INTEGRATIVE PROTEOMICS

Edited by **Hon-Chiu Eastwood Leung**

Integrative Proteomics
Edited by Hon-Chiu Eastwood Leung
Subject Editors: Tsz-Kwong Man and Ricardo J. Flores

Published by InTech

Janeza Trdine 9, 51000 Rijeka, Croatia

Notice

Statements and opinions expressed in the chapters are these of the individual contributors and not necessarily those of the editors or publisher. No responsibility is accepted for the accuracy of information contained in the published chapters. The publisher assumes no responsibility for any damage or injury to persons or property arising out of the use of any materials, instructions, methods or ideas contained in the book.

Publishing Process Manager Martina Durovic

Technical Editor Teodora Smiljanic

Cover Designer InTech Design Team

Additional hard copies can be obtained from orders@intechopen.com

Integrative Proteomics, Edited by Hon-Chiu Eastwood Leung;
Subject Editors: Tsz-Kwong Man and Ricardo J. Flores
p. cm.
ISBN 978-953-51-0070-6

For EU product safety concerns:
IN TECH d.o.o., Prolaz Marije Krucifikse Kozulić 3, 51000 Rijeka, Croatia,
info@intechopen.com or visit our website at intechopen.com.

Meet the editor

Dr Hon-Chiu Eastwood Leung finished his Ph.D. degree in microbiology and molecular genetics from University of Texas, Houston in 1997. He did his post-doctoral training in Baylor College of Medicine studying genomics and proteomics profiles of pediatrics medulloblastoma and osteosarcoma. Dr. Leung was a research scientist in Ciphergen Biosystems Inc. briefly, and then was recruited to Genome Institute of Singapore to lead the clinical proteomics section. In 2006 he returned to USA. He was the director of the genomics and proteomics core laboratory of the Texas Children's Hospital and director of genomics profiling of Baylor College of Medicine. At present, he is the director of the mass spectrometry-proteomics core facility of Baylor College of Medicine.

Contents

Preface

Proteomics has come of age. Two-dimensional gel electrophoresis (2D gel) was first used in the 1970s. With the advent of the human genome sequence and associated bioinformatics tools, and the advancement on protein separation and mass spectrometry in the 1990s and 2000s, the field Proteomics blossomed. The past decade has witnessed the development of clinical proteomics studies and fine tuning of proteomics studies to focus on a subset of proteome that are related to a biological phenotype. The rigorous bioinformatics standards developed along the way helped the data across platforms to be more compatible with each other.

The goal of this book is to provide insights for a broad audience base ranging from those who are new to those who are more experienced in the field of proteomics. This book covers the historical overview of proteomics, sample preparations, 2D gel separation of complex proteomes and associated databases, structural proteomics, and subproteome analyses with focus on i) glycoproteome, ii) phosphor-proteome, iii) protein:protein interactions, and iv) sub-proteomes captured by specific chemical groups. Surely, the data generated in the field of proteomics will only be making sense by the corresponding development of bioinformatics tools. The chapters in the section of bioinformatics will also touch on the pathway generation from the proteomics data.

Finally, I would like to thank the authors for sharing their experience within the chapters and to express my gratitude to Martina Durovic for the wonderful administrative assistance. Last but not least, I would like to thank the staff for taking the final steps to bring this publication to reality.

Hon-Chiu Eastwood Leung, Ph.D.
Assistant Professor, Baylor College of Medicine,
Houston, Texas,
USA

Part 1

Proteomics – Historical Review

1

Strategies for Protein Separation

Fernanda Salvato[1], Mayra Costa da Cruz Gallo de Carvalho[2] and Aline de Lima Leite[3]
[1]*Universidade de São Paulo, Escola Superior de Agricultura Luiz de Queiroz*
[2]*Empresa Brasileira de Pesquisa Agropecuária (EMBRAPA)*
[3]*Universidade de São Paulo, Faculdade de Odontologia de Bauru*
Brazil

1. Introduction

The proteome of a cell or tissue depends on cellular and environmental conditions, showing a dynamic system subject to large variations. To study these large changes of variability and quantity, proteomics has emerged, providing techniques dedicated to global characterization of all proteins simultaneously. The expectation is that this information will produce new insights into the biological function of proteins in different physiological states of a cell or tissue.

The proteome has a dynamic and complex nature that is the result of many post-translational modifications, molecular interactions, and a variety of proteins arising from alternative mRNA splicing. With this in mind, the number of modified and unmodified proteins found in any biological system is much bigger than the number of genes (Anderson et al., 2004), which is why mRNA expression may not correlate with protein content (Rogers et. al, 2008). In addition, not all proteins are expressed in the same or similar level in the proteome. For example, the enzyme Rubisco comprises 30–50% of leaf proteome (Feller et al., 2008), which is a big issue in the proteomic assessment of low-abundance proteins. In fact, the majority of proteins are in the low-abundance level. To overcome these challenges, the proteome must be fractionated for effective detection and quantification by mass spectrometry (MS). Consequently, the analysis of proteins on the large or small scale is dependent on separation methods.

As the ultimate goal in proteomics is to resolve all individual proteins in the cell, although it is quite difficult to find a separation method that could accommodate the diversity of proteins equally, protein separation methods directly affect the achievement of reliable results. Such methods are based on the physical or chemical properties of different proteins, such as their mass or net charge.

The combination of sequential methods exploiting different properties can provide high-resolution analysis of very complex protein mixtures. Then, current analytical strategies can reach different levels of resolution depending on the platform used. Two-dimensional gel electrophoresis (2DGE) and multidimensional liquid chromatography (MDLC) are the two methods that dominate the separation steps in proteomics. The differences of each strategy are basically related to sensitivity, automation, and high-throughput possibilities. In this chapter, the limitations and principles of these techniques will be discussed.

2. 2D-PAGE: Principles, advantages and limitations

The 2D-PAGE (two dimensional polyacrylamide gel electrophoresis) was developed by Patrick H. O'Farrell who successfully combined two known electrophoresis methods, isoeletric focusing (IEF) and sodium dodecyl sulfate electrophoresis (SDS-PAGE) (O'Farrell, 1975) with the objective of resolving more complex proteomes. The author was brilliant in his idea of combining both techniques once now proteins could be separated by two non-related properties given a uniform distribution throughout the gel. Surprisingly, the paper "High Resolution Two-dimensional Electrophoresis of Proteins" was firstly rejected by the JBC (Journal of Biological Chemistry) journal because of its "speculative character", as pointed out in the commemorative issue of the JBC (2006), but the power of 2D gel electrophoresis in resolving proteomes had already spread rapidly. Although the combination made by O'Farrell had immediately caused great impact on proteins separation, its commercial application in proteomics become possible only after a technical modification that made the 2D gel electrophoresis reproducible. In the mid1980s, was introduced to the 2D-PAGE system, commercial strips with immobilized pH gradients (IPG strip) and instruments for IEF (isoelectric focusing) (Bjellqvist et al., 1982) and, since then, the 2D-PAGE assume a central role in proteomics. Together, the 2D-PAGE and mass spectrometric techniques provided the characterization of thousands of proteins in single gels.

2.1 Principles of 2D-PAGE

To perform proteins separation, the two dimensional electrophoresis uses sequentially two non-related physical proprieties. In a first dimension, proteins are separated owing their migration in an immobilized pH gradient. Then, in a second dimension, proteins that occasionally took the same migration point after the first separation could now be separate in the polyacrylamide gel, according to their molecular weight, what guarantees to this technique a greater resolution power than achieved in one dimensional electrophoresis. Protein separation can be achieved as low as 0.1 isoelectric point (pI) unit and 1 kDa in molecular weight (MW) (Figeys, 2005). The spots visualized in a second dimension gel are unique proteins or simple mixture of proteins depending on certain factors that can influence technique resolution. To improve resolution, proteins should be completely denatured, reduced, disaggregated from protein complex and solubilized to disrupt macromolecular interactions (Chevalier, 2010).

In the 2D-PAGE protocol, preparation of protein samples is a fundamental and determining stage in electrophoresis efficiency. Usually, to solubilize samples, buffers containing chaotropic agents (urea and/or thiourea), nonionic or zwitterionic (CHAPS or Triton X-100) detergents, reducing agent (DTT) and proteases and phosphatase inhibitors are used. The chaotropic agents will act in the non-covalent macromolecular interactions, interfering in hydrophobic interactions; surfactants (CHAPS and Triton X-100) will act synergistically with chaotropics preventing the adsorption or aggregation of hydrophobic proteins which after the action of thiourea will have their hydrophobic domains exposed; the reducing agent will reduce protein disulfides breaking up intra and inter molecular interactions and proteases and phosphatase inhibitors will avoid modifications in the proteome. Also to optimize proteases and phosphatase inhibition, diluted TCA or TCA-acetone can be used in the solubilization process. One important aspect in solubilization process is to avoid salts accumulation through dialysis or precipitation. Salts can migrate through the pH gradient in the IPG strip and accumulate in the ends, inducing water accumulation and electric current reduction, what interferes in the focalization process.

The first dimension in 2D-PAGE, also called isoelectric focusing (IEF), is performed in acrylamide gel strips with immobilized pH gradient (IPG strips). The gel in the strip is formed through acrylamide polymerization with amphoteric acrylamide monomers named immobilins. Immobilins with different pKa are added to the acrylamide mixture and after gel polymerization; immobilins are immobilized in the strip generating the pH gradient, that's why strips used in IEF are called immobilized pH gradient or IPG strips. The IPG strips are commercially available in many pH ranges such as 6-9, 6-11 or 7-10. They are sold dried and should be rehydrated to be used. In this process, the rehydration solution must be composed by a commercial mixture of carrier ampholytes containing molecules corresponding to all pIs (isoelectric points) in the strip pH range and by the solubilized protein sample to be separated. Ampholytes act as good buffering agents next to their pIs, assisting proteins in the mixture to migrate in the gel.

Isoelectric focusing like the whole electrophoresis process is based on the migration of charged biomolecules under an electric field. The separation of a protein mixture in a pH gradient occurs because proteins are amphoteric molecules and thus can present negative or positive charges in their ionized groups depending on the pH medium. When an electric current is applied, proteins migrate in the gel while the balance between their charges is positive or negative until the difference between charges became equal to zero (isoelectric point - pI), in this point protein migration ceases and protein get focused. Proteins positively charged, i. e., the ones those are in the strip region where pH value is lower than their pI, keep migrating directly to the positive pole until reach their pI. In the other side, proteins negatively charged, i.e., the ones those are in the strip region where pH value is higher than their pI, keep migrating directly to the negative pole until reach their pI. Focusing process can last from 12 to 20 hours.

After IEF ends, the strip containing focused proteins must be equilibrated with the anionic detergent sodium dodecyl sulfate (SDS) solution that denatured these proteins and forms negatively charged protein/SDS complex. The amount of SDS linked into the protein should be directly proportional to its weight, thus proteins that are totally coupled to SDS will migrate in polyacrylamide gel (SDS-PAGE) only due to their weight. Other reagents in the reaction include Tris-HCl buffer, urea, glycerol, DTT, iodoacetamide and bromophenol blue.

The second dimension is performed by placing the IPG strip above and in direct contact with the gel in a system composed by two spaced glass. An electric current is applied and proteins migrate from the strip to a second dimension where they are solved due to their molecular weight. In the second dimension, gel can be heterogeneous: with a superior phase or stacking gel with acrylamide 6% and with an inferior phase or resolution gel containing 12 to 15%. In some cases, gel can be homogeneous with acrylamide 13% (Görg et al., 2000). The second dimension can be performed vertically or horizontally, but only the horizontal systems allow multiple runs simultaneously. Gels usually run with 1 or 2 W of current in the first hour, followed by 15 mA/gel overnight with temperature regulation (10°C to 18°C) (Chevalier, 2010).

To visualize the spots in the gel dyes visible to naked eye or fluorescent dyes can be used. In both cases, are necessary to fix the gel after the run, using an acid (phosphoric acid or acetic acid) or an alcoholic (ethanol or methanol) solution depending on the chosen dyeing protocol (Görg et al., 2000). Among the non-fluorescent dyes are Coomassie Brilliant Blue, Colloidal Coomassie Blue and silver nitrate which detect spots respectively with minimal protein of 50,

10 and 0.5 ng (Patton, 2002 & Smejkal, 2004). Usually, fluorescent dyes used are SYPRO dyes, Flamingo and Deep Purple. All these three dyes are sensitive enough to detect spots with up to 1 ng protein (Patton, 2002) however, because of their high costs, they are less used.

Once stained, gels are scanned and gel image can be analyzed using specific software available. It's always recommended to reproduce the proteome of the sample in at least three gels, representing identical technical repetitions. Software will search for a representative spots profile among all repetitions and, if desirable, compare this generated profile with others previously obtained. To perform the comparison, normally markers spots are designated in the gel and the position of all others spots is determinate using these spots as reference. It's also possible to estimate the volume of interested spots assigning them relative quantification values when the objective is to compare proteins differentially expressed. Through the use of these tools a proteomic map for a determined sample can be assembled and yet information on protein differential expression can be obtained. Among the software available are: Image Master, Progenesis, PDQuest, Samespots, the Melanie package from the Swiss Institute of Bioinformatics, the Phoretix 2D software from Phoretix and Gellab II from Scanalytics.

2.2 Why to use 2D-PAGE

The 2D-PAGE cannot be used alone to directly identity proteins through the visualization of resulting spots in the gel, even when proteomic maps and sequence information are available to the tested sample. That's because there's a great variation in the proteome of two identical samples, which beings in the protein extraction technique and solubilization and ends in the electrophoresis acrylamide gel. Thus, the identification of proteins in the spots depends on a sequencing stage performed through mass spectrometry (MS or MS/MS). This workflow is usually assumed in proteomics laboratories once can be easily conducted, is applied in many laboratories despite the structure and offers a resolution power enough to detect hundreds of proteins in one gel. Besides that, the 2D-PAGE system is unique about the possibility to visualize the protein profile of a studied sample, allowing immediate comparison with distinct profiles, interesting spots isolation to further studies or yet the enrichment of labeled proteins or specially stained. All these characteristics guarantee its massive application in proteomics characterization. Many other high throughput "gel free" strategies to perform protein separation are available today, but the 2D-PAGE system keep being an important toll in different workflows proposed to protein studies.

The proteomic map assembly is, until the present moment, realized merely by two-dimensional electrophoresis. Gel images are digitalized and made available in data banks what enables the *in silico* comparison between different profiles and the selection of interesting spots. The Japanese Bank containing rice proteomic maps shows for example, more than 13000 characterized spots to different tissues and development phases (http://gene64.dna.affrc.go.jp/RPD/). In humans, there are a great number of studies that report the generation of proteomic maps directed to protein identification that works as biomarkers to reproductive dysfunctions and tumor development (Guo et al., 2010 & Klein-Scory et al., 2010).

Another important contribution of 2D-PAGE system to proteomics is found in the identification and relative quantification of differentially expressed proteins between samples, i.e., differential-display proteomics. Until 1997, this assignment was not easy due to proteome variation in identical samples and gel-to-gel variation frequently observed in the repetition of runs from the same sample. In that time, it was necessary to obtain a great number of gels to reach the required reproducibility in the "average gel" and then perform the comparison between samples. Since DIGE technique or difference gel electrophoresis (Unlu et al., 1997) was developed, the reproducibility problem of 2D-PAGE gel was bypassed. The DIGE system consists in a modification in the conventional protocol of 2D-PAGE that make possible to analyze in a unique gel three different samples giving the electrophoresis system a "multiplex" character. Samples are pre-labeled with fluorescent markers such as Cy2, Cy3 and Cy5, pooled and separate in a single run. Therefore, in addition to solve the 2D-PAGE reproducibility problem, DIGE system allow the direct quantification of spots from different samples resolved in the same gel and is much more sensitive due to fluorescent dye labeling raising the gel resolution dynamic range up to 1,000 times (Chevalier et al., 2010).

The third sample used in the DIGE system is an internal running control composed by identical aliquots from each experimental sample. The mixture: internal standard and sample 01 and 02 are labeled, pooled and resolved in the same gel what avoid diversion on sample preparation. The internal standard control is normally labeled with Cy2 dye and the other samples with Cy3 and Cy5 dyes. The quantification of each protein is obtained from the signal Cy3:Cy2 and Cy5:Cy2 ratio. The Cy3:Cy2 and Cy5:Cy2 ratios for each protein are then normalized across all the gels in a large experiment, using the Cy2 signals for separate normalization of each protein under survey (Lilley & Friedman, 2004).

A problem in DIGE lies in the hydrophobicity of the cyanine dyes, which label the protein by reacting to a large extent, with surface-exposed lysines in the protein, and lead to removal of multiple charges from the protein. Consequently, this decreases the solubility of the labeled protein, and in some cases may lead to protein precipitation prior to gel electrophoresis. To address this problem, minimal labeling is generally employed in DIGE. In this reaction only 1-5% of total lysines in a given protein are labeled avoiding protein precipitation. Alternatively to minimal labeling the saturation labeling method can be done for Cy3 and Cy5 dyes by reacting to free cysteines in a protein (Shaw et al., 2003). This strategy circumvents the sensitivity problem of minimal labeling but limits the proteome analysis to proteins that show free-cystein residues (Chevalier, 2010).

The 2D-PAGE can also be very useful to identify post-translation modifications (PTMs). The affinity chromatography systems are usually used to enrich samples containing a specific PTM, but the 2D-PAGE visual character enable the direct selection on spots differentially expressed to a specific PTM. Proteins resolved in gel can be for example, specifically labeled to detect phosphorylations or glycosylations, and after visually selected; proteins are excised and identified using mass chromatography.

2.3 Limitations of 2D-PAGE

Technical characteristics related to gel reproducibility and others that can prevent or influence protein resolution are considered the main limitation of 2D-PAGE technique. However, gel reproducibility in 2D-PAGE method is also strongly influenced by own sample biology, what cannot be considered a limitation from the system *per se*. Some limitations associated to 2D-PAGE are pointed and discussed below.

2.3.1 Reproducibility

Protocols available for protein extraction can be applied to various types of biological samples, but the efficiency varies a lot depending on the biological characteristics of the sample. It's much simpler to reproduce the proteome from samples with unique cellular types, like a cell culture for example, than from samples containing many distinct types of cells or cells in different development phases, like for example, from an onion root. The cellular type also offers challenges to protein extraction and solubilization procedures, with a higher reproducibility to animal cell than to plant cells, which are cover by cellular walls and are rich in membranous compartments (plastids). Besides these intrinsic factors associated to sample biology, the proteome dynamism represents an important variation source in experimental repetitions from a same sample, especially when the objective is to perform a protein relative quantification. The proteome can be promptly modify by degradative pathways or by any of the hundreds post-translational modification that exists. Furthermore, small variations due to differences in the genetic backgrounds between sample repetitions can introduce relevant variations in the proteome.

Another factor that can reduce importantly gel reproducibility in 2D-PAGE is gel-to-gel variations which begin in the sample preparation and extend to focusing process and SDS-PAGE. Even in simultaneous runs that preserve exactly the same experimental conditions the gel-to-gel oscillations are present. To minimize or eliminate this effect, two alternatives are available: built an average gel from at least three replicates or use a multiplex run system (DIGE).

It's also important to emphasize that the reproducibility problem of 2D-PAGE system can restrict the applicability of proteomic maps databank that are being generated when no sequencing information to interesting spots are available.

2.3.2 Resolution

2.3.2.1 Proteins with high molecular weight

Proteins with molecular weight higher than 250kDa cannot be resolved in polyacrylamide gels. To realize this, the ideal is to use an agarose gel followed by an isoelectric focusing (Yokoyama et al., 2009).

2.3.2.2 Low abundance or rare proteins

Low abundance proteins operate in cellular activities of high interest, participating in signal reception, gene activity regulation and in signal transduction cascades. The detection of these proteins is masked in 2D gels by abundant proteins, which depending on the sample can be present in a magnitude concentration up to 12 times higher. That's the case for example, of albumin protein present in plasma samples. One possible strategy to avoid this problem is the depletion of abundant protein through methods such as affinity chromatography (Greenough et al., 2004). This is normally used to plasma sample, but is not yet possible to other many systems. In plant cells, the abundance of ribulose bisphosphate carboxylase/oxygenase (RuBisCo) enzyme mask low abundant proteins and the current used strategy to understate this effect is to reduce sample complexity through the use of IPG strips with overlapping narrow pH ranges (Görg et al., 2000).

2.3.2.3 Hydrophobic proteins and membrane proteins

In 1998, an important paper was published (Wilkins et al., 1998) in which was demonstrated that hydrophobic proteins were almost absent in 2D gels done using urea as the only chaotropic agent in the protein solubilization solution. This information was very valuable once hydrophobic proteins comprise the proteins present in cellular membranes and represent around 30% of total proteome (Molloy, 2000). After this observation, proteins solubilization began to be realized using a combination of higher concentration of urea and a lower concentration of thiourea, a chaotropic agent much more efficient than urea. The elevated concentration of urea was necessary to solubilize thiourea, which was used in lower concentration, because if in higher concentration, thiourea can interfere in protein focusing process (Molloy, 2000). This modification in the original protocol of protein solubilization resulted in a greater efficiency to solubilize hydrophobic proteins but yet the combination of urea-thiourea cannot keep the proteins in solubilized forms in the aqueous environment necessary to IEF. Other variations in the solubilization protocols combining urea and others nonionic or zwitterionic detergents were suggested, but all resulted in a additional identification of only some membrane proteins spots (reviewed by Rabilloud et al., 2008). It was clear that the solubilization of membrane proteins, mainly those with high hydrophobicity (multiple transmembrane domain), could not be achieved under IEF compatible conditions (reviewed by Tan et al., 2008). The gel systems intent to resolve membrane proteins should use strong detergents for solubilization of this kind of proteins and agents that can add charges to the proteins preventing their aggregation. Such gel-based systems (blue native-PAGE or BN-PAGE, clear-native-PAGE or CN-PAGE, benzyldimethyl-n-hexadecylammonium chloride or BAC, and SDS/SDS or dSDS-PAGE) exclude the IEF resulting in a severely impaired gel resolution (reviewed byTan et al., 2008). The resulting spots are generally composed by a misture of proteins that carry different post-translational modification and/or by complexes of membrane and soluble associated proteins (Rabilloud et al., 2008). Other strategies to detect membrane proteins are available using gel free systems, sample pre-fractionation through subcellular fractionation or affinity purification, and the avidin–biotin technology (Elia, 2008). However, there is still a great necessity of development of protocols that allow the high resolution detection of membrane proteins and the simultaneously detection of membrane and soluble proteins. This is especially important when we are looking for desease responses or physiological phenomena because membrane proteins play key functions in normal development, participating in cellular recognition and signal transduction. Identification of altered membrane proteins could lead to the discovery of novel biomarkers in the disease diagnosis (Adam et al., 2002 & Jang & Hanash, 2003) and targets to therapeutic approaches (Bianco et al., 2006).

2.3.2.4 Basic proteins

The basic proteins represent approximately one third to half of total cellular proteome. Among them, are ribosomal proteins and nucleases which exhibit pI superior to 10 and because of this reason are poorly resolved in pH ranges available for alkaline proteins (pH ranges 6–9, 6–11 or 7–10). This 2D-PAGE limitation began to be settle with the commercialization of IPG strips comprising pH ranges of 3-12, 6-12 and 9-12 which are successfully used in the resolution of strongly alkaline protein, with pI superior to 11 (Drews et al., 2004 & Görg et al., 1997).

3. Principles of liquid chromatography in proteomics

Chromatographic separation methods have been applied in different laboratories around the world to decipher the many complex problems in industry and science, involving, for example, amino acids and proteins, nucleic acids, carbohydrates, drugs, pesticides, etc. This method separates the components of a mixture by the distribution of these components into two phases, where an immiscible stationary phase remains fixed while the other moves through it. The sample components more strongly connected to the stationary phase move

very slowly in mobile phase flow, while those linked more weakly to the stationary phase move more quickly. This process results in differential migration of these components.
The main criteria for classification of chromatographic separation are related to the separation mechanism involved and the different types of stages used. Thus, the physical form of the system classifies the general technique as planar or column chromatography. In the former, the stationary phase is prepared on a flat surface, while in the latter, the stationary phase is arranged in a cylinder. In Gas Chromatography (GC), the mobile phase is an inert gas that does not contribute to the separation process, whereas in Liquid Chromatography (LC), the mobile phase is a liquid that can interact with the solutes, so their composition is very important in the separation process. Supercritical Fluid Chromatography (SFC) utilizes a substance with temperature and pressure higher than the critical temperature (Tc) and critical pressure (Pc) proper to fluids, with the advantage of having lower viscosity than the liquid while maintaining the properties of interaction with the solutes (Skoog et al., 2006).
The LC techniques may be further divided into classic liquid chromatography (LC) and High-Performance Liquid Chromatography (HPLC). LC utilizes glass columns at atmospheric pressure, and the flow rate is due to gravitational forces. HPLC is the automation of LC under conditions that provide for enhanced separations during a shorter time. It utilizes a metal column and the mobile phase flow rate is due to a high-pressure pump, which increases the efficiency achieved in the separation of compounds, thus making HPLC one of the main techniques used in the separation of proteins and peptides from a wide variety of synthetic or biological sources.
The HPLC equipment comprises a reservoir of mobile phase, which contains the solvents used as the mobile phase to achieve selectivity in HPLC; a pumping system; sample injector; columns; and detectors (Figure 1). The pumping system is required to pump the mobile phase and overcome the pressure exerted by the particles of the column. The major requirements for an efficient pumping system include the ability to generate pressure to

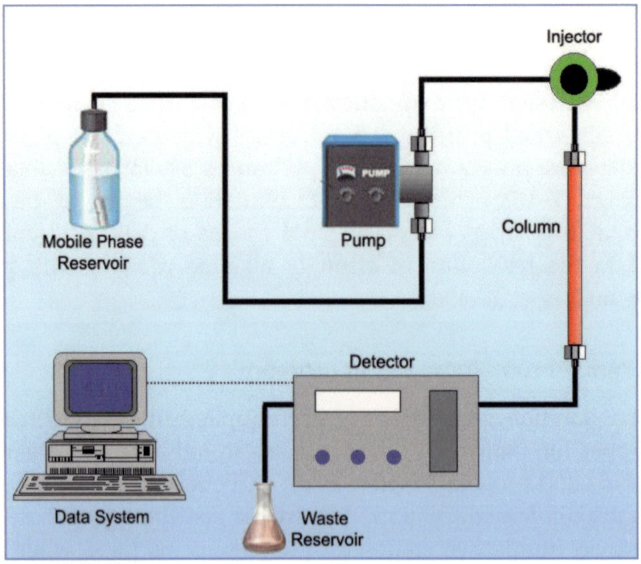

Fig. 1. Scheme of a HPLC system.

6000 psi, with no pulse output; flow rate ranging from 0.1 to 10 mL/min; constant solvent flow (with no variations greater than 0.5%); and corrosion-resistant components. There are two main pumps for HPLC: reciprocal pumps, which are employed in 90% of HPLC systems, consist of a small chamber in which the solvent is pumped by an oscillatory movement of the piston controlled by a motor. Because of this, the flow is not continuous, requiring a shock pulse. Syringe pumps consist of a large chamber equipped with a plunger that is activated by a screw mechanism. The rotation of the screw provides a continuous movement of the mobile phase that is free from pulsations from 0.1 to 5 mL/min. The most common injection system is sampling loops, which allow the introduction of samples up to 7000 psi with excellent precision. These loops can be manual or automated (Oliver, 1991; Meyer, 2010).

In a chromatograph, there are two types of columns: a guard column and separation column. The guard column has a length of 2 to 5 cm and is placed between the injector and separation column, allowing it to retain possible solids that can block the filters of the column and, in some cases, retain materials that can precipitate chemical reactions in the stationary phase. The separation columns (stationary phase) are the heart of a chromatograph, since they are responsible for the separation of the components present in the sample. They consist of a tube of inert material, usually stainless steel, and uniform internal diameter (i.d.), capable of resisting high pressures. They can be classified according to i.d (Saito et al., 2004).

Column designation	Internal Diameter [mm]
Conventional HPLC	3 - 5
Narrow-bore HPLC	2
Micro LC	0.5 - 1
Capillary LC	0.1 - 0.5
Nano LC	0.01 - 0.1
Open tubular LC	0.005 - 0.05

Table 1. Classification of liquid chromatography according to internal diameter (ID) of columns (Saito et al., 2004)

Silica is the most common stationary phase in HPLC because of advantages such as resistance to high pressures and physicochemical properties. Despite these advantages, silica has two limitations: the first restricts its use in a pH range of 2 to 8 because at pH below 2, bonds of Si-O-Si, which compose the silica and are responsible for maintaining the organic groups immobilized on the silica surface, become more susceptible to hydrolysis. On the other hand, at pH above 8, the hydroxyl groups (OH-) can easily react with the residual silanols, promoting silica dissolution that result in low efficiency and peak enlargement. The second limitation refers to the presence of residual silanol groups that can result on the asymmetry of the peak when basic samples are analyzed (Neue, 1997; Oliver, 1991; Meyer, 2010).

The particle structures are classified as porous, non-porous, and pellicular. The porous particles are most often used for HPLC, since it allows for greater surface area for interactions between the stationary phase and the analyte. Non-porous particles allow faster chromatography without losing efficiency because there is no diffusion of the analyte inside of the particles. However, to keep the sample capacity, it is necessary to use particles with diameters of 1 to 2 μm, as the capacity is 50 times less than that of porous particles. Pellicular particles are constituted of a solid nucleus coated with a thin layer (1–3 μm) of the stationary phase, and they have a good efficiency when analyzing macromolecules due to the fast mass transfer kinetics. The particle shape may be regular (spherical), irregular, or monolithic. The columns packed with spherical particles have a higher resistance to high pressures and good efficiency. The columns packed with irregular particles can have good efficiency when compared to regular particles; however, they have no mechanic stability and can result in higher pressures in the system (Meyer, 2010). Recently, monolithic particles have been introduced in HPLC. They are single pieces of porous silica or a highly intercrossed porous polymer such as polyacrylamide. The skeletons of monolithic particles contain macropores with diameters of approximately 2 μm and mesopores with diameters of approximately 13 nm. Because of those characteristics, they can provide higher flow rates without increasing the pressure, as well as great chemical stability and high permeability (Neue, 1997; Meyer, 2010).

In HPLC, there are different ways to detect the compounds eluting from the column. The ideal detectors are linear, selective and non-destructive and have adequate sensitivity, good stability and reproducibility, and a short response time. However, there are no detectors with all the features mentioned above, so the choice of the detector should be based on objective analysis as well as the type of sample to be analyzed. Liquid chromatographic detectors are basically of two types. Bulk property detectors respond to mobile-phase properties, such as refractive index, dielectric constant, or density. In contrast, solute property detectors respond to properties of solutes, such as UV absorbance, fluorescence, or diffusion current, which are not present in the mobile phase (Skoog et al., 2006; Meyer, 2010). Table 2 shows the major detectors used in HPLC.

Type of Detector	Limit of Detection	Commercial Available
Absorbance	10 pg	Yes
Conductivity	100 pg - 1 ng	Yes
Electrochemical	100 pg	Yes
Element Selective	1 ng	No
Fluorescence	10 fg	Yes
FTIR	1 μg	Yes
Light Scattering	1 μg	Yes
Mass Spectrometers	< 1 pg	Yes
Optical Activity	1 ng	No
Photoionization	< 1 pg	No
Refractive Index	1 ng	Yes

Source: Skoog et al., 2006

Table 2. The most common detectors used in HPLC.

High-performance chromatography supplanted gas phase chromatography because it is more versatile; it is not limited to volatile and thermally stable samples, thus allowing a wide choice of mobile and stationary phases. Because the mobile phase carries the solutes through the stationary phase, the correct choice of mobile phase is extremely important in the separation process, as it can completely change the selectivity of separations. The solvents used must be compatible with the stationary phase and detector and the high power of sample solubilization. The elution mode can be isocratic or gradient. In isocratic elution, the separation employs a single solvent or solvent mixture of constant composition, and the mobile phase remains constant with time. Gradient elution, in contrast, utilizes two or more solvent systems that differ significantly in polarity. In this case, when the elution process is begun, the ratio of the solvents varies with time, and separation efficiency is greatly enhanced by gradient elution (Skoog et all., 2006).

The major separation modes that are used to separate most compounds are normal-phase chromatography (NP), reverse-phase chromatography (RP), size-exclusion chromatography (SEC), ion-exchange chromatography (IEX), and affinity chromatography (AC).

In **normal-phase chromatography**, the stationary phase is polar while the mobile phase is non-polar. The retention of analytes occurs by the interaction of the stationary phase's polar functional groups with the polar groups on the particles' surfaces, and they elute from the column by addition of the low polarity compound followed by other compounds of increasing polarity (Figure 2). This method is widely used to separate analytes with low to intermediate polarity (Skoog et al., 2006).

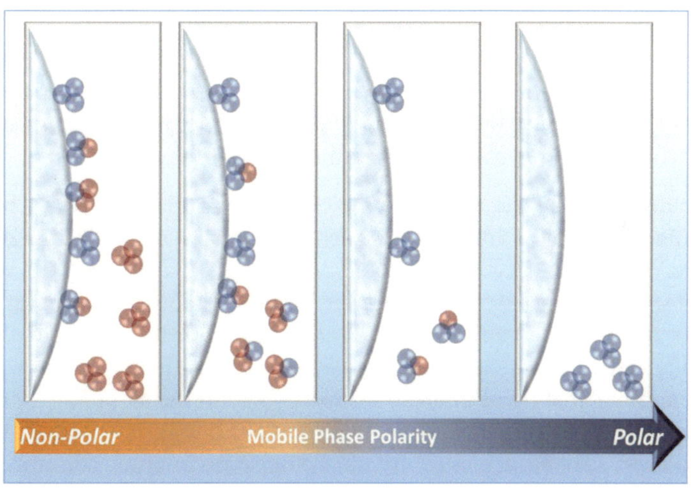

Fig. 2. Diagram of normal-phase chromatography separation. The stationary phase is polar and retains the polar molecule (blue) most strongly. The relatively non-polar molecules (red circles) are quickly eluted by the mobile phase, a non-polar solvent. An increase in mobile phase polarity will move polar molecules through the column.

Reverse-phase liquid chromatography has become a powerful tool widely used in the analysis and purification of biomolecules because of the high resolution provided by the technique. It is considered a very versatile technique because it can be used for non-polar, polar, ionizable, and ionic molecules. In RP-HPLC, the separation principle is based on the hydrophobic interaction between the analytes and non-polar groups bound on the stationary phase. Silica is the most common material used for column packing, which consists mainly of silicon dioxide (SiO_2) and has octadecyl (hydrocarbons having 18 carbon atoms) and octyl (hydrocarbons having 8 carbon atoms) groups chemically bound to the surface. The mobile phase composition is usually water or a water-miscible organic solvent (methanol, acetonitrile). The analytes adsorbed on the hydrophobic surface remain bound until the higher concentration of the organic solvent promotes the desorption of the molecules from the hydrophobic surface (Figure 3). More hydrophobic analytes are eluted slower than are the hydrophilic analytes (Skoog et al., 2006; GE Healthcare, 2006).

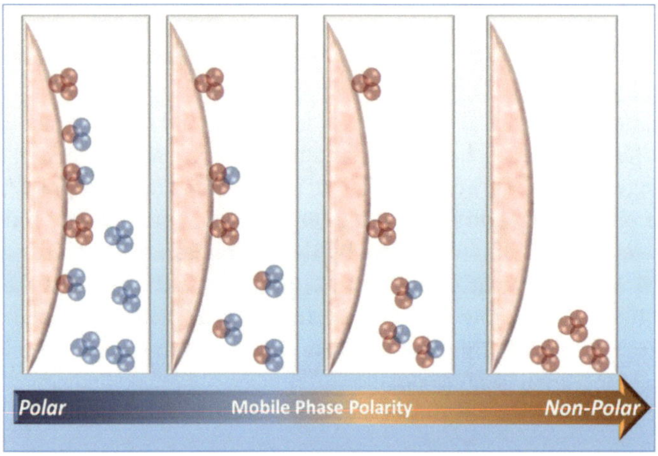

Fig. 3. Diagram of Reverse-phase chromatography separation. The stationary phase is non-polar and retains the non-polar molecule (red) most strongly. The relatively polar molecules (blue circles) are quickly eluted by the mobile phase, a polar solvent. A decrease in mobile phase polarity will move non-polar molecules through the column.

Affinity chromatography is the most specific chromatographic method. The separation is based on specific biochemical interactions such as enzyme-inhibitor, antigen-antibody or hormone-carrier. The stationary phase involves an inert matrix coupled with an *affinity ligand* specific for a binding site on the target molecule. The substance to be purified is specifically and reversibly adsorbed to a ligand, immobilized by a covalent bond to a chromatographic matrix. The samples are loaded in an affinity column containing the specific ligand, and the analyte of interest is adsorbed from the sample, while the molecules which have no affinity for the ligand pass through the column (Figure 4). Recovery of molecules of interest can be achieved by changing experimental conditions such as pH values, temperature, or ionic strength or by adding a stronger ligand to the mobile phase. For success in affinity chromatography, some important points have to be considered, such as finding a ligand specific enough and determining the ideal conditions for safe binding between analyte and ligand, as well as the ideal conditions for the retention and elution of the molecules involved (Skoog et al., 2006; GE Healthcare, 2007; Hage, 1999).

Ion-exchange chromatography (IEC) is based on the charge properties of the molecules. A stationary phase matrix constituted from a porous and inert material contains charged groups that interact with analyte ions of opposite charge. If these groups are acidic in nature, they interact with positively charged analytes and are called cation exchangers; however, if these groups are basic in nature, they interact with negatively charged molecules and are called anion exchangers. As the matrix material, they can be classified as organic (most common) and inorganic, natural or synthetic. Charged groups binding to the matrix

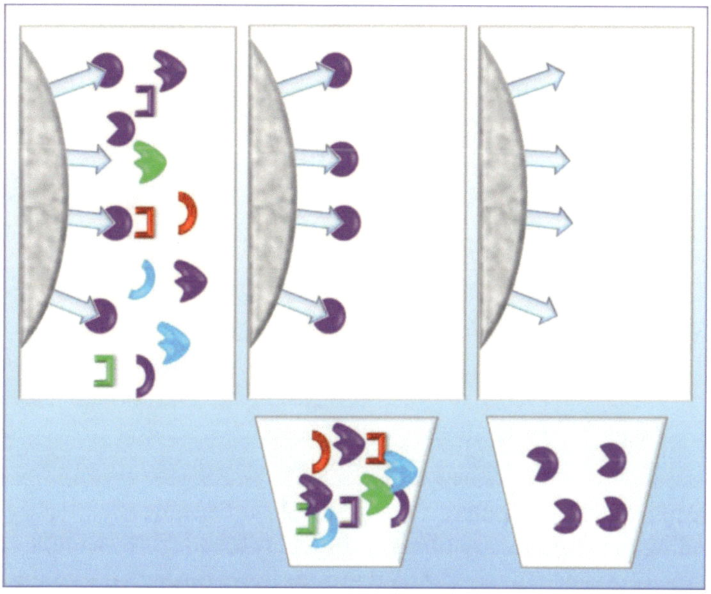

Fig. 4. Affinity chromatography column. The sample is loaded under ideal binding conditions. The target molecules bind specifically to the affinity ligands, while all other sample components, are not adsorbed.

are classified as strong ion exchangers that are completely ionized at a wide range of pH levels, while weak ion exchangers are ionized within a narrow pH range. Thus, weak exchangers offer more flexibility in selectivity than do strong ion exchangers, although the strong ion exchangers are used for initial development and optimization, because binding capacity does not change with pH. For the separation, the column is equilibrated with a start buffer, and then an analyte containing an opposite charge binds to the ionic groups of the matrix, whereas uncharged molecules, or those with the same charge as the ionic groups, are not retained. The adsorbed analyte of interest can be eluted by a gradient of ionic strength, pH values, or a combination of both in the mobile phase. The action mechanism of ion exchanger is shown in Figure 5 (Oliver, 1991; GE Healthcare, 2004; Meyer, 2010).

Size-exclusion chromatography (SEC) is a preparative and non-destructive analytical technique that, unlike other methods, is not based on interactions between molecules and the stationary phase, but on the size of molecules (Figure 6). The column is packed with inert material with pores of controlled size, within the stationary phase, such that the small molecules can enter most of the pores and therefore will be retained the longest time, while the larger molecules cannot penetrate and are kept for a shorter time period. The SEC can be classified according to the mobile phase used in *gel filtration chromatography* or *gel permeation chromatography*. Gel filtration chromatography (GFC) uses an aqueous mobile phase, which may contain organic modifiers or salts to change the ionic force or buffer solutions to change pH. Gel permeation chromatography (GPC) is a method used to separate high polymers, and it has become a prominent and widely used method for estimating molecular-weight distributions. Unlike GFC, GPC uses organic mobile phases such as tetrahydrofuran (THF), toluene, chloroform, dichloromethane, or dimethylformamide (Oliver, 1991; Meyer, 2004; GE Healthcare, 2010).

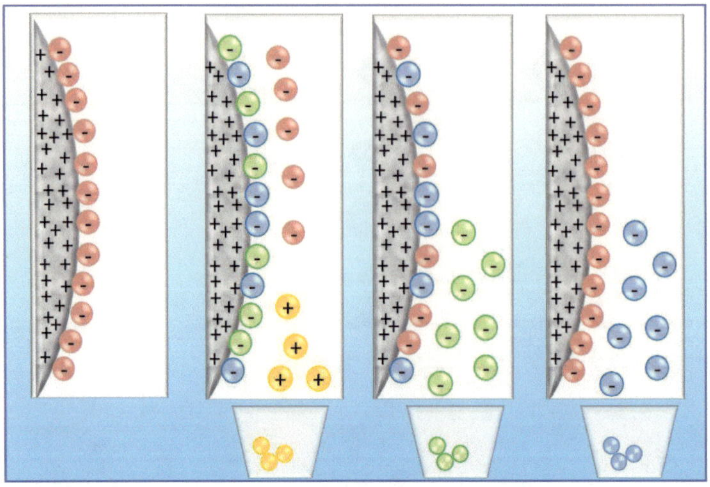

Fig. 5. The principle of IEC Separation. The mobile phase has ions negatively charged (red circles) that binding to the stationary phase (positively charged). The sample containing mixture of positively and negatively charged groups flows through the column. Those analytes containing negative charge are able to displace the mobile phase ions and bind to the stationary phase, while the positives groups (yellow) are eluted. The sample bound in the stationary phase can be eluted by increasing the concentration of a similarly charged species. The analyte that binds weakly (green) in the stationary phase will be eluted by buffer with salt ions at lower concentration. The analyte that binds strongly (blue) in the stationary phase will be eluted by buffer with salt ions at higher concentration.

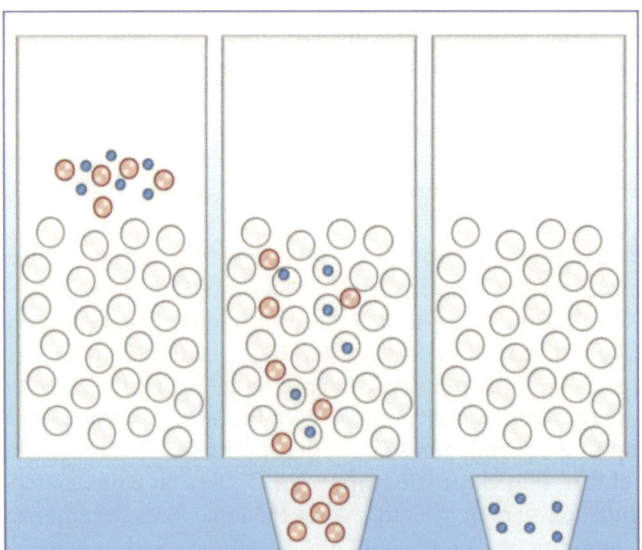

Fig. 6. Molecules smaller than pore will become trapped in the matrix. Those of larger molecular weight will not be trapped but will flow through column. Thus, larger molecules elute first, while smaller molecules are held longer inside the pores and will be eluted last.

3.1 Liquid chromatography coupled to MS

In proteomics, LC can be performed downstream of 2D gels to fractionate peptides from excised spots before MS analysis, upstream of 2D gels to prefractionate the protein sample, or instead of 2D gels as the main separation method in a multidimensional protein identification technology (MudPIT) experiment (item 4). Two strategies for protein identification and characterization by MS currently are employed in proteomics. In the bottom-up strategy, purified proteins or complex protein mixtures are subjected to enzymatic digestion, and the peptide products are analyzed by MS (Andersen et al., 1996; Pandley & Man, 2000) (Figure 7). In top-down proteomics, intact proteins or big protein fragments are subjected to fragmentation during MS analysis (Kelleher, 2004; Siuti & Kelleher, 2007). The major problem in bottom-up proteomics is that many peptides are generated for subsequent mass analysis, so it is not possible to get full protein sequence coverage and protein inference can be a problem. Different proteins can share some peptides, and consequently identification through bottom-up proteomics could be ambiguous. For these reasons, many efforts have focused on separation methods such as multidimensional chromatography to improve sensitivity and resolution. On the other hand, top-down proteomics permits high coverage sequence, thus overcoming one of the most important challenges in bottom-up strategy; however, it is a newer approach with several instrument limitations that will benefit from some hardware MS advances.

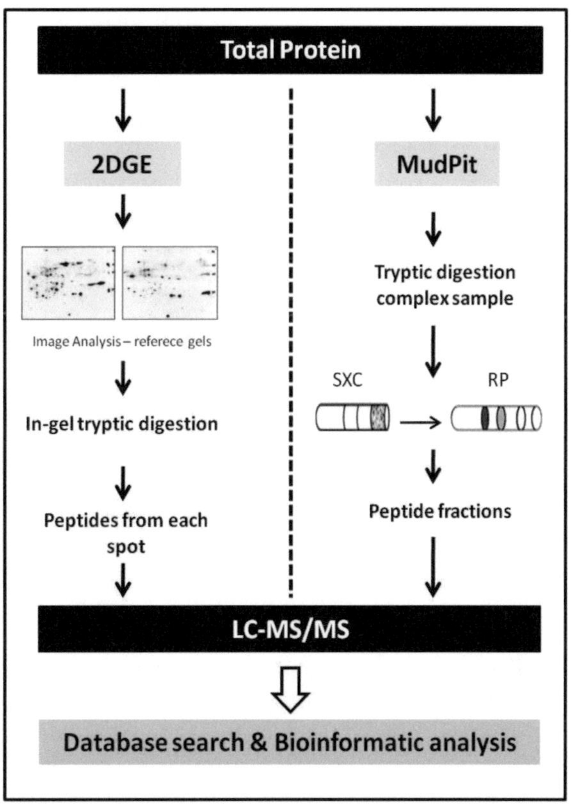

Fig. 7. Schematic representation of separation methods in bottom-up proteomics approaches (Adapted from Agrawal et al.,2008).

Also, the so-called "shotgun proteomics" is applied in a typical bottom-up approach and involves the utilization of HPLC coupled to tandem mass spectrometry (MS/MS) for identification of proteins on a large scale. This platform has been facilitated by the use of MudPIT, discussed below. As bottom-up proteomics is the most mature and most widely used approach for protein identification, in the following section we focus on chromatographic platforms for peptide separation.

4. Multidimensional liquid chromatography in proteomics

As previously mentioned, bottom-up proteomics offers some challenges to be overcome, such as sample complexity and large differences in protein concentration. As we know, these problems are far from being solved by single chromatographic or electrophoretic methods. Basically, two main approaches have been developed to face these difficulties: methods to separate abundant proteins from lower abundance proteins and multidimensional separation methods to maximize fractionation of peptides, thus increasing the proteome coverage analyzed by MS.

Abundant proteins can be a big problem when analyzing complex samples from certain tissues. Some tissues have high percentage (around 50% or more) of some classes of proteins. For example, leaves show approximately 50% of Rubisco (Feller et al., 2008) and the great majority of human blood serum proteome is comprised of albumin, fibrinogen, and immunoglobulins, transferrin, haptoglobin, and lipoproteins (Burtis & Ashwood, 2001; Turner & Hulme, 1970). If these abundant proteins are not removed from the sample, the peptides resulting from these proteins for proteomics analysis will overlap the peptides from the low-abundance proteins. To remove abundant proteins in complex samples, immunoaffinity separation or gel-based fractionation could be good options, although immunoaffinity separation can also remove low- abundance proteins that have any level of interaction with those abundant proteins (Granger et al., 2005).

Sequential chromatographic separations utilizing different chemical or physical properties have improved the assessment of classes of proteins that are difficult to handle in gel electrophoresis. MDLC can reach the same resolution of 2D gel electrophoresis with added advantages like automation, better sensitivity, and increased proteome coverage. MDLC was first described by Giddings in 1984 (Giddings, 1984) as a technique which combines two or more types of LC to increase the peak capacity and selectivity, contributing to a better fractionation of peptides that will enter the mass spectrometer.

The peak capacity is the number of peaks that can be resolved in a given time. To increase the peak capacity, the combination of two or more orthogonal separation dimensions is necessary. In other words, the properties affecting the separation in the first dimension do not affect the separation in the subsequent dimensions; thus, the process simplifies the sample complexity and improves the power of resolution, fractionating more components in a given time.

Many LC combinations have been reported utilizing chromatographic methods like strong cation exchange (SCX), strong anion exchange (SAX), size-exclusion (SEC) and reverse-phase chromatography (Zhang et al., 2007; Hynek et al., 2006; Moritz et al., 2005). Some factors are essential when considering an MDLC approach. The choice of different types of chromatographic columns to reach a satisfactory MDLC peak capacity is the most important point within this strategy. In order to separate a large variety of peptides in high-performance chromatography, the columns used in each dimension have to work with no

correlated properties. Other factors to be considered include the fact that the first column should have a larger loading capacity and have solvent compatibility with the subsequent columns. Thus, this compatibility between dimensions is essential in online systems. The last step of chromatography immediately before MS analysis should be compatible with electrospray ionization (ESI). For this purpose, the reverse-phase (RP) chromatography is frequently used, as it can desalt the samples and it is completely compatible with ESI, thus providing effective resolution and facilitating MS detection.
Several proteomic studies using different multidimensional configurations have been done along these years (reviewed by Gao et al., 2010) Here we point out some examples of the most applicable approaches.

4.1 Ion-exchange and reverse-phase chromatography (IEX-RPLC)

MUDPIT was first introduced by Yates and co-workers (Link et al., 1999; Wolters et al., 2001) using a biphasic column sequentially packed with SCX particles and then with C18 particles for peptide separation prior to MS analysis. Other works reported the use of an SCX column and an RP column connected to perform an online SCX-RP-MS/MS analysis (Tram et al., 2008; Gilar et al., 2005). Besides their orthogonality, SCX and RP columns show mobile phase compatibility between each other and with MS analysis.
In the SCX-RP configuration, peptides from a complex mixture are acidified and applied to an SCX column, in which the elution steps can be done using a salt gradient. Then, a fraction of peptides are absorbed in the RP column, and after washing away salts and buffers, peptides are eluted from the RP column into the mass spectrometer using a gradient of an organic solvent. Finally, the RP is reequilibrated and new fraction of peptides from the SCX column is eluted to be absorbed in the RP column, and the process repeats.
Several studies came out to improve the application of the MudPIT technique. One of the limitations of SCX-RPLC is related to the use of salt gradient SCX separation, resulting in limited resolution and peak capacity for peptide separation. Extensive salt usage could also cause ion suppression, thus reducing MS performance. Second, by using coupled SCX-RP columns, separation on both dimensions has to be performed at the same flow rate, which may sacrifice MS detection sensitivity for low-abundance components. As an alternative approach, offline SCX-RP separation has been used to identify more than 1200 proteins from zebrafish liver (Wang et al., 2007).However, the implementation of offline configurations is not always the best option, as extensive offline sample handling increases the overall analysis time and causes sample loss and sensitivity reduction.
Dai et al. (2005) reported an integrated column composed of SCX and RP where peptides were fractionated by a pH step gradient. The exclusion of salt removal steps could lead to fast 2D-LC separation and MS/MS detection. Some years later, the same research group developed a SCX and SAX combined pre-fraction strategy called Yin-Yang multidimensional chromatography (Dai et al., 2007). Peptides eluted from a SCX column at pH 2.5 were then separated by a SAX column using a pH gradient solution. Subsequently, SAX fractions were analyzed by RPLC-MS/MS. This approach revealed proteins from a broad range of different pH values.

4.2 Reverse-phase-reverse-phase chromatography (RP-RPLC)

The use of RP columns in both dimensions of a MudPIT analysis exhibited high-throughput, automatability, and performance comparable with that of SCX-RP. Despite all successful studies based on SCX-RP, these approaches are frequently encountering higher complexity samples than can be accommodated by their power of resolution. Accordingly, new strategies have been developed, including the RP-RPLC. Gilar et al. (2005) investigated the use of a pH gradient in a RP-RP platform. The RP-RP approach provides higher peak capacity in both dimensions, and the pH gradient has the most significant impact on the selectivity of this platform. The pH gradient modulates the peptide hydrophobicity during the elution of fractions of peptides.

Another factor that has been increasing the sensitivity of this approach is the integration of a high-pressure system in RP-RP platforms. This was first implemented in the Agilent 1100 2-D liquid chromatographic system and more recently by the 2D NanoACQUITTY system from Waters, frequently used with a pH gradient and two columns of C18 particles. Zhou et al. (2010) revealed that RP-RP fractionation outperforms SXC-RP. It was also demonstrated that the combination of RP-RP systems with the nanoflow format had good impact in the efficiency of electrospray ionization prior to MS/MS analysis.

4.3 Affinity chromatography

Another type of chromatography largely used in proteomics is affinity-based. The main usage of this technique is to enrich post-translationally modified (PTM) peptides or proteins subject to MS analysis. Due to their low stoichiometry and dynamic modification patterns, PTM materials have to be enriched in complex mixtures prior to MS analysis. Affinity chromatographic columns can be used online or offline to RP columns and the mass spectrometer.

4.4 Enrichment of phosphopeptides or phosphoproteins

In phosphoproteomics, the interest focuses on the identification and quantification of phosphorylated proteins or peptides. With this purpose, several selective enrichment methods are applied to increase the content of phosphopeptides or phosphoproteins in complex mixtures, thus preventing the suppression of ion signals by unphosphorylated molecules.

Immobilized metal affinity chromatography (IMAC) is one of the most popular techniques employed in phosphoproteomics. IMAC contains immobilized positive ions (Fe^{3+}, Ga^{3+} and Al^{3+}) interacting with the negative ions of phosphate groups. The elution of the interacted molecules makes the enrichment possible. Tsai et al. (2011) showed a pH/acid-controlled IMAC protocol for phosphopeptide purification with high specificity and lower sample loss. They characterized over 2,360 nondegenerate phosphopeptides and 2,747 phosphorylation sites in the H1299 lung cancer cell line, showing that a low pH buffer increases the specificity of IMAC for phosphopeptides. The buffer composition was directly associated to the specificity and selectivity of the IMAC technique (Tsai et al., 2008; Jensen et al., 2007).

Another important enrichment technique widely used is the metal oxide affinity chromatography (MOAC). MOAC uses a principle similar to that of IMAC, incorporating metal oxides, such as titanium dioxide TiO_2, zirconium dioxide ZrO_2, or aluminum hydroxide $Al(OH)_3$. Metal oxides tend to have higher selectivity for phosphopeptides, making it easier to trap them in the column (Zhou et al., 2007). During the last few years, TiO_2 has emerged as the main MOAC-based phosphopeptide enrichment method (Pinkse et al., 2004). The principle is the same as that of IMAC; however, when loading peptides with DHB (2,5-dihydroxybenzoic acid) non-specific binding is reduced, thus increasing the selectivity of TiO_2 (Larsen et al., 2005).

4.5 Enrichment of glycopeptides or glycoproteins

Glycosylation is the most complex PTM presented in eucaryotes. Like other PTMs, glycopeptides are often very minor constituents compared to peptides derived from proteolytic digestion. Therefore, enrichment of glycoproteins or glycopeptides is essential in a glycoproteomics study (Ito et al., 2009). The lectin affinity approach is the most common tool for glycoprotein/glycopeptide enrichment. Lectins are sugar-binding proteins that are highly specific to sugar moieties. For example, concanavalin A (conA) is a mannose-binding lectin largely used to study N-glycosylated proteins, while lectin from *Vicia villosa* (VVL), which preferentially binds to alpha- or beta-linked terminal GalNAc, has a different preference. All lectins have sugar specificity, and therefore serial lectin affinity columns (SLAC) have been developed to reduce the complexity of proteolytic digests by more than one order of magnitude. SLAC can be used to study O-glycosylation proteins, which are difficult to access because lectins for studying them are not specific enough. Jacalin, a lectin, is relatively specific for O-glycosylation but has the problem that it also selects mannose N-glycans. This problem can be overcome by first using a ConA affinity column to first remove mannose, and then using Jacalin columns. When used in a serial configuration, O-glycosylated peptides can be accessed (Durham & Regnier, 2006).

4.6 LC-MS repeatability and reproducibility

First of all, it is important to understand the difference between repeatability and reproducibility. The first, represents the variations in the measurements on the same sample, made in the same instrument and by the same operator (Bland & Altman, 1986). The second, is the variation observed for an analytical technique when operator, instrumentation, time, or location is changed (McNaught & Wilkinson, 1997). So in proteomics, it is possible to calculate variations in results from run to run estimating the repeatability of the analytical technique or estimating the reproducibility between different laboratories in completely differently instruments. These measurements are imperative to the clinical utility of biomarker candidates and must be expected (Baggerly et al.,2005).

During proteomics analysis there are many potential contributors to variability that can compromise the approach repeatability and reproducibility. The variations can begin in sample collection, specimen processing techniques, storage, and instrument performance (Hsieh at al, 2006; Banks et al., 2005; Pilny et al., 2006). All steps in the proteome analysis can offer a source of variations. The proteomics analysis by LC-MS begins with the digestion of complex mixtures followed by peptide fractionation in LC systems. Then MS/MS scans are acquired and spectra matches are resulted from bioinformatics analysis. The complexity of these steps leads to variations in peptide and protein identification and quantification. Minor differences in LC can result in modified peptide elution times (Prakash et al., 2006) or change which peptides are selected for MS/MS fragmentation (Liu et al., 2004).

In this context, the use of MudPIT also can introduce more variation. Delmotte et al. (2009) showed that an introduction of a separation dimension decreases the repeatability by approximately 25% upon 1D or 2D chromatographic separations. Slebos et al. (2008) reported superior reproducibility in isoelectric focusing (IEF) compared to SCX separations. They found that IEF more quickly reached maximal detection within three replicate analysis. In contrast, the SCX required six replicates. In this study, approximately 90% of all peptide identifications are found in a single fraction. In contrast, SCX is characterized by spread of peptides into adjacent fractions. Peptides at lower abundance or those generating lower signal intensity are more likely to be selected for MS/MS if they appear in multiple fractions. Thus SCX produces more chances of peptides detection which is consequence of the greater SXC sensitivity of peptide identifications. Though, greater sensitivity leads to much higher variability in peptide detection at each fractionation step from run to run, demanding a greater number of technical replicas, which compromises the overall throughput. Several groups have evaluated the number of replicates necessary to observe a particular percentage of the proteins in a sample (Liu et al., 2004; Slebos et al., 2008; Kislinger et al., 2005).

On the other hand, Tabb et al. (2009) concluded that a standardized platforms results a high degree of repeatability and reproducibility (around 70-80%) in protein identifications. This is an indication that LC-MS/MS platforms can generate consistent protein identifications. They showed that instrumentation can also increment variations. In this study, they observed that the high resolution of Orbitraps outperforms lower resolution of LTQs in repeatability and reproducibility. They also observed that reproducibility between different instruments of the same type is lower than repeatability of technical replicates on a single instrument by several percent.

Standardizations of methodologies and system configurations, as well as appropriate instrument tuning and maintenance would result in lower noise level from biological, chemical or instrumental sources between LC-MS/MS analysis. A comprehensive understanding about components that affect reproducibility and repeatability can help in the determination of the best alternative of experimental design to reach the reliability desired in proteomics studies.

5. Concluding remarks

Protein separation methods are vital for the characterization of proteomes. All these methods exploit one or more general properties of proteins and are directly linked to the effectiveness of any proteomic analysis. Traditionally, 2D gels have been the most frequently employed protein separation tool, capable of separating up to 10,000 components. However, 2D gels have some limitations, as discussed above, preventing efficient automation. In addition, LC methods coupled to MS analysis have been emerged as a promising strategy to achieve better automation and sensitivity. Combining orthogonal chromatographic columns increases peak capacity and permits a large dynamic range to be more efficiently measured in complex samples. Also, affinity chromatography allows the selection of a subgroup of proteins (proteins with certain PTMs), directing the study into specific biological pathways and increasing the generation of relevant information. Although its resolution has been improved resolution, LC fractioning has several concerns to be circumvented, such as the amount of time required, reproducibility, and protein inference from peptides. An experiment involving RP analysis can take 1.5 to 2 hours for a single analysis. Then, considering 15 fractions (taken from an IEX) from a complex mixture, there will be 22.5 hours required. Moreover, there is the analysis of biological (10 individual) and technical replicas (3 injections) that could reach 56 days of analysis, taking into account an experiment comparing control and affected. The required time increases instrument and maintenance costs and influences the experimental throughput. The second concern involves reproducibility during fractionating, which can be related to the number of dimensions employed and if they are inline coupled to the mass spectrometer or not. Another point that has received considerable attention is the development of data analysis pipelines to effectively process LC data from multiple peptide fractionation. The challenge is to reach better accuracy in protein inference, mainly in bottom-up proteomics. In parallel, top-down proteomics is developing rapidly, with significant progress in MDLC systems for protein separation and probably warranting more attention in the near future.

Finally, the achievement of better representation, resolution, sensitivity, and automation has been developed with the application of LC methods coupled to MS, but they are far from being considered as the perfect strategy and are subject to improvements. The use of 2D gels or LC as the central separation method or both in combination is debatable, depending on the study objective. However, the constant advancement of these platforms is resulting in successful proteomics studies.

6. References

Adam, B.L.; Qu, Y.; Davis, J.W.; Ward, M.D.; Clements, M.A.; Cazares, L.H.; Semmes, O.J.; Schellhammer, P.F.; Yasui, Y.; Feng, Z. & Wright, G.L.Jr. (2002). Serum protein fingerprinting coupled with a pattern-matching algorithm distinguishes prostate cancer from benign prostate hyperplasia and healthy men. *Cancer Research*, Vol.62, No.13, (July 2002), pp.3609–3614, ISSN 1541-7786

Agrawal, G.K.; Hajduch, M.;, Graham, K.; Thelen, J.J.(2008) In-Depth Investigation of the Soybean Seed-Filling Proteome and Comparison with a Parallel Study of Rapeseed. *Plant Physiology*, Vol. 148, pp. 504–518

Albrethsen, J.; Bogebo, R.; Olsen, J.; Raskov, H.; Gammeltoft, S. (2006). Preanalytical and analytical variation of surface-enhanced laser desorption-ionization time-of-flight mass spectrometry of human serum. *Clinical Chemistry and Laboratory Medicine*, Vol.44, No.10, pp. 1243-52

Andersen, J. S., Svensson, B., and Roepstorff, P. (1996) Electrospray ionization and matrix assisted laser desorption/ionization mass spectrometry: Powerful analytical tools in recombinant protein chemistry. *Nature Biotechnology*. Vol.14, pp. 449–457

Anderson, N.L.; Polanski, M.; Pieper, R.; Gatlin, T.; Tirumalai, R.S.; Conrads, T.P.; Veenstra, T.D.; Adkins, J.N.; Pounds, J.G.; Fagan, R. & Lobley, A. (2004). The Human Plasma Proteome: A Non-Redundant List Developed by Combination of Four Separate Sources, *Mol. Cell Proteomics*, Vol. 3, pp. 311–326

Baggerly, K. A.; Morris, J. S.; Edmonson, S. R.; Coombes, K. R. (2005) Signal in noise: evaluating reported reproducibility of serum proteomic tests for ovarian cancer. *Journal of the National Cancer Institute*, Vol.97, No.4, pp. 307-9

Banks, R. E.; Stanley, A. J.; Cairns, D. A.; Barrett, J. H.; Clarke, P.; Thompson, D.; Selby, P. J. (2005). Influences of blood sample processing on low-molecular-weight proteome identified by surface-enhanced laser desorption/ionization mass spectrometry. *Clinical Chemistry*, Vol. 51, No.9, pp. 1637-49

Bianco, C.; Strizzi, L.; Mancino, M.; Rehman, A.; Hamada, S.; Watanabe, K.; De Luca, A.; Jones, B.; Balogh, G.; Russo, J.; Mailo, D.; Palaia, R.; D'Aiuto, G.; Botti, G.; Perrone, F.; Salomon, D.S. & Normanno, N. (2006). Identification of cripto-1 as a novel serologic marker for breast and colon cancer. *Clinical Cancer Research*, Vol.12, No.17, (September 2006), pp.5158-64, ISSN 1557 -3265

Bjellqvist, B.; Ek, K.; Righetti, P.G.; Gianazza, E.; Görg, A.; Westermeier, R. & Postel, W. (1982).Isoelectric-focusing in immobilized ph gradients - principle, methodology and some applications. *Journal of Biochemical and Biophysical Methods*, Vol.6, No.4, (September 1982), pp.317-339 ISSN 0165-022X

Bland, J. M.; Altman, D. G. (1986). Statistical methods for assessing agreement between two methods of clinical measurement. *The Lancet*, Vol. 1, No. 8476, pp. 307–310

Burtis,C.A. & Ashwood E. R. (Eds.). (2001). *Tietz Fundamentals of Clinical Chemistry*, 5th Edition, WB Saunders, ISBN 0-7216-8634-6, Philadelphia

Chevalier, F. (2010). Highlights on the capacities of "Gel-based" Proteomics. In: *Proteome Science*, Vol.8, No.23, 29/07/2011, Available from: <http://www.proteomesci.com/content/8/1/23>

Dai, J.; Jin, W.H.; Sheng, Q.H.; Shieh, C.H.; Wu, J.R. & Zeng, R. (2007). Protein phosphorylation and expression profiling by Yin-yang multidimensional liquid chromatography (Yin-yang MDLC) mass spectrometry. *Journal of Proteome Research*, Vol.6, pp. 250-62

Dai, J.; Shieh, C.H.; Sheng, Q.; Zhou, H. & Zeng, R. (2005). Proteomic analysis with integrated multiple dimensional liquid chromatography/mass spectrometry based on elution of ion exchange column using pH steps, *Analytical Chemistry*, Vol.77, pp. 5793–5799

Delmotte, N.; Lasaosa, M.; Tholey, A.; Heinzle, E.; van Dorsselaer, A.; Huber, C. G. (2009). Repeatability of peptide identifications in shotgun proteome analysis employing off-line two-dimensional chromatographic separations and ion-trap MS, *Journal of Separation Science.*, Vol. *32, No.*8, pp. 1156–11164

Drews, O.; Reil, G.; Parlar, H. & Görg, A. Setting up standards and a reference map for the alkaline proteome of the Gram-positive bacterium *Lactococcus lactis*. (2004). *Proteomics*, Vol.4, No.5, (May 2004), pp.1293-1304, ISSN 1615-9853

Durham, M. & Regnier F. E. (2006). Targeted glycoproteomics: serial lectin affinity chromatography in the selection of O-glycosylation sites on proteins from the human blood proteome. *Journal of chromatography A*, Vol. 1132, pp. 165-73.

Elia, G. (2008). Biotinylation reagents for the study of cell surfasse Proteins. *Proteomics*, Vol. 8, pp. 4012–4024

Feller, U.; Anders, I. & Mae, T. (2008). Rubiscolytics: fate of Rubisco after its enzymatic function in a cell is terminated. *Journal of Experimental Botany*, Vol.59, pp. 1615–1624

Figeys, D. (2005). Proteomics: The Basic Overview. In: *Industrial Proteomics: Applications for Biotechnology and Pharmaceuticals*, edited by Daniel Figeys, pp.1-62, John Wiley & Sons Incorporated, ISBN 0471703753, San Francisco, USA

Gao, M.; Qi, D.; Zhang, P.; Deng, C. & Zhang, X. (2010). Development of multidimensional liquid chromatography and application in proteomic analysis, *Expert Review of Proteomics*, Vol.7, No.5, pp.665-678

GE Healthcare. (2004). *Ion Exchange Chromatography & Chromatofocusing,* Amersham Biosciences Limited, 11-0004-21, Edition AA, Sweden.

GE Healthcare. (2006). *Hydrophobic Interaction and Reversed Phase Chromatography, Principles and Methods,* GE imagination at work, 11-0012-69 AA, Sweden.

GE Healthcare. (2007). *Affinity Chromatography, Principles & Methods,* GE imagination at work, 18-1022-29 AE, Sweden.

GE Healthcare. (2010). *Gel Filtration, Principles and Methods,* GE imagination at work, 18-1022-18 AK, Sweden.

Giddings, J.C. (1984). Two-Dimensional Separations: Concept and Promise, *Anal. Chem.*, Vol.56, pp.1258A–1270A

Gilar, M.; Olivova; P., Daly; A. E. & Gebler, J. C. (2005). Two-dimensional separation of peptides using RP-RP-HPLC system with different pH in first and second separation dimensions, *Journal of Separation Science*, Vol.*28*, pp. 1694–1703

Görg, A.; Obermaier, C.; Boguth, G.; Csordas, A.; Diaz, J.J. & Madjar, J.J. Very alkaline immobilized pH gradients for two dimensional electrophoresis of ribosomal and nuclear proteins. (1997). *Electrophoresis*, Vol.18, No.3-4, (March-April 1997), pp.328-337, ISSN 1522-2683

Görg, A.; Obermaier, K.; Boguth,G.; Harder, A.; Scheibe, B.; Wildgruder, R. & Weiss, W.(2000). The current state of two-dimentional electrophoresis with immobilized pH gradients. *Electrophoresis*, Vol.21, No.6, (April 2000), pp.1037-1053, ISSN 1522-2683

Granger, J.; Siddiqui, S.; Copeland, S. & Remick D. (2005). Albumin depletion of human plasma also removes low abundance proteins including the cytokines *Proteomics*, Vol.5, pp. 4713–4718.

Greenough, C.; Jenkins, R.E.; Kitteringham, N.R.; Pirmohamed, M.; Park, B.K. & Pennington, S.R. (2004). A method for the rapid depletion of albumin and immunoglobulin from human plasma. *Proteomics*, Vol.4, No.10, (October 2004), pp.3107-3111, ISSN 1615-9853

Guo, X.; Zhao, C.; Wang, F.; Zhu, Y.; Cui, Y.; Zhou, Z.; Huo, R. & Sha, J. (2010). Investigation of human testis protein heterogeneity using 2-dimensional electrophoresis. *Journal of Andrology*, Vol31, No.(4), pp.419-429, (July-August 2010), ISSN 0196-3635

Hsieh, S. Y.; Chen, R. K.; Pan, Y. H.; Lee, H. L. (2006). Systematical evaluation of the effects of sample collection procedures on low-molecular-weight serum/plasma proteome profiling. *Proteomics*, Vol. 6, No.10, pp.3189-98

Hynek, R.; Svensson, B.; Jensen, O.N.; Barkholt, V. & Finnie, C. (2006). nrichment and identification of integral membrane proteins from barley aleurone layers by reversed-phase chromatography, SDS-PAGE and LC-MS/MS, *Journal Proteome Res.*, Vol.5, pp. 3105–3113

Ito, S.; Hayama, K. & Hirabayashi, J. (2009). Enrichment strategies for glycopeptides, *Methods in Molecular Biology, Vol.*534, pp.195-203

Jang, J.H. & Hanash, S. (2003). Profiling of the cell surface proteome. *Proteomics*, Vol.3, No.10, (October 2003), pp.1947-1954, ISSN 1615-9853

JBC Centennial 1905–2005 100 Years of Biochemistry and Molecular Biology. (2006). The Development of Two-dimensional Electrophoresis by Patrick H. O'Farrell. *The Journal of Biological Chemistry*, Vol.281, No.32, (August 2006), pp. e26, ISSN 1083-351X

Jensen S.S. & Larsen M.R. (2007). Evaluation of the impact of some experimental procedures on different phosphopeptide enrichment techniques. *Rapid Commun. Mass Spectrom*, Vol. 21, pp. 3635–45

Klein-Scory, S.; Kübler, S.; Diehl, H.; Eilert-Micus, C.; Reinacher-Schick, A.; Stühler, K.; Warscheid, B.; Meyer, H.E.; Schmiegel, W. & Schwarte-Waldhoff, I. (2010). Immunoscreening of the extracellular proteome of colorectal cancer cells. *BMC Cancer*, Vol.10, No.70, pp.2-19, (February 2010), ISSN 1471-2407

Larsen M.R.; Thingholm T.E.; Jensen O.N.; Roepstorff P. & Jorgensen TJ. (2005). Highly selective enrichment of phosphorylated peptides from peptide mixtures using titanium dioxide microcolumns. *Mol. Cell Proteomics*, Vol. 4, pp. 873–86

Lilley, K.S. & Friedman, D.B. (2004). All about DIGE: quantification technology for differential-display 2D-gel proteomics. *Expert Review of Proteomics*, Vol.1, No.4, (December 2004), pp.401-409, ISSN 1478-9450

Link, A.J.; Eng, J.; Schieltz, D.M.; Carmack, E.; Mize, G.J.; Morris, D.R.; Garvik, B.M. & Yates, J.R. III (1999). Direct analysis of protein complexes using mass spectrometry, *Nature Biotechnology*, Vol.17, pp. 676–682

Liu, H.; Sadygov, R. G.; Yates, J. R., III. (2004). A model for random sampling and estimation of relative protein abundance in shotgun proteomics. *Analytical Chemistry.*, Vol.*76, No.*14, pp.4193–4201

Kelleher, N. L. (2004). Top-Down Proteomics, *Analytical Chemistry*, Vol.76, No.11, pp.197A – 203A

Kislinger, T.; Gramolini, A. O.; MacLennan, D. H.; Emili, A. (2005). Multidimensional protein identification technology (MudPIT): technical overview of a profiling method optimized for the comprehensive proteomic investigation of normal and diseased heart tissue. *Journal of the American Society for Mass Spectrometry*, Vol. 16, No. (8), pp. 1207–1220

McNaught, A. D.; Wilkinson, A. (1997). *Compendium of Chemical Terminology (the "Gold Book").*, 2.02 ed.; Blackwell Scientific Publications: Oxford, ISBN 0-86542-684-8, North Carolina, United States of America

Meyer,V.R. (2010). *Practical High-Performance Liquid Chromatography (5th)*, Wiley, ISBN 978-0-470-68218-0, United Kingdom

Molloy, M.P. (2000). Two-dimensional electrophoresis of membrane proteins using immobilized pH gradients. *Analytical Biochemistry*, Vol.280, No.1, (April 2000), pp.1-10, ISSN 1096-03

Moritz, R.L. ; Clippingdale, A.B.; Kapp, E.A.; Eddes, J.S.; Ji, H.; Gilbert, S.; Connolly, L.M. & Simpson, R.J. (2005). Application of 2-D free-fl ow electrophoresis/ RP-HPLC for proteomic analysis of human plasma depleted of multi-high abundance proteins, *Proteomics*, Vol.5, pp. 3402–3413

Neue, U. D. (1997). *HPLC Columns, Theory, Technology and Practice*, Wiley, ISBN 978-0-471-19037-0, New York

O'Farrell, P. (1975). High Resolution Two-Dimensional Electrophoresis of Proteins. *The Journal of Biological Chemistry*, Vol.250, No.10, (May 1975), pp.4007–4021, ISSN 1083-351X

Oliver, R.W.A. (1991). *HPLC of macromolecules: a practical approach*, IRL press, ISBN 0199630208, England.

Pandey, A., and Mann, M. (2000) Proteomics to study genes and genomes. *Nature*, Vol.405, pp.837–846

Patton, W.F. (2002) Detection technologies in proteome analysis. *Journal of Chromatography B-Analytical Technologies in the Biomedical and Life Sciences*, Vol.771, No.(1-2), (May 2002), pp.3-31, ISSN 1873-376X

Pilny, R.; Bouchal, P.; Borilova, S.; Ceskova, P.; Zaloudik, J.; Vyzula, R.; Vojtesek, B.; Valik, D. (2006). Surface-enhanced laser desorption ionization/time-of-flight mass spectrometry reveals significant artifacts in serum obtained from clot activator-containing collection devices. *Clinical Chemistry*, Vol.52,No.11, pp.2115-6

Pinkse M.W.; Uitto P.M.; Hilhorst M.J.; Ooms B. & Heck A.J. (2004). Selective isolation at the femtomole level of phosphopeptides from proteolytic digests using 2D-NanoLC-ESI-MS/MS and titanium oxide precolumns. *Anal. Chem.* Vol.76, pp. 3935–43

Rabilloud, T.; Chevallet, M.; Luche, S. & Lelong, C. (2008). Fully denaturing two-dimensional electrophoresis of membrane proteins: A critical update. *Proteomics,* Vol. 8, pp. 3965–3973

Prakash, A.; Mallick, P.; Whiteaker, J.; Zhang, H.; Paulovich, A.; Flory, M.; Lee, H.; Aebersold, R.; Schwikowski, B. (2006). Signal maps for mass spectrometry-based comparative proteomics, *Mol. Cell.Proteomics,* Vol.*5, No.*3, pp.423–432

Rogers, S.; Girolami, M.; Kolch, W.; Waters, K.M.; Liu, T.; Thrall, B. & Wiley, H.S. (2008). Investigating the correspondence between transcriptomic and proteomic expression profiles using coupled cluster models *Bioinformatics,* Vol.24, pp. 2894–2900

Saito, Y.; Jinno, K.; Greibrokk, T. (2004). Capillary columns in liquid chromatography: between conventional columns and microchips, *Journal of Separation Science,* Vol. 27, No.17-18, (December 2004), pp. 1379-1390, ISSN 1615-9306.

Santoni, V.; Molloy, M. & Rabilloud, T. (2000). Membrane proteins and proteomics: Un amour impossible? *Electrophoresis,* Vol.21, No.6 , (April 2000), pp.1054-1070, ISSN 1522-2683

Shaw, J.; Rowlinson, R.; Nickson, J.; Stone, T.; Sweet, A.; Williams, K. & Tonge, R. (2003). Evaluation of saturation labelling two-dimensional difference gel electrophoresis fluorescent dyes. *Proteomics,* Vol.3, No.7, (July 2003), pp.1181-1195, ISSN 1615-9853

Siuti, N; Kelleher, N .L. (2007). Decoding protein modifications using top-down mass spectrometry, *Nature Methods,*Vol.4, pp.817-821

Slebos, R. J.; Brock, J. W.; Winters, N. F.; Stuart, S. R.; Martinez, M. A.; Li, M.; Chambers, M. C.; Zimmerman, L. J.; Ham, A. J.; Tabb, D. L.; Liebler, D. C. (2008). Evaluation of strong cation exchange versus isoelectric focusing of peptides for multidimensional liquid chromatography-tandem mass spectrometry. *J. Proteome Res.,* Vol.7, No.12, pp.5286–5294

Skoog, D.A.; Holler, J.F.; Crouch S.R. (2006). Liquid Chromatography, In: *Principles of Instrumental Analysis (6th),* Skoog, D.A.; Holler, J.F.; Crouch, Thomson Brooks/Cole (David Harris), pp. (816-855), ISBN 9780495012016, Canada.

Smejkal, G.B. (2004) The Coomassie chronicles: past, present and future perspectives in polyacrylamide gel staining. *Expert Review of Proteomics,* Vol.1, No.4, (December 2004), pp.381-387, ISSN 1478-9450

Tan, S.; Tan, H.T. & Chung, M.C.M. (2008). Membrane proteins and membrane proteomics. *Proteomics,* Vol. 8, pp. 3924–3932

Tabb, D. L.; Veja-Montoto, L.; Rudnick, P.A.; Variyath, A. M. et al. (2010). Repeatability and Reproducibility in Proteomic Identifications by Liquid Chromatography-Tandem Mass Spectrometry, *Journal of Proteome Research,* Vol.9, pp. 761–776

Tran, B. Q.; Loftheim, H.; Reubsaet, L.; Lundanes, E. & Greibrokk, T. (2008). On-Line multitasking analytical proteomics: How to separate, reduce, alkylate and digest whole proteins in an on-Line multidimensional chromatography system coupled to MS, *Journal of Separation Science,* Vol.31, pp. 2913–2923

Tsai, C.F.; Wang, Y.T.; Chen, Y.R.; Lai, C.Y.; Lin, P.Y.; et al. (2008). Immobilized metal affinity chromatography revisited: pH/acid control toward high selectivity in phosphoproteomics. *Journal Proteome Res.*, Vol. 7, pp. 4058–69

Tsai, C.F.; Wang, Y.T.; Lin, P.Y. & Chen, Y.J. (2011). Phosphoproteomics by Highly Selective IMAC Protocol, *Neuroproteomics*, Series Neuromethods, Vol.57, pp.181-196, Pub. Date: May-25-2011

Turner, M. W. & Hulme, B. (1970) *The Plasma Proteins: An Introduction*, Pitman Medical & Scientific Publishing Co., Ltd., London

Ünlü, M.; Morgan, M.E. & Minden, J.S. (1997). Difference gel electrophoresis: a single gel method for detecting changes in protein extracts. *Electrophoresis*, Vol.18, No.11, pp.2071-2077, (October 1997), ISSN 1522-2683

Wang, N., MacKenzie, L., De Souza, A. G., Zhong, H. Y., Goss, G., Li, L.. (2007), Proteome profile of cytosolic component of zebrafish liver generated by LC-ESI MS/MS combined with trypsin digestion and microwave-assisted acid hydrolysis, *Journal of Proteome Research,* Vol.6, pp. 263–272

Wilkins, M.R.; Gasteiger, E.; Sanchez, J.C.; Bairoch, A. & Hochstrasser, D.F. (1998). Two-dimensional gel electrophoresis for proteome projects: the effects of protein hydrophobicity and copy number. *Electrophoresis*, Vol.19, No.8-9, (June 1998), pp.1501-1505, ISSN 1522-2683

Wolters, D.A.; Washburn, M.P. & Yates, J.R.III (2001). An automated multidimensional protein identification technology for shotgun proteomics, *Analytical Chemistry*, Vol.73, pp. 5683–5690

Yokoyama, R.; Iwafune, Y.; Kawasaki, H. & Hirano, H. Isoelectric focusing of high-molecular-weight protein complex under native conditions using agarose gel. (2009). *Analytical Biochemistry*, Vol.387, No.1, (April 2009), pp.60-63, ISSN 1096-0309

Zhang, J.; Xu, X.; Gao, M.; Yang, P. & Zhang, X. (2007). Comparison of 2-D LC and 3-D LC with post- and pre-tryptic-digestion SEC fractionation for proteome analysis of normal human liver tissue, *Proteomics*, Vol.7, pp. 500–512

Zhou, F.; Cardoza, J.D.; Ficarro, S.B.; Adelmant, G.O.; Lazaro, J.B.; Marto, J.A. (2010). Online Nanoflow RP-RP-MS Reveals Dynamics of Multicomponent Ku Complex in Response to DNA Damage, *Journal of Proteome Research*, Vol. 9, No.12, pp. 6242-55

Zhou, H. J.; Tian, R. J.; Ye, M. L.; Xu, S. Y.; Feng, S.; Pan, C. S.; Jiang, X. G.; Li, X. & Zou, H. F. (2007). Highly specific enrichment of phosphopeptides by zirconium dioxide nanoparticles for phosphoproteome analysis. *Electrophoresis*, Vol.28, No.13, pp. 2201-2215

2

Evolution of Proteomic Methods for Analysis of Complex Biological Samples – Implications for Personalized Medicine

Amanda Nouwens and Stephen Mahler
The University of Queensland
Australia

1. Introduction

We are on the threshold of a paradigm shift for proteomics, moving from largely a qualitative discipline, to now being capable of quantification of a protein within a complex sample at great sensitivity. The potential application of such advanced proteomic technology is enormous as we will be able to detect and quantify low levels of expressed proteins in complex samples, and so move comparative proteomics to a new level.

The evolving practice of personalized medicine will be dependent on devising new techniques and methodologies that will allow the detection and quantification of proteins that are implicated in contributing to the diseased state. There is perhaps somewhere over 5000 genes that are linked to disease states and complex networks of interactions of these expressed genes ultimately lead to these disease states. The myriad of single nucleotide polymorphisms (SNPs) contribute to such phenomena, as an individual's SNP profile play a major role in susceptibility to disease and in adverse reactions to drugs, for example. Coupled with mutations that occur throughout life, the complex "disease state" proteome will contain mutant proteins at low levels that need to be identified and quantified, so that therapeutic intervention based on rational scientific hypotheses can be investigated.

Plasma and serum contain an unknown number of proteins with amounts ranging from pg-g/L levels (i.e. very high dynamic range). As we know one of the major problems faced by proteomic studies of plasma or serum, or indeed any complex protein sample, is that a relatively small number of abundant proteins accounts for the great majority of protein content of the sample. The upshot is that the proteins of interest, which may have regulatory function, are masked by these abundant proteins, and non-targeted methods of proteomic analysis bias at the top end of the abundance scale. The development of new methods for quantifying low abundance proteins has evolved rapidly, concomitant with the evolution of powerful mass spectrometers of increasing sensitivity. The use of antibodies for targeting peptides prior to mass spectrometry analysis is becoming prominent, as a means of partitioning low abundance peptides away from peptides in the bulk sample.

This review will provide a broad overview of the evolution of proteomic methods to analyse biological samples, including Differential In-Gel Electrophoresis (DIGE), Isotope-Coded Affinity Tag (ICAT), Isobaric tags for relative and absolute quantification (iTRAQ), Stable isotope labeling with amino acids in cell culture (SILAC), Unique ion signature Mass

Spectrometry, Selected reaction monitoring (SRM)- based targeted mass spectrometry and Stable Isotope Standards and Capture by Anti-Peptide Antibodies (SISCAPA). Examples of the application of these methods to the identification of proteins involved in a variety of disease states and their implication for personalized medicine will be provided.

2. Overview of personalized medicine

Personalized medicine is designed medicine based on the genotype, or more specifically the SNP profile of individuals. Personalized medicine facilitates the selection of treatments best matched to the individual and disease phenotype (Marko-Varga et al., 2007). What are the main factors which contribute to genotype diversity? SNP is the overriding factor, reflecting past mutation, and occurs wherever there is more than one nucleotide when comparing two sequences. It is the spread of SNPs within genomes which contribute to our individuality, with an estimated 93% of genes containing an SNP (Chakravati, 2001). The individual "fingerprint" of SNPs reflects differences in susceptibility to disease and our varying response to drugs. Pharmacogenetics is the study of how these differences in genotype are manifested in inter-individual variation in response to drugs. The convergence of traditional pharmacogenetics with the relatively new discipline of human genomics has resulted in the evolution of pharmacogenomics (Weinshilbaum et al., 2004). The Pharmacogenetics and Pharmacogenomics Knowledge Base (PharmGKB) is an entity associated with the cataloging genes involved in modulating the response to drugs. The Pharmacogenomics Research Network (PGRN) is a collaboration of scientists studying the effect of genes on responsiveness to a wide variety of medicines (Altman, 2007). The PGRN is linked to the PharmGKB integrative database containing genetic and clinical information on participants in studies (http:www.pharmgkb.org). Thus the integration and availability of data associated with information at the genomic and transcriptomic levels are well developed and a valuable resource for researchers involved in the development of personalized medicine.

The incorporation of proteomics in the further development of the concept of personalized medicine is a more recent phenomenon, and has given rise to the area of pharmacoproteomics, which in essence studies how the proteome changes in response to a drug, and is a logical extension of the pharmacogenetics and pharmacogenomics (Jain, 2004). While pharmacogenetics and pharmacogenomics provide information at the level of the genome and transcriptome, pharmacoproteomics yields information on function (i.e. translational level), although it should be pointed out that small nuclear (sn) RNAs have relatively recently been shown to contribute to regulation of cellular processes and thus have a functional role (Mercer et al., 2009; Mattick, 2009). To this end proteomics and proteomic profiling of individuals' serum and tissues are becoming increasingly important in patient diagnosis and assessment, and together with pharmacogenetics and pharmacogenomics, will provide a more complete picture of the status of an individual, particularly at the functional level.

Identification of disease states can be based on genomic analyses. For example, identification of mutations in breast cancer genes BRAC1 and BRAC2 can be used in the diagnosis of breast cancer (Miki et al., 1994; Wooster et al., 1995). However, at present DNA alone does not necessarily reflect the physiological state of functioning cells and thus analysis of gene products, both RNA and protein are required. RNA expression in comparison to protein, is easier to perform, but transcript levels in the cell do not always reflect protein levels, nor account for post-translational modifications on proteins or alternative splicing events.

2.1 Significance

There are a number ways in which proteomics may be utilized in personalized medicine and more broadly in drug discovery, research and development. In its simplest application, proteomics has contributed to the discovery of disease biomarkers, clinical entities that define and /or predict normal and pathogenic states (Krejsa et al., 2006). Furthermore, the clinical response to treatment can be monitored through proteome profiling of relevant biomarkers. The clinical use of a biomarker is contingent on whether it is a validated biomarker, which ultimately depends on its clinical reliability and utility. Combinations of validated markers biomarkers can indicate surrogate endpoints that can predict clinical outcomes. On a more global protein expression level, comparative proteomics can generate patterns of protein expression or expression profiles, which may be utilized to define a specific physiological state, or diseased state.

One area where proteomics can yield information not possible by other means is the identification and localization of proteins in various cellular compartments and extracellular space. The paradigm that proteins have fixed locations within cells has recently proven to be simplistic, and that proteins have diverse functions depending on their cellular location. The identification of a protein outside of its known functional zone in cellular preparations was once thought to be due to rupture of cells/cell organelles and leakage of the protein into other fractions. However it is now known that proteins translocate between intracellular and extracellular compartments (Butler et al., 2009). This has enormous implications in drug targeting as the presence of a target in multiple locations may complicate therapy. For example, chaperone proteins, including HSP 10, 70 and 90 have now been shown to exist in extracellular locations, where it was once thought that the chaperone proteins were exclusively located intracellularly, to aid protein folding and carry out chaperone function. Inhibitors of HSP90α are in clinical trials for treatment of cancer (Banerji, 2009), however inhibition of the extracellular wound-healing function of HSP90α21 could be an undesired adverse effect.

2.2 Personalized medicine for cancer

Due to the great diversity of cancer types, and individual variation within specific tumours, cancer perhaps shows the greatest potential for development of personalized therapy. Cancer accounts for about 13% of all deaths annually world wide, and is a major cause of morbidity and mortality (Krause-Heuer et al., 2009). Notwithstanding the emergence of new chemotherapeutic drugs and novel therapies for cancer, significant challenges remain for understanding tumourigenesis and tumour cell biology, and in developing new, effective strategies for cancer treatment. (Mozafari et al., 2009). Some of these challenges include;

- Identification of new tumour targets
- Drug potency, due to inadequate concentration at the cell surface.
- Non-selective nature of cytotoxic agents and a low therapeutic index.
- Development of multi-drug resistance (MDR)

The development of therapeutic monoclonal antibodies (mAbs) has shown promise in treatment of cancer amongst other indications. At present there are around 30 approved therapeutic mAbs predominantly for treating cancer and diseases associated with inflammation (Walsh, 2010). Anti-cancer antibodies are designed to target tumour cell surface antigens, with subsequent eliciting of an immune response on tumour binding. Most commonly this is termed antibody dependent cellular cytotoxicity (ADCC), and activated natural killer cells are recruited to attack the tumour. Currently approved anti-cancer therapeutic mAbs targeting the tumour cell surface are specific for antigens including EGFR, HER2/*neu*, CD20, CD52 and CD33. Other targets which are receiving much attention are the mucins, principally MUC1 (there is no existing approved antibody for MUC1), IGFR and CEA and cancer stem cell antigen CD44. The future development of monoclonal antibody (and other) cancer therapies will be contingent on the identification, development and validation of new tumour targets. However identification of new tumour biomarkers that reliably and accurately diagnose early stage cancer has not been met with great success. As an example, a group of researchers, as part of the Early Detection Research Network (EDRN), tested a group of recently discovered putative biomarkers for ovarian cancer, however none were superior to CA-125, which has been used extensively for 30 years. Notwithstanding the significant challenges in the discovery and development of clinically useful biomarkers, proteomics will be central for the discovery of new, novel biomarkers for early detection and diagnosis; some of these biomarkers may be suitable for development as novel drug targets (Pastwal et al., 2007).

Recently the International Cancer Genomes Consortium (ICGC) was formed, with a charter to co-ordinate and integrate large-scale cancer genome sequencing projects, focusing on 50 different types of cancers (The International Cancer Genome Consortium, 2010). The expanded studies will consist of investigating around 25,000 specific cancers (biopsy material from individuals). The primary objectives of the consortium were made public in April 2008 and were released in April 2008 (http://www.icgc.org/files/ICGC_ April_29_2008.pdf). However these studies will be at the genomic, epigenomic and transcriptomic levels. At the proteomic level, the Human Proteome Organization (HUPO, http://www.hupo.org/) instigated the Human Proteome Project (HPP), a co-ordinated global initiative to map the protein-based molecular architecture of the human body. This initiative, will aid in the discovery and cataloging of new tumour associated antigens and potential targets.

2.3 Evolution of methods of analysis using proteomics

The term proteomics, encompassing the analysis of and tools used to examine proteins expressed by a genome was coined only about 15 years ago (Wilkins et al., 1996). The progression and development of proteomics since this time however, has naturally afforded a refinement in techniques and methods for simultaneously detecting, identifying and quantifying proteins in biological samples. Fundamental to any identification or further characterization is the ability to first separate proteins from complex samples. This is particularly important for samples such as blood where it is estimated that the dynamic range of proteins is greater than 10exp9. With such a vast dynamic range, just 22 proteins account for 99% of protein content in blood (reviewed in Simpson et al., 2008). With respect to the search for protein biomarkers, the situation is also complicated by the notion that most biomarkers will be low abundance proteins. As protein function and levels of abundance are often altered in disease states, identification of such changes by comparison of healthy and disease samples will allow a greater understanding of the disease, provide new therapeutic targets, as well as identify markers of disease status. Establishment of what could be defined as the "healthy phenotype", will depend on detailed characterization of the proteomes of healthy and diseased states of cells/tissue. One school of thought suggests that by creating complex "proteomic fingerprints" of healthy and diseased states (and transitions thereof), one may recognize perturbations from the healthy state phenotype before manifestation of the disease state. Therapeutic intervention during the transition-to-disease state may instigate a reversion to the healthy phenotype. Thus a systems biology approach to studying the proteomes of cells in normal and diseased states, and also the network of protein-protein interactions, should enhance the opportunity for attaining this goal.

Proteomics and the tools to identify and quantify proteins have evolved substantially since its conception. The availability of genomic information, particularly the complete human genome sequence has in many ways pushed the bottleneck from the genomic to proteomic arena. Developments include tools for protein separation, protein identification, quantification and automated processes. The following sections provide a summary of some of the major approaches to protein separation, identification and quantification using both gel and gel-free proteomic methods. Separation of proteins can be based on one or more physical or biochemical parameter including size, pI, sub-cellular location, or other depletion / enrichment strategies, with separation involving two or more 'orthagonal' approaches providing greater separation than a single dimension alone.

2.3.1 Two-dimensional gel electrophoresis

One of the earliest approaches to protein separation was based on the use of two-dimensional gel electrophoresis (2DGE), in which proteins are first separated by charge in a pH gradient (isoelectric focusing), followed by separation based on size in SDS-PAGE gels (O'Farrell, 1975). The gel is subsequently stained to visualize proteins, with each protein spot-volume representative of abundance of the protein(s) within it. This approach, particularly when combined with other upstream fractionation steps (e.g. sub-cellular) can provide separation and resolution of a large number of proteins (Cordwell et al., 2000) and has been used in various projects aiming to identify differentially expressed proteins – potentially biomarkers, by comparing control vs. diseased samples via gel-analysis software and statistical tests. Using 2DGE, identification of candidate biomarkers has been achieved for a range of diseases including atherothrombotic ischemic stroke (Brea et al., 2008), pancreatic cancer (Park et al., 2011) and breast cancer (Lee, et al., 2011) with many studies identifying new potential markers.

2.3.2 Differential In-Gel Electrophoresis (DIGE)

Improvements to 2DGE include the addition of small fluorescent tags (CyDyes) on protein samples prior to separation, thus allowing multiple samples to be combined in the same physical gel (Tonge et al., 2001). This approach, known as Differential In-Gel Electrophoresis or DIGE, circumvents some of the problems of quantifying proteins across different samples in different gels as the distinct fluorescent tags on different samples allow the researcher to detect proteins from control vs. disease samples simultaneously. The inclusion of a pooled internal standard representing a mix of all samples can help circumvent some of the technical difficulties (e.g. gel warping and spot matching) that arise from single-sample gels. An advantage of DIGE is that due to the sensitive fluorescent nature of the CyDye labels low amounts of sample (as little as 10 μg) are required for analysis. However, a caveat is that although protein separation and quantification can be achieved, the identity of the proteins still remains unknown unless the protein spots are excised and further analysed, which can be difficult due to the low amounts of protein used in the analysis. Limitations also exist in the dynamic range of protein detection, estimated at 10exp4 (reviewed in Rabilloud, 2002) Typically a separate, unlabelled gel with larger amounts of protein loaded are required to be run and subsequently cross matched with the original DIGE experiment, as described by Matigian et al., 2010. Despite these difficulties, 2DGE has a proven track record in the separation and identification of proteins, with numerous differentially expressed proteins and potential biomarkers uncovered. 2DGE has been applied to a wide range of sample types including tissues (e.g. breast, skin, brain) and fluids such as cerebrospinal fluid, serum, urine, and tears targeting diseases such as various cancers, Alzheimers disease and dementia, cardiovascular diseases, infections such as HIV, conjunctivitis and toxoplasmosis. Overall, 2DGE with or without fluorescent labeling of samples does provide good separation of proteins, but is time consuming and a laborious process both in the gel procedures, and analysis of 2D protein spot profiles.

2.4 Mass spectrometry-based approaches

Mass spectrometry (MS) has formed the basis for standard protein identification for many years, typically in a 'bottom-up' approach (in which proteins are digested, usually with trypsin) and the resulting peptides analyzed to determine protein identity, but also in some cases by 'top-down' approaches where intact proteins form the basis of analysis. MS-based approaches hold some advantages over traditional 2DGE methods in that samples can potentially be analyzed and identified simultaneously through methods such as two dimensional LC-MS/MS using a combination of strong cation exchange followed by reverse phase separation of peptides. In comparison to gel-based approaches, MS analyses appear to be more effective at identifying low abundance proteins, as well as those with extreme physical properties such as molecular weights (low or high) or pI values, which are often difficult to resolve on gels. MS-based analyses also offer better prospects for automation of separation, analysis and identification of proteins.

With the ability to rapidly identify large numbers of proteins via MS, the emphasis has since shifted to also quantifying those proteins detected by MS. Broadly speaking, the approaches to quantifying proteins via mass spectrometry can be based on labeled or label-free methods. Each approach has its own advantages and disadvantages. Label-based methods include Isotope-Coded Affinity Tag or ICAT, and Isotope Tag for Relative and Absolute Quantitation or ITRAQ, in which multiple samples can be labeled, mixed and then analyzed simultaneously via MS to avoid technical issues relating to reproducibility that may be encountered with label-free approaches.

2.4.1 ITRAQ, ICAT and SILAC

ICAT and ITRAQ differ in their labeling chemistries and site of attachment. For ICAT, cysteine (cys) residues are targeted and selected for via avidin affinity chromatography. The enrichment of only those cys-containing peptides provides one avenue to quantify samples without the complexity of analyzing all peptides in a sample. However, ICAT becomes problematic for analysis of proteins which lack any cys residues. Furthermore, as reviewed in Patton et al., 2002, approximately 70% of proteins contain four or less cys residues thus limiting the usefulness of this approach.

ITRAQ utilizes lysine residues for labeling (Ross et al., 2004), thus avoiding the problem of limited cys residues encountered with ICAT. ITRAQ is a multiplexed approach, where tags are based on isobaric reagents. This means that up to eight different samples can be labeled with unique tags. The physical properties of the tags differ only in the isotopes used in their synthesis, meaning during LC separations, and in MS mode are identical. Only upon fragmentation in MS/MS mode are the isobaric tags distinguishable. The result is that proteins can be identified via MS/MS and due to unique reporter ions from the ITRAQ tag, the protein can also be quantified. The initial ITRAQ labels were designed to label up to four different samples, although tags to label and detect up to eight different samples are now available. Limitations for ITRAQ lie with the difficulty of identify proteins and quantify them when uniquely expressed in only one sample type eg a protein expressed only in the diseased state. Some technical difficulties, in particular with 8plex tags, resulting in a reduction in identification efficiency have also been reported (Thingholm et al., 2010).

Other label-based strategies such as Stable Isotope Labeling with Amino acids in Cell culture (SILAC) also exist. This method is based on labeling proteins in culture with heavy and light forms of amino acids. The approach is a useful way of comparing two samples, but is limited only to cells grown in culture or in some cases, animal models (Zanivan et al., 2012) and would not be applicable to human or clinical studies due to the use of radioactivity.

2.4.2 Selected reaction monitoring / multiple reaction monitoring

Selected reaction monitoring (SRM), also known as Multiple Reaction Monitoring (MRM) provides a targeted approach to quantifying proteins in a sample. It advances the 'global' approach to quantification by simply targeting those proteins specifically of interest to the researcher. Typically MS instruments such as triple quadrupoles are used, where a precursor mass representing a peptide (typically tryptic) from the protein of interest is selected, fragmented and specific product ions unique to that peptide, monitored. Generally for each protein of interest a number of precursor ions, and subsequent product ions are monitored to ensure specificity. This information is then used to identify and quantify the proteins present. The main limitation of SRM/MRM analyses is that it is essential to know the proteins of interest beforehand, so that appropriate precursor / product ion can be monitored.

SRM / MRM is a targeted approach, where identify of the protein(s) of interest must be known beforehand. The usefulness of SRM/MRM analyses is thus as a downstream technology after discovery-phase experiments have concluded, and candidate proteins of interest requiring quantification already identified. The advantage of SRM / MRM assays is the ability to simultaneously monitor numerous potential biomarkers in a single analysis and quantify protein levels and is thus currently a popular area of investigation. To help with SRM / MRM analyses, a consortium called SRMAtlas has been established (www.mrmatlas.org) to quantify proteins in complex samples by MS. As well as human entries, mouse and yeast information is also contained, and provides both web-interface and command line tools to search for assays. This readily available information means researchers can potentially circumvent some method development steps as optimal coordinates for SRM / MRM transitions of numerous proteins are available.

2.4.3 Stable isotope standards and capture by anti-peptide antibodies

Stable isotope standards and capture by antipeptide antibodies (SISCAPA) is a method which allows the quantification of peptides from complex digests. Originally described by Anderson et al., 2004 the method utilizes stable-isotope-labeled internal standards for comparison with (unlabelled) peptides that are enriched via anti-peptide antibodies, with subsequent quantification performed by electrospray mass spectrometry. The approach offers increased sensitivity over non-enriched methods, particularly when coupled with SRM / MRM assays. SISCAPA also offers potential in the verification of diagnostic protein panels from large samples as well as increased efficiency in assay time for the bind/elute process over conventional reverse phase separations. There are also distinct advantages over traditional techniques such as ELISA in development time for biomarker assays (Whiteaker et al, 2009). The main disadvantage of SISCAPA is the need for preselected targets, as well as the cost in producing the internal peptide standards and generation of the peptide binding antibodies. However, given the sensitivity of the assay (only low fmol - pmol amounts are required), and the fact the antibody itself can be recycled and used again means on-going costs can be reduced. Since the original design of SISCAPA, refinements in the assay have been developed to reduce loss of low abundance peptides, automated processing steps, and improvement in antibody sources i.e. from polyclonal to monoclonal (Anderson et al., 2009; Schoenherr 2009). As this method is only a relatively recent development, no biomarkers have as yet been published as validated with this approach, although proof-of-principle experiments have been performed with established biomarkers such as tropinin I (cTnI) (Kuhn, et al., 2009) and thus SISCAPA remains a promising tool.

2.4.4 Alternative strategies

In addition to the above technologies, other strategies have been developed to complement gel and MS approaches to detect biomarkers through improved sample preparation methods. For example, hexapeptide libraries, based on combinatorial peptide libraries offer a way to deplete samples of highly abundant proteins (Guerrier et al., 2008). In this innovative technique, a large collection of specific hexapeptides (hexapeptide library) is attached to beads. The complex protein sample of interest is mixed with the hexapeptide-bead library. The peptide library is of high diversity and so it would be expected that a specific peptide(s) in the library would have affinity for each individual protein in the complex sample. After separation of the beads from the mixture, the adsorbed proteins are eluted from the beads. As each hexapeptide is equally represented within the library, the end result is that the abundant proteins are depleted, while proteins of low abundance are concentrated. This approach is particularly useful for biological fluids (serum, saliva, urine etc) which have particularly large dynamic range of protein abundance. For example, hexapeptide enrichment of urine has uncovered an additional 251 proteins that were not previously known to be present in this fluid (Castagna et al., 2005). Although depletion/enrichment strategies may not, in their own right, uncover biomarkers, their usefulness lies in the ability to mine deeper into the proteome of these highly complex samples so that low abundance proteins can be identified.

Depletion / enrichment strategies can be problematic if the abundant protein is a carrier for low abundance molecules and the use of depletion strategies must be done with caution. For example it has been shown that the depletion of albumin from human plasma can also remove low abundance proteins such as cytokines (Granger et al., 2005). More recent studies (Bellei et al., 2010) have also concluded that removal of high-abundance proteins can result in a loss of non-targeted, less abundant proteins. Obviously unintentional and unknowing loss of low abundance proteins is a cause for concern in the search for biomarkers of disease.

Other approaches for analysis of samples include Surface Enhanced Laser Desorption Ionisation (SELDI) and Matrix Assisted Laser Desorption Ionisation (MALDI), which have both been utilised particularly in examination of body fluids such as serum for biomarker discovery. This is effectively a "protein pattern recognition" approach (reviewed in Zhan & Desiderio, 2010) which compares profiles from control versus disease samples to identify those proteins differentially expressed. This approach has been used in particular for analysis of cancer patients.

2.4.5 Post-translational modifications

The majority of the above technologies focus on protein expression and differential expression in control vs. disease states. However, greater emphasis in the future on protein post-translational modifications (PTMs) such as glycosylation and phosphorylation will be needed. Already, perturbations in modifications of proteins by the glycan O-linked B-N-acetylglucosamine (O-GlcNAc) has been implicated in a range of diseases, including Alzheimers and diabetes (reviewed in Dias & Hart 2007). Similarly, differential phosphorylation has been identified in diseased states such as cancer compared to control patients (Semaan et al., 2011).

2.4.6 The Human Proteome Project

Fundamental to rational design for disease treatment and prevention is the understanding of genes present, and the expression and function of gene products, including proteins involved in the disease process. The Human Genome Project (HGP) was established to map all genes encoded in the human genome. Surprisingly, the total number of protein coding genes, only approximately 20,300, was substantially lower than expected, with increased complexity presumably due to splice variants, and post-translational modifications. Following on from this ground-breaking work, is the recent establishment of the Human Proteome Project which aims to map the human proteome (Legrain et al., 2011). At present, of the protein-coding genes in humans identified in the human genome, approximately one third have not been detected at the protein level, while for many others, basic information such as abundance, sub-cellular localization, or function are unknown. Mapping of the proteome will be valuable in understanding human biology, and downstream applications in developing diagnostics, prognostics and new therapies to treat diseases. The HPP will have a 'gene-centric' approach to map information about proteins back to gene loci. HPP will aim to address three parts (HUPO Views, 2010):

- Identification and characterization of proteins from every gene.
- Distribution of proteins in all normal tissues and organs.
- Mapping of pathways and protein networks and interactions.

With respect to sample type, bodily fluids relatively easily attainable, such as urine, saliva, tears as well as those requiring more slightly more invasive methods for collection such as serum, plasma and CFS have all been analyzed for a variety of diseases. Fluids as opposed to solid tissues would generally form a better basis for determining personalized signatures and biomarker detection due to their ease of attainment. There has been some question over whether blood is the best choice for searching for biomarkers. The rationale has been that specific proteins are secreted by the body from different organs, and that these can represent a biological "fingerprint" of physiological state (reviewed in Simpson et al., 2008).

Using model systems such as mice, researchers have shown changes in the plasma proteome prior to any clinical evidence of breast cancer (Pitteri et al., 2011). A separate study in humans (Li, 2011) also suggests that it may be possible to observe proteome plasma changes prior to diagnosis. Plasma proteomics has also been used in the search for pre-diagnostic markers in other diseases such as coronary heart disease (Prentice et al., 2010). The ability to identify proteome changes prior to manifestation of disease phenotypically, will potentially improve patient outcome, particularly for those diseases such as cancer where early diagnosis strongly correlates with survival rates.

3. Conclusion

The ability to define proteomic "signatures"' for individuals will vastly enhance the ability of the medical community to diagnose and treat diseases, as well as potentially identify disease before symptoms appear. Early treatment will in turn prolong life, as well as potentially address healthcare costs through the application of more refined and defined therapies suitable for individual patients. The heterogeneity of some diseases such as breast cancer, in which specific proteins, e.g. progesterone receptor, estrogene receptor and HER-2 may or may not be expressed, make it difficult to broadly treat patients, as a 'one size-fits all' approach does not always apply, and emphasizes the need for individualized and personalized medicine. By examining the proteome, it is possible to gain a better understanding of the heterogeneity present in an individual and potentially can help determine best choice of therapies, as well as indicate disease status and progression. Given the complication of genetic factors and environmental influences on an individual, personalized medicine strategies will require complementation of proteomic data with other areas and strategies for analysis and compile this information to determine diagnostic approaches and tailor therapeutic strategies for the individual.

As yet, despite the excitement of biomarker discovery, and the vast number of publications claiming detection of biomarkers for a specific disease, the majority of candidate biomarkers are yet to be validated or used in clinical settings. However, once candidate biomarkers are confirmed, the emphasis will be on high-throughput approaches to expand analyses to greater numbers of samples. Clearly the proteomic tools available to detect and characterize samples, particularly in a high-throughput quantitative fashion are now a reality. Thus, personalized medicine is not far off the horizon. We anticipate a new era of therapeutic approaches and more refined medicinal treatments for diseases which will be more targeted and precise, not just for the disease, but for the individual, based on establishment of "proteomic fingerprints". In addition to greater confidence in diagnoses, proteome signatures would allow a more individualized and targeted approach to therapy. Potentially, such signatures may also provide better insight into future recurrences of the disease.

Besides the quest for discovery, research and development of new and unique biomarkers, other facets of biomarker research incorporate aims such as improving reliability, increasing the speed of detection and reducing the amount of sample needed for analysis. However, the search for biomarkers is particularly important for those diseases such as breast cancer, for which there are no current clinical biomarkers available, and for which mortality is tightly related to disease stage in the initial surgery (Bohm et al., 2011). Using proteomics however, a biomarker signature for non-metastatic breast cancer has been uncovered. This study (Bohm et al., 2011) found using serum samples and SELDI-TOF and MALDI-

TOF/TOF analyses, a combination of 14 biomarkers that can identify breast cancer patients from controls, with a specificity of 67%. It is unlikely at this stage that this panel or signature will entirely replace imaging diagnostics, but does have the potential to aid current diagnostic approaches, particularly when cancer survival rates are greatly improved with early detection, while tumors <5 mm are normally not detected.

One of the fundamental problems of assigning 'biomarker' status for a protein found to be differentially expressed in a disease is the overlap of these differentially expressed proteins across different diseases. A number of proteins have been implicated across a number of different diseases, making the notion of a single biomarker to indicate a specific disease more difficult. For example, serum amyloid A has been proposed as a prognostic marker for melanoma (Findeisen et al., 2009), breast cancer (Schaub et al., 2009), atherothrombotic ischemic stroke (Brea et a., 2009). Potentially, for greater confidence in disease diagnosis or prognosis, it may be required that a suite of biomarkers, be needed to provide greater specificity and confidence.

The significance of the future development of personalized medicine is far reaching, and will allow/facilitate the following:

- Predicting a patient's response to drugs.
- Development of customized' prescriptions.
- Minimizing, or in some cases eliminating adverse events.
- Improving rational drug development.
- Improving drug R&D and the approval of new drugs - better designed clinical trials based on genomic/proteomic information.
- Screening and monitoring certain diseases e.g. advanced diagnosis before disease symptoms.
- Reducing the overall cost of healthcare.

If the concept of routine personalized medicine is to become a reality in the future, the development of new proteomic techniques and methodologies will be vital, and will build on current methodologies now available.

4. References

Altman, R. (2007). PharmGKB: a logical home for knowledge relating genotype to drug response phenotype. *Nature Genetics*, Vol.39, pp. 426, ISSN 1061-4036

Anderson, N.L.; Anderson, N.G.; Haines, L.R.; Hardie, D.B.; Olafson, R.W.; & Pearson T.W. (2004). Mass Spectrometric Quantitation of Peptides and Proteins Using Stable Isotope Standards and Capture by Anti-peptide Antibodies (SISCAPA) *Journal of Proteome Research,* Vol.3, pp. 235-244, ISSN 1535-3893

Anderson, N.L.; Jackson, A.; Smith, D.; Hardie, D.; Borchers, C. & Pearson, T.W. (2009) SISCAPA Peptide Enrichment on Magnetic Beads Using an In-line Bead Trap Device. *Molecular & Cellular Proteomics,* Vol.8, pp.995-1005, ISSN 1535-9476

Banerji, U. (2009). Heat shock protein 90 as a drug target: some like it hot. *Clinical Cancer Research,* Vol.15, pp. 9–14, ISSN 1078-0432

Butler, G.S. & Overall, C.M. (2009). Proteomic Identification of Multitasking Proteins in Unexpected Locations Complicates Drug Targeting *Nature Reviews Drug Discovery,* Vol.8, pp. 935-948, ISSN 1474-1776

Bellei, E.; Bergamini, S.; Monari, E.;Fantoni, L.I.; Cuoghi, A.; Ozben, T.; & Tomasi, A. (2010). High-abundance proteins depletion from serum proteomic analysis: concomitant removal of non-targeted proteins. *Amino Acids* Vol.40, pp.145-156 ISSN 0939-4451

Bohm, D.; Keller, K.; Wehrwein, N.; Lebrecht, A.; Schmidt, M.; Kolbl, H. & Grus, F.-H. (2011). Serum proteome profiling of primary breast cancer indicates a specific biomarker profile. *Oncology Reports,* Vol.26, pp. 1051-1056, ISSN 1021-335X

Brea, D.; Sobrino, T.; Blanco, M.; Fraga, M.; Agulla, J.; Rodriguez-Yanez, M.; Rodriguez-Gonzalez, R.; Perez de la Ossa, N.; Leira, R.; Forteza, J.; Davalos, A. & Castillo, J. (2009) Usefulness of haptoglobin and serum amyloid A proteins as biomarkers for atherothrombotic ischemic stroke diagnosis confirmation. *Atherosclerosis,* Vol.205, pp. 561-567, ISSN 0021-9150

Buchen, L. (2011). Missing the Mark: Why Is It So Hard to Find a Test to Predict Cancer *Nature,* Vol.471, pp. 428-432, ISSN 0028-0836

Butler, G.S. & Overall, C.M. (2009). Proteomic Identification of Multitasking Proteins in Unexpected Locations Complicates Drug Targeting *Nature Reviews Drug Discovery,* Vol.8, pp. 935-948, ISSN 1474-1776

Castagna, A.; Cecconi, D.; Sennels, L.; Rappsilber, J.; Guerrier, L.; Fortis, F.; Boschetti, E.; Lomas, L.; & Righetti, P.G. (2005). Exploring the Hidden Human Urinary Proteome via Ligand Library Beads. *Journal of Proteome Research,* Vol.4, pp. 1917-1930, ISSN 1535-3893

Chakravati, A. (2001). To a Future of Genetic Medicine (2001) *Nature,* Vol.409, pp. 822-823, ISSN 0028-0836

Cordwell, S.J.; Nouwens, A.S.; Verrills, N.M.; Basseal, D.J.; & Walsh, B.J (2000). Sub-proteomics based upon protein cellular location and relative solubilities in conjuction with composite two-dimensional electrophoresis gels. *Electrophoresis,* Vol. 21, pp. 1094-1103, ISSN 0173-0835

Dias, W.B. & Hart, G.W. (2007) *O*-GlcNAc modification in diabetes and Alzheimer's disease. *Molecular Biosystems,* Vol.3, pp. 766-772, ISSN 1742-206X

Filiou, M.; Turck, C. & Martins-de-Souza, M. (2011). Quantitative proteomics for investigating psychiatric disorders. *Proteomics Clinical Applications,* Vol.5, pp. 38-49, ISSN 1862-8354

Granger, J.; Siddiqui, J.; Copeland, S.; & Remick, D. (2005). Albumin depletion of human plasma also removes low abundance proteins including the cytokines. *Proteomics* Vol.5, pp. 4713-4718, ISSN 1615-9853

Guerrier, L.; Righetti, P.G.; & Boschetti, E. (2008). Reduction of Dynamic Protein Concentration Range of Biological Extracts for the Discovery of Low-Abundance Proteins by Means of Hexapeptide Ligand Library. *Nature Protocols,* Vol.3, pp.883-890, ISSN 1750-2799

HUPO Views (2010). A Gene Centric Human Proteome Project. *Molecular and Cellular Proteomics,* Vol.9, pp. 427-429, ISSN 1535-9476

Jain, R. (2004). Role of Pharmacogenomics in the Development of Personalized Medicine *Pharmacogenomics,* Vol.5, pp. 331-336, ISSN 1462-2416

Krause-Heuer, A.M.; Grant, M.P.; Orkey, N. & Aldrich-Wright JR. (2008). Drug Delivery Devices and Targeting Agents for Platinum(ii) Anticancer Complexes. *Australian Journal of Chemistry,* Vol.61, pp. 675-81, ISSN 0004-9425

Krejsa, C.; Rogge, M. & Sadee, W. (2006). Protein therapeutics: new applications for pharmacogenetics. *Nature Reviews Drug Discovery,* Vol.5, pp. 507-521, ISSN 1474-1776

Kuhn, E.; Addona, T.; Keshishian, H.; Burgess, M.; Mani, D.R.; Lee, R.T.; Sabatine, M.S.; Gerszten, R.E.; Carr, S.A. (2009). Developing Multiplexed Assays for Troponin I and Interleukin-33 in Plasma by Peptide Immunoaffinity Enrichment and Targeted Mass Spectrometry. *Clinical Chemistry,* Vol.55, pp. 1108-1117, ISSN 0009-9147

Lee, S.; Terry, D.; Hurst, D.R.; Welch, D.R.; & Sang, Q-X.A. (2011). Protein signatures in human MDA-MB-231 breast cancer cells indicating a more invasive phenotype following knockdown of human endometase / matrilysin-2 by siRNA. *Journal of Cancer,* Vol.2, pp. 165-176, ISSN 1837-9664

Legrain, P.; Aebersold, R.; Archakov, A.; Bairoch A.; Bala, K.; Beretta, L.; Bergeron, J.; Borchers, C.H.; Corthals, G.L.; Costello, C.E.; Deutsch, E.W.; Domon, B.; Hancock, W.; He, F.; Hochstrasser, D.; Marko-Varga, G.; Salekdeh, G.H.; Sechi, S.; Snyder, M.; Srivastava, S.; Uhlén, M.; Wu, C.H.; Yamamoto, T.; Paik, Y.K. & Omenn, G.S. (2011) The human proteome project: current state and future direction. *Molecular and Cellular Proteomics,* Vol.10, M111.009993 ISSN 1535-9476

Li, C.I (2011). Discovery and Validation of Breast Cancer Early Detection Biomarkers in Preclinical Samples. *Hormones and Cancer,* Vol.2, pp. 125-131, ISSN 1868-8497

Marko-Varga, G.; Ogiwara, A.; Nishimura, T.; Kawamura, T.; Fujii, K.; Kawakami, K.; Kyono, Y.; Tu, H.; Anyoji, H.; Kanazawa, M.; Akimoto, S.; Hirano, T.; Tsuboi, M.; Nishio, K.; Hada, S.; Jiang, H.; Fukuoka, M.; Nakata, K.; Nishiwaki, Y.;Kunito, H.; Peers, I.; Harbron, C.; South, M.; Tim Higenbottam, T.; Nyberg, F.; Kudoh, S and I and Kato, H. (2007). Personalized Medicine and Proteomics: Lessons from Non-Small Cell Lung Cancer. *Journal of Proteome Research,* Vol.6, pp. 2925-2935, ISSN 1535-3893

Matigian, N.; Abrahamsen, G.; Sutharsan, R.; Cook, A.; Vitale, A.; Nouwens, A.; Bellette, B.; An, J.; Anderson, M.; Beckhouse, A.; Bennebroek, M.; Cecil, R.; Chalk, A.; Cochrane, J.; Fan, Y.; Féron, F.; McCurdy, R.; McGrath, J.; Murrell, M.; Perry, C.; Raju, J.; Ravishankar, S.; Silburn, P.; Sutherland, G.; Mahler, S.; Mellick G.; Wood, S.; Sue, C.; Wells, C. & Mackay-Sim, A. (2010). Disease-specific, neurosphere-derived cells as models for brain disorders. *Disease Models and Mechanisms,* Vol.3, pp. 785-798, ISSN 1754-8403

Mattick, J.S. (2009). Deconstructing the dogma: a new view of the evolution and genetic programming of complex organisms. *Annals of the New York Academy of Sciences,* Vol.1178, pp. 29-46, ISSN 0077-8923

Mercer, T.R.; Dinger, M.E.; Sunkin, S.M.; Mehler, M.F. & Mattick, J.S. (2009). Long non-coding RNAs: insights into functions. *Nature Reviews Genetics,* Vol.10, pp. 155-159, ISSN 1471-0056

Miki, Y.; Swensen, J.; Shattuck-Eidens, D.; Futreal, P.A.; Harshman, K,; Tavtigian S.; Liu, Q.; Cochran, C.; Bennett, L.M.; Ding, W.; et al., (1994). A strong candidate for the breast and ovarian cancer susceptibility gene BRCA1. *Science,* Vol.266, pp. 66-71, ISSN 0036-8075

Mozafari, M.R.; Pardakhty, A.; Azarmi, S.; Jazayeri, J.A.; Nokhodchi, A. & Omri, A. (2009). Role of nanocarrier systems in cancer nanotherapy. *Journal of Liposome Research,* Vol.19, pp. 310-21, ISSN 0898-2104

O'Farrell, P.H. (1975). High resolution two-dimensional electrphoresis of proteins. *Journal of Biological Chemistry,* Vol.250, pp. 4007-4021

Park, J.Y.; Kim, S.-A.; Chung, J.W.; Bang, S.; Park, S.W.; Paik, Y.K. & Song, S.Y. (2011). Proteomic analysis of pancreatic juice for the identification of biomarkers of pancreatic cancer. *Journal Cancer Research and Clinical Oncology,* Vol.137, pp. 1229-1238, ISSN 0171-5216

Pastwal, E.; Somiari, S.; Czyz, M. & Somiari, R. (2007). Proteomics in human cancer research. *Proteomics Clinical Applications,* Vol.1, pp. 4-17, ISSN 1862-8354

Patterson, S.; Van Eyk, J. & Banks, R. (2010). Report from the Wellcome Trust/EBI "Perspectives in Clinical Proteomics" retreat - A Strategy to Implement Next-Generation Proteomic Analyses to the Clinic for Patient Benefit: Pathway to Translation. *Proteomics Clinical Applications,* Vol.4, pp. 883-887, ISSN 1862-8354

Patton, W.F.; Schulenberg, B. & Steinberg, T.H. (2002) Two-dimensional gel electrophoresis: better than a poke in the ICAT? *Current Opinion in Biotechnology,* Vol.13, pp.321-328, ISSN 0958-1669

Pitteri, S.J,; Kelly-Spratt, K.S.; Gurley, K.E,; Kennedy, J.; Busald Buson, T.; Chin, A.; Wang, H.; Zhang, Q.; Wong, C.-H.; Chodosh, L.A ; Nelson, P.S.; Hanash, S.M.; & Kemp, C.J. (2011). Tumor Microenvironment-Derived Proteins Dominate the Plasma Proteome Response during Breast Cancer Induction and Progression. *Cancer Research,* Vol.71, pp. 5090-5100, ISSN 0008-5472

Prentice, R.L.; Paczesny, S.J.; Aragaki, A.; Amon, L.M.; Chen, L.; Pitteri, S.J.; McIntosh, M.; Wang, P. ; Buson Busald, T,; Hsia, J. ; Jackson, R.D. ; Rossouw, J.E. ; Manson, J.E. ; Johnson, K.; Eaton, C. & Hanash, S.M. (2010). Novel proteins associated with risk for coronary heart disease or stroke among postmenopausal women identified by in-depth plasma proteome profiling. *Genome Medicine,* Vol.2, pp. 48-60, ISSN 1756-994X

Rabilloud, T. (2002). Two-dimensional gel electrophoresis in proteomics : Old, old fashioned, but it still climbs up the mountains. *Proteomics* Vol.2 pp. 3-10, ISSN 1615-9853

Remily-Wood, E.; Liu, R.; Xiang, Y.; Chen, Y.; Thomas, C.; Rajyaguru1, N.; Kaufman, L.; Ochoa, J.; Hazlehurst, L.; Pinilla-Ibarz, J.; Lancet, J.; Zhang, G.; Haura, E.; Shibata, D.; Yeatman, T.; Smalley, K.; Dalton, W.; Huang, E.; Scott, E.; Bloom, G.; Eschrich, S. & Koomen, J. (2011). A database of reaction monitoring mass spectrometry assays for elucidating therapeutic response in cancer. *Proteomics Clinical Applications,* Vol.5, pp. 383-396, ISSN 1862-8354

Ross. P.L.; Huang, Y.N.; Marchese, J.N.; Williamson, B.; Parker, K.; Hattan, S.; Khainovski, N.; Pillai, S.; Dey, S.; Daniels, S.; Purkayastha, S.; Juhasz, P.; Martin, S.; Bartlet-Jones, M.; He, F.; Jacobson, A. & Pappin, D.J. (2004). Multiplexed Protein Quantitation in Saccharomyces cerevisiae using amine-reactive isobaric tagging reagents. *Molecular and Cellular Proteomics,* Vol.3, pp. 1154-1169, ISSN 1535-9476

Schoenherr, R.M.; Zhao, L.; Whiteaker, J.R.; Feng, L.-C.; Li, L.; Liu, L.; Liu, X. & Paulovich, A.G. (2010) Automated Screening of Monoclonal Antibodies for SISCAPA Assays using a Magnetic Bead Processor and Liquid Chromatography-selected Reaction Monitoring-mass Spectrometry. *Journal of Immunological Methods,* Vol.353, pp. 40-61, ISSN 0022-1759

Simpson, R.J.; Berhhard, O.K.; Greening, D.W. & Moritz, R.L. (2008). Proteomics-driven cancer biomarker discovery: looking to the future. *Current Opinion in Chemical Biology,* Vol.12, pp. 72-77, ISSN 1367-5931

Semaan, S.M. ; Wang, X. ; Stewart, P.A.; Marshall, A.G. & Sang, Q.X.A (2011). Differential phosphopeptide expression in a benign breast tissue, and triple-negative primary and metastatic breast cancer tissues from the same African-American woman by LC-LTQ/FT-ICR mass spectrometry. *Biochemical and Biophysical Research Communications,* Vol.412, pp. 127-131, ISSN 0006-291X

Sherman, J.; McKay, M.; Ashman, K. & Molloy, M.P. (2009). Unique Ion Signature Mass Spectrometry, a Deterministic Method to Assign Peptide Identity. *Molecular and Cellular Proteomics,* Vol.8.9, pp. 2051-2062, ISSN 1535-9476

The International Cancer Genome Consortium (2010). International Network of Cancer genome Projects. *Nature,* Vol.464, pp. 993-998, ISSN 0028-0836

Thingholm, T.E.; Palmisano, G.; Kjeldsen, F. and Larsen, M.R. (2010). Undesirable Charge-Enhancement of Isobaric Tagged Phosphopeptides Leads to Reduced Identification Efficiency. *Journal of Proteome Research,* Vol.9, pp. 4045-4052

Tonge, R.; Shaw, J.; Middleton, B.; Rowlinson, R.; Rayner, S.; Young, J,; Pognan, F.; Hawkins, E.; Currie, I. & Davison, M. (2001). Validation and development of fluorescence two-dimensional differential gel electrophoresis proteomics technology. *Proteomics,* Vol.1, pp. 377-396, ISSN 1615-9853

Walsh G. (2010). Biopharmaceutical Benchmarks. *Nature Biotechnology,* Vol.28, pp. 917-924.

Weinshilbaum, R. & Wang, L. (2004). Pharmacogenomics: Bench to Bedside. *Nature Reviews Drug Discovery,* Vol.5, pp. 38-49, ISSN 1474-1776

Whiteaker, J.R.; Zhao, L.; Anderson, L. & Paulovich, A.G. (2010). An Automated and Multiplexed Method for High Throughput Peptide Immunoaffinity Enrichment and Multiple Reaction Monitoring Mass Spectrometry-based Quantification of Protein Biomarkers. *Molecular and Cellular Proteomics,* Vol.9, pp. 184-196, ISSN 1535-9476

Wilkins, M.R.; Pasquali, C.; Appel, R.D,; Ou, K,; Golaz, O.; Sanchez, J.-C.; Yan, J.X.; Gooley, A.A.; Hughes, G.; Humphery-Smith, I,; Williams, K.L. & Hochstrasser, D.F. (1996). From Proteins to Proteomes: Large Scale Protein Identification by Two-Dimensional Electrophoresis and Amino Acid Analysis. *Biotechnology,* Vol.14, pp. 61-65, ISSN 0733-222X

Wooster, R.; Bignell, G.;, Lancaster J, Swift S, Seal S, Mangion J, Collins N, Gregory S, Gumbs C, Micklem G. et al., (1995). Identification of the breast cancer susceptibility gene BRCA2. *Nature,* Vol.375, pp. 789-792, ISSN 0028-0836

Zanivan, S.; Krueger, M. & Mann, M. (2012). In vivo Quantitative Proteomics: The SILAC mouse. *Methods in Molecular Biology,* Vol.757, pp. 435-450, ISSN 1064-3745

Zhan, X. & Desiderio, D.M. (2010). The use of variations in proteomes to predict, prevent, and personalize treatment for clinically nonfunctional pituitary adenomas. *EPMA Journal*, Vol.1, pp. 439-459, ISSN 1878-5077

Part 2

Sample Preparation

3

Proteomic Analyses of Cells Isolated by Laser Microdissection

Valentina Fiorilli[1], Vincent P. Klink[2] and Raffaella Balestrini[1*]

[1]*Istituto per la Protezione delle Piante del CNR and Dipartimento Biologia Vegetale dell'Università di Torino, Torino*

[2]*Department of Biological Sciences, Harned Hall, Mississippi State University - Mississippi State*

[1]*Italy*

[2]*USA*

1. Introduction

Living organisms conduct biological processes by transducing biotic and abiotic stimuli through gene regulation into well-orchestrated growth and development. Analyzing RNA from a transcribed genome (the transcriptome) is fairly easy due to the availability of various nucleotide sequencing technologies. The translation of a transcriptome provides a blueprint of tens of thousands of different proteins that is known as the proteome (Wasinger et al., 1995). The analysis of proteins from whole tissues, organs or organisms has been made easier thanks to various technologies, including the Edman degradation technique, which is used in sequencing polypeptides (Edman, 1950), and 2 dimensional gel electrophoresis (2-DE), which could resolve up to 10,000 polypeptides (Barrett & Gould, 1973; O'Farrell, 1975; O'Farrell et al., 1977). Thus, the problems of assaying a transcriptome and a proteome from samples isolated from tissues, organs or whole organisms have largely been overcome. However, although it is fairly easy to assay a transcriptome and a proteome at these higher levels of biological organization, assaying a proteome at a cellular resolution level involves a set of problems that in primarily centered on the ability to collect sufficient cells for meaningful studies. The central focus of this chapter is a discussion on the technologies that have allowed the proteomic analyses of cells, isolated from complex samples thanks to a procedure that was first called laser microdissection (Isenberg et al., 1976).

2. The diversity of genome activity

Thanks to the recent advances in high-throughput technologies, the past decade has witnessed an explosion of global transcriptome profiling studies, which have produced novel insights into many developmental, physiological and medicinal aspects. Although a great deal of information can be obtained from transcriptome profiling, it is however insufficient for a comprehensive delineation of biological systems. A single approach cannot fully

* Corresponding Author

unravel the complexity of living organisms (Persidis, 1998). In addition, enzymatic reactions and signaling pathways depend on the activity of proteins, and protein quantities are regulated by protein synthesis and degradation. These processes may be independent of transcriptional control or have only a weak correlation (Lu et al., 2007; Nie et al., 2007). By generating information on the proteome at cellular resolution, a greater understanding of biological complexity is gained, including post-translational modification, isoforms, and splice variants, which may lead to the identification of important cell-specific protein entities (Schulze & Usadel, 2010). The proteomics approach can shed light on a number of protein species that can be translated from a single gene as a result of alternative splicing (AS) or PTMs. Proteomics analyses can also provide the biological meaning of each variant (Kim et al., 2007; Witze et al., 2007). For example, the *Drosophila Dscam1* gene, which encodes a membrane receptor protein, has 115 exons. The various combinations permit the possibility of 38,016 different proteins to be produced and many have been identified (Schmucker et al., 2000; Chen et al., 2006; Meijers et al., 2007). On the basis of large-scale EST-cDNA alignments and bioinformatics analyses on the genomes of *Arabidopsis thaliana* (thale cress) and *Oryza sativa* (rice), it has been estimated that approximately 30–35% of the their genes are alternatively spliced (Cambpell et al., 2006; Xiao et al., 2005), while in humans up to 95% of multi-exon genes undergo alternative splicing (Pan et al., 2008). The number of alternatively spliced genes in plants is still likely to be underestimated because of the relatively low EST coverage and depth of sequencing of many plant transcripts (Simpson et al., 2008; Xiao et al., 2005). Extensive AS variation has been shown in some *Arabidopsis*-specific gene families, for example in genes encoding serine/arginine-rich proteins, and this results in a five-fold increase in transcriptome complexity (Palusa et al., 2007; Tanabe et al., 2007). In addition, stress conditions seem to dramatically alter the splicing pattern of many plant genes (Ali & Reddy, 2008). For these reasons, there is growing interest in complementing transcriptomic studies with proteomics, which should be considered as part of a multidisciplinary integrative analysis that extend from the gene to the phenotype through proteins.

3. Developmental plasticity of protein complexes

Many processes and structures are composed of protein complex aggregates. Protein complexes can vary in size and composition, and range from mega-Dalton assemblies of dozens of proteins (such as the ribosome and the spliceosome) to smaller clusters of just a few proteins. The composition and stability of protein complexes is highly regulated in both a context dependent manner, such as cell-type-specific differences, and a time-dependent manner (Michnick et al., 2004). This biological variability of proteins and their range of physicochemical properties reflect the difficulty of characterizing the structure and the function of protein complexes (Cravatt et al., 2007). In addition, in proteomics, the sample amount is often a limiting factor since, unlike transcript profiling, proteomic approaches cannot benefit from amplification protocols. It should be evident that sensitivity, resolution and speed in data capture are all significant problems with proteomics techniques. In order to circumvent these problems, methods have been developed to extract, separate, detect and identify a wide range of proteins from small sample amounts (Gutstein & Morris, 2007). Technical advances in mass spectometry have facilitated major progress in both the qualitative and quantitative analysis of proteins (Kaspar et al., 2010). Most of these improvements have occurred over the last decade and proteomics has developed a broad range of new protocols, platforms and workflows.

4. Proteomic workflow

The workflow of a standard proteomics experiment is crucial for the success of any experiment and it usually includes a good experimental design, an appropriate extraction/fractionation/purification protocol that considers the needs of different samples (tissue/cells or organelle), a suitable separation protocol, protein identification, statistical analysis and validation. The use of proteomics in plant biology research has increased significantly over the last few years with an improvement in both quality and quantitative analysis, inaugurating a new phase of "Second Generation Plant Proteomics" (Jorrín et al., 2009). This growing interest in plant proteomics has continually produced a large number of developmental studies on plant cell division, elongation, differentiation, and formation of various organs using various proteomics approaches (Hochholdinger et al., 2006; Takàč et al., 2011, Miernyk et al., 2011). Most of the studies published in the plant field concern the proteome of *Arabidopsis* and rice. The work has focused on profiling organs, tissues, cells, and/or subcellular proteomes (Rossignol et al., 2006; Komatsu et al., 2007; Jorrin et al., 2007; Jamet et al., 2008; Baerenfaller et al., 2008; Jorrin et al., 2009; Agrawal & Rakwal, 2011) and studying developmental processes and responses to biotic (Mehta et al., 2008) and abiotic stresses (Nesatyy & Suter 2008) using differential expression strategies. However, proteomics research results have recently appeared on several non-model herbaceous non-crop species, woody plants, fruit and forest trees (Table 1). Furthermore, over the past year, proteome analysis has increasingly been applied to the study of cereal grains with the aim of

model species		
Lycosersicon esculentum	tomato	Sheoran 2007
Hordeum vulgare	wheat	Song et al., 2007
Glycine max	soybean	Djordjevic et al., 2007
Zea mays	maize	Dembinsky et al., 2007
Medicago truncatula	alfalfa	de Jong et al., 2007
non-model species		
Elymus elongatum	wheatgrass	Gazanchian et al., 2007
Nicotiana alata	jasmine tobacco	Brownfield et al., 2007
Boea hygrometrica		Jiang et al., 2007
Xerophyta viscose		Ingle et al., 2007
Solanum chacoense	chaco potato	Vyetrogon et al., 2007
Citrullus lanatus	wild watermelon	Yoshimura et al., 2008
Citrus sinensis		Lliso et al., 2007
Pinus nigra	Australian pine	Wang et al., 2006
Pinus radiate	Californian pine	Fiorani Caledon et al., 2007
Eucalyptus grandis	rose gum eucalyptus	Lippert et al., 2007
Picea sitchenisis	sitka spruce	Valledor et al., 2008
Pyrus communis	conference pears	Pedreschi et al., 2007

Table 1. Proteomics analyses perfomed on model and non-model plants

providing knowledge that will facilitate the improvement of crop quality, either in terms of resistance to biotic and abiotic stress, or in terms of nutritional processing quality (Salekdeh & Komatsu, 2007; Finnie et al., 2011).

5. Proteomic approaches

Comparative plant proteomic approaches are still largely based on traditional two dimentional polyacrilamide gel electrophorsis (2D PAGE) with isoelectric focusing in the first dimension and SDS-PAGE in the second dimension. This technique was initially considered the most suitable method to visualize the differences between protein samples derived from samples grown under different conditions and/or from different tissues. Complex protein mixtures can be resolved efficiently, and the detection of differences in bands or spot intensities is intuitive. Currently, it is possible to visualize over 10,000 protein spots, corresponding to over 1,000 proteins, on single 2D gels (Görg et al., 2009). In many cases, however, individual spots may consist of more than one protein. The differences in spot composition can only be identified by means of mass-spectrometry. The quantitative mass-spectrometry-based proteomics field is constantly evolving, with continuous improvement in protocols, machines and software. Most of the early developments quantitative mass-spectrometry-based proteomic applications were driven by research on yeast and mammalian cell lines. However, in plant physiology analyses, mass spectrometry-based proteomics is no longer only used as a descriptive tool. Instead, well-designed quantitative proteomics has been applied to various aspects of organelle biology, growth regulation and signaling (Schulze & Usadel, 2010). These efforts have greatly improved our knowledge of protein diversity during complex processes. Encouraging pioneer studies on specific subproteomes in plants have revealed candidate proteins that are phosphorylated under specific stress conditions (Oda et al., 1999; Benshop et al., 2007; Niittylä et al., 2007) or during the light independent cycle of photosynthesis (Reiland et al., 2009). Protein abundance changes have been monitored in response to heat shock (Palmblad et al., 2008), during leaf senescence (Hebeler et al., 2008) and during the protein turnover of photosynthetic proteins, monitored using pulse-chase labeling in combination with mass-spectrometry (Nowaczyk et al., 2006). The combination of subcellular fractionation techniques and mass-spectrometry has led to the extensive characterization of the plant subcellular proteome which in turn has led to the discovery of new metabolic pathways (Dunkley et al., 2006). Organelle proteomes were also characterized, such as chloroplasts (Kleffmann et al., 2007; Mejaran et al., 2005; Peltier et al., 2000; Pevzner et al., 2001; Reiland et al., 2009) and plasma membranes and their microdomains (Kierszniowska et al., 2009; Nelson et al., 2006a).

6. Problems with proteomic analyses

Although quantitative methods and their results are desirable, the proteomics data that is usually produced is very complex and often variable in quality. The main problem is incomplete data, since the most advanced mass spectrometers cannot sample and fragment all the peptide ions that are present in complex samples. In fact, only a subset of the peptides and proteins present in a sample can be identified. The first step in primary data extraction is the manual validation of the identity of a peptide and quantification through the revision of the spectra assigned to each sequence. The identification of proteins through the use of algorithms has long been practiced and has been well documented (Eng et al., 1994; Pevzner et al., 2001; Craig & Beavis, 2004; Geer et al., 2004; Tanner et al., 2005). The development of robust algorithms to extract quantitative information from multidimensional proteomic experiments, based on mass spectrometry, is instead a more recent development (Schulze & Usadel, 2010 and references therein). Parallel investigations that provide complete genome sequences for several important agricultural crops will make proteomics-based analyses more useful and increase confidence in proteomic identification and characterization. Unfortunately, genome sequencing is still a relatively new approach and is still fairly expensive therefore most plant species of interest have not yet been sequenced, with consequent gaps in the databases. In such cases, it is possible to exploit the homology-driven proteomics for the characterization of proteomes (Junqueira et al., 2008). The availability of fairly large databases of genomic data from model systems has made it feasible to explore the proteomics of single cell types isolated from complex tissues through a procedure known as laser microdissection. The remainder of this chapter is focused on the use of laser microdissection-assisted proteomic analyses on plant tissues.

7. Laser microdissection in plant biology

Plants are considered to have about 40 different cell types (Martin et al., 2001). Therefore, the gene expression profiles, protein levels and chemical composition of these cell types are destined to be different, even when they are directly adjacent to each other. For this reason, it is important that the sampling and analysis of data are generated in an ever more spatio-temporal cognizant manner, to allow for a far greater resolution in gene expression (Moco et al., 2009). For many years, *in situ* hybridization and experiments with transgenic plants expressing promoter-gene reporter fusion constructs have been used to identify the expression of individual genes in specific cell types (Jefferson et al. 1987; reviewed in Balestrini & Bonfante, 2008). While these techniques cannot be developed with a high-throughput capability, there is a clear need to analyze a transcriptome and proteome at the specific cell-type level (Klink et al., 2007, 2009, 2010a, 2010b, 2011a, 2011b). It is well known that cell-type specific differences occur in gene expression. Identifying these differences in gene expression is complicated by the complexity of the cells that compose the tissues and organs. Thus, the primary reason for obtaining gene expression information from specific cell types is to minimize the dilution effect caused by the cellular complexity found in tissues and organs. This limitation has been overcome by the laser microdissection (LM) technique which was first described by Isenberg et al. (1976) and then developed at the NIH (National Institute of Health, U.S.) for the dissection of cells from histological tissue sections (Emmert-Buck et al., 1996). Laser microdissection permits the rapid procurement of selected cell populations from a section of heterogeneous tissue in a manner conducive to the extraction of DNA, RNA or proteins. Since it was re-designed for histological sections, LM technology has been used routinely in mammalian (Kamme et al., 2003; Kim et al., 2003; Mouledous et al., 2003) and, in more recent years, in plant systems (Asano et al., 2002; Nakazano et al., 2003; Kerk et al., 2003; Day et al., 2005; Klink et al., 2005). The LM apparatus is generally attached to a light microscope and the dissection of the region of interest is computer-controlled. Several instruments are commercially available to isolate individual cells or groups of cells from intact tissues and they are based on two major methods: laser capture microdissection (LCM) and laser cutting (Day et al., 2005; Nelson et al., 2006b). In LCM, the target cells are attached to a thermoplastic film, which covers an optically clear tube cap, using a pulsed infrared laser. The laser is manipulated so that it melts and fuses the film onto the desired cells. When the cap is removed, the target region is selectively pulled away from the surrounding tissues (Emmert-Buck et al., 1996). An alternative approach uses a UV laser to excise target regions from tissue sections. In the first system, the excised fragment is catapulted upwards into a tube cap (laser microdissection pressure catapulting, LMPC), whereas in the second, the sample falls into the collection tube without any extra forces (LMD). These two instruments allow the collection of a single cell and/or a group of cells or tissue regions. A new generation of LCM systems includes both an infrared laser and a UV laser that allow both laser excised microdissection and capture. Some recent reviews have highlighted the increasing interest of the scientific community in the application of this approach in plant biology (Day et al., 2005, 2006; Nelson et al., 2006b; Ramsay et al., 2006; Balestrini et al., 2009). The preparation of plant samples has been described extensively (Asano et al. 2002; Nakazono et al. 2003; Kerk et al., 2003; Inada & Wildermuth 2005; Klink et al. 2005; Tang et al., 2006; Yu et al., 2007; Balestrini et al., 2007; Klink et al. 2007) with additional details being provided in several reviews (Day et al., 2005, 2006; Nelson et al., 2006b).

8. Tissue processing for LM

The tissues for LM are first fixed and sectioned and then the target cells are isolated from the non-target cells under the LM microscope. Sample preparation for LM requires a balance between two contrasting aims: to preserve enough visual detail to identify specific cells during the harvest, and to allow the maximum subsequent recovery of the nucleic acids/proteins from the harvested cells (Figure 1). Two methods have been adopted to prepare sample sections for LM: cryosectioning and paraffin sectioning. Cryosectioning is commonly used in animal research, due to its speed, and it is better at preserving intact molecules, including RNAs and proteins. Although cryosectioning has been described in plant studies (Nakazono et al., 2003), its applicability should be judged on a case-by-case basis (V.K., unpublished observations). Freezing procedures can cause the formation of ice crystals inside vacuoles and air spaces between cells in mature plant tissues: both these features compromise tissue cytology, and eventually lead to the disassembly of cell structures. Cryosectioning of more mature or vacuolated plant material generally requires fixation as well as a cryoprotectant treatment using for example 10–15% sucrose, in order to alleviate the tissue damage caused by freezing. As an alternative, samples are embedded in paraffin after fixation when a more satisfactory preservation of tissue histology is required for target identification. Although this protocol provides excellent cytology, the RNA and protein yield is reduced compared with that from frozen samples. Therefore, it is clear that tissue fixation and paraffin embedding could result in a considerable loss in quality and quantity of the extracted material during RNA studies (Ramsay et al., 2006). Nevertheless, satisfactory amounts of RNA have been obtained from paraffin-embedded material (Kerk et al., 2003; Klink et al., 2005; Tang et al., 2006; Klink et al., 2007, 2009; Hacquard et al., 2010) and an improved morphology is sometimes essential to identify the appropriate cell types for collection purposes. The embedding of *Medicago truncatula* roots in Steedman's wax has recently been used as an alternative to paraffin, and sections of satisfactory morphology and improved RNA quality have been obtained (Gomez & Harrison, 2009). A method for preparing serial sections that reduces RNA degradation has been recently described by using a microwave method (Takahashi et al., 2010). As far as the analysis of nucleic acids is

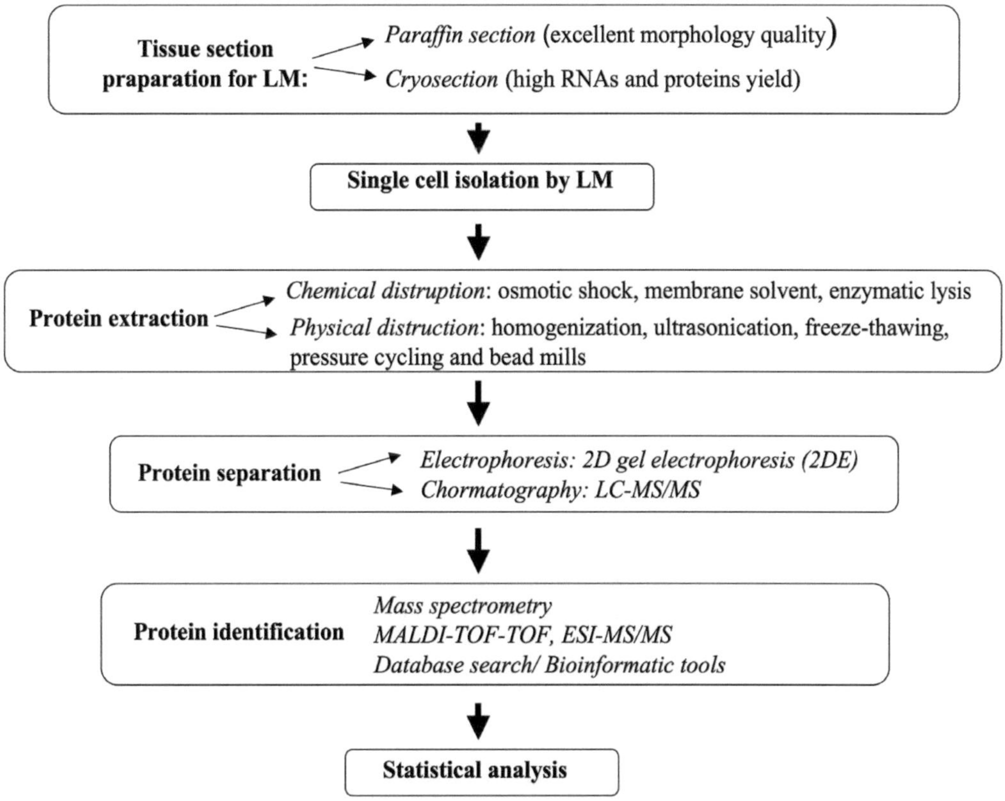

Fig. 1. Experimental proteomics workflow. The classical proteomics workflow has been adapted for a targeted analysis of microdissected samples.

concerned, the possibility of amplifying the RNA extracted from laser microdissected cells allows a transcriptome to be explored by means of microarrays (Nakazono et al., 2003, Casson et al., 2005; Jiang et al., 2006; Klink et al., 2007, 2009; Hacquard et al., 2010) or mRNA-seq techniques based on pyrosequencing platforms, such as 454 Roche and Illumina/Solexa (Graveley, 2008; Simon et al., 2009).

In recent years, LM technology has been applied to gene expression analysis on specific plant cell-types (Day et al., 2005; Nelson et al., 2006b; Ohtsu et al., 2007; Balestrini & Bonfante, 2008; Day et al., 2006; Nelson et al., 2008). The gene expression profile of a number of plant vegetative tissues or cell types, including root cortical cells, vascular bundles, parenchyma, meristem, incipient leaves, syncytia developed from nematode parasitism and abscission zones have been analyzed using the LM technique in several plants (Klink et al., 2005; 2007, 2009, 2010a, 2010b, 2011a, 2011b; Ramsay et al., 2006; Cai & Lashbrook, 2008; Augusti et al., 2009; Nelson et al., 2008 and reference therein). Recently, LM has also been used to provide new insight into fruit development and physiology through the collection of epidermal and subepidermal cells from green, expanding *Citrus clementina* fruit (Matas et al., 2010). A few studies have also focused on the application of LM to gene expression in plant-microbe interactions (Tang et al., 2006; Balestrini et al., 2007; Gomez et al., 2009; Guether et al., 2009a, 2009b; Fiorilli et al., 2009; Chandran et al., 2010; Hacquard et al., 2010).

9. Proteomics/metabolomics and LM

The proteome varies in different cells and various cells respond differently to physiological perturbations. Obtaining a better understanding of tissue complexity could be accomplished by isolating specific cells and analyzing them through proteomic analyses, that could compliment mRNA studies. Over the last few years, the combined use of LM and proteomic analysis has been widely adopted in animal biology and significant progress has been made in adapting the technology to the study of plant cellular processes (Gutstein & Morris, 2007). A list of papers on the application of LM in proteomic and metabolomic studies in plant biology is showed in Table 2. However, difficulties in upstream tissue processing, for example achieving cellular morphological integrity and extracting specific types of protein from cells have limited the efficiency of this approach. The most critical step involves extracting as many proteins as possible from the sample of interest. The wide range of chemical properties of proteins implies that the extraction of all the different types of proteins cannot occur with the same efficiency. Despite these difficulties, recent studies have shown that it is possible to obtain useful information from samples as small as those of single cells (Rubakhin et al., 2003; Hummon et al., 2006). Two general classes of fixatives are usually used in LM analysis: cross-linking and precipitating. Cross-linking fixatives generally have little effect on genomic DNA recovery, but have profound effects on RNA (Goldsworthy et al., 1999) and proteins (Rekhter et al., 2001). Therefore, precipitating fixatives such as ethanol and Methacarn are preferred for protein work (Shibutani et al., 2000; Ahram et al., 2003). It has been demonstrated that brief ethanol post fixation and LM using the IR-laser method does not adversely affect proteomic profiling by 2DE (Banks et al., 1999). In plant biology UV laser seems the most used for proteomic studies (Table 2). This could be probably related to the fact that in more recent years the UV-laser systems are the more widespread and also instruments with IR laser cell capture are combined with UV-laser tissue cutting (Balestrini et al., 2009; Nelson et al., 2006b). It has also been showed that paraffin embedding can have only a slight effect on proteomic profiling whether the tissue is processed properly (Ahram et al., 2003; Hood et al., 2006). This is an interesting observation because it opens the way towards the proteomics analyses of LM-collected cells, above all for plant tissues that are particularly prone to cell morphology damage during cryosectioning. Several studies on animal systems have suggested the staining of the tissue section with such dyes as hematoxylin and eosin to guide the dissection process. However, it has been demonstrated that conventional histological staining methods such as cresyl, hematoxylin/eosin and tolouidine blue, as well as some non-conventional methods such as chlorazol black E and Sudan black B, are incompatible with the 2DE-based proteomic analysis of samples isolated by LM (Banks et al., 1999; Craven & Banks, 2001; Moulédous et al., 2002; Craven et al., 2002; Sitek et al., 2005).

As previously mentioned, many efforts have been made to ensure that sample collection methods involving LM do not interfere with the subsequent proteomic analysis. Extractions can be performed both physically and chemically, or as a combination of mechanical disruption and chemical treatments. A wide range of methods has been described to physically disrupt cells for protein analysis: homogenization, ultrasonication, freeze-thawing, pressure cycling, and bead mills (Butt & Coorssen, 2006; Rabilloud et al., 1996). Cellular homogenization and ultrasonication methods are generally more applicable for a wide variety of biological samples. Chemical extraction and protein solubilization have improved substantially over the past few years. The used approaches include denaturation,

Subject	Tissue preparation	LM system	Technique	Reference
Optimization of several tissue fixing and embedding procedures, and of protein extraction methods from *Arabidopsis thaliana* stem microdissected vascular bundle	Fixation in - 70% ethanol - ethanol/acetic acid (75:25 v/v) Paraffin embedding (30 μm) Cryosectioning (30 μm)	LMPC (UV)	2-DE LC-MS/MS	Schad et al., 2005a
Comparison of gene expression and protein accumulation in pericycle cells of maize root	Fixation in ethanol/acetic acid 3:1 Cryosectioning (10 μm)	PixCell II LCM	2-DE ESI-MS/MS	Dembinsky et al., 2007
Analysis of tissue-specific differences in proteome profiles during barley grain development	Cryosectioning (20 μm)	LMPC (UV)	nanoUPLC combined with ESI-Q-TOF MS	Kaspar et al., 2010
Micromethod for the analysis of amino acid concentrations in NP and ETC cell-type populations from developed barley grain	Cryosectioning (15 μm)	LMPC (UV)	UPLC	Thiel et al., 2009
Metabolite measurement in microdissected vascular bundle samples from *A. thaliana* stem	Cryosectioning (30 μm)	LMPC (UV)	GC-TOF MS	Schad et al., 2005b
Analysis of cell wall carbohydrates from lignified and unlignified parenchyma cells, and xylem fibres of *Urtica dioica*	Fixation in 0.2% glutaraldehyde and 2% formaldehyde Paraffin embedding (4 μm)	LCM (UV)	GC-MS	Angeles et al., 2006
Identification of secondary plant metabolities in specific cells from Norway spruce	Cryosectioning (30 μm)	LMD (UV)	NMR MS	Li et al., 2007

Subject	Tissue preparation	LM system	Technique	Reference
Analysis of metabolite profiling in leaf and flower secretory cavities from fresh and dried sample of *Dilatris* plants	Cryosectioning (60 µm)	LMD (UV) LMPC (UV)	NMR HPLC	Schneider & Hölscher, 2007
Combined analysis of RNA transcripts abundance, enzyme activity and metabolite profiles in individual specialized tissues from white spruce stems	Cryosectioning (25 µm)	LMD (UV)	GC-MS	Abbot et al., 2010

Table 2. Application of LM in proteomic and metabolomic studies in plant biology

osmotic shock, the use of membrane solvents and enzymatic lysis (Asenjo & Andrews, 1990; Hopkins et al., 1991). When using chemical methods, it is important to reduce the interactions between the proteins, as well as the interactions between the proteins and other substances, including nucleic acids and lipids. It is also important to remove contaminants and interfering substances, and prevent protein precipitation during the separation process (Rabilloud et al., 1996, Gutstein & Morris, 2007). Once the proteins have been extracted, the resultant complex mixture needs to be separated for the subsequent detection, abundance and differential expression analyses.

10. Separation technologies used for proteins isolated from LM cells

One of most common methods used to perform protein quantification, which can be coupled with LM technology, is 2D gel electrophoresis (Table 2). At the same time, advances in high-efficiency liquid chromatography (LC), in conjunction with tandem mass spectrometry (MS/MS) have also been reported (Table 2). Although the application of LM to plant biology has been focused above all on cell-specific gene expression profiling, its application to protein analysis has rarely been reported for plant tissues (Nelson et al., 2008; Balestrini & Bonfante, 2008; Hölscher & Schneider, 2008).

This is probably because of the difficulties encountered due to the relatively large amount of proteins that are needed to achieve successful protein profiling (Schad et al., 2005a). As previously mentioned, unlike transcript profiling, which can be performed from very small sample amounts due to efficient amplification strategies, no *in vitro* amplification procedure is yet available for proteins. However, the applicability of 2-DE and high-efficiency liquid chromatography (LC), in conjunction with tandem mass spectrometry (MS/MS), to plant LM material has recently been demonstrated (Schad et al., 2005a). Schad and colleagues (2005a) have compared and optimized several tissue fixation and embedding procedures to obtain the cross sections of *Arabidopsis thaliana* stem tissue, which enabled the microdissetion of vascular bundles, as well as an efficient extraction of proteins. They demonstrated that cryosectioning retains a reasonable morphology and, at the same time, allows an efficient protein extraction. The analysis of proteins from 5000 vascular bundles (~ 250,000 cells yielding about 25 µg total protein) by means of analytical 2-DE has indicated that this tissue processing procedure does not lead to protein degradation/modification. Furthermore, they also optimized the LC-MS/MS approach, starting from a lower amount of material (400 vascular bundles, ~ 20,000 cells, about 2 µg total protein). This resulted in the identification of 131 proteins from 20 stem sections without vascular bundles and 33 specific proteins from 400 vascular bundles. The advantages of the LC-MS/MS approach include the possibility to use a lower amount of material, the capacity for high throughput, no bias against protein classes and high detection ability. The work of Schad et al. (2005a) has certainly increased interest in the application of this procedure, demonstrating that it is a very promising alternative for tissue-specific protein profiling. The number of studies that have employed LM techniques for protein identification and profiling in plant cells has increased significantly over the last years. For example, Dembinsky and colleagues (2007) have analyzed the transcriptome and proteome of pericycle cells in the primary root of maize (*Z. mays*) *versus* non-pericycle cells. For the proteomics experiments, about 1,000 rings of pericycle cells (200,000 cells) have been isolated from root cross sections, extracting approximately 30 µg of proteins, which were separated by 2-DE. The 56 most abundant protein spots were picked from a representative 2-D gel, digested with trypsin and the eluted peptides were subjected to liquid chromatography-tandem mass spectrometry (LC-MS/MS). The pericycle reference map was made in triplicate from indipendent protein preparations and all the identified proteins were detected in all the replicates. Twenty of the 56 proteins were identified by matching known plant proteins, thus defining a reference dataset of the maize pericycle proteome. In another study, Kaspar et al. (2010) focused their attention on tissue-specific differences in the proteome during barley grain development. In order to address this issue, nucellar projection (NP) and endosperm transfer cells (ETC) of barley grain were collected by LM. Proteins were subsequently extracted, digested with trypsin and analyzed through nanoLC separation combined with ESI-Q-TOF mass spectrometry. This procedure requires material from between 40 and 75 sections per sample. Three independent extractions showed highly reproducible chromatograms. Quantitative and qualitative protein profiling led to the identification of a number of proteins with tissue specific expression. For example, 137 proteins were identified from ETC and 44 from the NP. Among the identified proteins, 31 were identified in both tissues. The major differences between ETC and NP protein profiles concerned cell wall and protein synthesis (in the ETC but not in the NP) and the disease response (with a greater representation in NP), which is in agreement with previously published transcript analyses (Thiel et al., 2008). These experiments have shown that nanoLC-based separation in combination with MS detection can be considered a suitable platform for identifying proteins present in laser-microdissected samples, which contains only small quantities of proteins (Kaspar et al., 2010).

11. Metabolomic studies in cells isolated by LM

The last decade has seen an increase in metabolomic-based studies, which are crucial to understand cellular processes because they can connect metabolite profiles and metabolic changes to protein activity, and thus leading to a detailed and more comprehensive understanding of the phenotype of the organism of interest. So far, studies in this field have mainly been performed on whole plants, organs, such as fruits (Moco et al., 2006; Fraser et al., 2007), leaves (Kant et al., 2004; Glauser et al., 2008), tubers (Roessner et al., 2001; Sturm et al., 2007), flowers (Kazuma et al., 2003; Wang et al., 2004), and roots (Opitz et al., 2002; Hagel et al., 2008;). However, some studies have also been performed on specific tissues (Moco et al., 2007; Fait et al., 2008) and even on specific cells (Li et al., 2007; Schneider & Hölscher, 2007). Metabolite analysis at a microscale level from sectioned tissues or cells is a major challenge since metabolities (usually < 1500 Da) show an enormous chemical diversity and for this reason general multiple approaches are required for extraction, fractionation and analysis. Moreover, there is a higher turnover of metabolites than large biomolecules and there is a dynamic range of metabolite concentrations. Micromethods have been adapted from animal biology in order to determine the spatial distribution of small molecules in plant tissue (Schneider & Hölscher 2007; Fait et al., 2008; Hagel et al., 2008). Among the two different methods of LM, laser capture microdissection (LCM) and laser cutting, this last seems to be the most useful method for harvesting samples for metabolite analysis because, in contrast to LCM, it is contact-free and avoids potential contamination from the melting foil (Moco et al., 2009). In addition, most of the analyses have exploited the cryosection method, thus avoiding any further chemical treatment of the material (Table 2). Using standard tissue fixation and embedding protocols, metabolites can in fact either be extracted by means of dehydrating solvents, or washed out by embedding agents (Schad et al., 2005b). Paraffin embedding has been used for the carbohydrate analysis of the polysaccharides from the walls of lignified and unlignified parenchyma cells, and of xylem fibres of *Urtica dioica* (Angeles et al., 2006). The carbohydrate composition of different cell wall types was obtained by the combination of laser microdissection and GC-MS analysis.

For metabolite analyses based on LM, GC-TOF-MS, LC-MS, GC-MS and NMR-related strategies have been used (Schad et al., 2005b; Lisec et al., 2006; Moco et al., 2006). MS-based analytical methods probably ensure a higher identification power for small molecules than NMR measurements. In the first study in which LM was applied successfully to analyze the spatial distribution of metabolites in plant tissues, Schad and colleagues (2005b) used the GC-TOF MS technique to investigate vascular bundles obtained from *Arabidopsis thaliana* cross sections. Cryo-sectioned stem material of 30 μm section thickness was subjected to LMPC. Vascular bundles were dissected and catapulted into the collection device, which was filled with ethanol to inactivate the metabolic enzymes and protect the cell contents from undesired enzymatic modification. An ethanol extract of approximately 100 collected vascular bundles (~5,000 cells) was derivatized with N-methyl-N-(trimethylsilyl) trifluoroacetamide (MSTFA) and subjected to GC-time-of-flight (TOF) MS analysis to simultaneously detect compounds of different classes. Sixty-eight metabolites were detected in the vascular bundles; sixty-five metabolites were instead identified in control samples, which are sections without vascular bundles.

As an alternative, Thiel et al. (2009) used a combination of LMPC-based microdissection and liquid chromatography (UPLC) to analyze the amino acid concentrations in nucellar projections (NP) and endosperm transfer cells (ETC) from developing barley grains. In order to guarantee a sufficient amount of material to produce consistent values and detect the differences in the amino acid concentrations between the two tissues, the authors prepared 10-20 cryosections for one sample and analyzed 4-5 biological replicates/sample. UPLC technology was used to measure free amino acid concentrations from microdissected tissues and the sum of all the measured amino acids was 98 and 112 amol $m\mu^{-3}$ for NP and ETC, respectively. This metabolite approach based on LM was combined with a transcriptome analysis. On the basis of these studies, it has been concluded that combining metabolite data with a transcriptome approach leads to a better understanding of the metabolism, interconversion and transfer of amino acids at the maternal–filial boundary of growing barley seeds.

Methods have been also developed to analyze laser-microdissected samples by means NMR spectroscopy (http://www.ice.mpg.de/ext/769.html). For instance, high-resolution ^1H NMR spectroscopy has been used, in combination with LM, as a tool to analyze the contents of the secretory cavities from fresh leaves and herbarium specimen of *Dilatris* plants (Haemodoraceae) (Schneider & Hölscher, 2007). The secretory cavity sections show a typical storage cell surrounded by a thin layer of glandular epithelial cells. Their low water content makes them well accessible to LM (Moco et al., 2009). The dissected cavities were localized under a stereomicroscope. They were then picked up using an extremely sharp dissecting needle and transferred directly to a microcentrifuge tube containing the extraction solvent (acetone/water 20:1). In some experiments, the dissected material was transferred directly to the NMR tube without centrifugation, and extracted using the NMR solvent (deuterated acetone) in an ultrasonic bath. The extracts were subjected to cryogenic 1H NMR spectroscopy and reversed-phase high-performance liquid chromatography (HPLC). The results obtained from 180-year-old herbarium specimens of *Dilatris corymbosa* and *Dilatris viscosa* showed that phenylphenalenones, which are typical secondary metabolites of Hemodoraceae, were identified in secretory cells of leaves and flower petals (Schneider & Hölscher, 2007).

LM has not been widely applied to woody plant tissue. Cell-specific metabolic profilings have been conducted on special cells harvested from the bark of Norvegian spruce (*Picea abies*) (Li et al., 2007) by means a combination of LM, NMR, and MS. Sclereids (stone cells) were detected in cryosections of the bark taking advantage of their characteristic fluorescence and this was followed by laser microdissection. Non-fluorescing phloem tissue was microdissected from the same cryosections and used as a control sample. The collected samples were then transferred to NMR tubes to which deuterated methanol was added for extraction. ^1H and 2D NMR spectra were measured using a cryogenically cooled probehead. The results indicate that both sclereids and the adjacent parenchymatic tissue show similar phenolic components. Comparison with the spectra of reference compounds, together with MS analysis, revealed that astringin (major component) and dihydroxyquercetin 3′-O-β-**D**-glucopyranoside (minor component) are present in both the sclereids and the control cells. The control cells (sclereid-surrounding cells) showed higher levels of the two components.

Abbott and colleagues (2010) have recently reported, in a methodology article, the successful use of LMD technology to isolate individual specialized tissues from the stems of the woody perennial *Picea glauca* (white spruce), suitable for subsequent combined analysis of RNA transcripts abundance, enzyme activity and metabolite profiles. In agreement with previous papers, the authors underlined that sample preparation protocols for LM can vary substantially on the basis on the type of tissue and down-stream analysis. A tangential cryosectioning approach was essential to obtain large quantities of cortical resin ducts (CRD) and cambial zone (CZ) tissues using LM. Gene expression results showed a differential expression of genes involved in terpenoid metabolism between the CRD and CZ tissues, and in response to methyl jasmonate (MeJA). In addition, terpene synthase enzyme activity has been identified in CZ protein extracts and terpenoid metabolites were detected, by means of GC-MS, in both the CRD and CZ tissues. These analyses supported by LM seem to be very promising to improve the characterization of complex processes related to woody plant development, including cell differentiation and specialization associated with stem growth, wood development and the formation of defense-related structures such as resin ducts.

12. Bioinformatics

In 2002, Scheidler and colleagues demonstrated altered gene activity in *Arabidopsis* infected with *Phytopthora*. The work provided a meaningful context for the gene expression analyses that were performed, and resulted in the identification of the major shifts in physiology and metabolism that occur during the infection process. However, the analyses focused on gene expression in whole infected plants. Unlike Scheideler et al. (2002), Klink et al. (2011b) and

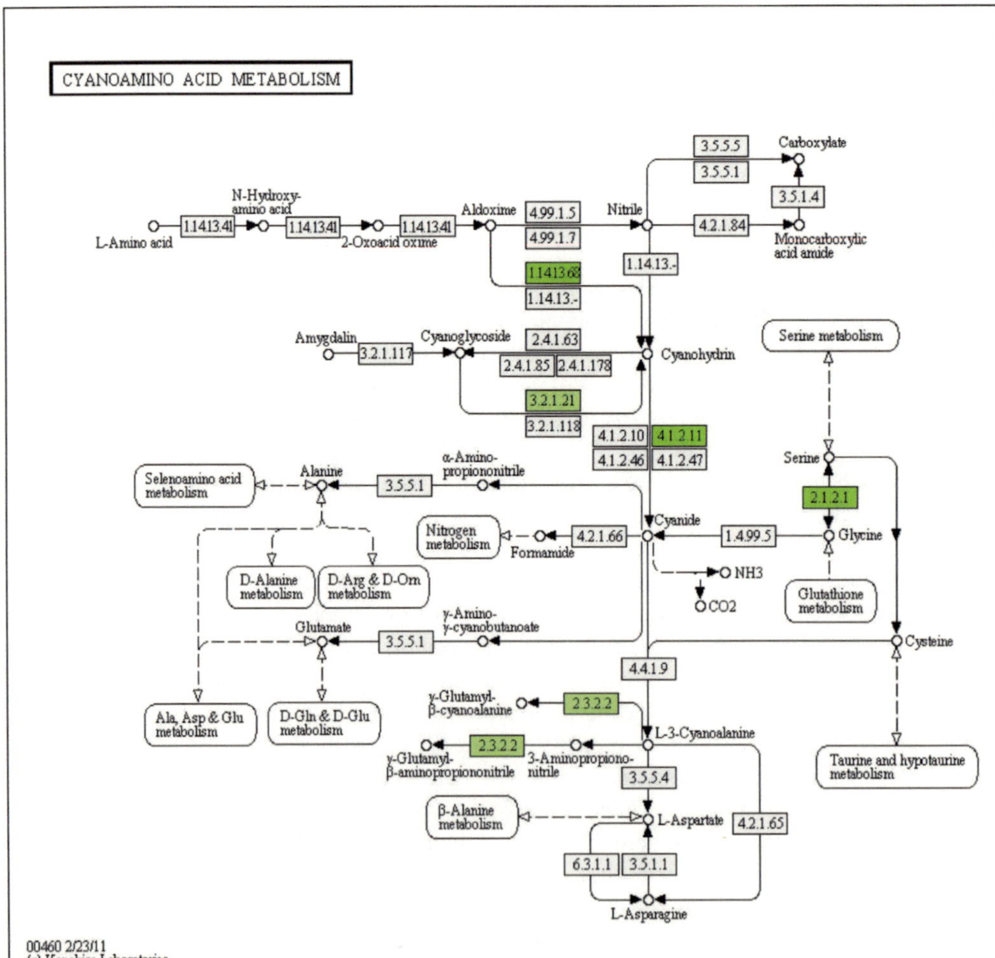

Fig. 2. A PAICE pathway for cyanoamino acid metabolism for 3 day post infection syncytia undergoing a resistant reaction in *G. max* as it is being infected by *Heterodera glycines* (soybean cyst nematode). The green boxes represents active genes (adapted from Matsye et al., submitted).

Matsye et al. (submitted) have used the same principle, adapting the publically available **K**yoto **E**ncyclopedia of **G**enes and **G**enomes (KEGG) (http://www.genome.jp/kegg/catalog/org_list.html) and modifying it so that gene expression can be visualized using a KEGG application called **P**athway **A**nalysis and **I**ntegrated **C**oloring of **E**xperiments (PAICE). PAICE was developed in the laboratories of Dr. Benjamin Matthews (USDA; Beltsville, MD) and Dr. Nadim Alkharouf (Towson University, Baltimore, MD) (Hosseini et al., unpublished) and is freely available (http://sourceforge.net/projects/paice/). PAICE has been used on LM cells infected with parasitic nematodes, and it provides a deeper understanding of the biochemical and metabolic activities during multiple defense reactions in multiple *G. max* genotypes compared to both pericycle control cell populations and the susceptible reaction (Klink et al., 2011; Matsye et al., submitted). However, the analyses were based on RNA isolated from the specific cell types and not on proteins or metabolites (Figure 2). It is believed that PAICE could be expanded to provide a comprehensive understanding of any cell isolated by LM and analyzed for its proteomic and metabolic content.

13. Conclusion

To have a better understanding of tissue and organ-defined processes and functions, it is necessary to study the biochemical activity at a cellular resolution level by analyzing the proteome. This has become increasingly important, since it has been demonstrated, in several comparative studies, that protein expression and abundance often poorly correlate with the mRNA levels in the same cell types (Schad et al., 2009a). Many proteins are the primary determinant molecules of physiological processes and are often restricted to specific tissues and cell types. Thus, the monitoring of protein expression at a very high spatial resolution could help enhance our understanding of the biological processes that control plant growth and development. At the same time, the use of different strategies and protocols for the characterization of a wide number of metabolites from a single cell or tissue have increased significantly over the last decade making the broad applicability of these analyses tractable. In order to address these issues, sampling methods, for example LM in plants, have been adopted to extract highly specific tissue regions and homogeneous cell-type populations with limited damage, and have led to the discovery of functions of genes/proteins/metabolites that contribute to cell specialization (Galbraith & Birnbaum 2006). Despite these considerable efforts, the current strategies used for protein/metabolite characterization still face significant obstacles. These challenges are mainly caused by the cellular complexity and spatial and temporal distribution of localized gene activity within living tissues, including metabolic processes. Other challenges concern the identification of the high degree of chemical diversity of the different cell types that can be affected by the analysis procedures. Technical improvements are still required to achieve reliable protein and metabolite profilings in small samples. The introduction of statistical analysis, applied to the handling and manipulation of data from proteomics and metabolomics, will lead to the development of promising strategies that can be used to extract precious information from large data sets and to identify new proteins and metabolites. Although most of these restrictions have already been solved in the field of genomics and transcriptomics, the problem still remains of adapting these computational strategies for proteomic and metabolic analyses.

14. Acknowledgements

Contributions to this chapter have been partly funded by CNR (Premio DAA 2009) to RB. VF was supported by a grant from BIOBIT-CIPE (Piedmont Region project). VPK would like to thank the Mississippi Soybean Promotion Board for the funding and critical reading of the manuscript by Prachi D. Matsye.

15. References

Abbott, E., Hall, D., Hamberger, B. & Bohlmann, J. (2010). Laser microdissection of conifer stem tissues: isolation and analysis of high quality RNA, terpene synthase enzyme activity and terpenoid metabolites from resin ducts and cambial zone tissue of white spruce (*Picea glauca*), *BMC Plant Biology* 10:106.

Ahram, M., Flaig, M.J., Gillespie, J.W., Duray, P.H., Linehan, W.M., Ornstein, D.K., Niu, S., Zhao, Y., Petricoin, E.F. 3rd & Emmert-Buck, M.R. (2003). Evaluation of ethanol-fixed, paraffin-embedded tissues for proteomic applications, *Proteomics* 3: 413–421.

Ali, G.S. & Reddy, A.S.N. (2008) Regulation of Alternative Splicing of Pre-mRNAs by Stresses, *Current Topics in Microbiology and Immunology* 326: 257-275.

Agrawal, G.K. & Rakwal, R. (2011). Rice proteomics: A move toward expanded proteome coverage to comparative and functional proteomics uncovers the mysteries of rice and plant biology, *Proteomics* 11: 1630–1649.

Angeles, G., Berrio-Sierra, J., Joseleau, J.P., Lorimier, P., Lefebvre, A. & Ruel, K. (2006). Preparative laser capture microdissection and single- pot cell wall material preparation: a novel method for tissue-specific analysis, *Planta* 224: 228–232.

Asano, T., Masumura, T., Kusano, H., Kikuchi, S., Kurita, A., Shimada, H. & Kadowaki, K. (2002). Construction of a specialized cDNA library from plant cells isolated by laser capture microdissection: toward comprehensive analysis of the genes expressed in the rice phloem, *The Plant Journal* 32: 401-408.

Asenjo, J.A. & Andrews, B.A. (1990) Enzymatic Cell Lysis for Product Release, in J.A. Asenjo, Marcel Dekker (eds), *Separation Processes in Biotechnology*, New York, pp. 143-175.

Augusti, J., Merelo, P., Cercós, M., Tadeo, F.R. & Talón, M. (2009) Comparative transcriptional survey between laser-microdissected cells from laminar abscission zone and petiolar cortical tissue during ethylene-promoted abscission in citrus leaves, *BMC Plant Biology* 9: 127.

Baerenfaller, K., Grossmann, J., Grobei, M.A., Hull, R., Hirsch-Hoffmann, M., Yalovsky, S., et al. (2008). Genome-scale proteomics reveals *Arabidopsis thaliana* gene models and proteome dynamics, *Science* 320: 938–41.

Balestrini, R., Gòmez-Ariza, J., Lanfranco, L. & Bonfante, P. (2007). Laser Microdissection reveals that transcripts for five plant and one fungal phosphate transporter genes are contemporaneously present in arbusculated cells, *Molecular Plant–Microbe Interactions* 20: 1055–1062.

Balestrini, R. & Bonfante, P. (2008). Laser Microdissection (LM): Applications to plant materials, *Plant Biosystems* 142: 331-336.

Balestrini, R., Gòmez-Ariza, J., Klink, V.P. & Bonfante P. (2009). Application of laser microdissection to plant pathogenic and symbiotic interactions, *Journal of Plant Interactions* 4: 81-92.

Banks, R.E., Dunn, M.J., Forbes, M.A., Stanley, A., Pappin, D., Naven, T., Gough, M., Harnden, P. & Selby, P.J. (1999). The potential use of laser capture microdissection to selectively obtain distinct populations of cells for proteomic analysis - Preliminary findings, *Electrophoresis* 20: 689–700.

Barrett, T.H. & Gould H. (1973) Tissue and species specificity of non-histone chromatin proteins, *Biochimica et Biophysica Acta* 294: 165-170.

Benschop, J.J., Mohammed, S., O'Flaherty, M., Heck, A.J., Slijper, M. & Menke, F.L. (2007). Quantitative phospho-proteomics of early elicitor signalling in *Arabidopsis, Molecular and Cellular Proteomics* 6: 1705–13.

Bligny, R. & Douce, R. (2001) NMR and plant metabolism. *Current Opinion in Plant Biology* 4: 191-196.

Brownfield, L., Ford, K., Doblin, M.S., Newbigin, E., Read, S. & Bacic, A. (2007). Proteomic and biochemical evidence links the callose synthase in *Nicotiana alata* pollen tubes to the product of the *NaGSL1* gene, *The Plant Journal* 52: 147–56.

Butt, R.H. & Coorssen, J.R. (2006). Pre-extraction Sample Handling by Automated Frozen Disruption Significantly Improves Subsequent Proteomic Analyses, *Journal of Proteome Research* 5: 437-448.

Cai, S. & Lashbrook, C.C. (2006). Laser capture microdissection of plant cells from tape-transferred paraffin sections promotes recovery of structurally intact RNA for global gene profiling, *The Plant Journal* 48: 628–637.

Campbell, M.A., Haas, B.J., Hamilton, J.P., Mount, S.M. & Buell, C.R. (2006). Comprehensive analysis of alternative splicing in rice and comparative analyses with *Arabidopsis, BMC Genomics* 7:327.

Casson, S., Spencer, M., Walker, K. & Lindsey, K. (2005). Laser capture microdissection for the analysis of gene expression during embryogenesis of *Arabidopsis, The Plant Journal* 42: 111-123.

Celedon, P.A.F., Andrade, A., Meireles, K.G.X., Carvalho, M.C.C.G., Caldas, D.G.G., Moon, D.H., Carneiro, R.T., Franceschini, L.M., Oda, S. & Labate, C.A. (2007). Proteomic analysis of the cambial region in juvenile *Eucalyptus grandis* at three ages, *Proteomics* 7: 2258–2274.

Chandran, D., Inada, N., Hather, G., Klindt, C.K. & Wildermuth M.C. (2010). Laser microdissection of *Arabidopsis* cells at the powdery mildew infection site reveals site-specific process and regulators, *Proceedings of the National Academy of Science USA* 107: 460-465.

Chen, B.E., Kondo, M., Garnier, A., Watson, F.L., Püettmann-Holgado, R., Lamar, D.R. & Schmucker, D. (2006). The molecular diversity of Dscam is functionally required for neuronal wiring specificity in *Drosophila, Cell* 125: 607-620.

Ciobanu, L. & Pennington, C.H. (2004). 3D micron-scale MRI of single biological cells, *Solid State Nuclear Magnetic Resonance* 25: 138-141.

Craig, R. & Beavis, R.C. (2004). TANDEM: matching proteins with mass spectra, *Bioinformatics* 20: 1466–67.

Cravatt, B.F., Simon, G.M. & Yates, III J.R. (2007). The biological impact of mass-spectrometry-based proteomics, *Nature* 450: 991-1000.

Craven, R. & Banks, R. (2001). Laser capture microdissection and proteomics: Possibilities and limitation, *Proteomics* 1: 1200-1204.

Craven, R.A., Totty, N., Harnden, P., Selby, P.J. & Banks, R.E. (2002). Laser capture microdissection and two-dimensional polyacrylamide gel electrophoresis:

evaluation of tissue preparation and sample limitations, *American Journal of Pathology* 160: 815–822.

Day, R.C., Grossniklaus, U. & Macknight, R.C. (2005). Be more specific! Laser-assisted microdissection of plant cells, *Trends in Plant Science* 10: 397–405.

Day, R.C., McNoe, L.A. & Macknight, R.C. (2006). Transcript analysis of laser microdissected plant cells, Technical Focus, *Physiologia Plantarum* 129: 267–282.

de Jong, F., Mathesius, U., Imin, N., Rolfe, B.G. (2007). A proteome study of the proliferation of cultured *Medicago truncatula* protoplasts, *Proteomics* 7: 722-36.

Dembinsky, D., Woll, K., Saleem, M., Liu, Y., Fu, Y., Borsuk, L.A., Lamkemeyer, T., Fladerer, C., Madlung, J., Barbazuk, B., Nordheim, A., Nettleton, D., Schnable, P.S. & Hochholdinger F. (2007). Transcriptomic and proteomic analyses of pericycle cells of the maize primary root, *Plant Physiology* 145: 575-88.

Djordjevic, M.A., Oakes, M., Li, D.X., Hwang, C.H., Hocart, C.H. & Gresshoff, P.M. (2007). The *glycine max* xylem sap and apoplast proteome, *Journal of Proteome Research* 6: 3771-9.

Dunkley, T.P., Hester, S., Shadforth, I.P., Runions, J., Weimar, T., Hanton, S.L., Griffin, J.L., Bessant, C., Brandizzi, F., Hawes, C., Watson, R.B., Dupree, P. & Lilley, K.S. (2006). Mapping the *Arabidopsis* organelle proteome, *Proceedings of National Academy Science USA* 103: 6518–23.

Edman, P. (1950). Method for determination of the amino acid sequence in peptides, *Acta Chemica Scandinavica* 4: 283-293.

Emmert-Buck, M.R., Bonner, R.F., Smith, P.D., Chuaqui, R.F., Zhuang, Z., Goldstein, S.R., Weiss, R.A. & Liotta, L.A. (1996). Laser capture microdissection, *Science* 274: 998–1001.

Eng, J., McCormack, A.L. & Yates, J.R.I. (1994). An approach to correlate tandem mass spectral data of peptides with amino acid sequences in a protein database, *Journal of the American Society for Mass Spectrometry* 5: 976–89.

Fait, A., Hanhineva, K., Beleggia, R., Dai, N., Rogachev, I., Nikiforova, V.J., Fernie, A.R. & Aharoni, A. (2008). Reconfiguration of the chene and receptacle metabolic networks during strawberry fruit development, *Plant Physiology* 148: 730–750.

Finnie, C., Sultan, A. & Grasser, K.D. (2011). From protein catalogues towards targeted proteomics approaches in cereal grains, *Phytochemistry* 72: 1145-1153.

Fiorilli, V., Catoni, M., Miozzi, L., Novero, M,. Accotto, G.P. & Lanfranco, L. (2009). Global and cell-type gene expression profiles in tomato plants colonized by an arbuscular mycorrhizal fungus, *New Phytologist* 184: 975-987.

Fraser, P.D., Enfissi, E.M.A., Goodfellow, M., Eguchi, T. & Bramley, P.M. (2007). Metabolite profiling of plant carotenoids using the matrix-assisted laser desorption ionization time-of-flight mass spectrometry, *The Plant Journal* 49: 552–564.

Galbraith, D.W. & Birnbaum, K. (2006). Global studies of cell type-specific gene expression in plants, *Annual Review of Plant Biology* 57:451-75.

Gazanchian, A., Hajheidari, M., Sima, N.K. & Salekdeh, G.H. (2007). Proteome response of *Elymus elongatum* to severe water stress and recovery, *Journal of Experimental Botany* 58: 291–300.

Geer, L.Y., Markey, S.P., Kowalak, J.A., Wagner, L., Xu M, Maynard, D.M., Yang, X., Shi, W. & Bryant, S.H.2004. Open mass spectrometry search algorithm, *Journal of Proteome Research* 3: 958–64.

Glauser, G., Guillarme, D., Grata, E., Boccard, J., Thiocone, A., Carrupt, P.A., Veuthey, J.L., Rudaz, S., & Wolfender, J.L. (2008). Optimized liquid chromatography-mass spectrometry approach for the isolation of minor stress biomarkers in plant extracts and their identification by capillary nuclear magnetic resonance, *Journal of Chromatography* 1180: 90–98.

Goldsworthy, S.M., Stockton, P.S., Trempus, C.S., Foley, J.F. & Maronpot, R.R. (1999) Effects of fixation on RNA extraction and amplification from laser capture microdissected tissue, *Molecular Carcinogenesis* 25: 86-91.

Gomez, S.K. & Harrison, M.J. (2009). Laser microdissection and its application to analyze gene expression in the arbuscular mycorrhizal symbiosis. Pest Management Sciences 65: 504-511

Gomez, K.S., Javot, H., Deewatthanawong, P., Torres-Jerez, I., Tang, Y., Blancaflor, B.E., Udvardi, M.K. & Harrison, J. M. (2009). *Medicago Truncatula* and *Glomus Intraradices* gene expression in cortical cells harboring arbuscules in the arbuscular mycorrhizal symbiosis, *BMC Plant Biology* 9:10.

Görg, A., Drews, O., Lück, C., Weiland, F. & Weiss, W. 2-DE with IPGs, *Electrophoresis* 30: S122-32.

Graveley, B.R. (2008) Molecular biology: power sequencing, *Nature* 453: 1197-8.

Guether, M., Balestrini, R., Hannah, M., He, J., Udvardi, M.K. & Bonfante, P. (2009a). Genome-wide reprogramming of regulatory networks, transport, cell wall and membrane biogenesis during arbuscular mycorrhizal symbiosis in *Lotus japonicus*, *New Phytologist* 182: 200–212.

Guether, M., Neuhauser, B., Balestrini, R., Dynowski, M., Ludewig, U. & Bonfante P. (2009b). A mycorrhizal-specific ammonium transporter from Lotus Japonicus acquires nitrogen released by arbuscular mycorrhizal fungi, *Plant Physiology* 105: 73-83.

Gutstein, H.B. & Morris, J.S. (2007). Laser capture sampling and analytical issue in proteomics, *Expert Review of Proteomics* 4: 627-37.

Hacquard, S., Delaruelle, C., Legué, V., Tisserant, E., Kohler, A., Frey, P., Martin, F. & Duplessis S. (2010) Laser capture microdissection of uredinia formed by *Melampsora larici-populina* revealed a transcriptional switch between biotrophy and sporulation, *Molecular Plant-Microbe Interactions* 23: 1275-1286.

Hagel, J.M., Weljie, A.M., Vogel, H.J. & Facchini, P.J. (2008). Quantitative H-1 nuclear magnetic resonance metabolite profiling as a functional genomics platform to investigate alkaloid biosynthesis in opium poppy, *Plant Physiology* 147: 1805-1821.

Hebeler, R., Oeljeklaus, S., Reidegeld, K.A., Eisenacher, M., Stephan, C., Sitek, B., Stühler, K., Meyer, H.E., Sturre, M.J., Dijkwel, P.P. & Warscheid, B. (2008). Study of early leaf senescence in *Arabidopsis thaliana* by quantitative proteomics using reciprocal 14 N/15 N labeling and difference gel electrophoresis, *Molecular and Cellular Proteomics* 7:108–20.

Hinse, C., Sheludko, Y.V., Provenzani, A., Stöckigt, J.H.H. (2001). *In vivo* NMR at 800 MHz to monitor alkaloid metabolism in plant cell cultures without tracer labeling, *Journal of America Chemical Society* 123: 5118-5119.

Hochholdinger, F., Sauer, M., Dembinsky, D., Hoecker, N., Muthreich, N., Saleem, M. & Liu, Y. (2006). Proteomic dissection of plant development, *Proteomics* 6: 4076–4083.

Hölscher, D., Schneider, B. (2008). Application of Laser-Assisted Microdissection for Tissue and Cell-Specific Analysis of RNA, Proteins, and Metabolites, *Progress in Botany* 69: 141-167.

Hood, B.L., Conrads, T.P. & Veenstra, T.D. (2006). Unravelling the proteome of formalin-fixed paraffin-embedded tissue, *Briefings in Functional Genomics and Proteomics* 5: 169–175.

Hopkins, T.R. (1991). Physical and chemical cell disruption for the recovery of intracellular proteins, *Bioprocess Technology* 12: 57–83.

Hummon, A.B., Amare, A. & Sweedler, J.V. (2006). Discovering new invertebrate neuropeptides using mass spectrometry, *Mass Spectrometry Reviews* 25: 77–98.

Inada, N. & Wildemuth, M.C. (2005). Novel tissue preparation method and cell-specific marker for laser microdissection of *Arabidopsis* mature leaf, *Planta* 221: 9–16.

Ingle, R.A., Schmidt, U.G., Farrant, J.M., Thomson, J.A. & Mundree, S.G. (2007). Proteomic analysis of leaf proteins during dehydration of the resurrection plant *Xerophyta viscosa, Plant Cell and Environment* 30: 435-46.

Isenberg, G., Bielser, W., Meier-Ruge, W. & Remy, E. (1976) Cell surgery by laser micro-dissection: a preparative method, *Journal of Microscopy* 107: 19–24.

Jamet, E., Boudart, G., Borderies, G., Charmont, S., Lafitte, C., Rossignol, M., Canut, H. & Pont-Lezica R.F. Isolation of plant cell wall proteins, *Methods Molecular Biology* 425: 187–201.

Jefferson, R.A., Kavanagh, T.A. & Bevan, M.W. (1987). GUS fusions: β-glucuronidase as a sensitive and versatile gene fusion marker in higher plants, *EMBO Journal* 6: 3901-3907.

Jiang, G., Wang, Z., Shang, H., Yang, W., Hu, Z., Phillips, J. Deng, X. (2007). Proteome analysis of leaves from the resurrection plant *Boea hygrometrica* in response to dehydration and rehydration, *Planta* 225: 1405-20.

Jiang, K., Zhang, S., Lee, S., Tsai, G., Kim, K., Huang, H., Chilcott, C., Zhu, T. & Feldman, L.J. (2006). Transcription profile analyses identify genes and pathways central to root cap functions in maize, *Plant Molecular Biology* 60: 343–363.

Jorrin, J.V., Maldonado, A.M. & Castillejo, M.A. (2007). Plant proteome analysis: a 2006 update, *Proteomics* 7: 2947–62.

Jorrín-Novo, J.V., Maldonado, A.M., Echevarría-Zomeño, S., Valledor, L., Castillejo, M.A., Curto, M., Valero, J., Sghaier, B., Donoso, G. & Redondo, I. (2009). Plant proteomics update (2007–2008): Second-generation proteomic techniques, an appropriate experimental design, and data analysis to fulfill MIAPE standards, increase plant proteome coverage and expand biological knowledge, *Journal of Proteomics* 72: 285-314.

Junqueira, M., Spirin, V., Santana Balbuenaa, T., Thomasa, H., Adzhubeib, I., Sunyaevb, S. & Shevchenko, A. (2008). Protein identification pipeline for the homology-driven proteomics, *Journal of Proteomics* 71: 346-356.

Kamme, F., Salunga, R., Yu, J., Tran, D.T., Zhu, J., Luo, L., Bittner, A., Guo, H.Q., Miller, N., Wan, J. & Erlander, M. (2003). Single-cell microarray analysis in hippocampus CA1: demonstration and validation of cellular heterogeneity, *Journal of Neuroscience* 23: 3607-3615.

Kant, M.R., Ament, K., Sabelis, M.W., Haring, M.A. & Schuurink, R.C. (2004). Differential timing of spider mite-induced direct and indirect defenses in tomato plants, *Plant Physiology* 135: 483- 495.

Kaspar, S., Weier, D., Weschke, W., Mock, H.P. & Matros, A. (2010). Protein analysis of laser capture micro-dissected tissues revealed cell-type specific biological functions in developing barley grains, *Analytical and Bioanalytical Chemistry* 398: 2883-93.

Kazuma, K., Noda, N. & Suzuki, M. (2003). Flavonoid composition related to petal color in different lines of *Clitoria ternatea, Phytochemistry* 64: 1133-1139.

Kerk, N.M., Ceserani, T., Tausta, S.L., Sussex, I.M. & Nelson, T.M. (2003). Laser capture microdissection of cells from plant tissues, *Plant Physiology* 132: 27–35.

Kierszniowska, S., Seiwert, B. & Schulze, W.X. (2009). Definition of *Arabidopsis* sterol-rich membrane microdomains by differential treatment with methyl-ß-cyclodextrin and quantitative proteomics, *Molecular and Cellular Proteomics* 8: 612–23.

Kim, E., Magen, A. & Ast, G. (2007). Different levels of alternative splicing among eukaryotes, *Nucleic Acids Research* 35: 125–31.

Kim, J.O., Kim, H.N., Hwang, M.H., Shin, H.I., Kim, S.Y., Park, R.W., Park, E.Y., Kim, I.S., van Wijnen, A.J., Stein, J.L., Lian, J.B., Stein, G.S. & Choi, J.Y. (2003). Differential gene expression analysis using paraffin-embedded tissues after laser microdissection, *Journal Cell Biochemistry* 90: 998-1006.

Kleffmann, T., von Zychlinski, A., Russenberger, D., Hirsch-Hoffmann, M., Gehrig, P, Gruissem, W. & Baginsky, S. (2007). Proteome dynamics during plastid differentiation in rice, *Plant Physiology* 143: 912–23.

Klink, V.P., MacDonald, M., Alkharouf, N. & Matthews, B.F. (2005). Laser capture microdissection (LCM) and expression analyses of *Glycine max* (soybean) syncytium containing root regions formed by the plant pathogen *Heterodera glycines* (soybean cyst nematode), *Plant Molecular Biology* 59: 969-983.

Klink, V.P., Overall, C.C., Alkharouf, N., MacDonald, M.H. & Matthews, B.F. (2007). Laser capture microdissection (LCM) and comparative microarray expression analysis of syncytial cells isolated from incompatible and compatible soybean roots infected by soybean cyst nematode (*Heterodera glycines*), *Planta* 226: 1389-1409.

Klink, V.P., Hosseini, P., Matsye, P., Alkharouf, N. & Matthews, B.F. (2009). A gene expression analysis of syncytia laser microdissected from the roots of the *Glycine max* (soybean) genotype PI 548402 (Peking) undergoing a resistant reaction after infection by *Heterodera glycines* (soybean cyst nematode), *Plant Molecular Biology* 71: 525-567.

Klink, V.P., Hosseini, P., Matsye, P., Alkharouf, N. & Matthews, B.F. (2010a). Syncytium gene expression in *Glycine max*$_{[PI\ 88788]}$ roots undergoing a resistant reaction to the parasitic nematode *Heterodera glycines*, *Plant Physiology and Biochemistry* 48: 176-193.

Klink, V.P., Overall, C.C., Alkharouf, N., MacDonald, M.H. & Matthews, B.F. (2010b). Microarray detection calls as a means to compare transcripts expressed within syncytial cells isolated from incompatible and compatible soybean (*Glycine max*) roots infected by the soybean cyst nematode (*Heterodera glycines*), *Genome* Article ID 491217: 1-30.

Klink, V.P., Matsye, P.D. & Lawrence, G.W. (2011a). Cell-specific studies of soybean resistance to its major pathogen, the soybean cyst nematode as revealed by laser capture microdissection, gene pathway analyses and functional studies in Aleksandra Sudaric (ed.) *Soybean - Molecular Aspects of Breeding*, Intech Publishers pp. 397-428.

Klink, V.P., Hosseini, P., Matsye, P.D., Alkharouf, N. & Matthews, B.F. (2011b). Differences in gene expression amplitude overlie a conserved transcriptomic program occurring between the rapid and potent localized resistant reaction at the syncytium of the *Glycine max* genotype Peking (PI 548402) as compared to the prolonged and potent resistant reaction of PI 88788, *Plant Molecular Biology* 75: 141-165.

Komatsu, S., Konishi, H. & Hashimoto, M. (2007). The Proteomics of plant cell membranes, *Journal Experimental Botany* 58: 103–12.

Krishnan, P., Kruger, N.J. & Ratcliffe, R.G. (2005). Metabolite fingerprinting and profiling in plants using NMR. *Journal of Experimental Botany* 56: 255-265.

Li, S.H., Schneider, B. & Gershenzon J. (2007). Microchemical analysis of laser-microdissected stone cells of Norway spruce by cryogenic nuclear magnetic resonance spectroscopy, *Planta* 225: 771-779.

Lippert, D., Chowrira, S., Ralph, S.G., Zhuang, J., Aeschliman, D., Ritland, C., Ritland, K., Bohlmann, J. (2007). Conifer defense against insects: proteome analysis of Sitka spruce (*Picea sitchensis*) bark induced by mechanical wounding or feeding by white pine weevils (*Pissodes strobi*), *Proteomics* 7: 248-70.

Lisec, J., Schauer, N., Kopka, J., Willmitzer, L. & Fernie, A.R. (2006). Gas chromatography mass spectrometry-based metabolite profiling in plants, *Nature Protocols* 1: 387–396.

Lliso, I., Tadeo, F.R., Phinney, B.S., Wilkerson, C.G. & Talón, M. (2007). Protein changes in the albedo of citrus fruits on postharvesting storage, *Journal of Agriculture Food Chemistry* 55: 9047–53.

Lu, P., Vogel, C., Wang, R., Yao, X. & Marcotte, E.M. (2007). Absolute protein expression profiling estimates the relative contributions of transcriptional and translational regulation, *Nature Biotechnology* 25: 117–24.

Martin, C., Bhatt, K. & Baumann, K. (2001). Shaping in plant cells, *Current Opinion in Plant Biology* 4: 540-9.

Matsye, P.D., Kumar, R., Hosseini, P., Jones, C.M., Alkharouf, N., Matthews, B.F.& Klink VP. Mapping cell fate decisions that occur during soybean defense responses. (Submitted)

Matas, A.J., Augusti, J., Tadeo, F.R., Talón, M. & Rose, J.K. (2010). Tissue-specific transcriptome profiling of the citrus fruit epidermis and subepidermis using laser capture microdissection, *Journal of Experimental Botany* 61: 3321-3330.

Mehta, A., Brasileiro, A.C., Souza, D.S., Romano, E., Campos, M.A., Grossi-de-Sá, M.F., Silva, M.S., Franco, O.L., Fragoso, R.R., Bevitori R. & Rocha, T.L. (2008). Plant–pathogen interactions: what is proteomics telling us?, *FEBS Journal* 275: 3731–3746.

Meijers, R., Puettmann-Holgado, R. Skiniotis, G., Liu, J.H., Walz, T., Wang, J.H. & Schmucker, D. (2007). Structural basis of Dscam isoform specificity, *Nature* 449: 487-491.

Michnick, S.W. (2004). Proteomics in living cells, *Drug Discovery Today* 9: 262–267.

Miernyk, J.A., Pret'ova, A., Olmedilla, A., Klubicova, K., Obert B., Hajduch M. et al. (2011). Using proteomics to study sexual reproduction in angioperms, *Sex Plant Reproduction* 24: 9–22.

Moco, S., Bino, R.J., Vorst, O., Verhoeven, H.A., de Groot, J., van Beek, T.A., Vervoort, J. & De Vos, R.C.H. (2006). A liquid chromatography-mass spectrometry-based metabolome database for tomato, *Plant Physiology* 141: 1205–1218.

Moco, S., Capanoglu, E., Tikunov, Y., Bino, R.J., Boyacioglu, D., Hall, R.D., Vervoort, J., De Vos, R.C.H. (2007). Tissue specialization at the metabolite level is perceived during the development of tomato fruit, *Journal Experimental Botany* 58: 4131-4146.

Moco, S., Forshed, J., De Vos, R.C.H., Bino, R.J. & Vervoort, J. (2008). Intra- and inter-metabolite correlation spectroscopy of tomato metabolomics data obtained by liquid chromatography-mass spectrometry and nuclear magnetic resonance, *Metabolomics* 4: 202–215.

Moco, S., Schneider, B. & Vervoort, J. (2009). Plant micrometabolomics: the analysis of endogenous metabolites present in a plant cell or tissue, *Journal of Proteome Research* 8: 1694-1703.

Moulédous, L., Hunt, S., Harcourt, R., Harry, J., Williams, K.L. & Gutstein, H.B. (2002). Lack of compatibility of histological staining methods with proteomic analysis of laser-capture microdissected brain samples, *Journal of Biomolecular Techniques* 13: 258-264.

Moulédous, L., Hunt, S., Harcourt, R., Harry, J., Williams, K.L. & Gutstein, H.B. (2003). Navigated laser capture microdissection as an alternative to direct histological staining for proteomic analysis of brain samples, *Proteomics* 3: 610-615.

Nakazono, M., Qiu, F., Borsuk, L.A. & Schable, P.S. (2003). Laser capture microdissection, a tool for the global analysis of gene expression in specific plant cell types: identification of genes expressed differentially in epidermal cells or vascular tissue of maize, *The Plant Cell* 15: 583-596.

Nelson, C.J., Hegeman, A.D., Harms, A.C. & Sussman, M.R. (2006a). A quantitative analysis of *Arabidopsis* plasma membrane using trypsin-catalyzed 18 O labeling. *Molecular Cell Proteomics* 5: 1382–95.

Nelson, T., Tausta, S.L., Gandotra, N. & Liu, T. (2006b). Laser microdissection of plant tissue: What you see is what you get, *Annual Review of Plant Biology* 57: 181–201.

Nelson, T., Gandotra, N. & Tausta, S.L. (2008). Plant cell types: reporting and sampling with new technologies, *Current Opinion in Plant Biology* 11: 567–573.

Nesatyy, V.J. & Suter, M.J. (2008). Analysis of environmental stress response on the proteome level, *Mass Spectrometry Reviews* 27: 556-74.

Nie, L., Wu, G., Culley, D.E., Scholten, J.C. & Zhang W. (2007). Integrative analysis of transcriptomic and proteomic data: challenges,solutions and applications, *Critical Reviews in Biotechnology* 27: 63–75.

Niittylä, T., Fuglsang, A.T., Palmgren, M.G., Frommer, W.B. & Schulze, W.X. (2007). Temporal analysis of sucrose-induced phosphorylation changes in plasma membrane proteins of *Arabidopsis, Molecular Cell Proteomics* 6: 1711–26.

Nowaczyk, M.M., Hebeler, R., Schlodder, E., Meyer, H.E., Warscheid, B. & Rögner, M. (2006). Psb27, a cyanobacterial lipoprotein, is involved in the repair cycle of photosystem II, *The Plant Cell* 18: 3121–3177.

Oda, Y., Huang, K., Cross, F.R., Cowburn, D. & Chait, B.T. (1999). Accurate quantitation of protein expression and site-specific phosphorylation. *Proceedings of the National Academy of Science USA* 96: 6591–96.

O'Farrell, P.H. (1975). High resolution two-dimensional electrophoresis of proteins, *The Journal of Biological Chemistry* 250: 4007-4021.

O'Farrell, P.Z., Goodman, H.M. & O'Farrell P.H. (1977). High resolution two dimensional electrophoresis of basic as well as acidic proteins, *Cell* 12: 1133-1142.

Opitz, S. & Schneider, B. (2002). Organ-specific analysis of phenylphenalenone-related compounds in *Xiphidium caeruleum*, *Phytochemistry* 61: 819–825.

Ohtsu, K., Takahashi, H., Schnable, P.S. & Nakazono, M. (2007). Cell type-specific gene expression profiling in plants by using a combination of laser microdissection and high-throughput technologies, *Plant Cell Physiology* 48: 3-7.

Palmblad, M., Mills, D.J. & Bindschedler, L.V. (2008). Heat-shock response in *Arabidopsis thaliana* explored by multiplexed quantitative proteomics using differential metabolic labeling, *Journal of Proteome Research* 7: 780–85.

Palusa, S.G., Ali. G.S. & Reddy A.S.N. (2007). Alternative splicing of pre-mRNAs of Arabidopsis serine/arginine-rich proteins: regulation by hormones and stresses, *The Plant Journal* 49: 1091–107.

Pan, H. & Lundgren, L.N. (1995). Phenolic extractives from root bark of *Picea abies*, *Phytochemistry* 39: 1423-8.
Pan, Q., Shai, O., Lee, L.J., Frey, B.J. & Blencowe, B.J. (2008). Deep surveying of alternative splicing complexity in the human transcriptome by high-throughput sequencing, *Nature* Genetic 40: 1413-5.
Pedreschi, R.,Vanstreels, E., Carpentier, S., Hertog, M., Lammertyn, J., Robben J., Noben, J.P., Swennen, R., Vanderleyden, J. & Nicolai, B. (2007). Proteomic analysis of core breakdown disorder in conference pears (*Pyrus communis* L.), *Proteomics* 7: 2083-99.
Peltier, J.B., Friso, G., Kalume, D.E., Roepstorff, P., Nilsson, F., Adamskaa, I. & van Wijk K.J. (2000). Proteomics of the chloroplast: systematic identification and targeting analysis of lumenal and peripheral thylakoid proteins, *The Pant Cell* 12: 319–342.
Persidis, A. (1998). Proteomics, *Nature Biotechnology* 4: 393-394.
Pevzner, P.A., Mulyukov, Z., Dancik, V. & Tang, C.L. (2001). Efficiency of database search for identification of mutated and modified proteins via mass spectrometry, *Genome Research* 11: 290–99.
Purea, A., Neuberger, T. & Webb, A.G. (2004). Simultaneous NMR micro-imaging of multiple single-cell samples, *Concepts Magn Reson Part B* 22B: 7–14.
Rabilloud, T. (1996). Solubilization of proteins for electrophoretic analyses, *Electrophoresis* 17: 813-829.
Ramsay, K., Jones, M.G.K. & Wang, Z. (2006). Laser capture microdissection: A novel approach to microanalysis of plant–microbe interactions, *Molecular Plant Pathology* 7: 429–435.
Reiland, S., Messerli, G., Baerenfaller, K., Gerrits, B., Endler, A., Grossmann, J., Gruissem W. & Baginsky, S. (2009). Large-scale *Arabidopsis* phosphoproteome profiling reveals novel chloroplast kinase substrates and phosphorylation networks, *Plant Physiology* 150: 889-903.
Rekhter, M.D. & Chen, J. (2001). Molecular analysis of complex tissues is facilitated by laser capture microdissection: Critical role of upstream tissue processing, *Cell Biochemistry and Biophysics* 35: 103–113.
Roessner, U., Willmitzer, L. & Fernie, A.R. (2001). High-resolution metabolic phenotyping of genetically and environmentally diverse potato tuber systems. Identification of phenocopies, *Plant Physiology* 127: 749–764.
Rossignol, M., Peltier, J.B., Mock, H.P., Matros, A., Maldonado, A.M. & Jorrin J.V. (2006). Plant proteome analysis: a 2004–2006 update, *Proteomics* 6: 5529-48.
Rubakhin, S.S., Greenough, W.T. & Sweedler, J.V. (2003). Spatial profiling with MALDI MS: distribution of neuropeptides within single neurons, Analytical Chemistry 75: 5374-5380.
Salekdeh, G.H. & Komatsu, S. (2007). Crop proteomics: aim at sustainable agriculture of tomorrow, *Proteomics* 7: 2976–96.
Schad, M,. Lipton, M.S., Giavalisco, P., Smith, R.D. & Kehr, J. (2005a). Evaluation of two-dimensional electrophoresis and liquid chromatography–tandem mass spectrometry for tissue-specific protein profiling of laser-microdissected plant samples, *Electrophoresis* 26: 2729–2738.
Schad, M., Mungur, R., Fiehn, O. & Kehr, J. (2005b). Metabolic profiling of laser microdissected vascular bundles of *Arabidopsis thaliana*, *Plant Methods* 1:2.

Schmucker, D., Clemens, J.C., Shu, H., Worby, C.A., Xiao, J., Muda, M., Dixon, J.E. & Zipursky, S.L. (2000). *Drosophila* Dscam is an axon guidance receptor exhibiting extraordinary molecular diversity, *Cell* 101: 671-684.

Schneider, B. & Hölscher, D. (2007). Laser microdissection and cryogenic nuclear magnetic resonance spectroscopy: an alliance for cell type-specific metabolite profiling, *Planta* 225: 763–770.

Schulze, W.X. & Usadel, B. (2010). Quantitation in mass-spectrometry-based proteomics, *Annual Review in Plant Biology* 61: 491–516.

Sheoran, I.S., Ross, A.R., Olson, D.J. & Sawhney, V.K. (2007). Proteomic analysis of tomato (*Lycopersicon esculentum*) pollen, *Journal of Experimental Botany* 58: 3525–35.

Shibutani, M., Uneyama, C., Miyazaki, K., Toyoda, K. & Hirose, M. (2000). Methacarn Fixation: A Novel Tool for Analysis of Gene Expressions in Paraffin-Embedded Tissue Specimens. *Lab Invest.* 80: 199–208.

Simon, S.A., Zhai, J., Nandety, R.S., McCormick, K.P., Zeng, J., Mejia, D. & Meyers, B.C. (2009). Short-Read Sequencing Technologies for Transcriptional Analyses, *Annual Review of Plant Biology* 60: 305–33.

Simpson, C.G., Lewandowska, D., Fuller, J., Maronova, M., Kalyna, M., Davidson, D., et al. (2008) Alternative splicing in plants, *Biochem Soc Trans* 36: 508–10.

Sitek, B., Luttges, J., Marcus, K., Kloppel, G., Schmiegel, W., Meyer, H. E., Hahn, S. A. & Stuhler, K. (2005). Application of fluorescence difference gel electrophoresis saturation labelling for the analysis of microdissected precursor lesions of pancreatic ductal adenocarcinoma, *Proteomics* 5: 2665–2679.

Song, X., Ni, Z., Yao, Y., Xie, C., Li, Z., Wu, H, Zhang, Y. & Sun, Q. (2007). Wheat (*Triticum aestivum* L.) root proteome and differentially expressed root proteins between hybrid and parents, *Proteomics* 27: 3538–57.

Sturm, S., Seger, C., Godejohann, M., Spraul, M. & Stuppner, H. (2007). Conventional sample enrichment strategies combined with high- performance liquid chromatography-solid phase extraction-nuclear magnetic resonance analysis allows analyte identification from a single minuscule *Corydalis solida* plant tuber, *Journal of Chromatography* 1163: 138–144.

Takáč, T., Pechan, T. & Samaj, J. (2011). Differential proteomics of plant development, *Journal of Proteomics* 74: 577-88.

Takahashi, H., Kamakura, H., Sato, Y., Shiono, K., Abiko, T., Tsutsumi, N., Nagamura, Y., Nishizawa, N.K., Nakazono, M. (2010). A method for obtaining high quality RNA from paraffin sections of plant tissues by laser microdissection, *Journal of Plant Research* 123: 807-813.

Tanabe, N., Yoshimura, K., Kimura, A., Yabuta, Y. & Shigeoka, S. (2007). Differential expression of alternatively spliced mRNAs of *Arabidopsis* SR protein homologs, atSR30 and atSR45a, in response to environmental stress, *Plant Cell Physiology* 48: 1036–1049.

Tang, W., Coughlan, S., Crane, E., Beatty, M. & Duvick, J. (2006). The application of laser microdissection to in planta gene expression profiling of the maize anthracnose stalk rot fungus *Colletotrichum graminicola, Molecular Plant–Microbe Interaction* 19: 1240–1250.

Tanner, S., Shu, H., Frank, A., Wang, L.C., Zandi, E., Mumby, M., Pevzner, P.A. & Bafna, V. (2005). InsPecT: identification of posttranslationally modified peptides from tandem mass spectra, *Analytical Chemistry* 77: 4626–39.

Thiel, J., Weier, D., Sreenivasulu, N., Strickert, M., Weichert, N., Melzer, M., Czauderna, T., Wobus, U., Weber, H., Weschke, W. (2008) Different hormonal regulation of cellular differentiation and function in nucellar projection and endosperm transfer cells—a microdissection-based transcriptome study of young barley grains, *Plant Physiology* 148: 1436–1452.

Thiel, J., Müller, M., Weschke, W. & Weber, H. (2009). Amino acid metabolism at the maternal-Wlial boundary of young barley seeds: a microdissection-based study, *Planta* 230: 205–213.

Valledor, L., Castillejo, M.A., Lenz, C., Rodríguez, R., Cañal, M.J. & Jorrín, J. (2008). Proteomic analysis of Pinus radiata needles: 2-DE map and protein identification by LC/MS/MS and substitution-tolerant database searching, *Journal of Proteome Research* 7: 2616–31.

van der Weerd, L., Claessens, M.M.A.E., Efde, C. & Van As, H. (2002). Nuclear magnetic resonance imaging of membrane permeability changes in plants during osmotic stress, *Plant Cell and Environment* 25: 1539-1549.

Vyetrogon, K., Tebbji, F., Olson, D.J., Ross, A.R. & Matton, D.P. (2007). A comparative proteome and phosphoproteome analysis of differentially regulated proteins during fertilization in the self-incompatible species *Solanum chacoense* Bitt., *Proteomics* 7: 232–47.

Wang, Y.L., Tang, H.R., Nicholson, J.K., Hylands, P.J., Sampson, J., Whitcombe, I., Stewart, C.G., Caiger, S., Oru, I., Holmes, E. (2004). Metabolomic strategy for the classification and quality control of phytomedicine: a case study of chamomile flower (*Matricaria recutita* L.), *Planta Medica* 70: 250-5.

Wang, D., Eyles, A., Mandich, D. & Bonello, P. (2006). Systemic aspects of host–pathogen interactions in Austrian pine (*Pinus nigra*): a Proteomics approach. *Physiological and Molecular Plant Pathology* 68: 149–57.

Ward, J.L., Harris, C., Lewis, J. & Beale, M.H. (2003). Assessment of H-1 NMR spectroscopy and multivariate analysis as a technique for metabolite fingerprinting of *Arabidopsis thaliana*, *Phytochemistry* 62: 949–957.

Wasinger, V.C., Cordwell, S.J., Cerpa-Poljak, A., Yan, J.X., Gooley, A.A., Wilkins, M.R., Duncan, M.W., Harris, R., Williams, K.L. & Humphery-Smith, I. (1995). Progress with gene-product mapping of the Mollicutes: *Mycoplasma genitalium*, *Electrophoresis* 16: 1090-1094.

Wenzler, M., Hölscher, D., Oerther, T. & Schneider, B. (2008). Nectar formation and floral nectary anatomy of *Anigozanthos flavidus*: a combined magnetic resonance imaging and spectroscopy study, *Journal of Experimental Botany* 59: 3425-3434.

Witze, E.S., Old, W.M., Resing, K.A. & Ahn N.G. (2007). Mapping protein post-translational modifications with mass spectrometry, *Nature Methods* 4: 798–806.

Xiao, Y.L., Smith, S.R., Ishmael, N., Redman, J.C., Kumar, N., Monaghan, E.L., Ayele, M., Haas, B.J., Wu, H.C. & Town C.D. (2005). Analysis of the cDNAs of hypothetical genes on *Arabidopsis* chromosome 2 reveals numerous transcript variants, *Plant Physiology* 139: 1323-37.

Yoshimura, K., Masuda, A., Kuwano, M., Yokota, A. & Akashi, K. (2008). Programmed proteome response for drought avoidance/tolerance in the root of a C3 xerophyte (wild watermelon) under water deficits, *Plant and Cell Physiology* 49: 226–41.

Yu, Y.Y., Lashbrook, C. & Hannapel, D. (2007). Tissue integrity and RNA quality of laser microdissected phloem of potato, *Planta* 226: 797–803.

4

A Critical Review of Trypsin Digestion for LC-MS Based Proteomics

Hanne Kolsrud Hustoft, Helle Malerod, Steven Ray Wilson, Leon Reubsaet, Elsa Lundanes and Tyge Greibrokk
University of Oslo
Norway

1. Introduction

Proteomics is defined as the large-scale study of proteins in particular for their structures and functions (Anderson and Anderson 1998), and investigations of proteins have become very important since they are the main components of the physiological metabolic pathways in eukaryotic cells. Proteomics increasingly plays an important role in areas like protein interaction studies, biomarker discovery, cancer prevention, drug treatment and disease screening medical diagnostics (Capelo et al. 2009).

Proteomics can be performed either in a comprehensive or "shotgun" mode, where proteins are identified in complex mixtures, or as "targeted proteomics" where "selective reaction monitoring" (SRM) is used to choose in advance the proteins to observe, and then measuring them accurately, by optimizing the sample preparation as well as the LC-MS method in accordance to the specific proteins (Mitchell 2010).

Whether "MS-based shotgun proteomics" has accomplished anything at all regarding clinically useful results was recently addressed by Peter Mitchell in a feature article (Mitchell 2010), and he states that the field needs to make a further step or even change direction. Referring to discussions with among others John Yates and Matthias Mann, Mitchell addresses the failure in the search for biomarkers as indicators of disease, the difficulties of protein arrays, the uncertainty of quantification in "shotgun proteomics" (due to among others the efficiency of ionization in the mass spectrometers), database shortcomings, the problems of detecting post translational modifications (PTMs), and finally the huge disappointment in the area of drug discovery. The field points in the direction of targeted proteomics, but targeted proteomics will not be the solution to all our questions and comprehensive proteomics will still be needed. In order to get as much information, with as high quality as possible, from a biological sample, both the sample preparation and the final LC-MS analyses need to be optimized.

The most important step in the sample preparation for proteomics is the conversion of proteins to peptides and in most cases trypsin is used as enzyme. Trypsin is a protease that specifically cleaves the proteins creating peptides both in the preferred mass range for MS sequencing and with a basic residue at the carboxyl terminus of the peptide, producing information-rich, easily interpretable peptide fragmentation mass spectra. Some other proteases can be used as well, such as Lys-C, which is active in more harsh conditions with 8 M urea, and give larger fragments than trypsin. Asp-N and Glu-C are also highly sequence-

specific proteases, but less active than the previously mentioned. Other less sequence-specific proteases are generally avoided since they create complex mixtures of peptides, difficult to interpret (Steen and Mann 2004). During a chromatographic separation of a complex mixture of peptides derived from a tryptic digestion, thousands of mass spectra are produced and sophisticated software is necessary to find matching proteins to the peptides identified. In complex proteomic samples, protein identification is performed by searching databases with search engines like Mascot, Sequest or Phenyx (IS 2011).
Protein identification traditionally follows two different workflows depending on the approach (Figure 1). In the gel electrophoresis-based approach the proteins are separated in one or two dimensions (1D/2D) on a gel and enzymatic digestion is performed in-gel, which is a time-consuming and tedious process (López-Ferrer et al. 2006). In the gel-free or in-solution based approach, the proteins or peptides, or both, are separated chromatographically using on-line LC systems and the proteins are digested in-solution

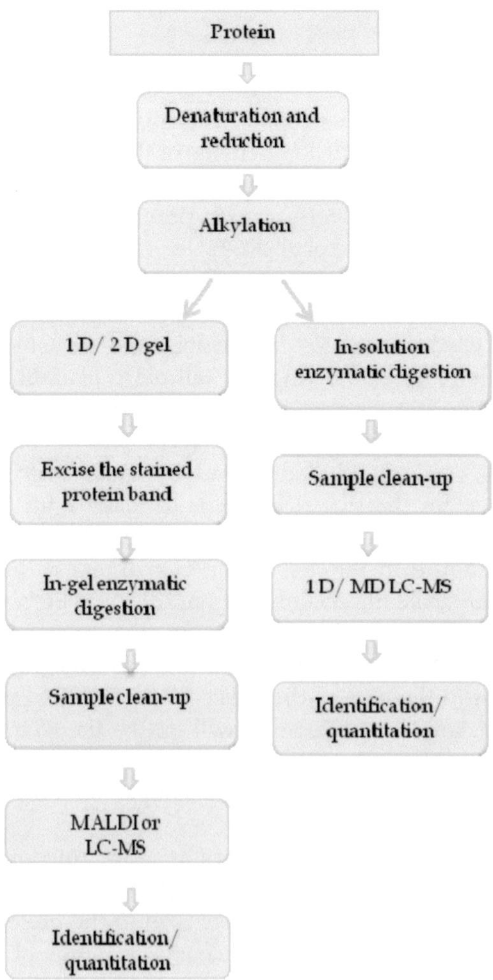

Fig. 1. Workflows of in-gel (left) and in-solution (right) digestion and subsequent LC-MS analysis of a protein sample.

(Capelo et al. 2009). The in-solution based approach tends to be the simplest in terms of sample handling and speed, but on the other hand it requires sophisticated LC-MS instrumentation which again requires constant maintenance.

The digestion step is the most time consuming step in the sample preparation workflow and different techniques to accelerate this procedure have been developed. Comparing these techniques, including some of our own experiments, the question of how we can evaluate the digestion efficiency materialized. The amino acid sequence coverage (SQ %) is often used as a measure for both the completeness of the protein digestion and the detection efficiency of the various tryptic peptides, and is a common way in proteomics to define the digestion rate (Xu et al. 2010). However, SQ % might be a misleading parameter to use, as different mass spectrometers and different search parameters in subsequent data analysis may reveal various SQ %. In addition it is of principal importance to relate SQ % to the degree of miss - cleavage peptides used to calculate this value: a high SQ % calculated from tryptic peptides without missed cleavages indicated a more complete digest than the same high SQ % calculated from tryptic peptides with many missed cleavages.

To get some information of the digestion efficiency, as a check before performing the data analysis, the possible presence of intact protein in the total ion chromatogram (TIC) may be used. However, this method can only be used for proteins small enough to be detected by the MS, such as cytochrome-C (cyt-C) (unless you have a MALDI MS available). On the other hand, evaluating the digestion rate this way, using an easily digested protein such as cyt-C, will give a good indication of the efficiency of the method; if an intact protein peak from cyt-C is detected in the chromatogram, then the digestion can be considered insufficient. Other non-protein reagents that are cleaved by trypsin might also be used as an internal standard when performing tryptic digestion of a complex sample, to have control over the digestion efficiency.

For quantitation of proteins it is necessary to find relevant indicators of their abundance in the mass spectrometer output. Several ways of protein quantitation have been suggested and they can be divided into two main categories; the isotope based and the label free methods. Two papers which give good overviews over the different labeling methods have been published recently (Capelo et al. 2010; Vaudel et al. 2010). In brief; the main modern strategies for isotopic labeling are divided into metabolic labeling at cell growth called SILAC, chemical labeling at protein level, called iCAT, enzymatic labeling at peptide level, after protein digestion like iTRAQ and labeling during protein digestion, such as ^{18}O labeling (Capelo et al. 2010). SILAC can only be used for samples which are produced using labeled amino acids, while the other methods can be used for all types of protein samples. Thiede *et al.* have recently introduced a promising new labeling method with relative or absolute quantification for identification and quantification of two differentially labelled states using MS/MS spectra, and which is called isobaric peptide termini labeling (IPTL) (Thiede and Koehler 2010). The method involve digestion of the protein samples and cross-wise labeling of N- and C-terminal ends of the obtained peptides, like the principle in ^{18}O labeling (Thiede and Koehler 2010).

The digestion efficiency in comprehensive proteomics is as important as the digestion repeatability in targeted proteomics. Everyone working in this field should strive to have control over these parameters during the sample preparation in proteomics, producing correctly identified proteins and reliable results. The focus in this review is on the in-solution based protocols in comprehensive proteomics, with emphasis on in-solution tryptic digestion and alternative methods to speed up the digestion, and also on how to evaluate the digestion efficiency of the used method.

2. Factors influencing proteolytic results

An issue that is little discussed in the literature of proteomics is the sample handling prior to the protein digestion. Some mention the need for enrichment, or elimination of interfering substances (López-Ferrer et al. 2006), but few focus on the steps prior to the enzymatic digestion of the protein fractions.

2.1 Protein concentration

A proper digestion procedure starts with the measurement of the protein content of the sample. This is necessary to determine for, among others, the needed amount of reduction and alkylation reagents, as well as the amount of enzyme in in-solution digestion. Quantifying the protein content of a sample separated on a gel is often relatively easy. In this case, guidelines of intensity of the stained gel-band can be used as a "semi-quantitative" measurement. The amount of the total protein content of gel-free samples can be measured with standard procedures like the NanoDrop (detection down to 10-15 μg/ml, using 2 μl sample) (NanoDrop 2011), the Bradford assay (detection down to 2.5 μg/ml, using 150 μl sample) (Bradford 1976) and the BCA assay (detection down to 20 μg/ml, using 25 μl sample) (Smith et al. 1985).

2.2 Keratin contamination

Avoiding keratin contamination, which is a problem common in both 1D or 2D gel and in-solution methods, but mostly in the gel-based analysis (Bell et al. 2009), is important. Keratins are naturally occurring structural proteins and appear more often in the sample as interference from the environment rather than from natural abundance. Fingerprints, hair, dead skin flakes, wool clothing, dust and latex gloves are common sources of contaminating keratins (Greenebaum 2011). If keratins are present at concentration levels greater than that of the protein of interest, their abundance will overwhelm the analytical capacity of the LC-MS system and obscure the protein of interest. This is particularly problematic when performing data dependent mass spectrometry, as the peptides from the more abundant keratins will be selected for tandem-MS analysis, providing little or no information about the actual proteins of interest. However, at low concentration levels, compared to the protein of interest, keratins are not a problem at all (Greenebaum 2011).

2.3 Detergents

Detergents are often used for total solubilisation of cells and tissues in biochemical studies, and sodium dodecyl sulphate (SDS) is often the choice. However, even at low concentrations, detergents can give rise to problems both concerning enzymatic digestion and in the subsequent LC-MS analysis. Hence it is most often necessary to deplete the detergents prior to the steps in the analytical method hampered by the detergent, or to find alternative ways to lyse the cells which are more compatible with the downstream steps in the analysis. This problem will be further discussed in section 5.2.2.

3. From proteins to peptides

3.1 Denaturation, reduction and alkylation

Prior to in-solution protein digestion the proteins in most samples need to be denatured, reduced and alkylated, using various reagents, for the proteolytic enzyme to be able to efficiently cleave the peptide chains of the proteins. A sample preparation workflow is presented in Figure 2 together with different suggested procedures to accelerate the tryptic digestion of proteins to peptides. These methods are presented in section 5.

Procedure	Intended effect	Experiment
Protein quantitation	To add correct amount of reagents	NanoDrop, Bradford assay, BCA assay
Denaturation and reduction	Reduces diculfide binds to avoid renaturation	1+50 DTT:protein, 15 min 95 °C
Alkylation	Alkylation of SH groups, avoid renaturation	1+10 IAM:protein, 15 min in the dark
Enzymatic digestion	Cleavage of proteins to peptides	1+20 E+S, pH 8.5, 37 °C overnight
Stop enzymatic digestion	To avoid autodigestion	Addition of 2.5 % formic acid (FA)
Desalting and enrichment	Clean-up and enrichment of peptides	C18 ZipTip procedure

Speed up

Immobilized trypsin acceleration
-Immobilized reactors
-Microspin columns
-FASP filters

Temperature related acceleration
-Heating
-Infrared radiation
-Microwave radiation

Other ways to accelerate
-Ultrasonication
-Solvent effects

Fig. 2. Procedure, intended effect and experimental conditions of a classical workflow for in-solution based sample preparation approaches in proteomics. To the very right different accelerating digestion techniques are presented.

In a study by Proc *et al.* the denaturation process of human plasma proteins was examined applying 14 different combinations of heat, solvents, chaotropic agents and surfactants for their effectiveness to improve tryptic digestion (Proc et al. 2010). The experiment was performed by quantifying the production of proteolytic tryptic peptides from 45 moderate-to-high-abundance plasma proteins which were grouped into rapidly digested proteins, moderately digested proteins and proteins resistant to digestion. Proc *et al.* did not find an "optimal" digestion method for all 45 proteins, but the denaturation procedure with the surfactant sodium deoxycholate (DOC), which is more compatible with MS than SDS, together with a digestion time of 9 hours, was found to be the most promising protocol for all proteins (Proc et al. 2010).

Denaturation and reduction can often be carried out simultaneously by a combination of heat and a reagent, like 1,4-dithiothreitol (DTT) (Choudhary et al. 2002), β-mercaptoethanol (Sundqvist et al. 2007) or tris(2-carboxyethyl)phosphine (Hale et al. 2004). Most used is DTT, which is a strong reducing agent, that reduce the disulfide bonds and prevent inter and intra-molecular disulfide formation between cysteines in the protein. By combining denaturation and reduction, renaturation of the proteins due to reduction of the disulfide bonds can be avoided (see Figure 3). Renaturation can be a problem using heat solely as the denaturation agent (Strader et al. 2005; Capelo et al. 2009).

Following protein denaturation and reduction, alkylation of cysteine is necessary to further reduce the potential renaturation (Figure 3), and the most commonly used agents for alkylation of protein samples prior to digestion are iodoacetamide (IAM) and iodoacetic acid (IAA) (López-Ferrer et al. 2006; Vukovic et al. 2008).

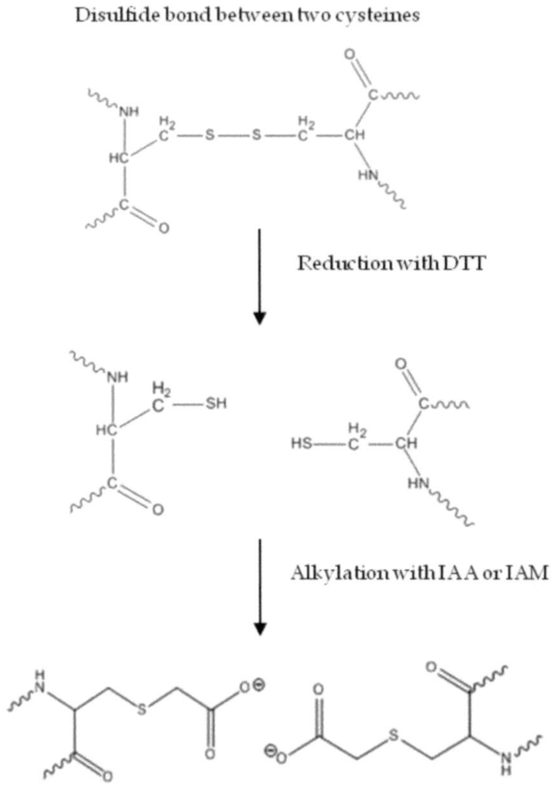

Fig. 3. The reduction and alkylation process: The breaking of disulfide bonds in proteins. Reduction by DTT to form cysteine residues must be followed by further modification of the reactive –SH groups (to prevent reformation of the disulfide bond) by acetylation by, in this case iodoacetic acid (adapted from (Nelson and Cox 2008)).

3.2 Trypsin digestion

Protein digestion with proteases is one of the key sample-preparation steps in proteomics, followed by LC-MS. As already mentioned, trypsin is the most commonly used protease for this purpose since it has a well defined specificity; it hydrolyzes only the peptide bonds in which the carbonyl group is followed either by an arginine (Arg) or lysine (Lys) residue, with the exception when Lys and Arg are N-linked to Aspartic acid (Asp). The cleavage will not occur if proline is positioned on the carboxyl side of Lys and Arg. Since trypsin is a protein it may digest itself in a process called autolysis. However, Ca^{2+}, naturally present in most samples, binds at the Ca^{2+}-binding loop in trypsin and prevents autolysis (Nord et al. 1956). With the modified trypsin presently used in most laboratories, autolysis is additionally reduced. Still addition of 1 mM $CaCl_2$ is recommended in the digestion medium, but not always absolutely necessary, when the contribution of Ca^{2+} from natural sources is low (Minnesota 2011). Tryptic digestion is performed at an optimal pH in the range 7.5-8.5 (Worthington 2011), and commonly at 37 °C for in-solution digestion. Thus prior to the addition of trypsin, a buffer is added (usually 50 mM triethyl ammonium bicarbonate (tABC) or 12.5 mM ammonium bicarbonate (ABC) buffer (López-Ferrer et al. 2006) to provide an optimal pH for the enzymatic cleavage. A 2-amino-2-hydroxymethyl-propane-1,3-diol (Tris) buffer may also be used for this purpose, but it should be taken into consideration that the Tris buffer is incompatible with the down stream MS analysis, such as MALDI and ESI-MS, and needs to be depleted through solid phase extraction (SPE) or ZipTips prior to such (Shieh et al. 2005; Sigma-Aldrich 2011).

Information about the enzyme to protein ratio needed for digestion of a protein sample is crucial to ensure an enzyme amount sufficient to perform the digestion, but not too high resulting in autolysis products from the trypsin used. Recent experiments indicate that a sufficient ratio of enzyme to substrate (E+S) is 1+20 (Hustoft et al. 2011). For targeted proteomics it may be beneficial to perform a pilot study on the necessary digestion time for the type of sample to be analyzed, to obtain an optimal digestion efficiency of the sample. For more comprehensive proteomics a longer digestion time, up to 9 hours is recommended to ensure the best overall digestion efficiency, as described by Proc *et al.* (Proc et al. 2010). Thus dealing with these long digestion times, an overnight digestion is often more convenient, starting with the post digestion sample preparation steps the following day.

Proteins may act differently in different environments and less effective digestions have been observed when model proteins were digested in a mixture as compared to being digested separately (Hustoft et al. 2011). One reason for these observations could be increased competition for the trypsin cleavage sites, when more proteins are digested together.

As mentioned in the introduction one of the main issues regarding digestion is how to measure the digestion efficiency of a method for a given complex sample of proteins. Examples from the literature show that different groups use various measures for the efficiency of their digestive method, where amino acid SQ % is the most common. However, based on our experience, we question whether the SQ % can serve as a reliable measurement of digestion efficiency, or not? Using a relatively high concentration of 250 ng/ ml of each of the model proteins, no significant difference in SQ % could be seen with a 5 min digestion versus an overnight digestion. Thus another measure for the digestion efficiency had to be evaluated. Since cyt-C was one of the model proteins, undigested intact protein could be detected by the Ion-Trap MS being used. The area of the intact protein peak decreased with increasing digestion time, - indicating better trypsination efficiency. The size of the intact protein peak could hence in some cases be used to compare the efficiency of digestion methods (Hustoft et al. 2011). When exploring the potential of microwave oven accelerated digestion of a mixture of proteins, different temperatures were examined both for the microwave oven and the Thermoshaker control samples. A decrease in the intact protein peak of cyt-C was detected indicating better efficiency at higher temperatures, in both cases. However, the peak area of four distinct tryptic peptides from cyt-C revealed that the decrease in cyt-C peak area was caused by denaturation of the sample as a function of higher temperatures, and not because of increased digestion. Hence, the area of the undigested protein peak is not necessarily a good measure for the digestion efficiency. Another way to describe the digestion efficiency is through the yield of peptides, used to study the effect of temperature, enzyme concentration, digestion time and surface area of the gel pieces in in-gel proteomics (Havliš et al. 2003).

4. Sample handling post digestion

4.1 Clean-up and enrichment of digests

Prior to LC-MS analysis, the digests must be purified to remove e.g. buffers and salts added during the sample preparation. This is most often carried out with ZipTips, which concentrate and purifies the samples for sensitive downstream analysis (Capelo et al. 2009). A C18 ZipTip is a 10 µl pipette tip with a 0.6 or 0.2 µl bed of C18 silica based medium fixed at its tip, used for single-step desalting, enrichment, and purification. Such ZipTips can be used for purification of, for instance peptides, proteins and nucleic acids. Purifying tryptic peptides with the C18 ZipTip results in high recovery, but noteworthy is that the capacity of the C18 ZipTips is limited; however, up to 10 µg digested protein could be loaded without losses (Hustoft et al. 2011). Another possible disadvantage of the C18 ZipTip procedure is the loss of small hydrophilic peptides which may be lost due to washing with an aqueous mobile phase containing 0.1 % trifluoroacetic acid (TFA) (Hustoft et al. 2011). Still, when the sample must be purified prior to LC-MS analysis the ZipTips are convenient to use because they are easy to handle and commercially available at a reasonable price, producing good recoveries.

5. Accelerating the protein digestion

An efficient proteolytic digestion, which is important to correctly identify proteins in comprehensive proteomics and to obtain low detection limits in targeted proteomics, requires the generation of peptides in a minimal amount of time. Conventional methods often involve up to 12-16 hours of incubation, but digestion times up to 24 hours are reported, due to protein heterogeneity in samples (López-Ferrer et al. 2006). Alternative methods have therefore been introduced in order to speed up the digestion method. Capelo *et al.* report eight ways to speed up the protein identification workflow (Capelo et al. 2009); heating, microspin columns, ultrasonic energy, high pressure, infrared (IR) energy, microwave energy, alternating electric fields and microreactors where the trypsin is immobilized on a solid support. The pros and cons of these methods were assembled in a table, including citations or validations of the methods from other research groups (Capelo et al. 2009). Capleo *et al.* found that heating, ultrasonication, microwave energy and microreactors (immobilized trypsin) are used in most applications, and recommend that the systems with microspin columns, high pressures, alternating electric fields and IR energy need to be further validated. In a recent study (Hustoft et al. 2011) we have evaluated some of these techniques; IR energy, microwave energy, solvent effects as well as a newly developed filter aided sample preparation (FASP) technique to perform both depletion of detergents like SDS and tryptic digestion of proteins on the same filter device. The different methods are grouped into "temperature related accelerated digestion", "immobilized trypsin accelerated digestion" and "other ways to accelerate digestion" in the following. The terminology used gives an indication of the acceleration method for enzymatic digestion.

5.1 Temperature related accelerated digestion

5.1.1 Heating

Enzymes perform best at a given temperature and for in-solution tryptic digestion, 37 °C has been suggested as the optimal temperature (Havliš et al. 2003), and is the temperature most commonly used both for overnight in-gel and in-solution based tryptic digestion. Havlis *et al.* showed that reductive methylation of trypsin decreases autolysis and shifts the optimum of its catalytic activity to 50-60 °C, with enzymatic digestion of bovine serum albumin (BSA) 12 times faster than in-solution at 37 °C, using the yield of peptides as a parameter of the digestion efficiency (Havliš et al. 2003). From time to time some approaches have been introduced regarding the use of elevated temperatures for trypsin digestion (Capelo et al. 2009), but no new papers have been published recently.

5.1.2 Ultrasonic assisted digestion

Among the different ways to speed up protein digestion, ultrasonic energy has been considered the most promising method of the techniques requiring specialized equipment (Capelo et al. 2009). Three different commercial devices are used for ultrasonication in laboratories today. The most available is the ultrasonic bath, but for the purpose of accelerating tryptic digestion this is not sufficiently powerful to shorten the digestion times (Capelo et al. 2009). Regardless of this, Li *et al.* claimed that an ultrasound bath-assisted method gave successful in-solution proteolysis of three model proteins; BSA, cyt-C and myoglobin, revealing higher SQ % than conventional overnight incubation at 37 °C (Li et al. 2010) . However, the experimental set up and type of samples used should be considered carefully. It would probably be more correct to compare the ultrasonic bath method to 37 °C incubation without the ultrasonic bath, using proteins denaturated, reduced and alkylated in the same fashion. Sonoreactors and ultrasonic probes are more effective, revealing a higher number of peptides and thus better SQ %, giving digestion in seconds as shown in the direct ultrasonic assisted enzymatic digestion of the soybean proteins (Domínguez-Vega et al. 2010). Carreira *et al.* proposed in another study a methodology that uses ultrasonic energy to speed up the protein digestion and throughput of ^{18}O labeling for protein quantification and peptide mass mapping through mass spectrometry based techniques (Carreira et al. 2010). This is a promising technique to accelerate the trypsin digestion of proteins, thus requiring specialized equipment, as mentioned in the start of this section.

5.1.3 Infrared radiation assisted digestion

In 2008 Wang *et al.* introduced a system where infrared (IR) energy was used to speed up the rate of trypsin digestion of proteins (Wang et al. 2008). The type of instrumentation used is presented in Figure 4, and the infrared light contributed, according to the authors, to shorter digestion times by increasing the excitation of the molecules and thus increasing the interaction between trypsin and the peptide bonds in the molecule.

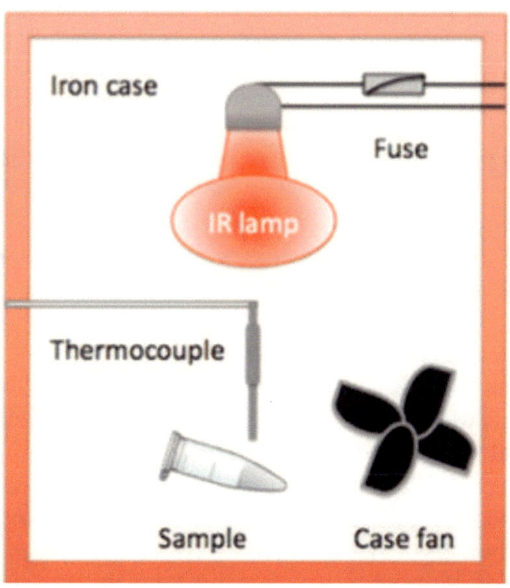

Fig. 4. Schematic diagram of the IR-assisted proteolysis system, adapted from (Wang et al. 2008).

In their first study IR assisted digestion was carried out for 5 min with trypsin in-solution and revealed almost a 100 % increase in SQ % of BSA and a 20 % increase in the SQ % of myoglobin compared to conventional trypsin digestion for 12 h at 37 °C. The method was considered repeatable when examined with a series of eight digestions giving myoglobin SQs of 90 % for all. Wang *et al.* later used the same system to study the digestion by another commonly used protease, α -chymotrypsin, which typically needs in-solution digestion times of 12-24 h (Wang et al. 2008). Using IR radiation the digestion time was reduced to 5 min for the digestion of BSA and cyt-C with SQs of 41 and 75 %, respectively. When the IR contribution was eliminated, the SQs were reduced to 11 and 56 % for BSA and cyt-C, respectively. For comparison the 12 h digestion at 37 °C yielded SQs comparable to those of 5 min IR radiation (37 % (BSA) and 75 % (cyt-C)). The same system was further examined three times in the years 2008-2009 (Bao et al. 2008; Wang et al. 2008; Bao et al. 2009) for the digestion of proteins on-plate MALDI-TOF-MS for in-gel proteolysis and one approach using trypsin-immobilized silica microspheres for peptide mapping. In 2010 another technique called photo thermal heating was introduced by Chen *et al.* A near infrared (NIR) diode laser was used to increase the reaction temperature during tryptic digestion on a Glass@AuNP slide, in a short period of time. The technique was used for four different proteins without the need for reduction and alkylation. The sequence coverages were in the range 43-95 % compared to 28-75 % with 12 h incubation at 37°C (Chen et al. 2010). Unfortunately no comparison of trypsin digestion efficiency with and without the NIR source was undertaken. We have found that, proteins can be digested in an IR oven, but compared to the traditional digestion procedure using 37 °C, there are no indications that the IR method has improved digestion efficiency for the commonly employed amount of proteins, at digestion times from 5 minutes up to 5 hours (Hustoft et al. 2011).

5.1.4 Microwave assisted digestion

Microwave assisted tryptic digestion was introduced in 2002 by Pramanik *et al.* as a tool to speed up the proteolytic cleavage of proteins (Pramanik et al. 2002). Other enzymes, as the endoproteinase Glu-C, has been reported to be inactivated by microwave induced denaturation, but trypsin digestion is accelerated according to the authors (Lill et al. 2007). In an attempt to investigate the acceleration of enzymatic cleavage, trypsin digestion with unmodified trypsin was performed at different microwave temperature settings, 37, 45 and 55 °C. The temperatures in the sample were found to be significantly higher than their microwave settings, and the authors emphasized that it was important to note the elevation of the reaction temperature which greatly enhanced the digestion reaction (Pramanik et al. 2002). Whether microwave accelerated digestion is a convenient way of heating, or whether the microwaves have a non-thermal positive effect on the digestion reaction, can be questioned. In a review on microwave-assisted proteomics, Lill *et al.* addressed the "heating principle" and stated that the kinetics in the microwave assisted incubation are different from the water bath incubation in that proteolysis was greatly enhanced when mediated by microwave radiation and that tightly folded proteins benefit the most from the microwave-assisted proteolysis (Lill et al. 2007). Two papers by Lin *et al.* (Lin et al. 2007; Lin et al. 2008) and one by Hahn *et al.* (Hahn et al. 2009) showed acceleration of digestion through a combination of immobilized trypsin and microwave radiation, when the digestion efficiency was measured as SQ % of different model proteins. In a short communication by Reddy *et al.* various solvents, temperatures and different enzyme: substrate (E+S) ratios were compared to see how they affected protein digestion under conventional heating and microwave-assisted digestion. Digestion efficiencies were referred to as the ratio of the abundance of the most abundant peptide product to that of this peptide plus the undigested protein. Optimal conditions were found to be microwave-assisted irradiation at 60 °C for 30 min in a 50 mM Tris buffer with a of 1:5 or 1:25 (Reddy et al. 2010). It should be noted that this method is incompatible with subsequent MS analysis when Tris is used as a buffer, without a buffer exchange. To make sure that no denaturation of the trypsin occurs at 60 °C, modified enzyme should be used.

The microwave approach has also been used in some recent papers (Hasan et al. 2010; Liu et al. 2010), for effective enrichment of phosphopeptides and ^{18}O labeling. High sensitivity and SQ % of phosphopeptides were obtained and explained by absorption of microwave radiation by accelerated activation of trypsin for efficient digestion of the phosphoproteins (Hasan et al. 2010). The microwave assisted ^{18}O labeling resulted in peptide mixtures with ^{18}O incorporation in less than 15 min with a low rate of back exchange (Liu et al. 2010). We have evaluated microwave assisted protein digestion using both a specialized temperature controlled microwave oven and a domestic microwave oven. No differences in SQ % (or area of intact protein peak of cyt-C) were found for microwave and temperature assisted protein digestion for four model proteins. As previously suggested, microwave irradiation seems to have no advantage over normal temperature induced digestion, within our experimental framework (Hustoft et al. 2011).

5.2 Immobilized trypsin accelerated digestion

5.2.1 Microreactors

The immobilization of enzymes onto solid materials can be traced back to the 1950s according to Ma *et. al* (Ma et al. 2009), and in the last decades numerous immobilization methods have been developed. Proteolytic enzymes can be covalently bonded or physically adsorbed onto different carriers, such as inorganic silica materials, and organic materials that display a great variability and good biocompatibility like polystyrene divinylbenzenes (PS-DVB), polyacrylamides and methacrylates (Ma et al. 2009). These types of reactors appear to have a promising future, and constitute the most used accelerating digestion techniques the last couple of years. Immobilized microreactors have a high enzymatic turnover rate, low reagent consumption, less contribution of enzyme autolysis and the possibility to be coupled on-line to nanoLC-MALDI or nanoLC-ESI (Capelo et al. 2009; Ma et al. 2009). In a review by Monzo *et al.* from 2009 the most important proteolytic enzyme-immobilization processes are summarized with emphasis on trypsin immobilized micro- and nanoreactors (Monzo et al. 2009). Another review on immobilized enzymatic reactors was published by Ma *et al.* in 2009 (Ma et al. 2009). Different inorganic and organic carriers for particle based, monolithic, open tubular capillaries and membranes with immobilized enzymes were included and the authors predicted that immobilized enzyme reactors might be one of the key points to combine the top-down and bottom-up strategies in the field of proteomics. Still, some characteristics like higher mechanical strength, larger surface area, lower backpressure, higher enzyme loading capacity and better biocompatibility, are needed.

In 2009 three papers concerning immobilized enzyme microreactors and LC-MS/MS were published (Krenkova et al. 2009; Yamaguchi et al. 2009; Yuan et al. 2009). Yuan *et al.* presented an integrated protein analysis platform based on column switch recycling size exclusion chromatography (SEC), a microenzymatic reactor and μLC-ESI-MS/MS. The system combines conventional SEC separation of intact proteins with on-line protein digestion on an immobilized enzymatic reactor (IMER) of conventional size and subsequent separation of peptides on a 300 µm (inner diameter) ID C18 column using ESI-MS/MS for identification (Yuan et al. 2009). The system requires large sample amounts and needs to be evaluated with real samples in order to be classified as a promising tool in proteomic studies. Monolithic enzymatic microreactors have been applied to digest, among others, immunoglobulin G at room temperature in only 6 minutes with reduced nonspecific adsorption of proteins and peptides to the stationary phase, as shown by Krenkova *et al.* (Krenkova et al. 2009). The SQ % was used as a measure for the digestion efficiency. Another microreactor was introduced by Yamaguchi *et al.*, using a PTFE microtube (500 µm ID, 13 cm length) with covalent binding of the enzyme. This tube was used to digest cyt-C and BSA, where the proteins (denatured in guanidine-HCl) were pumped into the immobilized microreactor and the tryptic peptides were subsequently purified on a C18 cartridge prior to LC or MS analysis. Yamaguchi *et al.* claimed that BSA could be digested without any reduction and alkylation procedures. Immobilized assisted digestion for 5.2 min at 30°C was compared with in-solution digestion of denatured BSA for 15 h at 37°C, however producing a rather low SQ % in both cases, 12 and 8 %, respectively (Yamaguchi et al. 2009). The authors claim that the low SQ % obtained was due to the stabilized tertiary structure by the 16 disulphide bonds that BSA contains, and probably better results would be obtained if reduction and alkylation of BSA had been performed in advance.

Xu *et al.* demonstrated a microporous reactor where polystyrene sulfonate and trypsin were adsorbed to a nylon membrane, to make a syringe based system for protein digestion. They used SQ % as the parameter for digestion efficiency claiming that the sequence coverage is a function of both the completeness of the protein digestion and the detection efficiency for the various tryptic peptides (Xu et al. 2010). The system showed improved SQ % of 84 % for BSA in only 6.4 seconds residence time compared to in-solution digestion for 16 h, and more promising cleavage in the presence of small amounts of SDS (Xu et al. 2010). Recently a critical overview of some highly efficient immobilized enzyme reactors termed IMERs, were presented (Ma et al.). This paper includes some newly developed IMERs and systems for protein-expression profiling, IMERs for characterization of proteins with PTMs and IMERs for protein quantification.

There are some drawbacks associated with the use of microreactors, like for instance the costs of the commercially available products of immobilized reactors. Self-fabrication requires adequate tools and experience in immobilization on different supports with enzymes. Automation is also still not easily achieved. However, as previously mentioned, on-going research can be expected to improve the techniques.

5.2.2 From microspin columns to filter-aided sample preparation

Commercial microspin columns or so called trypsin spin columns, where trypsin is immobilized at a high density on a solid support, - has been introduced by among others Sigma-Aldrich. It has been claimed that they reduce digestion times of proteins to 15 min, compared to conventional digestion times of 12 h, and give little autolysis fragments. However, the total microspin column method has been found to be both labor intensive and complex. The disappearance of these columns have been predicted because they do not give any apparent advantage over other types of immobilized trypsin which are commercially available and can be prepared in any lab (Capelo et al. 2009). This prediction can also be supported by the fact that Sigma-Aldrich's trypsin columns are no longer available. A kit intended for ^{18}O labeling called Prolytica ^{18}O labeling kit from Stratagene, based on trypsin immobilized spin columns, is also now out of production. Promega additionally had one product available, called "Immobilized trypsin", where, with the use of the spin column format, digested peptides could easily be separated from the immobilized trypsin, reducing enzyme interference during analysis (Wiśniewski et al. 2010). This product is also now withdrawn because of low demand.

Wisniewski *et al.* presented a "Universal sample preparation method for proteome analysis" based on a Filter-Aided Sample Preparation (FASP) (Wisniewski et al. 2009). The enzyme is not directly immobilized onto the ultrafiltration device, but the device acts as a "proteomic reactor" for detergent removal, buffer exchange, chemical modification and protein digestion, where trypsin is added in a dissolved form to the filter (Figure 5). Lately four other papers have been published based on this method, and it seems promising for both membrane proteins, brain phosphoproteins and the N-glycoproteins (Wiśniewski et al. 2009; Ostasiewicz et al. 2010; Wiśniewski et al. 2010; Zielinska et al. 2010).

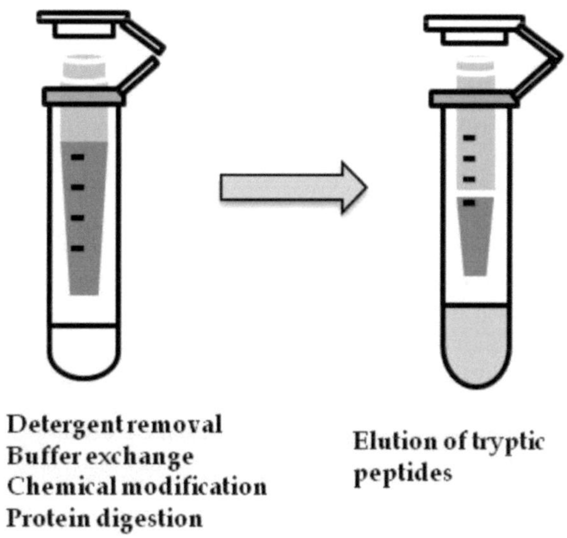

Fig. 5. The Amicon Ultra-0.5 mL Centrifugal Filters used in the FASP procedure. Adapted from, (Millipore 2011).

We, however, found that the filter device was not able to deplete all SDS, and this can lead to problems with the subsequent LC-MS analysis (Hustoft et al. 2011). The FASP procedure was found rather time consuming (using up to 3.5 h prior to the trypsin digestion) and the recommended 1:100 enzyme to protein ratio was not found satisfactory in our laboratory. Recently the FASP method was made commercially available through a FASP™ Protein Digestion Kit, from Protein Discovery. In this protocol the time of the centrifugation steps has been decreased, still it takes more than 2 hours to complete the protocol prior to 4-18 hours of trypsin digestion. Since the method has been made commercially available through a kit, and found to be convenient (Ostasiewicz et al. 2010; Wiśniewski et al. 2010; Zielinska et al. 2010; Hustoft et al. 2011) this method of trypsin digestion can be recommended when e.g. working with in-solution digestion of samples solubilised in detergents like SDS.

5.3 Solvent effects

The enzyme activity can also be improved in organic solvents as reported by Gupta and Roy (Gupta and Roy 2004). This was additionally shown by Strader *et al.* (Strader et al. 2005) who used an organic-aqueous system for digestion, containing 80 % acetonitrile (ACN), and which consistently provided the most complete digestion of microgram to nanogram quantities of proteins, by producing more peptide identifications at a shorter time (only 1 h compared to overnight). In a following paper Hervey *et al.* compared five different in-solution digestion protocols revealing that by adding 80 % ACN to the digestion solution the sequence coverages were as good as or in some cases better than using solvents with lower ACN % or chaotropes in the digests (Hervey IV et al. 2007). Addition of ACN to the digestion medium can cause (partial) denaturation of proteins and thus better accessibility to the cleavage sites of the protein. ACN can also improve digestion efficiency and enhance the solubility as well as elution of tryptic digests from e.g. a trypsin immobilized column (Tran et al. 2008). For the digestion of cyt-C, BSA, lysozyme and α-lactalbumin, addition of organic solvent up to 80 % did not increase the digestion efficiency regarding the area of the intact cyt-C peak or increased the sequence coverage (Hustoft et al. 2011). When more than 40 % ACN was used in combination with the tABC buffer, protein precipitation was seen. A solution to this problem is to use a buffer system with Tris-HCl/$CaCl_2$ when amounts of 40 % organic solvent or more are added to the sample solution (Hustoft et al. 2011). But, as before mentioned the Tris buffer is incompatible with the subsequent MS analysis and needs to be depleted prior to such.

6. Conclusions and recommended trypsin digestion procedure for LC-MS based proteomics

As has been pointed out, for some of the accelerating techniques used for tryptic digestion there is a need for more validation. We have thoroughly evaluated four of these techniques (Hustoft et al. 2011) finding no clear increase in the digestion efficiency (measured as SQ % or intact protein peak of cyt-C) of four model proteins when using IR energy, microwave energy, aqueous-organic solvent systems or FASP filters. What is of importance when comparing novel methods to established ones, is to include control experiments where the same treatment is used but without the accelerating factor for the control. When the digestion efficiency is measured based on amino acid sequence coverage, results have been found to be strongly dependent on the LC-MS data quality of the analyzed samples. Hence more replicates are strongly recommended for correct evaluation of the methods. The MS instrument available is of importance for examining the digestion efficiency and also the choice of model proteins are crucial because of their different response to tryptic digestion. Working with conventional shotgun (bottom-up) proteomic techniques the overall digestion efficiency is more intricate to study than when working with targeted proteomics. In targeted proteomics much more information can be found about the proteins to be determined, e.g. whether they have cysteines and need to be reduced and alkylated prior to enzymatic digestion. The literature can be searched in order to find relevant information and even established methods used for the targeted proteins, and selected reaction monitoring (SRM) can be used for targeted quantitative proteomics.

It should be kept in mind however, that many different variants of key words denoting the same method or process are used in the literature. One example is the method of trypsin digestion where different papers were found depending on which key word was entered

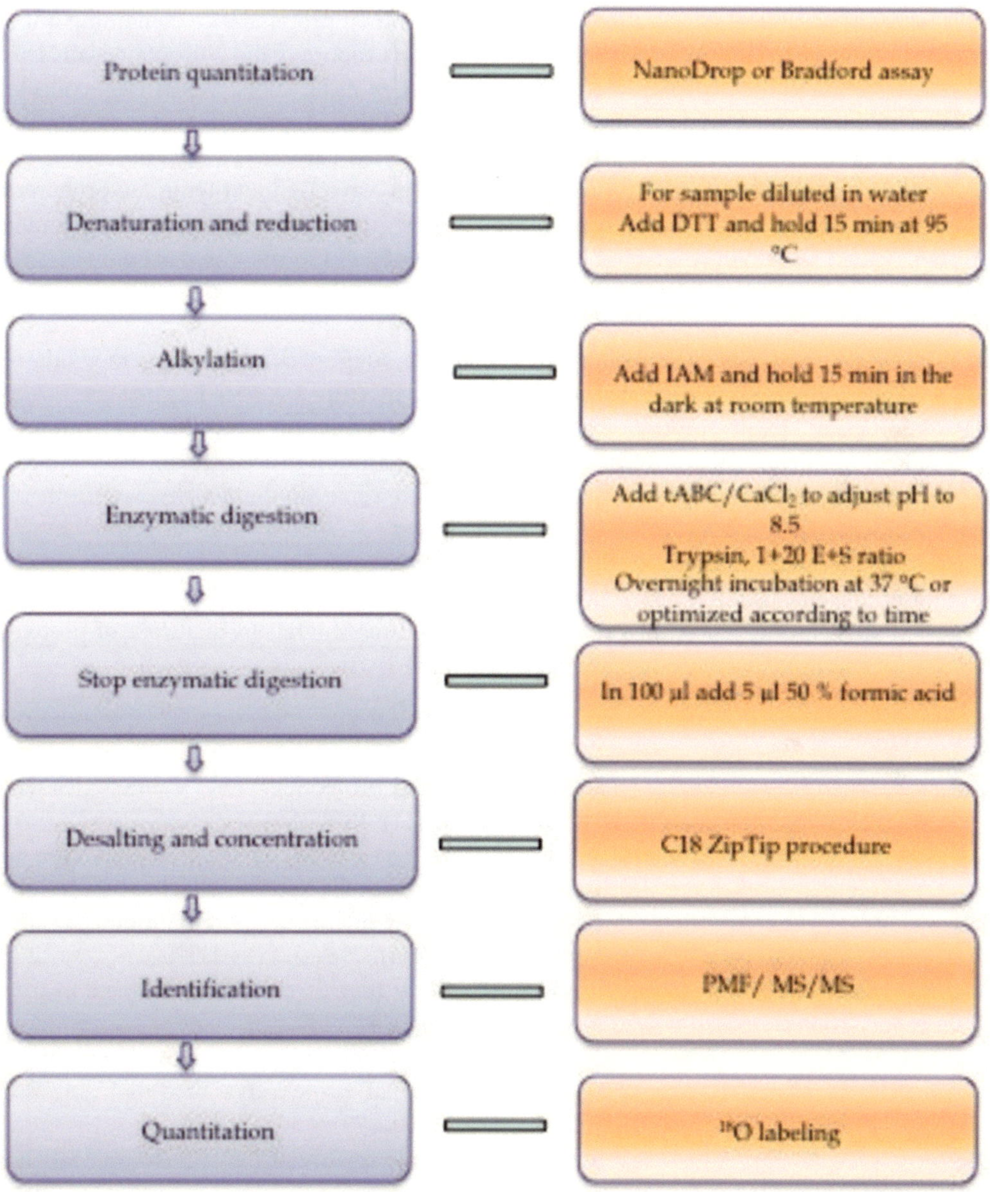

Fig. 6. A recommended procedure for in-solution based sample preparation and protein identification. PMF refers to peptide mass fingerprinting. (PFM).

into the search field of, in this case, SciFinder®: *Proteolysis, protein digestion, trypsin digestion, tryptic digestion, enzyme reaction, enzymatic digestion* or *enzymatic cleavage*, all produced hits for papers concerning trypsin/tryptic digestion (as we have chosen to use). Rounding up with Mitchell, he refers to a test sample study done in 2009, where 27 labs were included in reproducibility testing of standardized samples of 20 known proteins each containing one or more unique tryptic peptide. Only seven of the 27 labs reported the 20 proteins correctly, and only one identified all the proteolytic peptides (Bell et al. 2009). When they collected and analyzed the raw MS data from the labs, they found that all proteins and most peptides had been detected in all labs, but just not been interpreted correctly, indicating that it was the human element that failed. Due to the difficulties in correctly identifying the proteins in comprehensive proteomics the future of the field of proteomics will probably be more directed against targeted proteomics. However, as mentioned in the introduction, not every proteomic problem can be solved through targeted proteomics and it will still be a need for comprehensive analyses of complex samples.

Reviewing and in-house experiments of some of the suggested accelerating methods for trypsin digestion did not provide us with a better procedure for speeding up the sample preparation step in in-solution based proteomics, with the possible exception of ultrasonication. A complete recommended sample preparation procedure for newcomers in the field is presented in Figure 6, partially based on some of the conclusions from our investigations. The recommended procedure gives a robust and effective sample preparation guideline to comprehensive in-solution trypsin digestion of complex protein samples in proteomics. This procedure is more or less business as usual, since none of the suggested accelerating procedures revealed faster and more efficient digestion of proteins, than the inexpensive overnight in-solution digestion at 37 °C.

7. References

Anderson, N. and N. Anderson (1998). "Proteome and proteomics: new technologies, new concepts, and new words." *Electrophoresis* 19(11): 1853-1861.

Bao, H., et al. (2008). "Efficient In-Gel Proteolysis Accelerated by Infrared Radiation for Protein Identification." *Journal of Proteome Research* 7(12): 5339-5344.

Bao, H., et al. (2009). "Infrared-assisted proteolysis using trypsin-immobilized silica microspheres for peptide mapping." *PROTEOMICS* 9(4): 1114-1117.

Bell, A. W., et al. (2009). "A HUPO test sample study reveals common problems in mass spectrometry-based proteomics." *Nat Meth* 6(6): 423-430.

Bradford, M. M. (1976). "A rapid and sensitive method for the quantitation of microgram quantities of protein utilizing the principle of protein-dye binding." *Analytical Biochemistry* 72(1-2): 248-254.

Capelo, J., et al. (2010). "Latest developments in sample treatment for 18O-isotopic labeling for proteomics mass spectrometry-based approaches: A critical review." *Talanta* 80(4): 1476-1486.

Capelo, J. L., et al. (2009). "Overview on modern approaches to speed up protein identification workflows relying on enzymatic cleavage and mass spectrometry-based techniques." *Analytica Chimica Acta* 650(2): 151-159.

Carreira, R. J., et al. (2010). "Indirect ultrasonication for protein quantification and peptide mass mapping through mass spectrometry-based techniques." *Talanta* 82(2): 587-593.

Chen, J.-Y., et al. (2010). "Multilayer gold nanoparticle-assisted protein tryptic digestion in solution and in gel under photothermal heating." *Analytical and Bioanalytical Chemistry*: 1-9.

Choudhary, G., et al. (2002). "Multiple Enzymatic Digestion for Enhanced Sequence Coverage of Proteins in Complex Proteomic Mixtures Using Capillary LC with Ion Trap MS/MS." *Journal of Proteome Research* 2(1): 59-67.

Domínguez-Vega, E., et al. (2010). "First approach based on direct ultrasonic assisted enzymatic digestion and capillary-high performance liquid chromatography for the peptide mapping of soybean proteins." *Journal of Chromatography A* 1217(42): 6443-6448.

Greenebaum, C. C. (2011). From http://www.umgcc.org/research/proteomics_services.htm.

Gupta, M. N. and I. Roy (2004). "Enzymes in organic media." *European Journal of Biochemistry* 271(13): 2575-2583.

Hahn, H. W., et al. (2009). "Ultrafast Microwave-Assisted In-Tip Digestion of Proteins." *Journal of Proteome Research* 8(9): 4225-4230.

Hale, J. E., et al. (2004). "A simplified procedure for the reduction and alkylation of cysteine residues in proteins prior to proteolytic digestion and mass spectral analysis." *Analytical Biochemistry* 333(1): 174-181.

Hasan, N., et al. (2010). "Two-step on-particle ionization/enrichment via a washing- and separation-free approach: multifunctional TiO_2 nanoparticles as desalting, accelerating, and affinity probes for microwave-assisted tryptic digestion of phosphoproteins in ESI-MS and MALDI-MS: comparison with microscale TiO_2" *Analytical and Bioanalytical Chemistry* 396(8): 2909-2919.

Havliš, J., et al. (2003). "Fast-Response Proteomics by Accelerated In-Gel Digestion of Proteins." *Anal. Chem.* 75(6): 1300-1306.

Hervey IV, W., et al. (2007). "Comparison of digestion protocols for microgram quantities of enriched protein samples." *J. Proteome Res* 6(8): 3054-3061.

Hustoft, H. K., et al. (2011). "Critical assessment of accelerating trypsination methods." *J. Pharmaceut. Biomed. Anal.* 10.1016/j.jpba.2011.08.013.

IS, M. S. R. (2011). from http://ionsource.com/.

Krenkova, J., et al. (2009). "Highly Efficient Enzyme Reactors Containing Trypsin and Endoproteinase LysC Immobilized on Porous Polymer Monolith Coupled to MS Suitable for Analysis of Antibodies." *Anal. Chem.* 81(5): 2004-2012.

Li, Y.-P., et al. (2010). "Ultrasound-assisted urea-free chemical denaturants combined with thermal denaturation to accelerate enzymatic proteolysis." *Fenxi Huaxue* 38(Copyright (C) 2010 American Chemical Society (ACS). All Rights Reserved.): 663-667.

Lill, J. R., et al. (2007). "Microwave-assisted proteomics." *Mass Spectrometry Reviews* 26(5): 657-671.

Lin, S., et al. (2007). "Development of microwave-assisted protein digestion based on trypsin-immobilized magnetic microspheres for highly efficient proteolysis followed by matrix-assisted laser desorption/ionization time-of-flight mass spectrometry analysis." *Rapid Communications in Mass Spectrometry* 21(23): 3910-3918.

Lin, S., et al. (2008). "Fast and Efficient Proteolysis by Microwave-Assisted Protein Digestion Using Trypsin-Immobilized Magnetic Silica Microspheres." *Anal. Chem.* 80(10): 3655-3665.

Liu, N., et al. (2010). "Microwave-Assisted 18O-Labeling of Proteins Catalyzed by Formic Acid." *Anal. Chem.* 82(21): 9122-9126.

López-Ferrer, D., et al. (2006). "Sample treatment for protein identification by mass spectrometry-based techniques." *TrAC Trends in Analytical Chemistry* 25(10): 996-1005.

Ma, J., et al. "Immobilized enzyme reactors in proteomics." *TrAC Trends in Analytical Chemistry* In Press, Accepted Manuscript.

Ma, J., et al. (2009). "Recent advances in immobilized enzymatic reactors and their applications in proteome analysis." *Analytica Chimica Acta* 632(1): 1-8.

Millipore. (2011). from http://www.millipore.com.

Minnesota, U. (2011). from http://www.cbs.umn.edu/msp/protocols/insolution.shtml.

Mitchell, P. (2010). "Proteomics retrenches." *Nat Biotech* 28(7): 665-670.

Monzo, A., et al. (2009). "Proteolytic enzyme-immobilization techniques for MS-based protein analysis." *TrAC Trends in Analytical Chemistry* 28(7): 854-864.

NanoDrop (2011).

Nelson, D. N. and M. M. Cox, Eds. (2008). *Lehninger: Principles of Biochemistry*, Freeman and Company.

Nord, F. F., et al. (1956). "On the mechanism of enzyme action. LXI. The self digestion of trypsin, calcium-trypsin and acetyltrypsin." *Archives of Biochemistry and Biophysics* 65(1): 120-131.

Ostasiewicz, P., et al. (2010). "Proteome, Phosphoproteome, and N-Glycoproteome Are Quantitatively Preserved in Formalin-Fixed Paraffin-Embedded Tissue and Analyzable by High-Resolution Mass Spectrometry." *Journal of Proteome Research* 9(7): 3688-3700.

Pramanik, B. N., et al. (2002). "Microwave-enhanced enzyme reaction for protein mapping by mass spectrometry: A new approach to protein digestion in minutes." *Protein Science* 11(11): 2676-2687.

Proc, J. L., et al. (2010). "A Quantitative Study of the Effects of Chaotropic Agents, Surfactants, and Solvents on the Digestion Efficiency of Human Plasma Proteins by Trypsin." *Journal of Proteome Research* 9(10): 5422-5437.

Reddy, P. M., et al. (2010). "Digestion Completeness of Microwave-Assisted and Conventional Trypsin-Catalyzed Reactions." *Journal of the American Society for Mass Spectrometry* 21(3): 421-424.

Shieh, I. F., et al. (2005). "Eliminating the Interferences from TRIS Buffer and SDS in Protein Analysis by Fused-Droplet Electrospray Ionization Mass Spectrometry." *Journal of Proteome Research* 4(2): 606-612.

Sigma-Aldrich. (2011). from http://www.sigmaaldrich.com/etc/medialib/docs/Sigma/Bulletin/t6567bul.Par.0001.File.tmp/t6567bul.pdf.

Smith, P. K., et al. (1985). "Measurement of protein using bicinchoninic acid." *Analytical Biochemistry* 150(1): 76-85.

Steen, H. and M. Mann (2004). "The abc's (and xyz's) of peptide sequencing." *Nat Rev Mol Cell Biol* 5(9): 699-711.

Strader, M. B., et al. (2005). "Efficient and Specific Trypsin Digestion of Microgram to Nanogram Quantities of Proteins in Organic−Aqueous Solvent Systems." *Anal. Chem.* 78(1): 125-134.

Sundqvist, G., et al. (2007). "A general, robust method for the quality control of intact proteins using LC-ESI-MS." *Journal of Chromatography B* 852(1-2): 188-194.

Thiede, B. and C. Koehler (2010). Mass spectrometry-based quantitative proteomics using isobaric peptide termini labeling, Universitetet i Oslo: 47pp.

Tran, B. Q., et al. (2008). "On-Line multitasking analytical proteomics: How to separate, reduce, alkylate and digest whole proteins in an on-Line multidimensional chromatography system coupled to MS." *J. Sep. Sci.* 31(16-17): 2913-2923.

Vaudel, M., et al. (2010). "Peptide and protein quantification: A map of the minefield." *PROTEOMICS* 10(4): 650-670.

Vukovic, J., et al. (2008). "Improving off-line accelerated tryptic digestion:: Towards fast-lane proteolysis of complex biological samples." *Journal of Chromatography A* 1195(1-2): 34-43.

Wang, S., et al. (2008). "Infrared-Assisted On-Plate Proteolysis for MALDI-TOF-MS Peptide Mapping." *Anal. Chem.* 80(14): 5640-5647.

Wang, S., et al. (2008). "Efficient Chymotryptic Proteolysis Enhanced by Infrared Radiation for Peptide Mapping." *Journal of Proteome Research* 7(11): 5049-5054.

Wang, S., et al. (2008). "Infrared-assisted tryptic proteolysis for peptide mapping." *PROTEOMICS* 8(13): 2579-2582.

Wisniewski, J., et al. (2009). "Universal sample preparation method for proteome analysis." *Nat Meth* 6(5): 359-362.

Wiśniewski, J. R., et al. (2010). "Brain Phosphoproteome Obtained by a FASP-Based Method Reveals Plasma Membrane Protein Topology." *Journal of Proteome Research* 9(6): 3280-3289.

Wiśniewski, J. R., et al. (2009). "Combination of FASP and StageTip-Based Fractionation Allows In-Depth Analysis of the Hippocampal Membrane Proteome." *Journal of Proteome Research* 8(12): 5674-5678.

Worthington, B. C. (2011). from http://www.worthington-biochem.com/try/default.html.

Xu, F., et al. (2010). "Facile Trypsin Immobilization in Polymeric Membranes for Rapid, Efficient Protein Digestion." *Anal. Chem.*: null-null.

Yamaguchi, H., et al. (2009). "Rapid and efficient proteolysis for proteomic analysis by protease-immobilized microreactor." *Electrophoresis* 30(18): 3257-3264.

Yuan, H., et al. (2009). "Integrated protein analysis platform based on column switch recycling size exclusion chromatography, microenzymatic reactor and [mu]RPLC-ESI-MS/MS." *Journal of Chromatography A* 1216(44): 7478-7482.

Zielinska, D. F., et al. (2010). "Precision Mapping of an In Vivo N-Glycoproteome Reveals Rigid Topological and Sequence Constraints." *Cell* 141(5): 897-907.

5

Simple and Rapid Proteomic Analysis by Protease-Immobilized Microreactors

Hiroshi Yamaguchi[1,2], Masaya Miyazaki[1,3] and Hideaki Maeda[1,3]
[1]Measurement Solution Research Center, National Institute of Advanced Industrial Science and Technology
[2]Liberal Arts Education Center, Aso campus, Tokai University
[3]Interdisciplinary Graduate School of Engineering Science, Kyusyu University
Japan

1. Introduction

Proteomics is the large-scale study of proteins, particularly their stuructures and functions. One of the most important point is to develop efficient and rapid approachehes to identify the target proteins. Peptide mass mapping and tandem mass spectrometry (MS/MS)-based peptide sequencing are key methods in current protein identification for proteomic studies. Proteins are usually digested into peptides that are subsequently analyzed by MS. Therefore, proteolysis by sequence-specific proteases is the key step for positive sequencing in proteomic analysis integrated with MS (Aebersold & Mann, 2003). The conventional method of in-solution digestion by proteases is a time-consuming procedure (overnight at 37 °C). The substrate/protease ratio must be kept high (generally > 50) in order to prevent excessive sample contamination by the protease and its auto-digested products. But this leads to a relatively slow digestion. In addition, obtaining reliable peptide maps and meaningful sequence data by MS analysis requires not only the separation of the digested peptides but also strictly defined proteolysis conditions (Domon & Aebersold, 2006; Witze et al., 2007). In addition, the ionization efficiency of the digested fragments including a modified peptide such as a phosphopeptide is dependent on peptide-size or peptide-sequence, which directly correlates with sequence coverage of the target protein by MS analysis. Furthermore, peptide recovery from in-solution digestion is highly dependent on the structural properties of the target proteins because proteins with rigid structures, *e.g.* by disulfide bonds tend to be resistant to complete digestion. In fact, the typical preparation of a sample for proteolysis includes denaturation, reduction of disulfide bonds, and alkylation procedures lead to a decrease the conformational stability. It is obvious that insufficient sequence coverage could compromise the accuracy of proteome characterization. Therefore, it is important to develop novel digestion methods to achieve a highly efficient proteolysis for MS-based peptide mapping (Park & Russell, 2000; Slysz et al., 2006).

A microreactor is a suitable reaction system for handling small-volume samples in a microchannel to perform chemical or enzymatic reactions. Enzyme-immobilized microreactors have been widely utilized in chemical and biotechnological fields (Liu et al., 2008; Ma et al., 2009; Miyazaki et al., 2008; Asanomi et al., 2011). The protease-immobilized microreactor provides several advantages for proteolysis (Ma et al., 2009); *e.g.* low degree of auto-digestion even at high protease concentrations and a large surface and interface area that leads to rapid proteolysis. Furthermore, the immobilized proteases on the microchannel walls can be easily isolated and removed from the digested fragments prior to MS, which means elimination of the requirement to stop the reaction by chemical or thermal denaturation after digestion. These features can contribute to higher sequence coverage compared to the approach based on in-solution digestion. High sequence coverage is important to enhance the probability of identification of the protein and increase the likelihood of detection of structural variants generated by processes such as post-translational modifications.

Several methods for protease immobilization have been reported, wherein the protease, usually trypsin, has been immobilized in microchips by sol-gel encapsulation (Sakai-Kato et al., 2003; Wu et al., 2004), covalently bounded (Lee et al., 2008; Fan & Chen, 2007) and physically adsorbed onto different supports (Liu et al., 2006). In addition, trypsin-immobilized magnetic particles have been developed to carry out proteolysis with a short digestion time (Li et al., 2007a; Chen & Chen, 2007). However, preparations of these protease-immobilized microreactors require multi-step procedures consuming considerable amounts of time and effort. Therefore, a facile preparation method of the enzyme-immobilized microreactor is desirable for the routine proteolysis step in proteomic analysis. In addition, reusability is also an important feature required for laboratory use.

We developed the procedure for immobilizing enzymes on the internal surface of the poly-tetrafluoroethylene (PTFE) microtube by forming an enzyme polymeric membrane through a cross-linking reaction in a laminar flow between lysine (Lys) residues on the protein surfaces (Honda et al., 2005) or between the mixture of proteins with isoelectric point p*I* < 7.0 and poly-Lys (Honda et al., 2006). A typical sample preparation for proteolysis involves multi-steps (denaturation, reduction and alkylation) that are expected to produce enhancement of digestion efficiency. Because enzyme-immobilized microreactors prepared by our cross-linking method have excellent reaction performance and stability against high temperature and high concentration of denaturant (Honda et al., 2005), the microreactors are expected to achieve efficient digestion during the denaturation process. This idea inspired us to apply protease-immobilized microreactor for rapid and accurate proteolysis in proteomic analysis.

This chapter addresses the use of protease-immobilized microreactors with MS for proteomic applications. Preparation of the microreactors and examples of a simple and rapid analysis of protein sequence and protein post-translational modification are presented.

2. Preparation of protease-immobilized microreactors

A typical procedure for cross-linking enzyme involves activation of the primary amine groups of enzyme with cross-linker to create aldehyde groups that can react readily with other primary amine groups of enzymes (Honda et al; 2005; Ma et al., 2009; Miyazaki & Maeda, 2006). Cross-linking yields depend on the number of the Lys residues of enzyme. Therefore, the acidic or neutral enzyme (p*I* < 7.0) cannot be cross-linked efficiently merely by the use of cross-linker such as glutaraldehyde (GA). Although high concentration of cross-linker can increase the cross-linking yield, it often causes a change of its conformation, engendering a reduction of its catalytic activity (Wang et al., 2009). To overcome this difficulty, poly-Lys was used as the amine donor in this study. It is expected that the large number of primary amine groups of poly-Lys can improve the cross-linking yield with lower concentration of cross-linker than those in reported procedures (typically 5–10% of GA, v/v) (Ma et al., 2008; Fan & Chen, 2007).

2.1 Protease

We prepared two protease-immobilized microreactors for proteomic analysis; trypsin- (TY) and chymotrypsin- (CT) immobilized microreactors. TY hydrolyses peptide bonds after Arg and Lys residues. Because these basic residues are usually located on the surface of protein, especially in soluble proteins, the digested peptides generally fit the range (< 2 kDa) required for analysis by MS. However, if Pro residue is located at the C-terminal side of Arg or Lys, hydrolysis will not occur. Moreover, it is possibility that the conformational stability of the protein *e.g.* by disulfide bond has a resistance for proteolysis. These possibilities will cause the digested peptides to become too large to be detected by MS. Therefore, aside from TY, other endoproteases can be used for MS-based analysis to cover the whole sequence of the target protein. CT hydrolyses peptide bonds after aromatic residues (Phe, Trp and Tyr) and after Leu and Met in a less specific way and is often used for proteolysis of hydrophobic proteins such as membrane proteins (Fischer & Poetsch, 2006; Temporini et al., 2009).

2.2 Preparation techniques for the protease-immobilized microreactors

A microfluidics-based enzyme-polymerization technique (Honda et al., 2005; Honda et al., 2006) was used for the preparation of protease-immobilized microreactors. The procedure for immobilizing protease on the internal surface of the PTFE microtube by forming an enzyme polymeric membrane through a cross-linking reaction between Lys residues on the protein surface are presented in Fig. 1. The solutions were introduced to the PTFE microtube from gas tight syringes by syringe pumps. The combination of GA (0.25%, v/v) and paraformaldehyde (PA) (4%, v/v) provided better cross-linking yields between the proteases maintaining their activities (Honda et al., 2005; Honda et al., 2006). Because the p*I* value of CT (8.6) is close to the pH value of the reaction buffer (8.0), this probably leads to low polymerization yield. Thus, poly-Lys supporting polymerization procedure was used for the preparation of CT-immobilized microreactor. The molecular weight of poly-Lys was 62 kDa. Resulting Schiff base was reduced by $NaCNBH_3$. On the other hand, for TY-immobilized microreactor, poly-Lys was omitted because the p*I* value of trypsin was 10.5 and poly-Lys was a substrate for trypsin. To avoid autolysis of protease in bulk solution during the cross-linking reaction, the preparation of microreactor was conducted at 4 °C (Yamaguchi et al., 2009).

Protease immobilization on PTFE tube was analyzed by the Bradford method in order to measure the total polymerized enzyme. For example, 50 μg CT was formed by polymerizing on a 1 cm long PTFE tube. Other enzymes were also immobilized on PTFE tubes with similar concentration. Because these concentrations of the immobilized proteases are higher than those used in the experimental conditions of in-solution digestion, it can be suggested that our microreactors can perform rapid digestion compared with in-solution digestion.

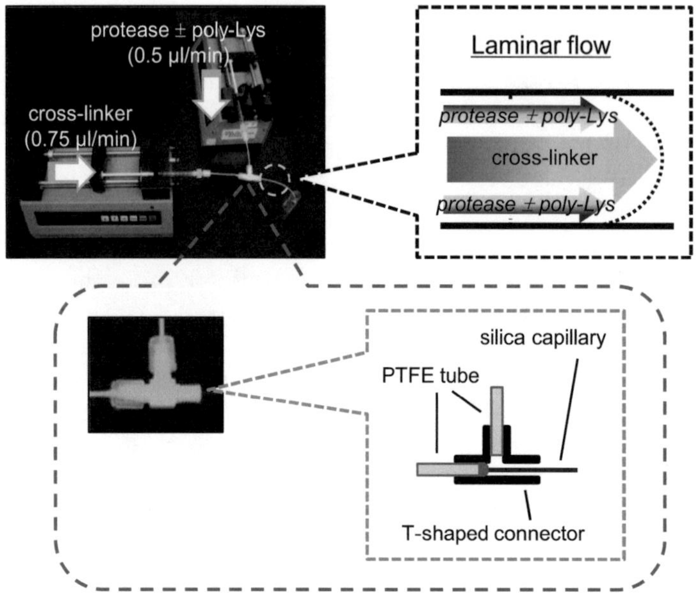

Fig. 1. The assembled microflow system for the preparation of protease-immobilized microreactor (Honda et al., 2006; Yamaguchi et al., 2009). The cross-linker solution was supplied to the substrate PTFE microtube through a silica capillary, corresponding to a central stream in the concentric laminar flow. A solution of proteases or a protease/poly-Lys mixture was supplied from another PTFE microtube connected to the T-shaped connector. Both solutions were introduced by syringe pumps.

In addition to these protease-immobilized microreactors, we also prepared alkaline phosphatase- (AP) immobilized microreactor for analysis of protein phosphorylation. Because the p*I* value of AP is 5.9, the poly-Lys supported immobilization procedure was used for the AP-immobilized microreactor. When high molecular weight of poly-Lys (62 kDa) which was intended for CT-microreactor was used for the AP-microreactor, an aggregation of protein was readily observed, suggesting that high positively charged poly-Lys (62 kDa) was quickly interacted with the acidic AP protein by electrostatic interaction. Because our enzyme-polymeric membrane is formed on the inner wall of the microchannel (500 µm inner diameter) through cross-linking polymerization in a laminar flow (Fig. 1), the quickly aggregated enzyme and poly-Lys can be stucked on the microchannel during cross-linking polymerization. Therefore, it is suggested that the large poly-Lys (62 kDa) molecular is not appropriate for preparation of AP-microreactor. To overcome this problem, a low molecular weight poly-Lys (4 kDa) was used for the polymerization of AP. As expected, with the use of poly-Lys (4 kDa), quick aggregation was suppressed and AP-microreactor was successfully prepared.

3. Characterization of the protease-immobilized microreactors

3.1 Kinetic characterization

In our digestion procedure, the substrate solution was pumped through the microreactor using a syringe pump (Fig. 2A). A reaction time is correlated with a flow rate of the substrate. In the present microreactors, the hydrolysis reaction at a flow rate of 5.0 µl/min

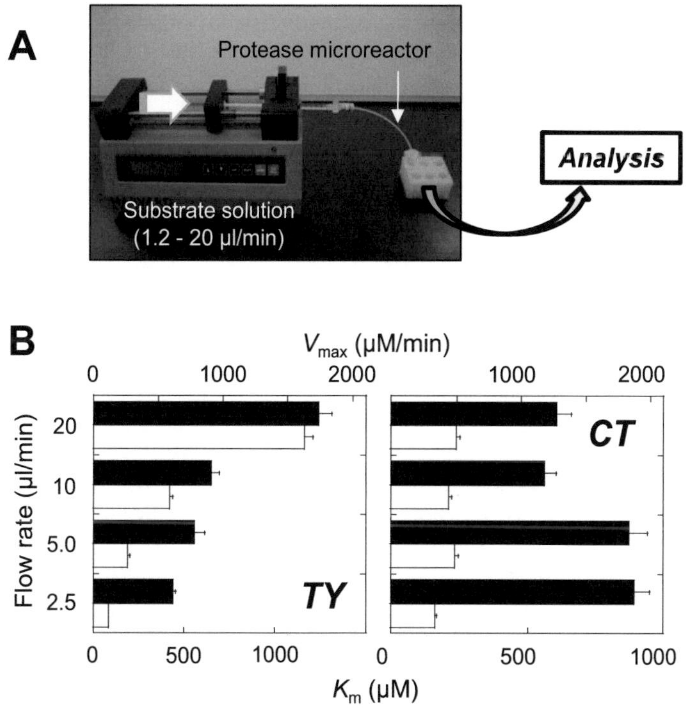

Fig. 2. (A) Schematic representation of the hydrolytic reaction by the protease-immobilized microreactor. Reaction temperature was kept in an incubator. (B) Kinetic parameters of hydrolysis activity of the protease-immobilized microreactors at different flow rates of the substrates. Substrates: BAPA for TY; GPNA for CT. Open bars, K_m (µM); closed bars, V_{max} (µM/min). The graph shows the mean ± standard error for at least three experiments. All assays were performed in 50 mM Tris−HCl (pH 8.0) at 30 °C.

yields a reaction time of 5.2 min (PTFE microtube volume of 26 µl). The individual hydrolysis activity of the microreactors was evaluated using synthetic small compounds; benzoyl-L-arginine *p*-nitroanilide (BAPA) for TY-microreactor and *N*-glutaryl-L-phenylalanine *p*-nitroanilide (GPNA) for CT-microreactor. Digestions by both microreactors showed a similar order of K_m values (hundred micromolar) to the reported K_m values for TY in a microtube (Yamashita et al., 2009) and CT-microreactor (Honda et al., 2005), suggesting that both immobilized proteases maintain their own hydrolysis activity after polymerization on PTFE surface.

It is known that the flow rate of substrate can affect the efficiency of the immobilized enzyme activity (Fan & Chen, 2007; Honda et al., 2005; Ma et al., 2008; Nel et al., 2008; Wu et al., 2004; Dulay et al., 2005). Therefore, we studied the effect of different delivery speeds of substrates on hydrolysis activities (Yamaguchi et al., 2009). As shown in Fig. 2B, the estimated K_m values for TY-microreactor decreased with increase in flow rate, while no significant change in K_m values was observed for CT-microreactor at different flow rate, indicating that the diffusion limitation of the substrate in the immobilized TY is more influenced than that in the immobilized CT. Because both proteases have similar conformational structures and catalytic-sites as well-known, the difference between TY-microreactor and CT-microreactor could be attributed to the difference in polymerization procedure. In contrast to K_m values, V_{max} values for both microreactors increased with increase in flow rate. Therefore, both microreactors showed higher V_{max}/K_m values at lower flow rate (longer reaction time) but lower V_{max}/K_m values at higher flow rate (shorter reaction time), suggesting that the hydrolysis activity was more efficient at lower flow rate. This indicates that efficient digestion of substrate by immobilized protease was achieved using an alternative substrate mobilization approach that involved incubation of substrate with immobilized enzyme. In contrast, the calculated K_m value for free TY was 806 μM, a K_m value that was 4.3-fold higher than TY-microreactor with the same reaction time. Free CT also showed much higher K_m value (over 1 mM) compared with that of CT-microreactor. It is known that K_m value represents the binding affinity between enzyme and substrate, and lower K_m value means higher affinity therefore, suggesting that the enhanced mass-transfer in the microchannel induced hydrolysis reaction (Honda et al., 2005; Honda et al., 2007).

3.2 Operational stabilities of the protease-immobilized microreactors

The operational stability and reusability of the protease-immobilized microreactor are important for its application in the proteomics analysis. Therefore, the stability of immobilized proteases against temperature in the hydrolysis reaction were tested (Yamaguchi et al., 2009). The relative hydrolysis activity against BAPA or GPNA of the protease-immobilized microreactors and free proteases at different temperatures were measured. The results showed that both immobilized proteases were more stable at high temperature than free proteases. At 50 °C, free TY and free CT showed 15 and 52% of hydrolysis activities respectively, while the immobilized proteases kept at 30 °C retained their activities. It is suggested that due to multipoint interactions between the TY molecules or between CT and poly-Lys, the immobilized proteases increased the thermal stabilities of the enzymes. Similar thermal stability of immobilized enzyme was previously reported (Kim et al., 2009; Sheldon 2007).

Next, the stability of the protease-microreactors against high concentration of denaturant was tested. When the substrate solutions in 3 M guanidine hydrochloride (Gdn-HCl) were delivered to both protease-immobilized microreactors, we observed almost the same hydrolysis activities compared with those in buffer containing no denaturant. In contrast, free proteases in 3 M Gdn-HCl were denaturated consequently and did not show any hydrolysis activities. Similar stability of CT-microreactor against 4 M urea and 50% dimethyl sulfoxide were previously observed (Honda et al., 2005). Although both immobilized proteases did not show any hydrolysis activities in 5 M Gdn-HCl, over 90% recovered activities were observed after washing the microreactors with buffer containing no Gdn-HCl. Moreover, the reusability of the microreactors was also investigated by storing them at pH 8.0 at 4 °C. After over 60 days (after over 20 times reuse), both microreactors retained over 90% of their hydrolysis activities against synthetic substrates while free proteases had very little to no activities. More recently, the immobilized subtilisin by poly-Lys supported cross-linking was also more stable than free protease at high temperature, in the presence of a chemical denaturant or in an organic solvent and was recycled without appreciable loss of activity (Yamaguchi et al., 2011). These results indicate that the stability of proteases was improved after formation of enzyme-polymerization. This enhancement in the efficiency of activity of the immobilized protease compared to free protease can be ascribed to minimization or elimination of the auto-digestion of proteases (Dulay et al., 2005; Shui et al., 2006) and possible stabilization of the structure of proteases by cross-linking, thus resulting in higher accessibility of the substrate to the catalytic-site of the enzyme (Honda et al., 2005; Honda et al., 2006; Dulay et al., 2005; Shui et al., 2006).

The operational stability of the protease-immobilized microreactors was also tested based on the digestions of cytochrome *c* (Cyt-C) for TY-microreactor and β-casein for CT-microreactor (Yamaguchi et al., 2010a). Between each digestion, both microreactors were washed with buffer solution and stored at 4 °C for over 60 days. The obtained 10 MS spectra (not shown) were identical with the similar sequence coverage of 93% (Cyt-C by TY-microreactor) and 57% (β-casein by CT-microreactor). In contrast, free proteases almost completely lost their activities at 25 °C within a couple of days, as reported previously (Sakai-Kato et al., 2002). These results indicate that the stability of proteases was increased by the prevention of auto-digestion after an enzyme-immobilization.

3.3 Proteolysis by the protease-immobilized microreactors

Our previous proteolysis procedure using microreactors was carried out in 50 mM Tris–HCl (pH 8.0) solution (Yamaguchi et al., 2009). In this buffer system, an additional purification step using reversed-phase micropipette tips prior to MS measurement could remove excessive amounts of buffer salt, but could lead to sample losses especially of hydrophobic peptides due to their inherent affinity to reversed-phase surfaces, leading to lower sequence coverage. To avoid this, proteolysis was carried out in a 10 mM ammonium acetate buffer (pH 8.5) that easily evaporated during ESI-TOF MS measurement without the need for any desalting procedure. The digested peptides were collected in a test tube and then directly analyzed by ESI-TOF MS using Mariner mass spectrometer (Applied Biosystems Inc.) and reverse-phase high performance liquid chromatography (HPLC) (Yamaguchi et al, 2010a).

Digestion methods	Temperature /°C	Reaction time	Identified amino acids	Sequence coverage / %
TY-microreactor	30	10.4 min	97/104	93
CT-microreactor	30	10.4 min	40/104	38
TY (in-solution)[a]	37	18 hours	99/104	95
CT (in-solution)[a]	37	18 hours	68/104	65

[a]In-solution digestion was carried out in 10 mM ammonium acetate buffer, pH 8.5. Concentrations of substrate and free proteases were 100 and 2 µg/ml, respectively.

Table 1. Summary of ESI-TOF MS results of the digests of Cyt-C

To study the effect of flow rate on the proteolysis efficiency, we first investigated the digestion of Cyt-C by TY-microreactor at several flow rates. Under our experimental conditions, with flow rate increase from 1.2 to 15 µl/min, auto-digestion peak of protease was not observed by MS and HPLC analyses. In addition, MALDI-TOF MS analysis using Bruker Autoflex (Bruker Daltonics) indicated that no free protease or cross-linked aggregation came off from the PTFE tubes, demonstrating good mechanical stability. Proteolysis was carried out at 30 °C. The digests by TY-microreactor were analyzed by ESI-TOF MS spectra and reverse-phase HPLC profiles. With a flow range of 2.5–15 µl/min, the intact Cyt-C was observed while below 1.2 µl/min over 90% of Cyt-C was digested by TY-microreactor. The results indicate that digestion at a lower flow rate (longer reaction time) is more efficient as shown in Fig. 2B (against small compound). Although intact Cyt-C remained in the samples above 2.5 µl/min flow rate, the matched peptides covered 93%

(97/104 amino acids) of the Cyt-C sequence (Table 1). This value was the same as that of the digested sample at 1.2 μl/min. For comparison, in-solution digestions of Cyt-C with an incubation time of 18 hours at 30 and 37 °C were performed. Although the reaction at 37 °C showed complete digestion, only 40% Cyt-C was digested at 30 °C, which is of the same proteolysis condition as that of the TY-microreactor. Thus, these results indicate that digestion of Cyt-C by TY-microreactor was much faster and efficient than in-solution digestion and that there is no need to perform the digestion at higher temperature. In contrast to TY-microreactor, CT-microreactor showed lower sequence coverage (38%) than that by in-solution digestion (65%). Because Cyt-C was not denaturated by denaturant in this digestion condition, a possible reason for the lower digestion ability of CT-microreactor could be mass transfer limitation of folded Cyt-C in the cross-linked CT and poly-Lys complex matrix.

The residues of Lys and Arg which are recognized by TY, are hydrophilic residue that usually locate the surface of protein, while aromatic residues for CT are buried inside the protein. It is possible that immobilized TY could easily recognize the residues of Lys and Arg; on the other hand, immobilized CT could not interact with aromatic residues. As described above, the microreactors have stability against high concentration of denaturant and the denaturated Cyt-C with its aromatic residue exposed to the outside is expected to be digested by CT-microreactor. To confirm this possibility, Cyt-C digestion by CT-microreactor was carried out in 3 M Gdn-HCl. Circular dichroism (CD) spectrum revealed that Cyt-C (100 μg/mL) in 3 M Gdn-HCl was denaturated. Under this condition, free CT did not show any proteolysis activity. In contrast to in-solution digestion, the digests by CT-microreactor showed a different HPLC profile compared with the HPLC profile of intact Cyt-C in 3M Gdn-HCl, indicating that the immobilized CT in 3M Gdn-HCl digested the denaturated Cyt-C (Yamaguchi et al., 2009). Furthermore, MS analysis of the digests also showed the efficient digestion by CT-microreactor in 3M Gdn-HCl. These results suggest that the stability of immobilized proteases is superior to that of free proteases.

A typical sample preparation for proteolysis before digestion involves multi-steps including denaturation, reduction of disulfide bond, and alkylation of free thiol group to reduce the conformational stability of protein; steps which are expected to produce enhancement of digestion efficiency (Ma et al., 2008; Li et al., 2007a; Ethier et al., 2006; Lin et al., 2008). However, the multi-step procedure is time-consuming. In contrast, our digestion method by protease-immobilized microreactor can be carried out with high concentration of denaturant and high sequence coverage. Moreover, it can directly use the denaturated protein as a substrate. In addition, it does not need any complicated reduction and alkylation steps, therefore, exhibiting superior advantages of our digestion procedure over other reported protease-immobilized microreactor in achieving rapid proteomics analysis.

3.4 Proteolysis by the tandem protease-immobilized microreactors

The improved sequence coverage is important to enhance the probability of identification and increase the likelihood of detection of structural variants generated by processes such as alternative splicing and post-translational modifications. The identification of the protein sequence is a first and important step in proteome analysis. Post-translational modification such as phosphorylation is also important information in understanding the role of target protein in the regulation of fundamental cellular processes. Because disregulations of mechanisms of modifications are implicated in various diseases, including cancer (Hunter, 2009), the characterization of protein sequence is useful for biological and clinical researches.

The conventional approach using multi-digestion by different proteases for improved sequence coverage is based on parallel digestions of the same samples and analyzes overlapping peptides. However, this approach takes long time and multi-step procedure. To overcome this difficulty, we prepared the tandem microreactor that was connected by different protease-immobilized microreactors using a Teflon connector. Connection was made easy because the present microreactors were made of PTFE microtube. It is expected that the combination of MS results obtained with the tandem microreactor that carries out multi-digestion may give significantly higher sequence coverage than that obtained with individual digestion by the single microreactor. Based on a similar idea, a reactor which was a bonded mixture of TY and CT to an epoxy monolithic silica column was reported (Temporini et al., 2009). In contrast to the mixture of TY and CT immobilized reactor (Temporini et al., 2009), an interesting feature of our microreactor is the ease in linking each microreactor by using a connector. In our system, we can easily change the order of each microreactor (for example, CT-TY or TY-CT) according to our preference (Yamaguchi et al., 2010a).

Cyt C (non-phosphoprotein), β-casein (phosphoprotein) and pepsin A (phosphoprotein) were used to test the performance of the tandem microreactors. The enzymatic reaction at a flow rate of 2.5 μl/min yielded a reaction time of 10.4 min (single microreactor) or 20.8 min (tandem microreactor). The digestion efficiency by the microreactors was evaluated by analyzing the sequence coverage and the identified peptide. Some of the digested peptides by the microreactors were the expected peptides that have one or two missed cleavage site. The p*I* value of Cyt-C is 9.6 (horse residue 2-104), suggesting that Arg or Lys residues locate the protein surface with high possibility. As shown above, Cyt-C was digested by TY-microreactor with higher sequence coverage (93% in Table 1) at 30 °C. MALDI-TOF MS analysis also showed high sequence coverage (89%). The value of sequence coverage was higher than the other trypsin-immobilized reactors reported (Li et al., 2007a; Liu et al., 2009) and the same as that performed by in-solution digestion (37 °C for 18 hours), suggesting that the immobilized TY showed rapid and efficient proteolysis.

The multi-digestion of Cyt-C by the tandem microreactor that was connected by using TY-microreactor and CT-microreactor was evaluated. TY-CT tandem microreactor also showed rapid Cyt-C digestion with the similar sequence coverage as that of the single TY-microreactor. As expected, the multi-digested peptides such as VQK, TGPNLHGLF or TGQAPGF were identified (Yamaguchi et al., 2010a). Because some digested peptides (< 300 Da) by tandem microreactor were too small to be identified by TOF-MS analysis, therefore based on comparison with the single TY microreactor the sequence coverage for Cyt-C by the tandem microreactor was found not to improve (88%, 91/104 amino acids). These results indicate that the peptide fragments by TY-microreactor were also digested by CT-microreactor, as expected.

Bovine β-casein (residue 16-224) is a phosphoprotein with well-characterized phosphorylated sites (Han et al., 2009; Li et al., 2007b). MS measurement of an intact protein revealed that β-casein in this study has five phosphorylated sites. By measuring the decrease in the peak area at 220 nm of HPLC profiles of the digests, it could be estimated that over 90% β-casein was digested by CT-microreactor for 10.4 min at 30 °C. ESI-TOF MS analysis revealed that 13 peptides containing 120 out of the 209 possible amino acids of β-casein were obtained, producing the sequence coverage of 57% (Yamaguchi et al., 2010a). This value was higher than that by in-solution digestion (45%, 95/209 amino acids). Moreover, the phosphopeptide containing four phosphoserine (pS) residues (**pSpSpS**EEpSITRINKKIEKF) was detected despite the low ionization efficiency of the phosphopeptide. To confirm this detection of the phosphopeptide by other MS system, MALDI-TOF MS analysis was performed. In addition to the **pSpSpS**EEpSITRINKKIEKF phosphopeptide, the Q**pS**EEQQQTEDEL phosphopeptide was also identified by MALDI-TOF MS analysis, thereby showing that all phosphorylation sites on β-casein in this study were detected from the digests by CT-microreactor. In contrast, HPLC analysis of digested β-casein by TY-microreactor showed a broad profile and was different with that by CT-microreactor. In addition to the p*I* value of β-casein (5.1) that was estimated from the primary structure and which did not take into account the number of phosphorylation sites, the total number of Phe, Trp, Tyr, Leu and Met residues (42) was larger than that of Arg and Lys residues (15). Therefore, it is suggested that the sequence coverage of β-casein by TY-microreactor (14%) or in-solution digestion by free trypsin (21%) is lower than that by CT-microreactor. We next studied the feasibility of enzyme-immobilized tandem microreactor. As expected, the multi-digestion by CT-TY tandem microreactor showed 20 digested peptides (Yamaguchi et al., 2010a). It is noteworthy that GVSK, VKEAMAPK, HKEMPFPK and YPVEPF peptides that were not identified by the single CT-microreactor were identified by the tandem microreactor. The results indicate that an improvement of the sequence coverage in digestion by tandem microreactor in comparison to the single microreactor and in-solution digestion can be expected.

3.5 Phosphorylation site analysis by the tandem microreactor

Further analysis of protein phosphorylation was carried out using AP-microreactor. HPLC profile of the digested peptides by CT-AP tandem microreactor revealed the disappearance of two peaks compared to that by the single CT-microreactor. This suggested that the phosphopeptides containing **pSpSpS**EEpSITRINKKIEKF were dephosphorylated by AP. MS analysis also revealed the dephosphorylation of the **pSpSpS**EEpSITRINKKIEKF phosphopeptide. The results indicate that the tandem microreactor which was made by using the protease-microreactor and the phosphatase-microreactor showed the feasibility of the identification of phosphorylation site in phosphoproteins without any enrichment strategies and radioisotope labeling. β-Casein has another well-known phosphorylated site (Ser35). Because the phosphopeptide containing Ser35 (Q**pS**EEQQQTEDEL) was detected by MALDI-TOF MS analysis, it is possible that the one peak that disappeared in HPLC profile may be the phosphopeptide containing pS35.

Pepsin A (porcine residue 60-385) is an acidic protease (p*I* of 3.2) and a phosphoprotein which has one phosphoserine residue (Kinoshita et al., 2006). An optimum pH of pepsin A for its protease activity is around 2.0, meaning that pepsin A does not have any activity under our digestion condition at pH 8.5. Therefore, pepsin A was used as a substrate without any denaturation procedure. Similar to the digestion of β-casein, pepsin A was efficiently digested by CT-microreactor but not by TY-microreactor thus, explaining the difference between the total number of cleavage sites by CT (65 residues) and those by TY (3 residues). The sequence coverage of 55% (179/326 amino acids) by CT-microreactor was lower than that by in-solution digestion (60%, 196/326 amino acids) (Yamaguchi et al., 2010a). In addition, the phosphopeptide (EAT**pS**QELSITY) was detected in the digestion by in-solution but not by CT-microreactor. When HPLC profile of the digests by the single CT-microreactor was compared with CT-AP tandem microreactor digest, it was found out that one peak disappeared after passing through the tandem microreactor. This suggests that the digests by CT-microreactor also contain the phosphopeptide but the size of the peptide was bigger than the EAT**pS**QELSITY phosphopeptide from in-solution digestion. In addition, it is possible that our MS system was not able to detect the phosphopeptide from CT-microreactor. A possible reason for the lower digestion ability of CT-microreactor could be mass transfer limitation of folded pepsin A in the cross-linked CT and poly-Lys complex matrix. Similar lower digestion ability of the immobilized-protease was observed in the digestion of Cyt-C by CT-microreactor as described above.

The present analytical method of protein phosphorylation is much simpler than the other conventional methods (Han et al., 2009; Kinoshita et al., 2006; Zhao & Jensen, 2009), for example, the phosphoprotein is just flowed through the microreactor and it eliminates purification of digests from the reaction system without any enrichment strategies. These interesting features are superior advantages of our approach using the enzyme-immobilized microreactors over the conventional method.

3.6 Analysis of disulfide bond using the protease-immobilized microreactor at several temperatures

Disulfide bond is covalent cross-linking between side-chains of two Cys residues of the same or different peptide chains. It is an important factor for stabilizing the protein conformation. In addition, oxidation-reduction of Cys residues in protein is significant to biological functions (Lee et al., 2004; Mieyal et al., 2008; Yano & Kuroda, 2008). Therefore, the assignment of disulfide bonding patterns will not only provide insights into its three-dimensional structure and contribute to the understanding of its structure-function relationship but will also play a role in the regulation of fundamental cellular processes. However, the conventional method for the assignment of disulfide bond by chemical cleavage and/or proteolysis is a time-consuming multi-step procedure. In addition, due to higher conformational stability of protein by disulfide bond(s), the conventional in-solution digestion by protease was usually carried out using reduced protein which was prepared by the reduction and alkylation procedure. Although this approach has provided us with information on the protein sequence (primary structure of protein), the information for disulfide bond such as Cys-Cys pair(s), the number of disulfide bond, and a distinction between Cys residues involving disulfide bond and free Cys residue are not from the conventional reduced sample.

As shown above, the protease-immobilized microreactor showed a rapid and efficient digestion compared with in-solution digestion. Furthermore, the immobilized-protease can be easily isolated and removed from the digests prior to MS without any purification step. Moreover, the immobilized proteases by our methods were more stable at high temperature than free proteases (Yamaguchi et al., 2009). Based on these interesting features, we tested whether the substrate proteins that maintain their disulfide bonds can be efficiently digested by the protease-immobilized microreactor at high temperature without any chemical modification and purification step, and whether the position of disulfide bond(s) in the resultant digests can be analyzed by MS (Yamaguchi et al., 2010b).

It was reported that thermally denatured proteins were efficiently digested by in-solution digestion (Park & Russell, 2000) or by protease-microreactors using free proteases (Liu et al., 2009; Sim et al., 2006). Our approach is based on the concept that proteins of stable conformation owing to their disulfide bonds(s) were thermally denatured at high temperature and were directly digested by the protease-immobilized microreactor. It is known that thermal denaturation at higher temperature (~ 90 °C) would form the protein aggregation (Park & Russell, 2000) although the mechanism of this formation has not yet been fully elucidated. Therefore, we performed the proteolysis between 30 to 50 °C. Substrate proteins (50 μg/ml) which were not treated with the reduction and alkylation procedure were pumped through the microreactor at flow rate of 1.2 μl/min (reaction time: 21.7 min). The digests were collected in a test-tube and directly analyzed by ESI-TOF MS without any purification or concentration procedure.

Lysozyme (chicken residue 19-147: MW 14,304 Da) has well-characterized four disulfide bonds. As shown in Fig. 3, the digests by both microreactors showed that increasing reaction temperature increased the number of digested fragments, which were correlated with the sequence coverage (Table 2). Most of the digested peptides by the microreactors were the expected peptides which did not have the missed cleavage site. At 50 °C by TY-microreactor, 10 peptides containing 126 out of the 129 possible amino acids of lysozyme were obtained, producing the sequence coverage of 98%. This value was higher than those by TY-microreactor at 30 °C (5%, 7/129 amino acids) and in-solution digestion at 37 °C (77%, 99/129 amino acids). Moreover, all four disulfide bonds (Cys24-Cys145, Cys48-Cys133, Cys83-Cys98, and Cys94-Cys112) were detected in MS spectrum from the digests at 50 °C but 3 of 4 disulfide bonds (Cys48-Cys133, Cys83-Cys98, and Cys94-Cys112) were from in-solution digestion at 37 °C (Yamaguchi et al., 2010b). The sequence coverage of 22% (28/129) by the TY-microreactor at 40 °C was lower than that by in-solution digestion at 37 °C. A possible reason for the lower digestion ability of TY-microreactor could be that the substrate protein has different exposure of the digestion site by TY between in the microchannel and in batch method (in-solution digestion).

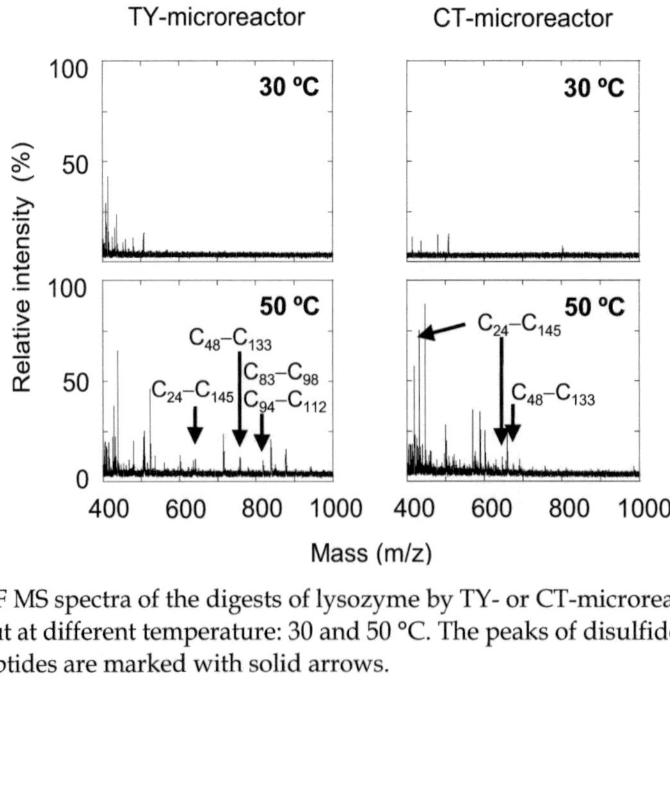

Fig. 3. ESI-TOF MS spectra of the digests of lysozyme by TY- or CT-microreactor. Digestion was carried out at different temperature: 30 and 50 °C. The peaks of disulfide bond(s)-containing peptides are marked with solid arrows.

Digestion methods	Temperature / °C	Reaction time	Identified amino acids	Sequence coverage / %
TY-microreactor	30	21.7 min	7/129	5
TY-microreactor	50	21.7 min	126/129	98
CT-microreactor	30	21.7 min	11/129	9
CT-microreactor	50	21.7 min	54/129	42
TY (in-solution)[a]	37	18 hours	99/129	77
CT (in-solution)[a]	37	18 hours	66/129	51

[a]In-solution digestion was carried out in 10 mM ammonium acetate buffer, pH 8.5. Concentrations of substrate and free proteases were 100 and 2 μg/ml, respectively.

Table 2. Summary of ESI-TOF MS results of the digests of lysozyme

In contrast, sequence coverage by CT-microreactor at 50 °C and in-solution digestion at 37 °C by free CT were lower than that by TY-microreactor (Table 2), although the total number of Phe, Trp, Tyr, Leu, and Met residues (22) for CT was larger than that of Arg and Lys residues (17) for TY. CD and fluorescence spectra measurements indicated that lysozyme at 50 °C was thermally denatured but partly formed some secondary structures. Basic residues are usually located on the surface of protein than hydrophobic residues. In addition, the p*I* value of lysozyme is 9.3. Therefore, it is possible that TY easily recognizes and hydrolyses peptide bonds after Arg and Lys residues that locate on the surface of lysozyme. Because sequence coverage by CT was lower than that by TY, the number of identified disulfide bonds was also low. The sequence coverage of lysozyme by in-solution digestion at 50 °C (18% for free TY and 22% for free CT) were lower than those by the microreactors at 50 °C, indicating that free proteases decreased their hydrolysis activities at 50 °C.

To further investigate the efficiency of our approach for the assignment of disulfide bond, the digestion of bovine serum albumin (BSA) by the microreactor was carried out (Yamaguchi et al., 2010b). BSA (bovine residue 25-607: MW 66,390 Da) has 17 disulfide bonds. Similar to the digestion of lysozyme, the digestion of BSA by TY-microreactor at 50 °C efficiently occurred but not at 30 °C. The sequence coverage of 37% (201/583 amino acids) was higher compared with that of in-solution digestion (26%, 151/583 amino acids) and was better or comparable to those of a thermal (Liu et al., 2009; Sim et al., 2006) or chemical denatured BSA (Li et al., 2007a; Chatterjee et al., 2010) or the reduced BSA (Ma et al., 2008) by the reported trypsin-microreactors (24-46%). In addition, the number of identified disulfide bonds was 10 of 17, which is superior to 6 obtained by in-solution digestion. The digests of BSA by the TY-microreactor at 50 °C was also analyzed by MALDI-TOF MS. In addition to the information on disulfide bonds from ESI-TOF MS, we identified the TCVADESHAGCEK peptide with intramolecular disulfide bond (Cys77-Cys86) that was not identified by ESI-TOF MS. Although all disulfide bonds in BSA were not identified under the present condition, these results indicate that proteolysis approach at high temperature by the microreactors can be useful not only for large proteins but also for the proteins having many disulfide bonds. After each proteolysis procedure at several temperatures, the microreactors were washed with buffer solution and digestions were carried out at 30 °C. The residual activities of both microreactors after proteolysis procedure even at 50 °C were identical with those of prior to proteolysis, indicating that both immobilized proteases maintained their activities after several use at different temperatures. We also tested the hydrolysis activity of TY-microreactor against Cyt-C. MS spectra of Cyt-C by TY-microreactor before and after proteolysis at 50 °C were the same. In addition, both sequence coverage of Cyt-C by TY-microreactor were 89% (93/104 amino acids). This value was higher than those of the reported trypsin-immobilized reactors (Liu et al., 2007a; Liu et al., 2009). In contrast, free proteases showed lower proteolysis activities at 50 °C than those at 37 °C. Once again, these results verified that the present protease-immobilized microreactors were thermally stable at high temperature but this was not observed in the case of free proteases. Similar stability of immobilized TY on polymer nanofibers was previously reported (Kim et al., 2009).

From these results, the procedures of proteolysis by the microreactor and the MS measurement took less than 2 hours. This is much faster and easier than the conventional procedure (multi-days). Therefore, our proteolysis approach by the thermostable microreactor is a simple and rapid analytical method for the assignment of disulfide bond without any chemical modification or purification procedure.

4. Conclusion

The microreactors showed efficient proteolysis with high sequence coverage, long-term stability, and good reusability. It is known that in-solution digestion by trypsin can induce artificial modifications such as asparagine deamidation (Krokhin et al., 2006) and *N*-terminal glutamine cyclization (Bongers et al., 1992; Dick et al., 2007) on target protein due to the elevated temperature and alkaline pH buffers used during digestion for overnight. Proteolysis by our protease-immobilized microreactors was achieved within a short period of time (~ 20 min) at 30 °C therefore suggesting these artificial modifications as a remote possibility. In addtion, proteolysis by the tandem microreactors showed higher sequence coverage, which is a remarkable result compared with those of the single microreactor or in-solution digestion. The tandem microreactor comprising a protease-microreactor and a phosphatase-microreactor also showed the capability to localize phosphorylation site(s) in phosphoproteins. Several protease-immobilized microreactors were developed for proteolysis. Most of these studies have focused on rapid digestion and reduction of sample volume. So far, there is no study yet on multi-enzymatic reaction system and analysis of post-translational modification in protein. Furthermore, proteolysis at 50 °C by the microreactors showed higher sequence coverage and assignment of disulfide bonds, which is a remarkable result compared with that of in-solution digestion.

The present procedure is much simpler than the other conventional methods, for example, the protein is just flowed through the microreactor and it is not necessary to purify the digests from the reaction system. These interesting features are superior advantages of our proteolysis approach over the conventional method. The enzyme-immobilization method using poly-Lys can be applied to proteins with wide-range p*I* values hence, the strategy based on multi-enzymatic reaction using the tandem microreactor provides a useful approach for other post-translational modification analysis (*e.g.* acetylation, methylation, ubiquitination or glycosylation). Coupling the protease-immibilized microreactor with MS and/or HPLC (on-line) can be also applied for high throughput proteomic analysis systems.

5. Acknowledgment

This work was supported by Grant-in-Aid for Basic Scientific Research (B: 23310092) and for Young Scientists (B: 23710153), from the Japan Society for the Promotion of Science (JSPS).

6. References

Aebersold, R. & Mann, M. (2003) Mass spectrometry-based proteomics. *Nature, 422*, 198-207.

Asanomi, Y., Yamaguchi, H., Miyazaki M. & Maeda, H. (2011) Enzyme-immobilized microfluidic process reactors. *Molecules*, 16, 6041-6059.

Bongers, J., Heimer, E. P., Lambros, T., Pan, Y. C., Campbell, R. M. & Felix, A. M. (1992) Degradation of aspartic acid and asparagines residues in human growth hormone-releasing factor. *Int. J. Pept. Protein Res.*, 39, 364-374.

Chatterjee, D., Ytterberg, A. J., Son, S. U., Loo, J. A. & Garrell, R. L. (2010) Integration of protein processing steps on a droplet microfluidics platform for MALDI-MS analysis. *Anal. Chem.*, 82, 2095-2101.

Chen, W. Y. & Chen Y. C. (2007) Acceleration of microwave-assisted enzymatic digestion reactions by magnetite beads. *Anal. Chem.*, 79, 2394-2401.

Dick Jr., L. W., Kim, C., Qiu, D. & Cheng, K. C. (2007) Determination of the origin of the N-terminal pyro-glutamate variation in monoclonal antibodies using model peptides. *Biotechnol. Bioeng*. 97, 544-553.

Domon, B. & Aebersold, R. (2006) Mass spectrometry and protein analysis. *Science*, 312, 212 217.

Dulay, M. T., Baca, Q. J. & Zare, R. N. (2005) Enhanced proteolytic activity of covalently bound enzymes in photopolymerized sol gel . *Anal. Chem.*, 77, 4604-4610.

Ethier M., Hou, W., Duewel, H. S. & Figeys, D. (2006) The proteomic reactor: a microfluidic device for processing minute amounts of protein prior to mass spectrometry analysis. *J. Proteomie Res.*, 5, 2754-2759.

Fan, H. & Chen, G. (2007) Fiber-packed channel bioreactor for microfluidic protein digestion. *Proteomics*, 7, 3445-3449.

Fischer, F. & Poetsch, A. (2006) Protein cleavage strategies for an improved analysis of the membrane proteome. *Proteome Sci*, 4, 2.

Han, G., Ye, M., Jiang, X., Chen, R., Ren, J., Xue, Y., Wang, F., Song, C., Yao, X. & Zou, H. (2009) Comprehensive and reliable phosphorylation site mapping of individual phosphoproteins by combination of multiple stage mass spectrometric analysis with a target-decoy database search. *Anal. Chem.*, 81, 5794-5805.

Honda, T., Miyazaki, M., Nakamura, H. & Maeda, H. (2005) Immobilization of enzymes on a microchannel surface through cross-linking polymerization. *Chem. Commun.*, 5062-5064.

Honda, T., Miyazaki, M., Nakamura, H. & Maeda, H. (2006) Facile preparation of an enzyme-immobilized microreactor using a cross-linking enzyme membrane on a microchannel surface. *Adv. Synth. Catal*. 348, 2163-2171.

Honda, T., Miyazaki, M., Yamaguchi, Y., Nakamura, H. & Maeda, H. (2007) Integrated microreaction system for optical resolution of racemic amino acids. *Lab. Chip.*, 7, 366-372.

Hunter, T. (2009) Tyrosine phosphorylation: thirty years and counting. *Curr. Opin. Cell Biol.*, 21, 140-146.

Kim, B. C., Lopez-Ferrer, D., Lee, S. -M., Ahn, H. -K., Nair, S., Kim, S. H., Kim, B. S., Petritis, K., Camp, D. G., Grate, J. W., Smith, R. D., Koo, Y. M., Gu, M. B. & Kim, J. (2009) Highly stable trypsin-aggregate coatings on polymer nanofibers for repeated protein digestion. *Proteomics*, 9, 1893-1900.

Kinoshita, E., Kinoshita-Kikuta, E., Takiyama, K. & Koike, T. (2006) Phosphate-binding tag, a new tool to visualize phosphorylated proteins. *Mol. Cell. Proteomics*, 5, 749-757.

Krokhin, O. V., Antonovici, M., Ens, W., Wilkins, J. A. & Standing, K. G. (2006) Deamidation of -Asn-Gly- sequences during sample preparation for proteomics: Consequence for MALDI and HPLC-MALDI analysis. *Anal. Chem.*, 78, 6645-6650.

Lee, K., Lee, J., Kim, Y., Bae, D., Kang, K. Y., Yoon, S. C. & Lim, D. (2004) Defining the plant disulfide proteome. *Electrophoresis*, 25, 532-541.

Lee, J., Musyimi, H. K., Soper, S. A. & Murray, K. K. (2008) Development of an automated digestion and droplet deposition microfluidic chip for MALDI-TOF MS. *J. Am. Soc. Mass Spectrom.*, 19, 964-972.

Li, Y., Xu, X. Q., Deng, C. H., Yang, P. Y. & Zhang, X. M. (2007a) Immobilization of trypsin on superparamagnetic nanoparticles for rapid and effective proteolysis. *J. Proteome Res.*, 6, 3849-3855.

Li, Y., Xu, X., Qi, D., Deng, C., Yang, P. & Zhang, X. (2007b) Novel Fe_3O_4 @ TiO_2 core-shell microspheres for selective enrichment of phosphopeptides in phosphoproteome analysis. *J. Proteome Res.*, 7, 2526-2538.

Lin, S., Yao G., Qi, D., Li, Y., Deng, C., Yang, P. & Zhang X. (2008) Fast and efficient proteolysis by microwave-assisted protein digestion using trypsin-immobilized magnetic silica microspheres. *Anal. Chem.*, 80, 3655-3665.

Liu, Y., Lu, H., Zhong, W., Song, P., Kong, J., Yang, P., Girault, H. H. & Liu B. (2006) *Anal. Chem.*, 78, 801-808.

Liu, Y., Liu, B., Yang, P. & Girault, H. H., (2008) Microfluidic enzymatic reactors for proteome research. *Anal. Bioanal. Chem.*, 390, 227-229.

Liu, T., Bao, H., Zhang, L. & Chen, G. (2009) Integration of electrodes in a suction cup-driven microchip for alternating current-accelerated proteolysis. *Electrophoresis*, 30, 3265-3268.

Ma, J., Ziang, Z., Qiao, X., Deng, Q., Tao, D., Zhang L. & Zhang, Y. (2008) Organic-inorganic hybrid silica monolith based immobilized trypsin reactor with high enzymatic activity. *Anal. Chem.*, 80, 2949-2956.

Ma, J., Zhang, L., Liang, Z., Zhang, W. & Zhang, Y. (2009) Recent advance in immobilized enzymatic reactors and their applications in proteome analysis. *Anal. Chim. Acta*, 632, 1-8.

Mieyal, J. J., Gallogly, M. M., Qanungo, S., Sabens, E. A. & Shelton, M. D. (2008) Molecular mechanisms and clinical implications of reversible protein *S*-glutathionylation. *Antioxid. Redox Signal.*, 10, 1941-1988.

Miyazaki M. & Maeda H. (2006) Microchannel enzyme reactors and their applications for processing. *Trends Biotechnol.*, 24, 463-470.

Miyazaki, M., Honda, T., Yamaguchi, H., Briones, M. P. P. & Maeda, H. (2008) Enzymatic processing in microfluidic reactors., *Biotechnol. Genet. Eng. Rev.*, 25, 405-428.

Nel, A. L., Krenkova, J., Kleparnik, K., Smadja, C., Taverna, M., Viovy, J. L. & Foret, F. (2008) On-chip tryptic digest with direct coupling to ESI-MS using magnetic particles. *Electrophoresis*, 29, 4944-4947.

Park Z. Y. & Russell D. H. (2000) Thermal denaturation: A useful technique in peptide mass mapping. *Anal. Chem.*, 72, 2667-2670.

Sakai-Kato, K., Kato, M. & Toyooka, T. (2002) On-line trypsin-encapsulated enzyme reactor by the sol-gel method integrated into capillary electrophoresis. *Anal. Chem.*, 74, 2943-2949.

Sakai-Kato, K., Kato M. & Toyooka, T. (2003) Creation of an on-chip enzyme reactor by encapsulating trypsin in sol gel on a plastic microchip. *Anal. Chem.*, 75, 388-393.

Sheldon R. A. (2007) Enzyme immobilization: The quest for optimum performance. *Adv. Synth. Catal.* 349, 1289-1307.

Sim, T. S., Kim, E. -M., Joo, H. S., Kim, B. G. & Kim, Y. -K. (2006) Application of a temperature-controllable microreactor to simple and rapid protein identification using MALDI-TOF MS. *Lab. Chip*, 6, 1056-1061.

Slysz, G. W., Lewis, D. F. & Schriemer, D. C. (2006) Detection and identification of sub-nanogram levels of protein in nanoLC-trypsin-MS system. *J. Proteome Res.*, 5, 1959-1966.

Shui, W., Fan, J, Yang, P., Liu, C., Zhai, J., Lei, J., Yan, Y., Zhao, D. & Chen, X. (2006) Nanopore-based proteolytic reactor for sensitive and comprehensive proteomic analyses. *Anal. Chem.*, 78, 4811-4819.

Temporini, C., Calleri, E., Cabrera, K., Felix, G. & Massolini, G. (2009) On-line multi-enzymatic approach for improved sequence coverage in protein analysis. *J. Sep. Sci.* 32, 1120-1128.

Wang, L. S., Khan, F. & Micklefield, J. (2009) Selective covalent protein immobilization: strategies and applications. *Chem. Rev.*, 109, 4025-4053.

Witze, E. S., Old, W. N., Resing, K. A. & Ahn, N. G. (2007) Mapping protein post-translational modifications with mass spectrometry. *Nat. Methods*, 10, 798-806.

Wu H., Tian Y., Liu, B., Lu, H., Wang, X., Zhai, J., Jin, H., Yang, P., Xu, Y. & Wang H. (2004) Titania and alumina sol–gel-derived microfluidics enzymatic-reactors for peptide mapping: design, characterization, and performance. *J. Proteome Res.*, 3, 1201-1209.

Yamaguchi, H., Miyazaki, M., Honda, T., Briones-Nagata, M. P., Arima, K. & Maeda, H. (2009) Rapid and efficient proteolysis for proteomic analysis by protease-immobilized microreactor. *Electrophoresis*, 30, 3257-3264.

Yamaguchi, H., Miyazaki, M., Kawazumi, H. & Maeda, H. (2010a) Multidigestion in continuous flow tandem protease-immobilized microreactors for proteomic analysis. *Anal. Biochem.*, 407, 12-18.

Yamaguchi, H., Miyazaki, M. & Maeda, H. (2010b) Proteolysis approach without chemical modification for a simple and rapid analysis of disulfide bonds using thermostable protease-immobilized microreactors. *Proteomics*, 10, 2942-2949.

Yamaguchi, H., Miyazaki, M., Asanomi Y. & Maeda, H. (2011) Poly-lysine supported cross-linked enzyme aggregates with efficient enzymatic activity and high operational stability. *Catal. Sci. Technol.*, DOI: 10.1039/c1cy00084e.

Yamashita, K., Miyazaki, M., Nakamura, H. & Maeda, H. (2009) Nonimmobilized enzyme kinetics that rely on laminar flow. *J. Phys. Chem. A*, 113, 165-169.

Yano, H. & Kuroda, S. (2008) Introduction of the disulfide proteome: Application of a technique for the analysis of plant storage proteins as well as allergens. *J. Proteome Res.*, 7, 3071-3079.

Zhao, Y. & Jensen, O. N. (2009) Modification-specific proteomics: strategies for characterization of post-translational modifications using enrichment techniques. *Proteomics*, 9, 4632-4641.

6

Labeling Methods in Mass Spectrometry Based Quantitative Proteomics

Karen A. Sap and Jeroen A. A. Demmers
Erasmus University Medical Center
The Netherlands

1. Introduction

Proteomics is loosely defined as the description of sets of proteins from any biological source, which have in most cases been identified by using mass spectrometry. However, only the mere identity of proteins present in a certain sample does not give any information about the dynamics of the proteome, involving relevant cellular events such as protein synthesis and degradation, or the formation of protein assemblies. In order to retrieve information on proteome dynamics, relative protein abundances between different protein samples should be assessed. Comparative or differential proteomics aims to identify *and* quantify proteins in different samples, to study *e.g.* differences between healthy and diseased states, mutant and wildtype cell lines, undifferentiated and differentiated cells, etc. Since mass spectrometry is in itself only a qualitative technique, various methods to obtain quantitative information of the proteome have been developed over the past decade and will be described in this Chapter.

We will focus on post-digestion labeling methods in the field of functional proteomics. Functional proteomics focuses on characterizing the composition of protein complexes, and generally involves the affinity purification of a protein of interest followed by the identification of co-purifying proteins by mass spectrometry (AP-MS). Generally, proteins in a negative control sample and those identified in the sample containing the protein of interest and its interacting partners are directly compared to determine which of the proteins interact in a specific manner. However, the mere presence or absence of a certain protein in protein data sets as a measure for either overlap or specificity is generally not sufficient, as this gives no information about the relative abundances of the present proteins. A generally recognized problem is the presence of contaminating proteins that are identified in the mass spectrometric screen, but do not really make part of the protein complex. Often, these background proteins are highly abundant proteins that stick to the complex or to beads to which the antibody is conjugated in a non-specific manner. A more accurate and correct approach would therefore involve a strategy in which protein abundance differences between sample and control can be assessed in a quantitative manner and which helps in discriminating *bona fide* interaction partners from such background proteins. Ideally, a differential mass spectrometric method would allow for an unbiased, sensitive, and high-throughput screening for protein-protein interaction networks.

1.1 SDS-PAGE based methods for protein quantitation

Two-dimensional sodium dodecylsulfate polyacrylamide gel electrophoresis (2D-SDS-PAGE) has traditionally been a popular method for differential-display proteomics on a global scale, although recently the popularity and applicability of stable isotope LC-MS based methods has exceeded those presented by gel based methods. 2D-SDS-PAGE based methods enable the separation of complex protein mixtures on a single gel. Proteins are separated in two dimensions: in the first dimension, they are fractionated according to their isoelectric point using a pH gradient gel, which is subsequently placed on a polyacrylamide gel slab for further separation based on their molecular weight using SDS-PAGE. Proteins are then visualized by staining the gel with a dye such as Coomassie, silver or Sypro Ruby. In principle, for comparative issues, samples are loaded on separate gels and protein spot patterns are compared visually. Proteins that differ in abundance can then be punched out of the gel, digested with a suitable protease and analyzed by mass spectrometry. In a variation of this technique, difference gel electrophoresis (DIGE), proteins from two samples are first labeled with different fluorescent dyes and then mixed, making it possible to compare two different samples on a single gel. Two fluorescence images are recorded and overlayed, and differentially expressed proteins appear in only one of the images (Unlu et al., 1997). Limitations of this method include the manual selection of proteins to be analyzed, making it a time-consuming technique, as well as the limited sensitivity, as a consequence of which that proteins with a low concentration may be failed to be selected. Nowadays, in many laboratories there is a tendency to replace 2D-SDS-PAGE based methods by more powerful, LC-MS based methods for relative protein quantitation.

1.2 Protein and peptide quantitation using LC-MS based methods

Rather than by comparing protein spot intensities on a gel, quantification of proteins in LC/MS based methods is based on the peak height or area of the proteolytic peptide peaks in the mass spectrum and/or chromatogram. As mentioned before, mass spectrometry is not an inherently quantitative analytical technique, meaning that the peak height or area in a mass spectrum in itself does not accurately reflect the abundance of a peptide in the sample. The main reasons for this are the differences in ionization efficiency and detectability of peptides because of their different physicochemical characteristics, as well as the limited reproducibility of an LC-MS experiment. Altogether, this makes it difficult to compare peptide peak intensities between different mass spec runs. In principle however, peak intensity differences of the same analyte within one LC-MS run do accurately reflect the abundance difference. One way to distinguish the same analyte from different sample sources within one LC-MS experiment is by using stable heavy isotope labeling. When different stable isotope labels are used for proteins or peptides which are derived from different samples, the same analyte can in principle be quantified in one experiment. Such heavy stable isotope labels should in principle not affect the biophysical and chemical properties of peptides and proteins, but solely the mass, designating one of the samples as 'light' and the other sample 'heavy' according to the mass introduced by the label. The heavy and light peptides co-elute from the LC column at the same retention time and the heavy stable isotope leads to a mass shift in the mass spectrum, resulting in the observation of peak pairs. The peak heights or areas of such pairs can be compared and give an accurate reflection of the difference in abundance of this peptide between both samples. Heavy stable isotope labels can be introduced at different stages in the sample treatment protocol. Below, we will give an overview of the most widely used labeling techniques.

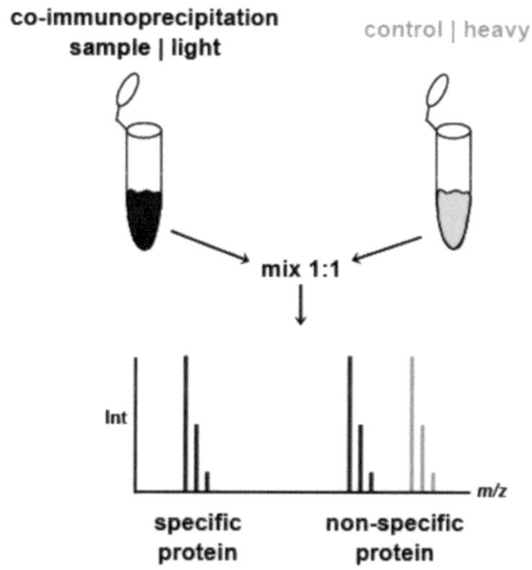

Fig. 1. In a differential labeling AP-MS experiment, proteins in a control sample are labeled with a heavy stable isotope label, whereas proteins in the experimental sample are labeled with a light label. Incorporation of the heavy label results in a shift of the *m/z* value and allows one to differentiate between the sources of the protein of interest.

2. LC-MS-based quantitation methods

2.1 Incorporation of stable isotopes by metabolic labeling

Heavy stable isotope labels can be introduced *in vivo* by growing cells or even whole organisms in the presence of amino acids or nutrients carrying such stable isotopes. Metabolic labeling is often the preferred labeling technique, since incorporation occurs at the earliest possible moment in the sample preparation process, thereby minimizing the error in quantification (see Figure 2). Several methods based on metabolic labeling have been developed and here we will give a brief overview. The first metabolic labeling studies were performed utilizing ^{15}N-enriched media to grow *S. cerevisiae* (Oda et al., 1999) and *E. coli* (Conrads et al., 2001). Next, the method was extended towards multicellular organisms which were ^{15}N labeled, such as *D. melanogaster* and *C. elegans* by feeding them on labeled yeast or bacteria, respectively (Krijgsveld et al., 2003). Even a higher eukaryote like a rat has been labeled with ^{15}N (McClatchy et al., 2007). Plants, as they are autotrophic organisms, can easily be labeled metabolically through feeding of labeled inorganic compounds in the form of ^{15}N-nitrogen-containing salts, as first demonstrated in NMR studies (Ippel et al., 2004), and later in MS-based proteomics (Engelsberger et al., 2006; Lanquar et al., 2007). ^{15}N atoms are incorporated into the sample during cell growth, eventually replacing all natural isotopic (*i.e.*, ^{14}N) nitrogen atoms. The corresponding mass shift depends on the number of nitrogen atoms present in each of the resulting proteolytic peptides. However, this variable mass shift complicates data analysis to a large extent and requires high resolution mass spectrometry for the analysis (Conrads et al., 2001). Specific software for the analysis of ^{15}N labeled samples has been developed (Mortensen et al., 2010).

Stable isotope labeling in cell culture (SILAC) is a metabolic labeling approach first published in 2002 by the lab of Matthias Mann (Ong et al., 2002). During cell growth, essential amino acids that carry heavy stable isotopes and which have been added to the

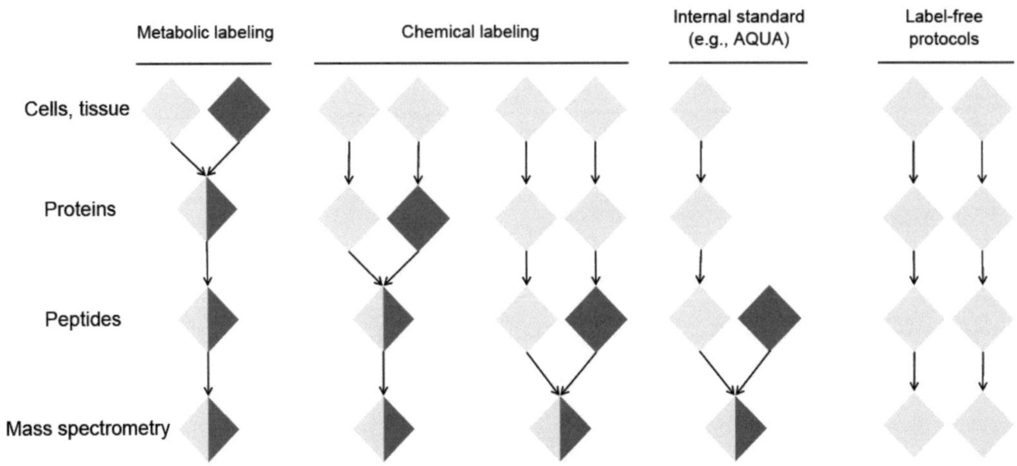

Fig. 2. Stages of incorporation of stable isotope labels in typical labeling workflows in quantitative proteomics. The light and dark grey diamonds represent the two protein samples to be differentially labeled and compared. Figure adapted from (Ong & Mann, 2005).

culturing medium are introduced in all newly synthesized proteins. After several cell doublings, the complete cellular proteome will have incorporated the supplied labeled amino acid(s). This results in a shift of the proteolytic peptide mass after protein digestion and subsequent MS analysis. When labeled and non-labeled cell cultures are now mixed and analyzed in the same experiment, peptides will be represented by peak pairs in the mass spectrum, where the mass difference will depend on the number and nature of the labeled amino acid(s). Usually, labeled lysine and arginine are used, with the result that every peptide will carry a label except for the carboxyl-terminal peptide of the protein, when digested with trypsin, as does labeling with lysine when digested with Lys-C (Ibarrola et al., 2003). In contrast to ^{15}N labeling, the number of incorporated labels in SILAC is defined and not dependent on the peptide sequence, thus facilitating data analysis. SILAC has been successfully applied in global proteome studies (de Godoy et al., 2006), for functional proteomics assays, as well as for the study of post-translational modifications (Blagoev et al., 2003; Blagoev & Mann, 2006).

Because of the label incorporation at early stages in the sample preparation protocol, SILAC is generally the preferred choice of labeling method. However, SILAC is limited in sample applicability, for example, not every cell line can grow in an efficient manner in media optimized for SILAC, often due to the requirement of dialyzed serum in the medium to prevent contamination with natural amino acids. Besides, the method may be hampered by *in vitro* conversion of labeled arginine to proline (Van Hoof et al., 2007). SILAC has been used to label higher organisms, for instance flies (Sury et al., 2010) and mice (Kruger et al., 2008), by feeding them with labeled food. In general though, this is a time consuming and expensive process. In the plant, SILAC has only yielded label incorporation of approximately 70% (Gruhler et al., 2005), which is not satisfying for many proteomics applications. Moreover, there are practical and moral limitations to SILAC labeling of human tissue. For these cases, methods for stable isotope label incorporation at a later stage in the sample preparation protocol are required. Chemical and enzymatic labeling techniques have been developed that can introduce the heavy stable isotope label only after sample collection and proteolytic digestion at the peptide level.

2.2 Incorporation of stable heavy isotope labels by chemical or enzymatic labeling

In general, the advantage of chemical labeling over metabolic labeling is the possibility to label a wide range of different sample types, since incorporation of the label is performed only after harvesting cells and subsequent purification of proteins. Chemical labeling is essentially based on similar mechanisms as metabolic labeling, except that the label is introduced into proteins or peptides by a chemical reaction, *e.g.*, with sulfhydryl groups or amine groups, or through acetylation or esterfication of amino acid residues. Alternatively, the heavy stable isotope label can be introduced into the peptide during an enzymatic reaction with heavy water ($H_2{}^{18}O$). Below, several of the most widely applied chemical and enzymatic labeling approaches are described.

2.2.1 Isotope-Coded Affinity Tags (ICAT)

Isotope-Coded Affinity Tagging (ICAT) is a chemical labeling method that was first described by the Aebersold lab in 1999 (Gygi et al., 1999). In chemical modification-based approaches, stable isotope-bearing chemical reagents are targeted towards reactive sites on a protein or peptide. The ICAT reagent consists of a reactive group that is cysteine-directed, a polyether linker region with eight deuteriums, and a biotin group that allows purification of labeled peptides. In an ICAT experiment, two pools of proteins are denatured and reduced, and the cysteine residues of the proteins are subsequently derivatized with either the 'heavy' or 'light' ICAT reagent. The labeled pools are then combined, cleaned up to remove excess reagent, and digested with an appropriate protease. The cysteine-containing peptides, carrying 'heavy' and 'light' isotope tags, are then captured on an avidin column via the biotin moiety present at the incorporated label. Peptides are then eluted from the column and analyzed by mass spectrometry. Since only cysteine-containing peptides are isolated, the peptide mixture complexity is in general limited, which in principle would enable identification of lower abundant proteins. On the other hand, some proteins contain no cysteines, while others would have to be quantified on the basis of just a single peptide. Additionally, the large biotin tag significantly increases the complexity of fragmentation spectra, complicating peptide identification, and, besides that, it has been demonstrated that deuterium atoms that are associated with the tag can cause a shift in retention time between the light and heavy peptides in reverse phase chromatography (Zhang et al., 2001). Subsequent iterations of the ICAT approach by substituting a cleavable and co-eluting tag have improved the method (Hansen et al., 2003; Li et al., 2003).

2.2.2 Dimethyl labeling

An alternative method based on chemical labeling is dimethylation of peptides. In this workflow, samples are first digested with proteases such as trypsin and the derived peptides of the different samples are then labeled with isotopomeric dimethyl labels. The labeled samples are mixed and simultaneously analyzed by LC-MS whereby the mass difference of the dimethyl labels is used to compare the peptide abundance in the different samples. Stable isotope labeling by dimethylation is based on the reaction of peptide primary amines (peptide N-termini and the epsilon amino group of lysine residues) with formaldehyde to generate a Schiff base that is rapidly reduced by the addition of cyanoborohydride to the mixture. These reactions occur optimally between pH 5 and 8.5. Dimethyl labeling can be used as a triplex reagent, making it possible to quantitatively analyze three different samples in a single MS run. Labeling with the light reagent generates a mass increase of 28 Da per primary amine on a peptide and is obtained by using regular formaldehyde and cyanoborohydride. Using deuterated formaldehyde in combination with regular cyanoborohydride generates a mass increase of 32 Da per primary amine; this is referred to as the intermediate label (Hsu et al., 2003) . Incorporation of the heavy label can be achieved through combining deuterated and ^{13}C-labeled formaldehyde with cyanoborodeuteride, resulting in a mass increase of 36 Da (Boersema et al., 2008). These reactions are visualized in Figure 3.

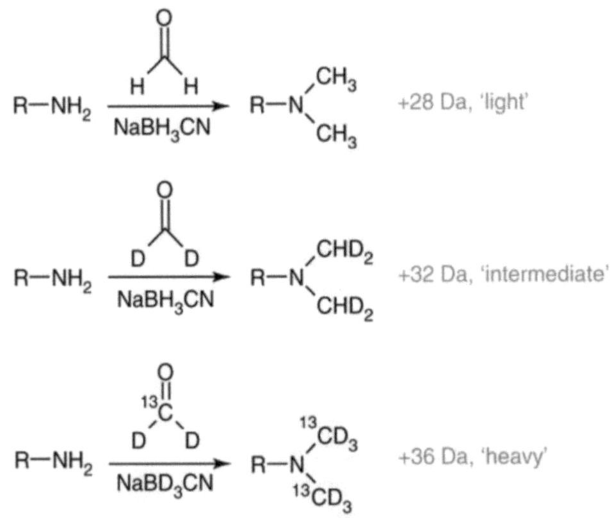

Fig. 3. Labeling schemes of triplex stable isotope dimethyl labeling. R: remainder of the peptide. Figure adapted from (Boersema et al., 2008).

One drawback of the incorporation of deuterium is that deuterated peptides show a small but significant retention time difference in reversed phase chromatography compared to their non-deuterated counterparts (Zhang et al., 2001). This complicates data analysis because the relative quantities of the two peptide species cannot be determined accurately from one spectrum but requires integration across the chromatographic time scale. As the stable isotope dimethyl labeling is performed at the peptide level, the method is not subjected to restrictions on the origin of the biological sample. Stable isotope dimethyl labeling can be performed in up to 8M urea, as well as after in-gel digestion protocols. It should be noted that during the sample preparation workflow, no buffers and solutions containing primary amines (such as ammonium bicarbonate and Tris) ought to be used, as formaldehyde would react with these, which would affect the labeling efficiency. This can be circumvented by desalting the peptide sample before the labeling reaction or by performing the digestion in buffers without primary amines (*e.g.*, triethyl ammonium bicarbonate (TEAB)). Since both the peptide N-termini and lysine side chain amino groups are labeled in this protocol, it is compatible with the peptide products of virtually any protease, such as trypsin, Lys-C, Lys-N, Arg-C, and V8 (Boersema et al., 2008). Typically, for proteomics experiments trypsin is used, which cleaves C-terminal of lysine and arginine residues. Labeling of tryptic peptides using the method described here results in a mass shift of either 4 Da (when cleaved after an arginine residue) or 8 Da (when cleaved after a lysine residue) between the light and intermediate and between the intermediate and heavy label.

Differential labeling of peptides resulting from digestion with Lys-C or Lys-N (cleaving respectively C- and N-terminal of lysine residues) will result in a mass difference of mainly 8 Da (both the N-terminus and the lysine residues are labeled), whereas peptide products from Arg-C and V8 will result in varying mass differences as the number of lysine residues per peptide will typically vary. After proteolytic digestion, the samples are labeled separately by incubation with CH_2O and $NaBH_3CN$ (light), CD_2O and $NaBH_3CN$ (intermediate) or $^{13}CD_2O$ and $NaBD_3CN$ (heavy).
Boersema and co-workers have described three different experimental protocols for dimethyl labeling, *i.e.* in-solution, online, and on-column (Boersema et al., 2009). In-solution labeling (Boersema et al., 2008; Hsu et al., 2003) can be used for sample amounts from 1 μg to several milligrams of sample and is most suitable for experiments in which large sample numbers have to be labeled since labeling can be performed in parallel here. Online stable isotope labeling is the optimal method for the labeling of small quantities (<< 1 μg) of sample, because the sample loss is diminished by combining sample clean-up and labeling and by performing LC-MS directly after labeling. Finally, the on-column stable isotope labeling method is most suited for larger (up to milligrams) sample amounts, as sample clean-up and labeling steps are combined and the quenching step is avoided. After labeling, the samples are mixed and analyzed by mass spectrometry. Finally, quantification is performed by comparing the signal intensities of the differentially labeled peptides (see section on Data Analysis).
Protein quantitation by dimethyl labeling has been applied in a variety of studies, *e.g.* for the investigation of tyrosine phosphorylation sites in Hela cells upon EGF stimulation (Boersema et al., 2010). Proteins in a HeLa cell extract were dimethyl labeled and subsequently enriched for phosphorylated-tyrosine-containing peptides using immunoaffinity assays. Several tens of unique phosphotyrosine peptides were found to be regulated by EGF, illustrating that such a targeted quantitative phosphoproteomics approach has the potential to study signaling events in detail. Furthermore, the method has been applied to unravel differences in composition between highly related protein complexes, such as tissue-specific bovine proteasomes (Raijmakers et al., 2008) and the yeast nuclear and cytoplasmic exosome protein complex (Synowsky et al., 2009).
In conclusion, dimethyl labeling is a reliable, cost-effective and undemanding procedure that can be easily automated and applied in high-throughput proteomics experiments. It is applicable to virtually any sample, including tissue samples derived from animals or humans and up to three samples can be analyzed simultaneously. Like other chemical labeling methods though, stable isotope dimethyl labeling is performed in one of the final steps of a typical proteomics workflow and is therefore more prone to errors in the quantitative analysis as compared to workflows in which the label is added at an earlier stage.

2.2.3 ^{18}O labeling

^{18}O labeling relies on class-2 proteases, such as trypsin, to catalyze the exchange of two ^{16}O atoms for two ^{18}O atoms at the C-terminal carboxyl group of proteolytic peptides, resulting in a mass shift of 4 Da between differently labeled peptides, as illustrated in Figure 4.
Hydrolysis of a protein in $H_2{}^{18}O$ by a protease results in the incorporation of one ^{18}O atom into the carboxyl terminus of each proteolytically generated peptide. This mechanism involves a nucleophilic attack by a solvent water molecule on the carbonyl carbon of the scissile peptide bond (reaction 1). Following this hydrolysis reaction, the protease

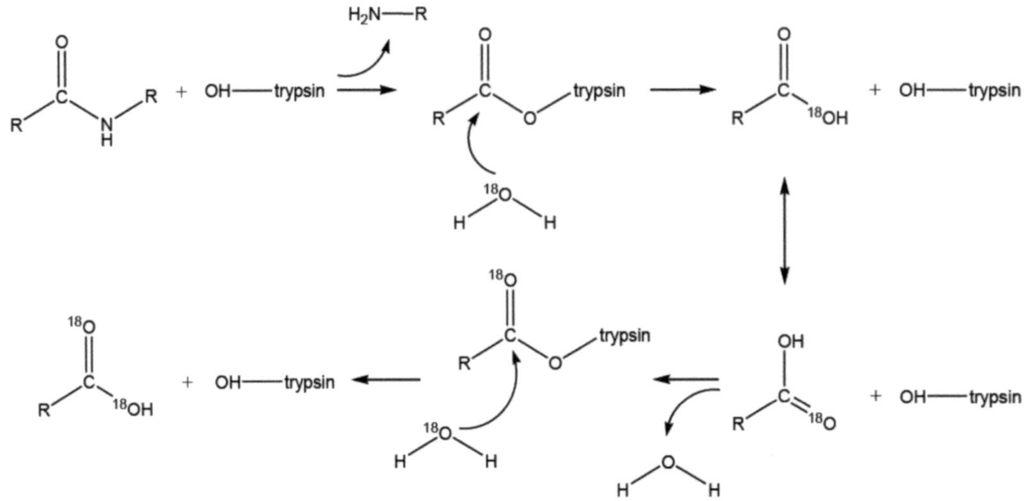

Fig. 4. Principle of trypsin catalyzed ^{18}O labeling. Incorporation of two ^{18}O labels at the C-terminus of a tryptic peptide takes place in a two-step reaction.

incorporates one more ^{18}O atom into the carboxyl terminus of the proteolytically generated peptide. This second incorporation results in two ^{18}O atoms being incorporated into the carboxyl terminus of the peptides (reaction 2). The second ^{18}O atom-incorporation is essentially the reverse reaction of peptide-bond hydrolysis or the peptide-bond formation reaction (Miyagi & Rao, 2007).

Proteolytic ^{18}O labeling has shown to be a useful tool in the field of comparative proteomics. A number of studies have been published, involving among others relative protein quantitation of the virus proteome (Yao et al., 2001), proteomes of cultured cells (Blonder et al., 2005; Brown & Fenselau, 2004; Rao et al., 2005) and proteins in serum (Hood, Lucas et al., 2005; Qian et al., 2005) and tissues (Hood, Darfler et al., 2005; Zang et al., 2004). In addition, ^{18}O labeling has been used for the relative quantitation of post-translational modification, *e.g.* changes of protein phosphorylation in response to a stimulus (Bonenfant et al., 2003). In the latter study, pools of differentially labeled phosphorylated proteins were enriched by using immobilized metal-affinity chromatography. Peptides were then dephosphorylated by alkaline phosphatase in order to quantify the changes in phosphorylation by mass spectrometry. A similar approach has been used for the global phosphoproteome analysis of human HepG2 cells (Gevaert et al., 2005).

Despite its relatively simple mechanism and low costs, ^{18}O labeling has not become the preferred method for differential proteomics based on heavy stable isotope labeling. The practical difficulties involved, most importantly the occurrence of incomplete incorporation of two ^{18}O atoms into the proteolytic peptide, and, as a consequence, the difficulties in data analysis and interpretation are the most likely reasons for this. Several factors are responsible for the variable degree of ^{18}O incorporation, including variable enzyme substrate specificity, oxygen back exchange, pH dependency and peptide physicochemical properties. To overcome inefficient labeling, algorithms for the correction of ^{18}O labeling efficiency have been developed (Ramos-Fernandez et al., 2007), while other studies have focused on minimizing back exchange of ^{18}O to ^{16}O. It was found that the latter can be achieved by either decreasing the pH value for trypsin catalyzed incorporation reactions

(Hajkova et al., 2006; Staes et al., 2004; Zang et al., 2004), or by using immobilized trypsin for the exchange reaction (Chen et al., 2005; Fenselau & Yao, 2007; Sevinsky et al., 2007). Trypsin immobilization allows the investigator to significantly increase the molar ratio of protease-to-substrate ratio, which subsequently increases the labeling efficiency. Another advantage of using immobilized proteases is that no protease-catalyzed oxygen back exchange reaction occurs, because the immobilized proteases are completely removed from the peptides after the labeling reaction.

Our lab has developed a two-step approach in order to completely label all proteolytic peptides (Bezstarosti et al., 2010). In this method, proteins are first digested with soluble trypsin. Subsequently, proteolytic peptides are incubated with $H_2{}^{18}O$ at pH 4.5 in the presence of immobilized trypsin. Clearly, no singly ^{18}O labeled variants were observed in any of the peptide mass spectra (see Figure 5), indicating that no partial labeling whatsoever occurred, nor did any back exchange from ^{18}O to ^{16}O take place during sample treatment or analysis. Thus, complete incorporation of two ^{18}O labels into each of the tryptic peptides in a mixture can be achieved routinely.

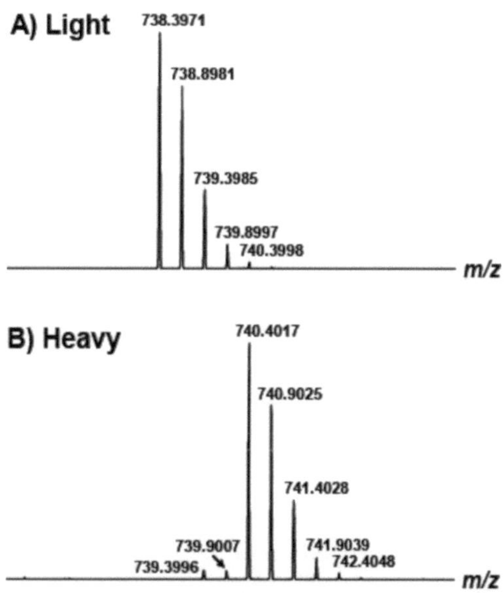

Fig. 5. Doubly charged tryptic peptide FLEQQNQVLQTK A) in the absence of ^{18}O label and B) after incorporation of the label. The two-step labeling reaction in the presence of immobilized trypsin as described here ultimately results in the complete incorporation of two ^{18}O labels, with no intermediary products present. The peptide isotope peaks in B) at *m/z* 739.39 and 739.90 are due to impurities of commercial $H_2{}^{18}O$, containing only 97% ^{18}O.

It was shown in this study that ^{18}O labeling can be applied in a functional proteomics assay to discriminate background proteins from specific interactors of a protein of interest. Generally, controls are heavy labeled and the coimunoprecipitation (co-IP) sample is labeled light. Specific interactors are expected not to be present in the control and would thus have a ratio of (close to) zero, whereas background proteins would show heavy-to-light (H:L) ratios of (close to) 1. ^{18}O labeling was used in order to differentiate between non-specific background proteins and specific, *bona fide* interactors of the Cyclin dependent kinase 9

(Cdk9) purified from nuclear extracts of murine erythroleukemia (MEL) cells. Biotinylated Cdk9 was expressed in MEL cells and purified using streptavidin beads under relatively mild conditions (de Boer et al., 2003). The proteins that co-purified with Cdk9 were washed and digested with trypsin while still bound to the beads and subsequently identified by tandem mass spectrometry. A control sample was taken following the same procedure from an equal number of cells, but using non-transfected MEL cells. Proteolytic peptides from the control sample were then labeled using $H_2{}^{18}O$ in the two-step approach mentioned earlier, while proteolytic peptides from the Cdk9 pulldown sample underwent the same procedure with unlabeled H_2O. The peptide mixtures were dissolved in equal volumes of buffer, mixed in a 1:1 volume ratio and identified by LC-MS/MS. H:L ratios were calculated for all proteins identified from the mixed sample.

As expected, H:L ratios of close to 1 were observed for typical background proteins, such as ribosomal, housekeeping, and structural proteins, which were present as non-specific background proteins (see Figure 6). In contrast, among the proteins that were quantified with H:L ratios close to 0, indicating specificity for the Cdk9 co-immunopurification sample, the far majority of interacting proteins that have been described in different studies in the literature were identified in a single experiment, as well as several novel interaction partners of diverse functionalities, suggesting putative additional roles for Cdk9 in various nuclear events such as transcription and cell cycle control (Bezstarosti et al., 2010). It was shown in this study that complete ^{18}O labeling of peptides in complex mixtures can be routinely achieved. This greatly simplifies the analysis of peak intensity ratios, since only two components (*i.e.*, 'light' and 'heavy') need to be considered and no correction algorithms have to be applied to convert peak intensities of intermediately labeled peptide species.

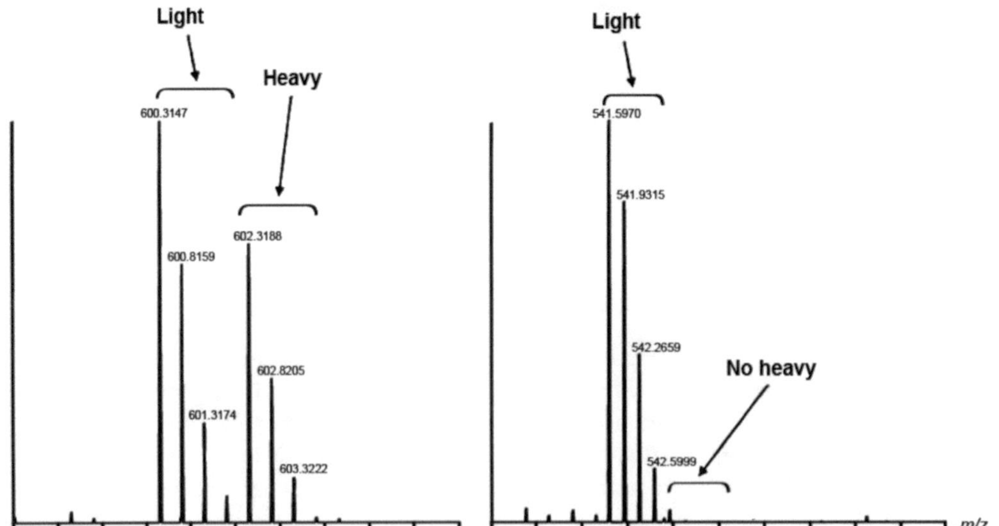

Fig. 6. MS spectra of two tryptic peptides from a 1:1 mixture of a digest of a Cdk9 co-IP experiment ($H_2{}^{16}O$) and a control sample (in $H_2{}^{18}O$). (A) Doubly charged peptide LGTPELSPTER of the contaminant acetyl-CoA carboxylase shows both the "light" and "heavy" forms and is therefore marked as a nonspecific protein. (B) Triply charged peptide GPPEETGAAVFDHPAK of cyclin T1 is only present in the "light" form and is therefore specific for the Cdk9 sample (see (Bezstarosti et al., 2010)).

2.2.4 Labeling with isobaric tags

Metabolic labeling, ICAT, enzymatic labeling and most other chemical labeling approaches for relative quantification are based on the mass difference between differentially labeled peptides. There are, however, some limitations imposed by mass difference labeling. The mass difference concept of many practical purposes is limited to a binary (2-plex) or ternary (3-plex) set of reagents; higher order multiplexing would increase the complexity of MS^1 spectra too much. This limitation makes comparison of multiple states difficult to undertake. Therefore, multiplexed sets of reagents for quantitative protein analysis have been developed. The isobaric tag for relative and absolute quantitation (iTRAQ) (Ross et al., 2004) and tandem mass tag (TMT) (Thompson et al., 2003) technologies are commercially available isobaric mass tagging reagents and protocols (Figure 7).

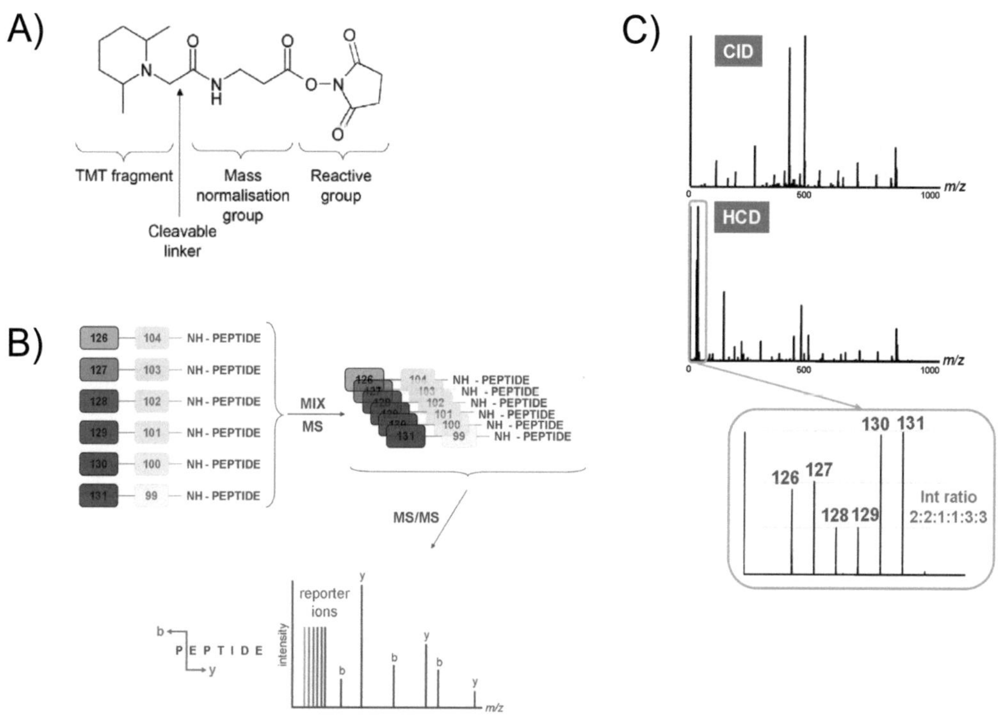

Fig. 7. A) Chemical structure of the TMT tag. The 6-plex tags have different distributions of the stable heavy isotopes of carbon and nitrogen in the molecule, resulting in different fragmentation spectra. B) A peptide that is present in 6 different samples is differentially labeled with a 6-plex TMT tag, containing reporter-balancer combinations, resulting in all conjugated peptides having the same *m/z* value. Upon high energy collision dissociation (HCD), the differentially labeled peptides show identical b and y fragment ions, but the reporter ion masses in the low *m/z* region are different. C) As an example, a protein was labeled in a 6-plex (TMT-126 through TMT-131) protocol and mixed in a 2:2:1:1:3:3 ratio. The resulting reporter ion intensity ratios show an excellent correlation with the mixing ratios. Panels A and B were adapted from (Thompson et al., 2003).

In these procedures, both N-termini and lysine side chains of peptides in a digest mixture are labeled with different isobaric mass reagents in such a way that all derivatized peptides are isobaric and chromatographically indistinguishable. Only upon peptide fragmentation can the different mass tags be distinguished. As each tag adds the same total mass to a given peptide, each peptide species produces only a single peak during liquid chromatography, even when two or more samples are mixed. Thus, there will be only one peak in the MS^1 scan, and, therefore, only a single *m/z* will be isolated for fragmentation. The different mass tags only separate upon fragmentation, when reporter ions that are typical for each of the different labels are generated. These reporter ions are in the low mass range, which usually is not covered by typical peptide fragment ions. The intensity ratio of the different reporter ions is used as a quantitative readout. Thus, quantitation in combination with isobaric mass tagging is based on peptide fragmentation (MS^2) spectra rather than on the survey scans and quantitative accuracy will depend on the isolation width of precursor ions for fragmentation, since all ions isolated in that window will contribute to fragments in the reporter ion mass ranges. One drawback of such a method is that often only a single fragmentation spectrum per peptide is available, while in quantitation based MS^1 scans, usually several data points across the eluting peptide peak are sampled, which may result in a lower overall sensitivity.

2.3 Label free quantitation

Over the past few years, mainly as a result of constantly improving LC-MS equipment, there has been growing interest in the use of label-free approaches for quantitative proteomic analysis (see (Neilson et al., 2011) for a recent review). In a label free quantitative proteomic analysis, protein mixtures are analyzed directly and samples are compared to each other after independent analyses. As a result, there is no mixing of samples, so that higher proteome coverage can be achieved and there is no limit to the number of experiments that can be compared (Bantscheff et al., 2007). The disadvantage of this approach is a lack of a formal internal standard, which can lead to greater error in individual datasets but is minimized through the analysis of several biological replicates.

Label-free approaches may be divided into two main groups by the way that the abundance of a peptide is measured. The first group comprises methods that are based on the ion count and compare either maximum abundance or volume of ion count for peptide peaks at specific retention times between different samples (Chelius & Bondarenko, 2002; Listgarten & Emili, 2005; Silva et al., 2005; Wiener et al., 2004). As ionized peptides elute from a reversed-phase column into the mass spectrometer, their ion intensities can be measured within the given detection limits of the experimental setup. Although this method is relatively straightforward conceptually, several considerations must be taken into account to ensure reproducible and accurate detection and quantitation between individual sample runs. Concerns with LC signal resolution can arise when peptide signals are spread over a large retention time range causing overlap with co-eluting peptides. Similar concerns include biological variations resulting in multiple signals for the same peptide as well as technical variations in retention time, MS intensity, and sample background noise from chemical interference. These aspects of quantitation based on 'area under the curve' necessitate a computational 'clean up' of the raw LC-MS data (Neilson et al., 2011).

The second group is based on the identification of peptides by MS/MS and uses sampling statistics such as peptide count, spectral counts (Lundgren et al., 2010), or sequence coverage to quantify the differences between samples (Choi et al., 2008; Liu et al., 2004; Old et al.,

2005; Rappsilber et al., 2002). For protein quantification based on spectrum counting, the data processing steps are basically identical to the general protein identification workflow in proteomics, which is one of the reasons why this approach has become so popular. The rationale behind this quantitation method is that more abundant proteins are sampled more often in fragment ion scans than are low abundance peptides or proteins. Obviously, the outcome of spectrum counting depends on the settings of data-dependent acquisition on the mass spectrometer. In particular the linear range for quantitation and the number of proteins to be quantified are influenced by different settings for dynamic exclusion (Wang & Li, 2008); the optimal settings will depend on sample complexity. The most significant disadvantage of spectrum counting is that it behaves very poorly with proteins of low abundance and few spectra. The accuracy of the spectrum count method, especially for low abundance proteins, suffers from the fact that each spectrum is scored independently of its ion intensities.

3. Comparison of different methods for quantitation

With the existence of a wide variety of LC-MS based quantitation methods, it may be hard to decide which approach to utilize for a certain application. As described earlier, each approach has its own strengths and limitations, and, additionally, other factors may play a role, such as available equipment, level of experience and budget. In the following section we summarize the pros and cons of earlier described quantitation methods which might serve as a guidance to decide which approach is most suitable in a specific situation.

3.1 Metabolic labeling

If it is possible to label samples metabolically, this would be the most advantageous option to quantitate proteins. The most important reason for this is that different samples can be combined at the level of intact cells, which, as a result, excludes all sources of quantitation error introduced by biochemical and mass spectrometric procedures, as these will affect both protein populations in the same way. Metabolic labeling is therefore the most sensitive MS based labeling technique to date, making it possible to study small protein abundance differences as small as 1.5-fold changes or even smaller. Despite a number of cases that demonstrate the feasibility of metabolic labeling of higher organisms using ^{15}N sources *in vivo,* such as *C. elegans, D. Melanogaster* (Krijgsveld et al., 2003) and the rat (Wu et al., 2004), it is not practical to apply this strategy routinely. The most important reason for this is that labeling with ^{15}N complicates data analysis to a large extent, as discussed in section 2.1. Nowadays, the most widely applied method to metabolically label material of eukaryotic origin is SILAC in immortalized cell lines. SILAC based MS has been extensively applied for the study of global proteomes, in the field of functional proteomics, and for the analysis of post-translational modifications. Additionally, SILAC can be applied to whole organisms, such as *E. coli, S. cerevisiae,* and *D. melanogaster*. Even metabolic labeling of higher eukaryotes like the mouse (Kruger et al., 2008) has shown to be possible.

Although SILAC is the most accurate MS based quantitation approach, it might not always be possible or preferable to use SILAC. As mentioned earlier, not every cell type might grow well in the SILAC medium. Some cell lines readily convert arginine to proline, which complicates data analysis, and require adaptation of the protocol such as titration of arginine in the medium (Ong et al., 2003). Otherwise, computational approaches to correct

arginine-to-proline conversion may be applied (Park et al., 2009). Finally, cell lines that are sensitive to changes in media composition or are otherwise difficult to grow or maintain in culture may not be amenable to metabolic labeling at all. When it is not possible to label a cell culture in SILAC medium, post-digestion incorporation methods may serve as an alternative. Moreover, post-digestion labeling might be the preferred method for affinity purification mass spectrometry (AP-MS) applications, as the starting material for co-IP assays is typically several milligrams of proteins. The use of stable isotope labeling by SILAC can be cost-prohibitive, whereas post-digestion labeling approaches such as stable isotope dimethyl labeling and ^{18}O labeling are performed with inexpensive generic reagents and do not pose severe financial restrictions to the amount of sample to be labeled.
In conclusion, SILAC can be applied in almost all sorts of proteomic applications since it is very sensitive, and limitations are mainly biological applicability or involve practical issues such as time, cost, or available equipment.

3.2 Chemical and enzymatic post-digestion labeling

One of the advantages of a chemical modification approach over metabolic labeling is the ability to label proteins after cell lysis and in a post-digestion manner. This makes the approach generically applicable, since it allows the quantitative analysis of biological samples that cannot be grown in culture, such as human body fluids or human tissue. ICAT was one of the first chemical labeling methods introduced for quantitative mass spectrometry. Although often and successfully applied, its main drawbacks are adverse side reactions and its inability to label peptides that do not contain cysteine residues. As a result, in many laboratories, ICAT has been substituted by other approaches, such as chemical dimethyl labeling or enzymatic ^{18}O labeling. Compared to ICAT, both ^{18}O labeling and dimethyl labeling are simple, free of extensive sample manipulations, virtually free of side reactions, and amenable to all protein species (*i.e.,* proteins that contain no cysteine residues). In contrast to ICAT, there is no lower limit of the protein amount that can be labeled for ^{18}O and dimethyl. Another advantage of the latter two labeling approaches is that they are cost-effective. This, together with the fact that proteins for any species can be labeled and the ease of sample preparation, makes chemical labeling the preferred method for the quantitative analysis of for instance size-limited human tissue specimens. Also, post-digestion labeling is practical for tissue samples of higher organisms such as mice, or cell lines that cannot be metabolically labeled.
One drawback of dimethyl labeling is that deuterated peptides show a small but significant retention time difference in reversed-phase HPLC compared to their non-deuterated counterparts (Zhang et al., 2001). This complicates data analysis because the relative quantities of the two peptide species cannot be determined accurately from one spectrum but requires integration across the chromatographic time scale. Retention time shifts are far less pronounced for labels such as ^{13}C, ^{15}N, or ^{18}O isotopes (Zhang & Regnier, 2002), so that the additional signal integration step over retention time can generally be omitted in approaches based on incorporation of these labels. However, compared to iTRAQ and TMT, dimethyl labeling is performed with inexpensive generic reagents and do not pose severe financial restrictions to the amount of sample to be labeled.
Multiplex labeling using TMT or iTRAQ has turned out to be particularly useful for following biological systems over multiple time points or, more generally, for comparing multiple treatments in the same experiment. With dimethylation labeling, iTRAQ and TMT

labeling, multiplexing can be achieved, which is not possible for ^{18}O labeling. iTRAQ is capable of simultaneously analyzing eight samples (Pierce et al., 2008), whereas with TMT labeling, six samples can be measured together (Thompson et al., 2003). It should be noted that the use of commercial isobaric iTRAQ or TMT labels can be cost-prohibitive. In terms of equipment, TMT and iTRAQ labeling approaches are limited to mass spectrometers which are capable of efficiently detecting ions that are present at a relatively low *m/z* and peptide quantification is based on a single fragmentation mass spectrum. An advantage of isobaric tags is that the labeled peptides co-elute from the chromatographic column which means that the MS signal is not split into different peaks, as in conventional isotope labeling, improving sensitivity in the MS mode.

In conclusion, chemical and enzymatic labeling can be applied to virtually any biological sample since incorporation is performed after cell lysis and generally also after digestion. Therefore, post-digestion labeling is specifically useful for the study of mammalian and human tissue or body fluids. Importantly, compared to metabolic labeling, label incorporation in chemical and enzymatic labeling approaches takes place at a later stage in the sample treatment protocol and are therefore in general less accurate. For absolute protein quantitation, peptides have to be labeled with stable heavy isotopes, which is usually done by synthesizing them with labeled amino acids, in order to serve as an internal standard.

3.3 Label free approaches

Since no labels are used whatsoever in label free quantitative proteomics, these approaches are inexpensive, they can be applied to any kind of biological material and the proteome coverage of quantified proteins is high because basically every protein that is identified by at least one peptide spectrum can in principle be quantified. In addition, the complexity of the sample is not increased by mixing different samples. Label free methods therefore usually have a high analytical depth and dynamic range, giving this method an advantage when large, global protein changes between treatments are expected. Also, since the samples are not mixed and quantification is done after MS analysis, the obtained data is not fixed and can be used in other contexts as well. These advantages make label free quantification an attractive approach for *e.g.* clinicians who have large patient material-derived datasets and want to compare multiple datasets, and have no wet lab available.

Despite the many advantages of label free quantitation, it is probably the least accurate among the mass spectrometric quantification methods when considering the overall experimental process because all the systematic and non-systematic variations between experiments are reflected in the obtained data. Consequently, the number of experimental steps should be kept to a minimum and every effort should be made to control reproducibility at each step.

There has been growing interest in the use of label-free approaches for quantitative proteomic analyses over the recent years, particularly because of ever increasing accuracy and reproducibility of high-resolution LC-MS equipment. Most MS analysis is performed with data dependent analysis (DDA) where the mass spectrometer runs a parent ion scan and selects the most abundant ions on which to conduct fragmentation scans, typically 4-10 scans, before returning to a parent ion scan. There may be a bias in this type of data for co-eluting peptides towards omitting the lower abundant peptides from MS/MS (Venable et al., 2004). This bias creates a subset of proteins effectively unseen due to the resultant level of detection limit. An experimental setup has been developed in which the mass spectrometer no longer cycles between MS and MS/MS mode but aims to detect and fragment all peptides in a chromatographic window simultaneously by rapidly alternating between high- and low-energy conditions in the mass spectrometer (Silva et al., 2006). Obviously, there are challenges with analyzing such data from complex samples as many fragmentation spectra will be populated with sequence ions from multiple peptides each contributing differently to the overall spectral content.

Also, there is evidence that label-free methods provide higher dynamic range of quantification than any stable isotope labeling approach (*i.e.*, 2-3 orders of magnitude) and therefore may be advantageous when large and global protein changes between experiments are observed (Old et al., 2005).

4. Data analysis

No matter the choice of quantitative method, quantitative proteomic data are typically very complex and often of variable quality. The main challenge stems from incomplete data, since even today's most advanced mass spectrometers cannot sample and fragment every peptide ion present in complex samples. As a consequence, only a subset of peptides and proteins present in a sample can be identified. Over the past years, a series of experimental strategies for mass spectrometry based quantitative proteomics and corresponding computational methodology for the processing of quantitative data have been generated (reviewed in (Matthiesen et al., 2011; Mueller et al., 2008). Conceptually different methods to perform quantitative LC-MS experiments demand different quantification principles and available software solutions for data analysis. Quantification can be achieved by comparing peak intensities in differential stable isotopic labeling, via spectral counting, or by using the ion current in label-free LC-MS measurements. Numerous software solutions have been presented, with specific instrument compatibility and processing functionality and which can cope with these basically different quantitation methods. It is important for researchers to choose an appropriate software solution for quantitative proteomic experiments based on their experimental and analytical requirements. However, it goes beyond the scope of this Chapter to discuss all of the available software tools separately. For an extensive and up-to-date overview of software solutions including links to websites for downloads, the reader is referred to http://www.ms-utils.org.

5. Concluding remarks

As we have discussed in this Chapter, all of the mass spectrometry based quantification methods have their particular strengths and weaknesses. The researcher has to choose the best method from the multitude of methods that have emerged for the analysis of simple and complex (sub-)proteomes using quantitative mass spectrometry for his or her specific research; a choice that depends on the financial aspects involved, the availability of high-resolution mass spectrometer and LC equipment and the available expertise present in the lab. Quantitative proteomics methods are now starting to mature to an extent that they can be meaningfully applied to the study of proteomes and their dynamics. Using the labeling methods described in this Chapter, it is now possible to identify and quantitate several thousands of proteins in a single experiment. However, there is still room for significant improvements to the experimental strategies that are required for the quantitative analysis of very complex mixtures and of post-translational modifications, with the ultimate aim to generate quantitative proteomic data at a scale which would allow the comprehensive investigation of a biological phenomenon. At the same time, the recent exponential increase in data volume and complexity demands the development of appropriate bioinformatic and statistical approaches in order to arrive at meaningful interpretations of the results. This can only be achieved if the influence of the employed technologies on the results obtained is well understood.

6. References

Bantscheff, M., Schirle, M., Sweetman, G., Rick, J. & Kuster, B. (2007). Quantitative Mass Spectrometry in Proteomics: A Critical Review. Anal Bioanal Chem, 389(4), 1017-1031.

Bezstarosti, K., Ghamari, A., Grosveld, F. G. & Demmers, J. A. (2010). Differential Proteomics Based on 18o Labeling to Determine the Cyclin Dependent Kinase 9 Interactome. J Proteome Res, 9(9), 4464-4475.

Blagoev, B., Kratchmarova, I., Ong, S. E., Nielsen, M., Foster, L. J. & Mann, M. (2003). A Proteomics Strategy to Elucidate Functional Protein-Protein Interactions Applied to Egf Signaling. Nat Biotechnol, 21(3), 315-318.

Blagoev, B. & Mann, M. (2006). Quantitative Proteomics to Study Mitogen-Activated Protein Kinases. Methods, 40(3), 243-250.

Blonder, J., Hale, M. L., Chan, K. C., Yu, L. R., Lucas, D. A., Conrads, T. P., Zhou, M., Popoff, M. R., Issaq, H. J., Stiles, B. G. & Veenstra, T. D. (2005). Quantitative Profiling of the Detergent-Resistant Membrane Proteome of Iota-B Toxin Induced Vero Cells. J Proteome Res, 4(2), 523-531.

Boersema, P. J., Aye, T. T., van Veen, T. A., Heck, A. J. & Mohammed, S. (2008). Triplex Protein Quantification Based on Stable Isotope Labeling by Peptide Dimethylation Applied to Cell and Tissue Lysates. Proteomics, 8(22), 4624-4632.

Boersema, P. J., Foong, L. Y., Ding, V. M., Lemeer, S., van Breukelen, B., Philp, R., Boekhorst, J., Snel, B., den Hertog, J., Choo, A. B. & Heck, A. J. (2010). In-Depth Qualitative and Quantitative Profiling of Tyrosine Phosphorylation Using a Combination of Phosphopeptide Immunoaffinity Purification and Stable Isotope Dimethyl Labeling. Mol Cell Proteomics, 9(1), 84-99.

Boersema, P. J., Raijmakers, R., Lemeer, S., Mohammed, S. & Heck, A. J. (2009). Multiplex Peptide Stable Isotope Dimethyl Labeling for Quantitative Proteomics. Nat Protoc, 4(4), 484-494.

Bonenfant, D., Schmelzle, T., Jacinto, E., Crespo, J. L., Mini, T., Hall, M. N. & Jenoe, P. (2003). Quantitation of Changes in Protein Phosphorylation: A Simple Method Based on Stable Isotope Labeling and Mass Spectrometry. Proc Natl Acad Sci U S A, 100(3), 880-885.

Brown, K. J. & Fenselau, C. (2004). Investigation of Doxorubicin Resistance in Mcf-7 Breast Cancer Cells Using Shot-Gun Comparative Proteomics with Proteolytic 18o Labeling. J Proteome Res, 3(3), 455-462.

Chelius, D. & Bondarenko, P. V. (2002). Quantitative Profiling of Proteins in Complex Mixtures Using Liquid Chromatography and Mass Spectrometry. J Proteome Res, 1(4), 317-323.

Chen, X., Cushman, S. W., Pannell, L. K. & Hess, S. (2005). Quantitative Proteomic Analysis of the Secretory Proteins from Rat Adipose Cells Using a 2d Liquid Chromatography-Ms/Ms Approach. J Proteome Res, 4(2), 570-577.

Choi, H., Fermin, D. & Nesvizhskii, A. I. (2008). Significance Analysis of Spectral Count Data in Label-Free Shotgun Proteomics. Mol Cell Proteomics, 7(12), 2373-2385.

Conrads, T. P., Alving, K., Veenstra, T. D., Belov, M. E., Anderson, G. A., Anderson, D. J., Lipton, M. S., Pasa-Tolic, L., Udseth, H. R., Chrisler, W. B., Thrall, B. D. & Smith, R. D. (2001). Quantitative Analysis of Bacterial and Mammalian Proteomes Using a Combination of Cysteine Affinity Tags and 15n-Metabolic Labeling. Anal Chem, 73(9), 2132-2139.

de Boer, E., Rodriguez, P., Bonte, E., Krijgsveld, J., Katsantoni, E., Heck, A., Grosveld, F. & Strouboulis, J. (2003). Efficient Biotinylation and Single-Step Purification of Tagged Transcription Factors in Mammalian Cells and Transgenic Mice. Proc Natl Acad Sci U S A, 100(13), 7480-7485.

de Godoy, L. M., Olsen, J. V., de Souza, G. A., Li, G., Mortensen, P. & Mann, M. (2006). Status of Complete Proteome Analysis by Mass Spectrometry: Silac Labeled Yeast as a Model System. Genome Biol, 7(6), R50.

Engelsberger, W. R., Erban, A., Kopka, J. & Schulze, W. X. (2006). Metabolic Labeling of Plant Cell Cultures with K(15)No3 as a Tool for Quantitative Analysis of Proteins and Metabolites. Plant Methods, 2, 14.

Fenselau, C. & Yao, X. (2007). Proteolytic Labeling with 18o for Comparative Proteomics Studies: Preparation of 18o-Labeled Peptides and the 18o/16o Peptide Mixture. Methods Mol Biol, 359, 135-142.

Gevaert, K., Staes, A., Van Damme, J., De Groot, S., Hugelier, K., Demol, H., Martens, L., Goethals, M. & Vandekerckhove, J. (2005). Global Phosphoproteome Analysis on Human Hepg2 Hepatocytes Using Reversed-Phase Diagonal Lc. Proteomics, 5(14), 3589-3599.

Gruhler, A., Schulze, W. X., Matthiesen, R., Mann, M. & Jensen, O. N. (2005). Stable Isotope Labeling of Arabidopsis Thaliana Cells and Quantitative Proteomics by Mass Spectrometry. Mol Cell Proteomics, 4(11), 1697-1709.

Gygi, S. P., Rist, B., Gerber, S. A., Turecek, F., Gelb, M. H. & Aebersold, R. (1999). Quantitative Analysis of Complex Protein Mixtures Using Isotope-Coded Affinity Tags. Nat Biotechnol, 17(10), 994-999.

Hajkova, D., Rao, K. C. & Miyagi, M. (2006). Ph Dependency of the Carboxyl Oxygen Exchange Reaction Catalyzed by Lysyl Endopeptidase and Trypsin. J Proteome Res, 5(7), 1667-1673.

Hansen, K. C., Schmitt-Ulms, G., Chalkley, R. J., Hirsch, J., Baldwin, M. A. & Burlingame, A. L. (2003). Mass Spectrometric Analysis of Protein Mixtures at Low Levels Using Cleavable 13c-Isotope-Coded Affinity Tag and Multidimensional Chromatography. Mol Cell Proteomics, 2(5), 299-314.

Hood, B. L., Darfler, M. M., Guiel, T. G., Furusato, B., Lucas, D. A., Ringeisen, B. R., Sesterhenn, I. A., Conrads, T. P., Veenstra, T. D. & Krizman, D. B. (2005). Proteomic Analysis of Formalin-Fixed Prostate Cancer Tissue. Mol Cell Proteomics, 4(11), 1741-1753.

Hood, B. L., Lucas, D. A., Kim, G., Chan, K. C., Blonder, J., Issaq, H. J., Veenstra, T. D., Conrads, T. P., Pollet, I. & Karsan, A. (2005). Quantitative Analysis of the Low

Molecular Weight Serum Proteome Using 18o Stable Isotope Labeling in a Lung Tumor Xenograft Mouse Model. J Am Soc Mass Spectrom, 16(8), 1221-1230.

Hsu, J. L., Huang, S. Y., Chow, N. H. & Chen, S. H. (2003). Stable-Isotope Dimethyl Labeling for Quantitative Proteomics. Anal Chem, 75(24), 6843-6852.

Ibarrola, N., Kalume, D. E., Gronborg, M., Iwahori, A. & Pandey, A. (2003). A Proteomic Approach for Quantitation of Phosphorylation Using Stable Isotope Labeling in Cell Culture. Anal Chem, 75(22), 6043-6049.

Ippel, J. H., Pouvreau, L., Kroef, T., Gruppen, H., Versteeg, G., van den Putten, P., Struik, P. C. & van Mierlo, C. P. (2004). In Vivo Uniform (15)N-Isotope Labelling of Plants: Using the Greenhouse for Structural Proteomics. Proteomics, 4(1), 226-234.

Krijgsveld, J., Ketting, R. F., Mahmoudi, T., Johansen, J., Artal-Sanz, M., Verrijzer, C. P., Plasterk, R. H. & Heck, A. J. (2003). Metabolic Labeling of C. Elegans and D. Melanogaster for Quantitative Proteomics. Nat Biotechnol, 21(8), 927-931.

Kruger, M., Moser, M., Ussar, S., Thievessen, I., Luber, C. A., Forner, F., Schmidt, S., Zanivan, S., Fassler, R. & Mann, M. (2008). Silac Mouse for Quantitative Proteomics Uncovers Kindlin-3 as an Essential Factor for Red Blood Cell Function. Cell, 134(2), 353-364.

Lanquar, V., Kuhn, L., Lelievre, F., Khafif, M., Espagne, C., Bruley, C., Barbier-Brygoo, H., Garin, J. & Thomine, S. (2007). 15n-Metabolic Labeling for Comparative Plasma Membrane Proteomics in Arabidopsis Cells. Proteomics, 7(5), 750-754.

Li, J., Steen, H. & Gygi, S. P. (2003). Protein Profiling with Cleavable Isotope-Coded Affinity Tag (Cicat) Reagents: The Yeast Salinity Stress Response. Mol Cell Proteomics, 2(11), 1198-1204.

Listgarten, J. & Emili, A. (2005). Statistical and Computational Methods for Comparative Proteomic Profiling Using Liquid Chromatography-Tandem Mass Spectrometry. Mol Cell Proteomics, 4(4), 419-434.

Liu, H., Sadygov, R. G. & Yates, J. R., 3rd. (2004). A Model for Random Sampling and Estimation of Relative Protein Abundance in Shotgun Proteomics. Anal Chem, 76(14), 4193-4201.

Lundgren, D. H., Hwang, S. I., Wu, L. & Han, D. K. (2010). Role of Spectral Counting in Quantitative Proteomics. Expert Rev Proteomics, 7(1), 39-53.

Matthiesen, R., Azevedo, L., Amorim, A. & Carvalho, A. S. (2011). Discussion on Common Data Analysis Strategies Used in Ms-Based Proteomics. Proteomics, 11(4), 604-619.

McClatchy, D. B., Dong, M. Q., Wu, C. C., Venable, J. D. & Yates, J. R., 3rd. (2007). 15n Metabolic Labeling of Mammalian Tissue with Slow Protein Turnover. J Proteome Res, 6(5), 2005-2010.

Miyagi, M. & Rao, K. C. (2007). Proteolytic 18o-Labeling Strategies for Quantitative Proteomics. Mass Spectrom Rev, 26(1), 121-136.

Mortensen, P., Gouw, J. W., Olsen, J. V., Ong, S. E., Rigbolt, K. T., Bunkenborg, J., Cox, J., Foster, L. J., Heck, A. J., Blagoev, B., Andersen, J. S. & Mann, M. (2010). Msquant, an Open Source Platform for Mass Spectrometry-Based Quantitative Proteomics. J Proteome Res, 9(1), 393-403.

Mueller, L. N., Brusniak, M. Y., Mani, D. R. & Aebersold, R. (2008). An Assessment of Software Solutions for the Analysis of Mass Spectrometry Based Quantitative Proteomics Data. J Proteome Res, 7(1), 51-61.

Neilson, K. A., Ali, N. A., Muralidharan, S., Mirzaei, M., Mariani, M., Assadourian, G., Lee, A., van Sluyter, S. C. & Haynes, P. A. (2011). Less Label, More Free: Approaches in Label-Free Quantitative Mass Spectrometry. Proteomics, 11(4), 535-553.

Oda, Y., Huang, K., Cross, F. R., Cowburn, D. & Chait, B. T. (1999). Accurate Quantitation of Protein Expression and Site-Specific Phosphorylation. Proc Natl Acad Sci U S A, 96(12), 6591-6596.

Old, W. M., Meyer-Arendt, K., Aveline-Wolf, L., Pierce, K. G., Mendoza, A., Sevinsky, J. R., Resing, K. A. & Ahn, N. G. (2005). Comparison of Label-Free Methods for Quantifying Human Proteins by Shotgun Proteomics. Mol Cell Proteomics, 4(10), 1487-1502.

Ong, S. E., Blagoev, B., Kratchmarova, I., Kristensen, D. B., Steen, H., Pandey, A. & Mann, M. (2002). Stable Isotope Labeling by Amino Acids in Cell Culture, Silac, as a Simple and Accurate Approach to Expression Proteomics. Mol Cell Proteomics, 1(5), 376-386.

Ong, S. E., Kratchmarova, I. & Mann, M. (2003). Properties of 13c-Substituted Arginine in Stable Isotope Labeling by Amino Acids in Cell Culture (Silac). J Proteome Res, 2(2), 173-181.

Ong, S. E. & Mann, M. (2005). Mass Spectrometry-Based Proteomics Turns Quantitative. Nat Chem Biol, 1(5), 252-262.

Park, S. K., Liao, L., Kim, J. Y. & Yates, J. R., 3rd. (2009). A Computational Approach to Correct Arginine-to-Proline Conversion in Quantitative Proteomics. Nat Methods, 6(3), 184-185.

Pierce, A., Unwin, R. D., Evans, C. A., Griffiths, S., Carney, L., Zhang, L., Jaworska, E., Lee, C. F., Blinco, D., Okoniewski, M. J., Miller, C. J., Bitton, D. A., Spooncer, E. & Whetton, A. D. (2008). Eight-Channel Itraq Enables Comparison of the Activity of Six Leukemogenic Tyrosine Kinases. Mol Cell Proteomics, 7(5), 853-863.

Qian, W. J., Monroe, M. E., Liu, T., Jacobs, J. M., Anderson, G. A., Shen, Y., Moore, R. J., Anderson, D. J., Zhang, R., Calvano, S. E., Lowry, S. F., Xiao, W., Moldawer, L. L., Davis, R. W., Tompkins, R. G., Camp, D. G., 2nd & Smith, R. D. (2005). Quantitative Proteome Analysis of Human Plasma Following in Vivo Lipopolysaccharide Administration Using 16o/18o Labeling and the Accurate Mass and Time Tag Approach. Mol Cell Proteomics, 4(5), 700-709.

Raijmakers, R., Berkers, C. R., de Jong, A., Ovaa, H., Heck, A. J. & Mohammed, S. (2008). Automated Online Sequential Isotope Labeling for Protein Quantitation Applied to Proteasome Tissue-Specific Diversity. Mol Cell Proteomics, 7(9), 1755-1762.

Ramos-Fernandez, A., Lopez-Ferrer, D. & Vazquez, J. (2007). Improved Method for Differential Expression Proteomics Using Trypsin-Catalyzed 18o Labeling with a Correction for Labeling Efficiency. Mol Cell Proteomics, 6(7), 1274-1286.

Rao, K. C., Palamalai, V., Dunlevy, J. R. & Miyagi, M. (2005). Peptidyl-Lys Metalloendopeptidase-Catalyzed 18o Labeling for Comparative Proteomics: Application to Cytokine/Lipolysaccharide-Treated Human Retinal Pigment Epithelium Cell Line. Mol Cell Proteomics, 4(10), 1550-1557.

Rappsilber, J., Ryder, U., Lamond, A. I. & Mann, M. (2002). Large-Scale Proteomic Analysis of the Human Spliceosome. Genome Res, 12(8), 1231-1245.

Ross, P. L., Huang, Y. N., Marchese, J. N., Williamson, B., Parker, K., Hattan, S., Khainovski, N., Pillai, S., Dey, S., Daniels, S., Purkayastha, S., Juhasz, P., Martin, S., Bartlet-

Jones, M., He, F., Jacobson, A. & Pappin, D. J. (2004). Multiplexed Protein Quantitation in Saccharomyces Cerevisiae Using Amine-Reactive Isobaric Tagging Reagents. Mol Cell Proteomics, 3(12), 1154-1169.

Sevinsky, J. R., Brown, K. J., Cargile, B. J., Bundy, J. L. & Stephenson, J. L., Jr. (2007). Minimizing Back Exchange in 18o/16o Quantitative Proteomics Experiments by Incorporation of Immobilized Trypsin into the Initial Digestion Step. Anal Chem, 79(5), 2158-2162.

Silva, J. C., Denny, R., Dorschel, C., Gorenstein, M. V., Li, G. Z., Richardson, K., Wall, D. & Geromanos, S. J. (2006). Simultaneous Qualitative and Quantitative Analysis of the Escherichia Coli Proteome: A Sweet Tale. Mol Cell Proteomics, 5(4), 589-607.

Silva, J. C., Denny, R., Dorschel, C. A., Gorenstein, M., Kass, I. J., Li, G. Z., McKenna, T., Nold, M. J., Richardson, K., Young, P. & Geromanos, S. (2005). Quantitative Proteomic Analysis by Accurate Mass Retention Time Pairs. Anal Chem, 77(7), 2187-2200.

Staes, A., Demol, H., Van Damme, J., Martens, L., Vandekerckhove, J. & Gevaert, K. (2004). Global Differential Non-Gel Proteomics by Quantitative and Stable Labeling of Tryptic Peptides with Oxygen-18. J Proteome Res, 3(4), 786-791.

Sury, M. D., Chen, J. X. & Selbach, M. (2010). The Silac Fly Allows for Accurate Protein Quantification in Vivo. Mol Cell Proteomics, 9(10), 2173-2183.

Synowsky, S. A., van Wijk, M., Raijmakers, R. & Heck, A. J. (2009). Comparative Multiplexed Mass Spectrometric Analyses of Endogenously Expressed Yeast Nuclear and Cytoplasmic Exosomes. J Mol Biol, 385(4), 1300-1313.

Thompson, A., Schafer, J., Kuhn, K., Kienle, S., Schwarz, J., Schmidt, G., Neumann, T., Johnstone, R., Mohammed, A. K. & Hamon, C. (2003). Tandem Mass Tags: A Novel Quantification Strategy for Comparative Analysis of Complex Protein Mixtures by Ms/Ms. Anal Chem, 75(8), 1895-1904.

Unlu, M., Morgan, M. E. & Minden, J. S. (1997). Difference Gel Electrophoresis: A Single Gel Method for Detecting Changes in Protein Extracts. Electrophoresis, 18(11), 2071-2077.

Van Hoof, D., Pinkse, M. W., Oostwaard, D. W., Mummery, C. L., Heck, A. J. & Krijgsveld, J. (2007). An Experimental Correction for Arginine-to-Proline Conversion Artifacts in Silac-Based Quantitative Proteomics. Nat Methods, 4(9), 677-678.

Venable, J. D., Dong, M. Q., Wohlschlegel, J., Dillin, A. & Yates, J. R. (2004). Automated Approach for Quantitative Analysis of Complex Peptide Mixtures from Tandem Mass Spectra. Nat Methods, 1(1), 39-45.

Wang, N. & Li, L. (2008). Exploring the Precursor Ion Exclusion Feature of Liquid Chromatography-Electrospray Ionization Quadrupole Time-of-Flight Mass Spectrometry for Improving Protein Identification in Shotgun Proteome Analysis. Anal Chem, 80(12), 4696-4710.

Wiener, M. C., Sachs, J. R., Deyanova, E. G. & Yates, N. A. (2004). Differential Mass Spectrometry: A Label-Free Lc-Ms Method for Finding Significant Differences in Complex Peptide and Protein Mixtures. Anal Chem, 76(20), 6085-6096.

Wu, C. C., MacCoss, M. J., Howell, K. E., Matthews, D. E. & Yates, J. R., 3rd. (2004). Metabolic Labeling of Mammalian Organisms with Stable Isotopes for Quantitative Proteomic Analysis. Anal Chem, 76(17), 4951-4959.

Yao, X., Freas, A., Ramirez, J., Demirev, P. A. & Fenselau, C. (2001). Proteolytic 18o Labeling for Comparative Proteomics: Model Studies with Two Serotypes of Adenovirus. Anal Chem, 73(13), 2836-2842.

Zang, L., Palmer Toy, D., Hancock, W. S., Sgroi, D. C. & Karger, B. L. (2004). Proteomic Analysis of Ductal Carcinoma of the Breast Using Laser Capture Microdissection, Lc-Ms, and 16o/18o Isotopic Labeling. J Proteome Res, 3(3), 604-612.

Zhang, R. & Regnier, F. E. (2002). Minimizing Resolution of Isotopically Coded Peptides in Comparative Proteomics. J Proteome Res, 1(2), 139-147.

Zhang, R., Sioma, C. S., Wang, S. & Regnier, F. E. (2001). Fractionation of Isotopically Labeled Peptides in Quantitative Proteomics. Anal Chem, 73(21), 5142-5149.

Part 3

2D Gel Electrophoresis and Databases

7

Preparation of Protein Samples for 2-DE from Different Cotton Tissues

Chengjian Xie[1], Xiaowen Wang[2], Anping Sui[2*] and Xingyong Yang[1,2*]
[1]Chongqing Normal University, Chongqing
[2]Southwest University, Chongqing
China

1. Introduction

Cotton today is one of the most important economic crops. Cotton fiber is the most used material for the textile industry, and takes an important strategic status in the world economy. Proteomics is one of the most important techniques in the post-genome era, and two-dimensional electrophoresis (2-DE) is one key technology for proteomics. Protein extraction and sample preparation are of prime importance for optimal 2-DE results (Isaacson et al., 2006). The cotton is a highly recalcitrant plant material and rich in compounds such as polysaccharides, polyphenols, nucleic acids, cellulose, and other secondary metabolites, which interfere with protein extraction, produce highly diluted protein extracts, and affect protein migration in 2-DE (Görg et al., 2000). Moreover, modified or different protein extraction method should be used for different cotton tissues (e.g., leaves, roots, seeds, and stems), which contain different secondary metabolites.
Here, several protocols to extract total proteins for different cotton tissues are described based on methods routinely used in our laboratory.

2. Materials

Ultrapure water (doubly distilled, deionized, > 18 MΩ) is used for all reagent preparation. Reagent grades should be of the highest quality.

1. 1 M Tris-saturated phenol (pH 8.0)
2. Extraction buffer: 0.1 M Tris-HCl, pH 8.0, containing 30% w/v sucrose, 2% w/v SDS, 1 mM phenylmethanesulfonyl fluoride (PMSF), 2% v/v thioglycol 2-mercapitoethanol (2-ME).
3. Lysis buffer: 7 M urea, 2 M thiourea, 4% w/v CHAPS, 65 mM DTT, and 0.5% v/v carrier ampholytes.
4. Equilibration buffer: 6 M urea, 20% w/v glycerol, 2% w/v SDS, and 50 mM Tris-HCl, pH 8.8.
5. Staining solution: 0.12% w/v Coomassie brilliant blue G-250, 10% w/v ammonium sulfate, 10% w/v phosphoric acid, 20% v/v methanol.

---

* Corresponding Authors

3. Methods

All plant tissue samples should be ground to fine powder with a pre-chilled mortar and pestle in liquid nitrogen. Before grinding, silicon dioxide (SiO_2) and polyvinylpolypyrrolidone (PVPP, 10% w/w of sample weight) were added into mortar. The finely ground powder (ca. 0.5 g per tube) was immediately transferred into a 10-mL centrifuge tube **(see Note 1)** precooled in liquid nitrogen. The powder sample can be immediately used or stored in a -80 °C freezer until protein extraction.

3.1 Extraction from leaves

- The powder sample was resuspended in 4 mL 10% v/v trichloroacetic acid (TCA) in acetone **(see Note 2)** and extensively homogenized.
- Centrifuge (12, 000*g* at 4 °C) for 5 min, the pellet of proteins was washed once with 5 mL 0.1 M ammonium acetate in 80% v/v methanol and once with cold 80% v/v acetone.
- The pellet was dried in vacuum **(see Note 3)**.
- It was resuspended in 3 mL of extraction buffer.
- An equal volume of 1 M Tris-saturated phenol (pH 8.0) was added, and then the mixture was homogenized on ice for 5 min. The upper phenol phase was collected after centrifuge (12, 000*g* at 4 °C) for 5 min, and this extraction step was repeated once **(see Note 4)**.
- The total phenol phase was transferred into a new tube, and an equal volume of extraction buffer was added into the total phenol phase, and then the mixture was homogenized on ice for 5 min. The upper phenol phase was collected after centrifuge (12, 000*g* at 4 °C) for 5 min **(see Note 4)**.
- The proteins were precipitated with five volumes of 0.1 M ammonium acetate in methanol overnight at -20 °C.
- Centrifuge (12, 000*g* at 4 °C) for 10 min. The collected protein pellets were washed once with 3 mL methanol, and then washed once with 3 mL cold 80% v/v acetone in water. The pellets were dried in a freeze vacuum dryer for 10 min and stored at -80 °C.

3.2 Extraction from roots (see Note 5)

- The powder sample was resuspended in 3 mL of extraction buffer.
- An equal volume of 1 M Tris-saturated phenol (pH 8.0) was added, and then the mixture was homogenized on ice for 5 min. The upper phenol phase was collected after centrifuge (12, 000*g* at 4°C) for 5 min, and this extraction step was repeated once **(see Note 4)**.
- The total phenol phase was transferred into a new tube, and an equal volume of extraction buffer was added into it. The mixture was homogenized on ice for 5 min, and the upper phenol phase was collected after centrifuge (12, 000*g* at 4°C) for 5 min **(see Note 4)**.
- Precipitation the phenol phase with 5 volumes of 0.1 M ammonium acetate in methanol overnight at -20°C.
- Centrifuge (12, 000 g at 4°C) for 10 min. The protein pellets were washed once with 3 mL methanol, and then washed once with 3 mL 80% v/v acetone in water. The pellets were dried in a freeze vacuum dryer for 10 min and stored at -80 °C.

3.3 Extraction from fibers (see Note 6)

- The ground powder was resuspended in 4 mL acetone and extensively homogenized. The sample was keeped at -20 °C overnight.
- Centrifuge (8, 000*g* at 4 °C) for 5 min **(see Note 6)**. Discard the supernatant and the pellets were washed once with 4 mL acetone containing 2% v/v 2-ME.
- Centrifuge (10, 000 g at 4°C) for 5 min, and the wash was repeated once.
- Discard the supernatant and the pellet was dried in vacuum.
- Resuspend the pellet with 4 mL of extraction buffer.
- An equal volume of 1 M Tris-saturated phenol (pH 8.0) was added and homogenized on ice for 5 min. Centrifuge (12, 000*g* at 4 °C) for 5 min.
- The upper phenol phase was collected. The phenol extraction procedure was repeated once.
- The collected phenol phase was precipitated with 5 volumes of 0.1 M ammonium acetate in methanol overnight at -20 °C.
- Centrifuge (10, 000*g* at 4 °C) for 10 min. Discard the supernatant and the pellet was washed twice with cold 0.1 M ammonium acetate in methanol. Wash twice with cold 80% acetone in water. The pellet was dried in a freeze vacuum dryer and stored at -80 °C.

3.4 Proteins pellet resuspension

Proteins pellet was resuspended in lysis buffer and shaked for 1 h (IKA Vortex Genius 3, Staufen, Germany). After centrifugation at 15, 000*g* for 20 min to remove debris, the supernatant could be used immediately for first-dimensional IEF gels. Protein concentration was determined using the Bradford method (Bradford, 1976) with bovine serum albumin as a standard.

3.5 Two-dimensional gel electrophoresis

The 2-D gel electrophoresis (2-DE) protocol was adapted by O'Farrell (1975). The first electrophoresis was performed using immobilized pH gradient (IPG) strips on an IPGphor isoelectric focusing (IEF) system (Amersham Pharmacia, San Francisco, CA). For example in our experiment, the IPG strips (13 cm, 3–10 nonlinear pH gradient; GE Healthcare, Piscataway, NJ) were rehydrated with 250 µl of rehydration buffer (containing 370 µg proteins). Focusing was then performed at 20 °C as follows: active rehydration at 30 V for 12 h, 200 V for 2 h, 500 V for 3 h, 1,000 V for 4 h, 8,000 V for 5 h, with a gradient increase in voltage between 8,000 V and 40,000 V. After IEF, the proteins in the strips were reduced with 1% w/v DTT in 10 ml of equilibration buffer for 15 min and alkylated with 2.5% w/v iodoacetamide in 10 ml of equilibration buffer for 15 min. The strips were transferred onto vertical 10.5% w/v SDS-PAGE selfcast gels. The second electrophoresis (SDS-PAGE) was performed on an Amersham Hoefer SE 600 system (Amersham Pharmacia) at 10 mA for 1 h and 20 mA for 6 h at 15 °C.

3.6 Protein visualization

The 2-DE gel was stained with blue silver (Candiano et al., 1975). The gel was fixed in a solution of 40% v/v methanol and 10% v/v acetic acid for 30 min, washed in distilled water 4 times for 15 min, and finally incubated in a staining solution **(see Note 7)** overnight with gentle shaking. The gel was decolorized in distilled water.

As shown in Figure 1, a great many of protein spots were detected on the 2-DE image. The crude protein yield and the number of protein on 2-DE gels are also summarized in Table 1.

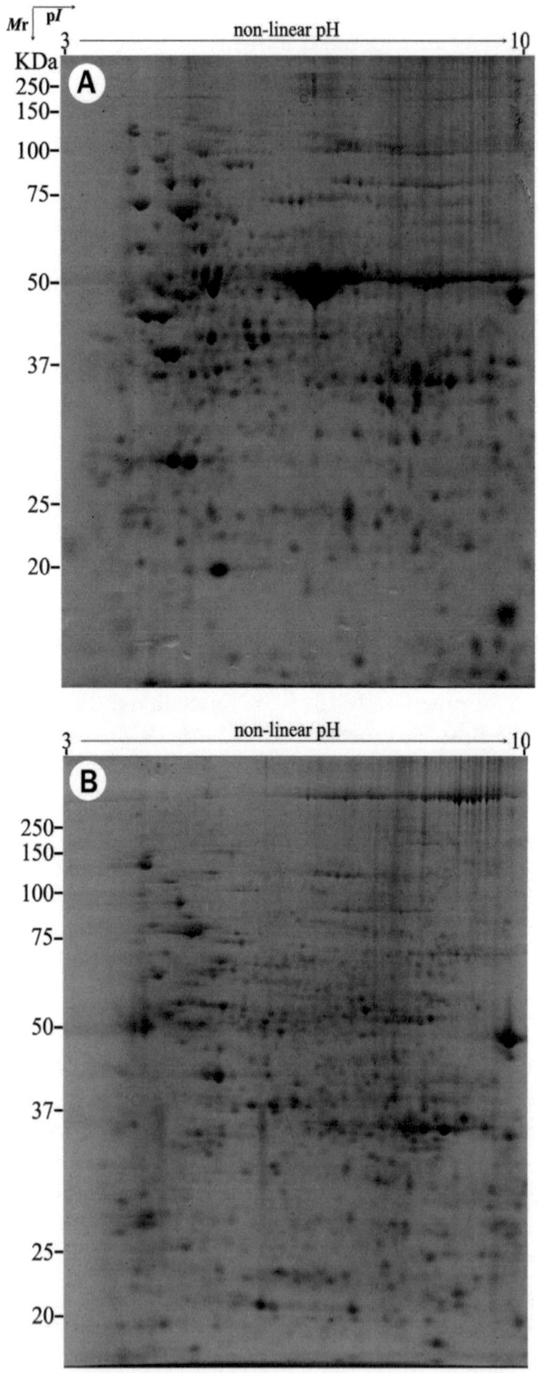

Fig. 1. Two-dimensional gel electrophoresis gel of proteins extracted from cotton leaves (A) and roots (B). Proteins (370 µg) were separated on a 13-cm pH 3–10 nonlinear gradient immobilized pH gradient strip and on 10.5% SDS-PAGE gel. The gels were stained using the blue-silver method. Mr, molecular mass; pI, isoelectric point.

Tissues	Protein yield (mg/g; crude protein/powdered tissue)	The number of protein on 2-DE gels
Leaves	8.7	About 900
Roots	5.1	About 830
Fibers	4.6	About 850

Table 1. The crude protein yield and the number of protein on 2-DE gels for different extraction methods. Proteins (370 μg) were separation on 13cm pH 3-10 non-linear gradient IPG strip and 10% SDS-PAGE gel. The gel was stained using Blue silver [4]. The above results are from different experiments and represent initial results. It can provide preliminary reference for selecting extraction method.

These results demonstrate that above protein extraction methods could be compatible with cotton different tissues.

4. Notes

1. Small plant samples can yield a sufficient amount of protein for 2-DE. Furthermore, using small plant samples easily extract the high purity of protein.
2. These mentioned solutions except for lysis buffer were stored at 4 °C and cooled on ice in advance for use.
3. A good principle for drying sample is that the edge of protein pellets turned white, and too dry samples is bad for resuspension.
4. To reduce loss of protein, this phenol extraction step was repeated once. However, some water soluble organic (such as polysaccharides and nucleic acids) was involved in collected total phenol phase and resulted in horizontal streaking visible on two dimensional gels. Once more wash step with an equal volume of extraction buffer was performed to remove water soluble organic in phenol phase.
5. TCA-acetone precipitation can effectively remove pigments (Xie et al., 1975). There are not too many pigments such as chlorophyll in cotton roots, so we could get the high purity protein samples if step 1-3 of "**Extraction from leaves**" was removed.
6. The method described here was adapted from published paper with minor modifications. Centrifugation at lower speeds (step 2) is beneficial to resuspend protein pellets. It, sometimes, is the necessary to use auxiliary tool for scraping pellets. Using TCA-acetone instead of acetone (step 1) is helpful for pellets' resuspension, although TCA-acetone precipitation could increase the loss of protein (Görg et al., 2004).
7. To prepare blue silver staining solution, phosphoric acid and distilled water (10% of the final volume) firstly were mixed, and then ammonium sulfate fine powder was added and completely dissolved on magnetic stirrer. Next, 0.12% (w/v) coomassie brilliant blue G-250 was added and stirred at least 2 hour. Fill to 80% final volume with distilled water, sequentially add 20% volume methanol and mix thoughly. These undissolved particles in staining solution were not needed to filter out and they can increase sensitivity of staining solution.

5. Acknowledgments

"Extraction from leaves" was adapted from Wang et al. method with modifications (Wang et al., 2003, 2006) and "Extraction from fibers" was based on the method previously described by Yao et al. (2006), with slight modifications. Furthermore, this work was supported by the National Natural Science Foundation of China (Grant no. 30771388), the Natural Science Foundation Project of CQ CSTC (Grant no. 2009BB1123) and the Fundamental Research Funds for the Central Universities (Grant no. XDJK2011C016).

6. References

Bradford, M.M. (1976) A rapid and sensitive method for the quantitation of microgram quantities of protein utilizing the principle of protein-dye binding. *Analytical Biochemistry*, Vol. 72, No. 1-2, (May 7), pp. 248-254, ISSN 0003-2697.

Candiano, G.; Bruschi, M.; Musante, L.; Santucci, L.; Ghiggeri, G.M.; Carnemolla, B.; Orecchia, P.; Zardi, L.; Righetti, P.G. (2004) Blue silver: A very sensitive colloidal Coomassie G-250 staining for proteome analysis. *Electrophoresis*, Vol.25, No.9, (May 15), pp.1327-1333, ISSN 0173-0835.

Görg, A,; Obermaier, C.; Boguth, G.; Harder, A.; Scheibe, B.; Wildgruber, R.; Weiss, W. (2000). The current state of two-dimensional electrophoresis with immobilized pH gradients. *Electrophoresis*, Vol. 21, No.6, (April 1), pp. 1037-1053, ISSN 0173-0835.

Görg, A.; Weiss, W.; Dunn, MJ. (2004) Current two-dimensional electrophoresis technology for proteomics. *Proteomics*, Vol. 4, No. 12, (December 20), pp. 3665-3685, ISSN 1615-9853.

Isaacson, T.; Damasceno, C.M.B.; Saravanan, R.S.; He, Y.; Catala, C.; Saladie, M.; Rose, J.K.C. (2006). Sample extraction techniques for enhanced proteomic analysis of plant tissues. *Nature Protocols*, Vol.1, No.2, (July 13), pp. 769-774, ISSN 1754-2189

O'Farrell, P.H. (1975) High resolution two-dimensional electrophoresis of proteins. *Journal of Biology Chemistry*, Vol.250, No.10, (May 25), pp. 4007-4021, ISSN 0021-9258

Wang, W.; Scali, M.; Vignani, R.; Spadafora, A.; Sensi, E.; Mazzuca, S.; Cresti, M. (2003) Protein extraction for two-dimensional electrophoresis from olive leaf, a plant tissue containing high levels of interfering compounds. *Electrophoresis*, Vol. 24, No.14, (July 15), pp. 2369-2375, ISSN 0173-0835.

Wang, W.; Vignani, R.; Scali, M.; Cresti, M. (2006) A universal and rapid protocol for protein extraction from recalcitrant plant tissues for proteomic analysis. *Electrophoresis*, Vol. 27, No.13, (July 1), pp. 2782-2786, ISSN 0173-0835.

Xie, C.; Wang, D.; Yang, X. (2009) Protein Extraction Methods Compatible with Proteomic Analysis for the Cotton Seedling. *Crop Science*, Vol. 49, No.2, (March 15), pp. 395-402, ISSN 1435-0653.

Yao, Y.; Yang, Y.; Liu, J. (2006) An efficient protein preparation for proteomic analysis of developing cotton fibers by 2-DE. *Electrophoresis*, Vol. 27, No. 22, (November 15), pp. 4559-4569, ISSN 0173-0835.

8

2D-PAGE Database for Studies on Energetic Metabolism of the Denitrifying Bacterium *Paracoccus denitrificans*

Pavel Bouchal[1,2], Robert Stein[3], Zbyněk Zdráhal[4], Peter R. Jungblut[5] and Igor Kučera[1]
[1]*Department of Biochemistry, Faculty of Science, Masaryk University, Brno*
[2]*Regional Centre for Applied Molecular Oncology, Masaryk Memorial Cancer Institute, Brno*
[3]*I&B Informatics and Biology, Berlin*
[4]*Core Facility – Proteomics, Central European Institute of Technology, Masaryk University, Brno*
[5]*Max Planck Institute for Infection Biology, Core Facility Protein Analysis, Berlin*
[1,2,4]*Czech Republic*
[3,5]*Germany*

1. Introduction

The gram-negative soil bacterium *Paracoccus denitrificans* is a chemoorganotroph and a facultative chemolithotroph, capable of using the oxidation of molecular hydrogen, methanol or thiosulphate as sole source of energy for autotrophic growth. Many different organic compounds serve as sole carbon source, the metabolism is, however, always respiratory and never fermentative. *P. denitrificans* synthesizes three distinct terminal oxidases (*aa_3*-type and *cbb_3*-type cytochrome *c* oxidases and *ba_3*-type quinol oxidase) during aerobic growth (Fig. 1). Under limited oxygen concentration, it can produce four additional terminal oxidoreductases for stepwise anaerobic conversion of nitrate to nitrogen gas (denitrification): nitrate reductase, nitrite reductase, nitrous oxide reductase and nitric oxide reductase (Fig. 2). Synthesis of these enzymes is tightly controlled at the transcription level: (i) globally according to an energetic hierarchy and (ii) on the level of the individual genes. As a result, a proper balance in the concentration and activity of these reductases is achieved and the cytotoxicity of the toxic intermediates of denitrification, nitrite and nitric oxide, is eliminated (Zumft 1997). The major players in the mentioned regulatory network are three members of the FNR (fumarate and nitrate reductase regulatory) protein family of transcription regulators. Upon activation by their corresponding signals, they bind to specific sites (FNR boxes) in target promoters upstream/downstream of the σ factor binding site and destabilize/stabilize the RNA-polymerase transcription initiation complex. The first regulatory protein is FnrP which has a [4Fe-4S] cluster for oxygen sensing, the second is

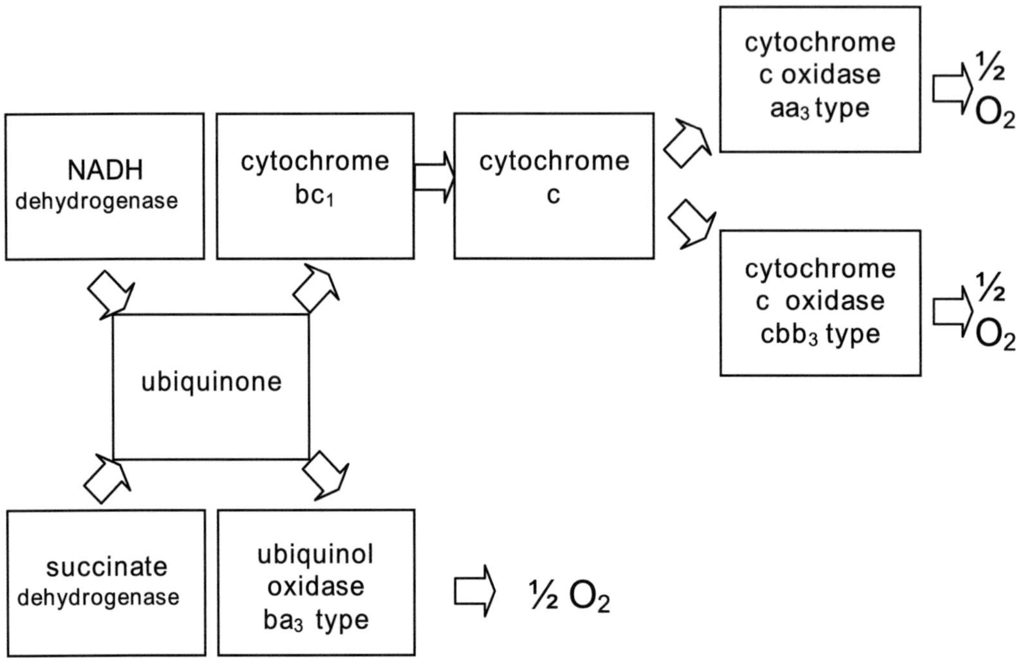

Fig. 1. Major components of *P. denitrificans* respiratory chain. Three distinct terminal oxidases are synthesized depending on environmental factors (*e.g.*, oxygen tension). The arrows indicate the electron flow. Inhibition of terminal oxidases with azide led to elevation of superoxide dismutase as revealed using a proteomic approach and confirmed at transcript and enzyme activity level (Bouchal *et al.* 2011).

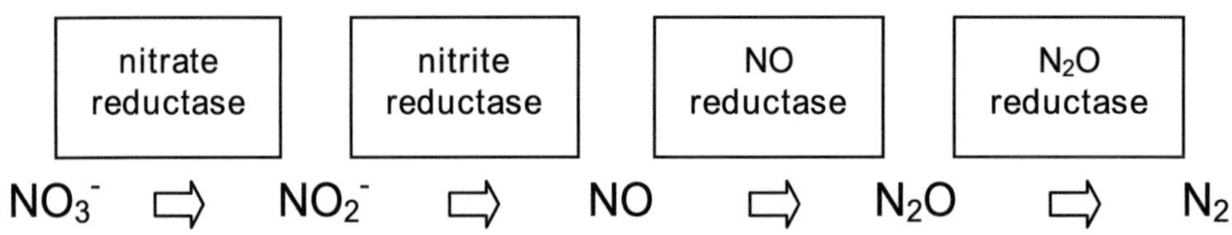

Fig. 2. Schema of *P. denitrificans* anaerobic denitrification pathway. Nitrate reductase β-subunit, nitrite reductase and nitrous oxide reductase can be detected and quantified using a proteomic approach (Bouchal *et al.* 2004; Bouchal *et al.* 2011). The expression of the denitrification enzymes is tightly controlled by FnrP, NNR and NarR transcription regulators at transcription level.

NNR, which has a heme for NO sensing and the third one is NarR which is poorly characterized and likely to be a nitrite sensor (Van Spanning *et al.* 1997; Wood *et al.* 2001). In response to oxygen deprivation, FnrP controls expression of the *nar* gene cluster encoding nitrate reductase, the *cco*-gene cluster encoding a *cbb*$_3$-type oxidase for respiration at low oxygen concentrations and the *ccp* gene encoding cytochrome *c* peroxidase. NNR specifically controls expression of the gene clusters encoding the nitrite (*nirS*), and nitric oxide (*norCB*) reductases and, to a certain extent, nitrous oxide (*nosZ*) reductase. NarR is required for transcription of the *nar* gene cluster in an unknown interplay with the FnrP protein (Wood *et al.* 2001; Veldman *et al.* 2006). These properties have been deduced from a number of studies on each of these transcriptional activators, but knowledge on the interplay between these regulators along with their position in the complete regulatory network are scarce.

Given its metabolic versatility, this bacterium becomes an excellent model system to study the mechanisms of cellular responses to different environments.

2. *P. denitrificans* proteome analysis: Method development

Our group gained extensive experience in studying *P. denitrificans* physiology, reaching from measurement of enzyme activities to characterization of the role of individual proteins, *e.g.*, pseudoazurin (Koutny & Kucera 1999), nitrate transporter (Kucera 2003) and ferric iron reductases (Mazoch *et al.* 2004; Sedlacek *et al.* 2009). New proteomic technologies have been employed in our research since 2001 when we used a two-dimensional gel electrophoresis (2D-PAGE) with carrier ampholytes for the first time. Using this technique, we obtained good separations of about 150 proteins present in membrane fraction, allowing a comparison of its protein composition with the periplasmic fraction (Bouchal & Kucera 2004), see also Fig. 3. However, the capacity of original sample preparation procedure (Bouchal & Kucera 2004) led to difficulties with identification of less intensive protein spots.

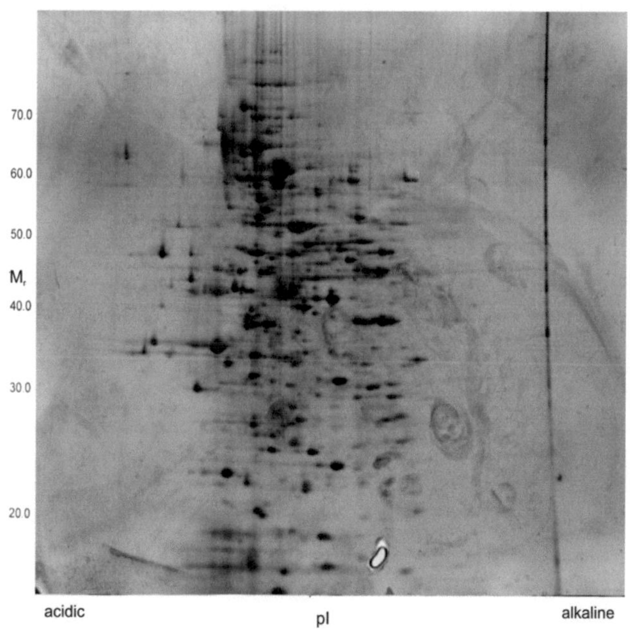

Fig. 3. A separation of *P. denitrificans* periplasmic proteins with 2D-PAGE based on carrier ampholytes–isoelectric focusing in the first dimension (Bouchal & Kucera 2002), with permission.

After establishing the mass spectrometry laboratory in 2002, we optimized new methods for the proteome analysis of *P. denitrificans* using immobilized pH gradients. See next paragraphs for complete protocols and Fig. 4 for a typical 2-D proteome map. Namely, the mass spectrometric analysis opened the way towards high-throughput and precise protein identification and valid conclusions made based on proteomics data. Matrix-Assisted Laser

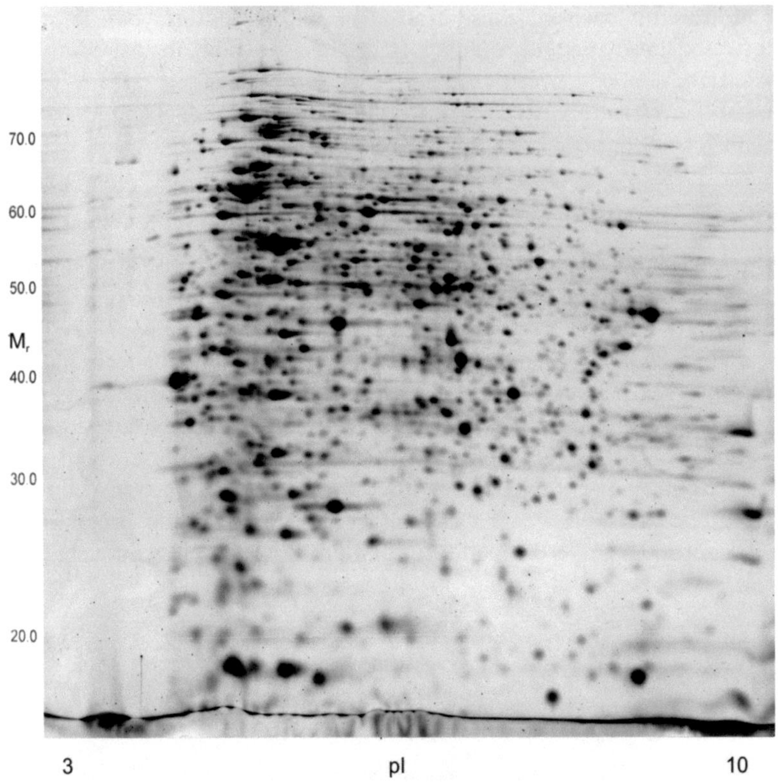

Fig. 4. Current stage of *P. denitrificans* proteomics approach: 2D-PAGE map (20 cm x 20 cm size) prepared with non-linear immobilized pH gradients (pH 3-10 NL) and Sypro Ruby staining has been used for both quantitative analysis and mass spectrometry protein identification.

Desorption-Ionization Mass Spectrometry (MALDI-MS) was used in initial proteomic studies identifying proteins exclusively by peptide mass fingerprinting. In addition, sensitivity of our MALDI-MS instrumentation of that time was not sufficient in case of weak protein spots. Since 2007, we have started to identify proteins using tandem mass spectrometric techniques, MALDI-MS/MS and ESI-MS/MS (concretely, capillary liquid chromatography - ion trap mass spectrometry with electrospray ionization), which resulted in more reliable protein identification based on MS/MS data. To improve the sensitivity of our LC-MS/MS system, we introduced nano-scale LC separation in 2008. At present, practically each analysis of protein spots leads to positive identification, involving sensitive fluorescent staining (Sypro Ruby). Subsequently, several comprehensive proteomic experiments were performed during the years using our proteomic platform in order to study the differences in protein composition caused by the growth on different terminal electron acceptors in both total cell lysates and membrane fractions (Bouchal *et al.* 2004). An additional large proteomic study with data confirmation at transcript level was performed to describe the regulons of three FNR-type transcription regulators FnrP, NNR and NarR at protein level (Bouchal *et al.* 2010). Quantitative and statistical image analysis primarily resulted in creation of local database files in a PDQUEST format. Subsequently, we decided to publish the 2-D maps in a web form to make all details accessible to other researchers online *via* the "Proteome Database System for Microbial Research at Max Planck Institute for Infection Biology" in Berlin, Germany as described below.

3. Optimized methods used for *P. denitrificans* proteomics

3.1 Bacteria and culture conditions

Four strains of *P. denitrificans* were used in the published studies: Pd1222 (wild type), Pd2921 (FnrP mutant (Van Spanning *et al.* 1997)), Pd7721 (NNR mutant (Van Spanning *et al.* 1995)) and Pd11021 (NarR mutant, unpublished data). These four strains were cultivated at 30 °C in 1 l bottles filled with 0.5 l cultures with a starting optical density at 600 nm of 0.01, under the three following growth conditions: (i) aerobically at 250 rpm up to an optical density of 0.6, (ii) semiaerobically at 100 rpm up to an optical density of 1.0 and (iii) semiaerobically with nitrate at 100 rpm up to an optical density of 1.0. The minimal medium was composed of NH_4Cl (30 mM), sodium succinate (25 mM), Na_2MoO_4 (0.6 mM), $MgSO_4$ (0.4 mM), EDTA (0.25 mM), Lawford trace solution (1 mL/L) and potassium phosphate (65 mM, pH 7.0); KNO_3 (100 mM) was added in the case of cultivations in the presence of nitrate. Each culture was grown in three biological replicates, and as such we availed of a set of 36 independently grown *P. denitrificans* cultures. Cells were harvested by centrifugation (6 200 x g, 30 min), washed with 50 mM tris(hydroxymethyl)aminomethane/HCl (Tris/HCl) pH 7.3 and stored as a pellet at -80 °C.

3.2 Sample preparation and two-dimensional gel electrophoresis

After cultivation, the cells were disrupted by sonicating 15 mg (wet weight) of pellet for 30 x 0.1 s (50 W output) in 300 µL of lysis buffer containing 7 M urea, 2 M thiourea, 1% (w/v) (3-((4-Heptyl)phenyl-3-hydroxypropyl)dimethylammoniopropanesulfonate) (C7BzO), 40 mM Tris-base, 70 mM dithiothreitol (DTT), 2% (v/v) Pharmalyte 3/10, 5 mM NaF, 0.2 mM $NaVO_3$, CompleteMini Protease Inhibitor Cocktail (Roche, Penzberg, Germany, one tablet per 10 mL of lysis buffer) and 150 U of benzonase (Sigma-Aldrich, St. Louis, MO, USA). The cell extracts were incubated for 1.5 h at 20 °C. Cellular debris was then removed by centrifugation (16000 x g, 20 min, 15 °C) and the supernatant (total cell lysate) was stored at -80 °C.

For preparation of membrane fraction, cells were disintegrated and converted into membrane vesicles as previously described (Burnell *et al.* 1975) with several modifications. Briefly, the suspension of harvested cells was diluted with 5.7 volumes of a solution containing 0.5 M sucrose, 200 mM Tris/HCl pH 7.3, 0.5 mM EDTA and lysozyme (1 mg/ml of the total volume). After 45 min of enzymatic lysis at 30 °C, an osmotic lysis (45 min, 30 °C) was initiated by addition of equal volume of ice-chilled water. After centrifugation (4600 x g / 30 min/4 °C), the pellet (spheroplasts) was further lyzed (30 min/4 °C) in 7.5 volumes of chilled water containing a trace of DNAse and 14 mM $MgSO_4$. The unbroken spheroplasts were sedimented at 4600 x g (30 min/4 °C) and the supernatant from this step was subjected to ultracentrifugation at 184000 x g for 40 min (4°C) using Beckman L8-55M Ultracentrifuge with 45 Ti rotor (Beckman, USA). The collected membranes were resuspended in 50 mM Tris/HCl pH 7.3, utracentrifuged again and resuspended in the same buffer. 150 µg of protein for analytical gels or 1 mg of protein for micropreparative separations, respectively, were extracted using sample solution containing 7 M urea, 2 M thiourea, 1 % (w/v) 3-[N,N-dimethyl(3-myristoylaminopropyl)ammonio]propanesulfonate (ASB 14), 1 % (v/v) TRITON X-100, 2 mM tributylphosphine, 15 mM Tris base, 1 % (v/v) Pharmalyte 3/10 and 0.5 % (v/v) Pharmalyte 8/10 for 1.5 h at 20 °C.

The protein content was determined by RC-DC Protein Assay (Bio-Rad, Hercules, CA) with BSA as a standard. Bio-Rad 2-D standards were added for determination of approximate M_r and p*I*.
Aliquots containing 150 μg of protein for analytical purposes or 400 μg of protein for micropreparative separation, respectively, were precipitated overnight with 7.5 volumes of acetone containing 0.2 % (w/v) DTT at -20 °C. After washing the pellets again in the same solution, the samples were resolubilized in 350 μL of rehydration solution containing 7 M urea, 2 M thiourea, 1 % (w/v) C7BzO, 40 mM Tris-base, 70 mM DTT and 2 % (v/v) Pharmalyte 3/10 by incubating at 20 °C for 1 h. The samples were centrifuged again (16000 x g, 20 min, 15 °C) before loading on 18 cm nonlinear immobilized pH gradients (IPG) 3-10 (Bio-Rad, Hercules, CA) by in-gel rehydration.
Proteins were separated by isoelectric focusing using PROTEAN IEF Cell (Bio-Rad). The voltage was varied from 100 V (100 Vh, rapid), 500 V (500 Vh, linear), 1000 V (1000 Vh, linear) to 8000 V (95000 Vh, rapid), subsequently. The paper electrode wicks were changed 10 times during the first 10 kVh (the anodic wicks were soaked with water and the cathodic ones with 50 mM DTT). The IPGs were stored frozen at -80 °C.
The IPG strips containing total cell proteins were equilibrated for 12 min in a solution containing 6 M urea, 30 % (v/v) glycerol, 2 % (w/v) SDS, 50 mM Tris/HCl pH 8.8, trace of bromphenol blue and 1 % (w/v) DTT and then for a further 12 min in the same buffer except that DTT was replaced with 2.5 % (w/v) iodoacetamide. The IPGs were then embedded onto SDS-PAGE gels (20 cm x 20 cm in size, 1 mm thick) using 0.5 % (w/v) low-melting agarose in Laemmli electrode buffer. In the second dimension, homogenous (12 % T, 1.07 % C) SDS-PAGE gels, Laemmli buffer system (Laemmli 1970) and PROTEAN Plus Dodecacell were used. After an initial ramp up period of 2 h at 50 V, the gels were run at 100 V for about 20 h at 4 °C. The gel patterns were visualized by tetrathionate-silver nitrate staining (Rabilloud 1992) for analytical purposes or by SYPRO Ruby (Molecular Probes) in the case of micropreparative separations according to manufacturer's instructions. GS-800 and Pharos FX Pro instruments (Bio-Rad) were used for gel scanning. Spot detection, background subtraction, spot matching and data normalization using a local regression model method were performed using PDQUEST 8.0 software.

3.3 Statistical analysis of 2D-PAGE data

The normalized data exported from PDQUEST 8.0 were analyzed as follows: Values estimated by threshold level were excluded from the analysis. To reveal differences between groups, significance analysis of microarrays (SAM) (Tusher *et al.* 2001) was performed if there were at least 3 replicates in each of the compared groups. Proteins were considered as significantly differentially regulated if the false discovery rate (FDR) did not exceed 10 % and if the mean quantitative change was higher than 2 (up-regulation) or lower than 0.5 (down-regulation). In order to visualize the effect of selected proteins, hierarchical clustering based on Spearman correlation was performed. Data analysis was performed in a R-2.8.1 environment for statistical computing (R_Development_Core_Team 2008). For SAM the "samr" package was used, and clustering was performed using the package "cluster".

3.4 Mass spectrometry analyses

Sypro Ruby-stained protein spots selected for MS analysis were excised from 2D-PAGE gels. After destaining, the proteins in the gel pieces were incubated with trypsin (sequencing grade, Promega) at 37 °C for 2 h (Havlis *et al.* 2003). Peptide mass fingerprinting and tandem mass spectrometry (MS/MS) analyses were performed by matrix-assisted laser desorption-ionization mass spectrometry (MALDI-MS) with an Ultraflex III mass spectrometer (Bruker Daltonik, Bremen, Germany).

Sample preparation protocol for MALDI-MS employing α-cyano-4-hydroxycinnamic acid solution prepared according to Havlis (Havlis *et al.* 2003) used as the matrix in combination with AnchorChip target was used to enhance measurement sensitivity. The sample (1 µl) was mixed with matrix solution on the target in a 2:1 ratio. Peptide maps were acquired in reflectron positive mode (25 kV acceleration voltage) with 800 laser shots. Twelve dominant peaks within 700 - 3600 Da mass range and minimum S/N 10 were picked for MS/MS analysis employing laser induced dissociation - "LIFT" arrangement with 600 laser shots for each peptide. Known autoproteolytic products of trypsin were used for internal calibration of digested peptides. In absence of these products, an external calibration procedure was employed, using a mixture of seven peptide standards (Bruker Daltonik) covering the mass range of 1000 - 3100 Da. The Flex Analysis 3.0 and MS Biotools 3.1 (Bruker Daltonik) software were used for data processing.

In case of insignificant or negative results of the MS/MS ion search, tryptic digests were subjected to electrospray ionization liquid chromatography-tandem mass spectrometry (ESI-LC-MS/MS) analysis. LC-MS/MS experiments were accomplished on a high performance liquid chromatography system consisting of a gradient pump (Ultimate), autosampler (Famos) and column switching device (Switchos; LC Packings, Amsterdam, The Netherlands) on-line coupled with an HCTultra PTM Discovery System ion trap mass spectrometer (Bruker Daltonik). The column used for LC separation was filled according to a previously described procedure (Planeta *et al.* 2003). Prior to LC separation, tryptic digests were concentrated and desalted using PepMap C18 trapping column (300 µm x 5 mm, LC Packings). Sample volume was 15 µl. After washing with 0.1 % formic acid, the peptides were eluted from the trapping column using an acetonitrile/water gradient (4 µL/min) onto a fused-silica capillary column (320 µm x 180 mm), on which peptides were separated. This column was filled with 4-µm Jupiter Proteo sorbent (Phenomenex, Torrance, CA). The mobile phase A consisted of acetonitrile/0.1 % formic acid (5/95 v/v) mixture and the mobile phase B consisted of acetonitrile/0.1 % formic acid (80/20 v/v) mixture. The gradient elution started at 5 % of mobile phase B, and after 4 minutes, it was increased linearly from 5 % to 50 % during 55 minutes. The analytical column outlet was connected to the electrospray ion source via a 50-µm-inner diameter fused-silica capillary. Nitrogen was used as nebulizing as well as drying gas. The pressure of nebulizing gas was 15 psi. The temperature and flow rate of drying gas were set to 300 °C and 6 L/min, respectively, and the capillary voltage was 4.0 kV. The mass spectrometer was operated in the positive ion mode in an m/z range of 300 - 1500 for MS and 100 - 3000 for MS/MS scans. Extraction of the mass spectra from the chromatograms, mass annotation and deconvolution of the mass spectra were performed using DataAnalysis 4.0 software (Bruker Daltonik).

3.5 Mass spectrometry data processing

MASCOT 2.0 (MatrixScience, London, UK) search engine was used for processing the MS and MS/MS data. Database searches were done against the translated genome sequence data of *P. denitrificans* downloaded from http://genome.ornl.gov/microbial/pden/, the last sequence version released in 2006). A mass tolerance of up to 30 ppm was accepted during processing MALDI-MS data for PMF and 0.6 Da during processing laser-induced dissociation -"LIFT" data for MS/MS ion searches. For ESI-MS/MS data, mass tolerances of peptides and MS/MS fragments for MS/MS ion searches were 0.5 Da. Oxidation of methionine and carbamidomethylation of cysteine as optional and fixed modifications, respectively, and one enzyme miscleavage were set for all searches. Gene annotations are consistent with *P. denitrificans* genome database at http://genome.ornl.gov/microbial/pden/. Note: In the time of preparation of this chapter, the *Paracoccus denitrificans* genome database was just in the process of moving to a new address: http://genome.jgi-psf.org/parde/parde.download.ftp.html.

4. Web accessible 2D-PAGE dataset of *P. denitrificans* proteome

PDQUEST 8.0 software was used for image analysis of 2D-PAGE gels and handling and keeping all operational data in a local file. This software also served for gel calibration, spot numbering and quantitation. Local files in PDQUEST format contain 2-D maps of total cell lysates and membrane fractions annotated with spot numbers, experimental Mr/pI, gel ID numbers, identification status, MS identification mode, protein name and UniProt accession number.

These data have been submitted to the "Proteome Database System for Microbial Research at Max Planck Institute for Infection Biology" in Berlin, Germany where they are now stored in the "Proteome 2D-PAGE Database" subsection (http://www.mpiib-berlin.mpg.de/2D-PAGE/) (Mollenkopf *et al.* 1999). The "Proteome 2D-PAGE Database" currently contains 11146 protein identifications from 10975 spots and 3124 mass peaklists in 55 reference maps representing experiments from 26 different organisms and strains. The data were submitted by 104 submitters from 30 institutes from 13 nations. The aim of the PDBS is to share proteomics information in a readily manner with the scientific community as an invitation for data mining. Showing experimental data like MS peaklists and raw spectra leads to more transparency of the results. In addition, protein identification data are integrated with genomic, metabolic and other biological knowledge sources to increase the value of the primary data.

The frontend of the "Proteome 2D-PAGE Database", i.e. the website, is dynamically generated mainly by a combination of PERL and CGI, but also JAVA, PHP and R (http://www.r-project.org/) are used. The data in the backend are organized in a relational database under the control of MySQL (http://www.mysql.com) as database management system.

The user of the *P. denitrificans* 2D-PAGE dataset can find three 2-D maps of *P.denitrificans* proteome: (i) Coomassie-stained 2-D map of total cell lysate (Fig. 2, 26 proteins identified) , (ii) silver-stained map of membrane fraction (Fig. 3, 14 proteins identified) and (iii) Sypro Ruby-stained total cell lysate map (Fig. 4, 640 proteins identified). The third Sypro Ruby total cell lysate map was prepared with the latest technologies and protocols and contains the most comprehensive annotation, including the possibility of downloading the complete list of ORFs identified at proteome level.

The user of the database has several possibilities to highlight the MS analyzed spots according to MS analysis results: He/she can highlight (i) only identified spots (ii) only non-identified spots, (iii) all spots, and (iv) none spots. The spots with significant identification are marked with a red cross while the spots without identification are labeled with a blue cross. A zoom function is available; a detailed map view can be thus obtained. If the user moves the cursor over a cross, spot number and protein name appears. If he clicks on the

Fig. 5. The web browser window of *P. denitrificans* dataset in the 2D-PAGE database: total cell lysate 2D-PAGE map. The cursor-selected spot is annotated with its protein name.

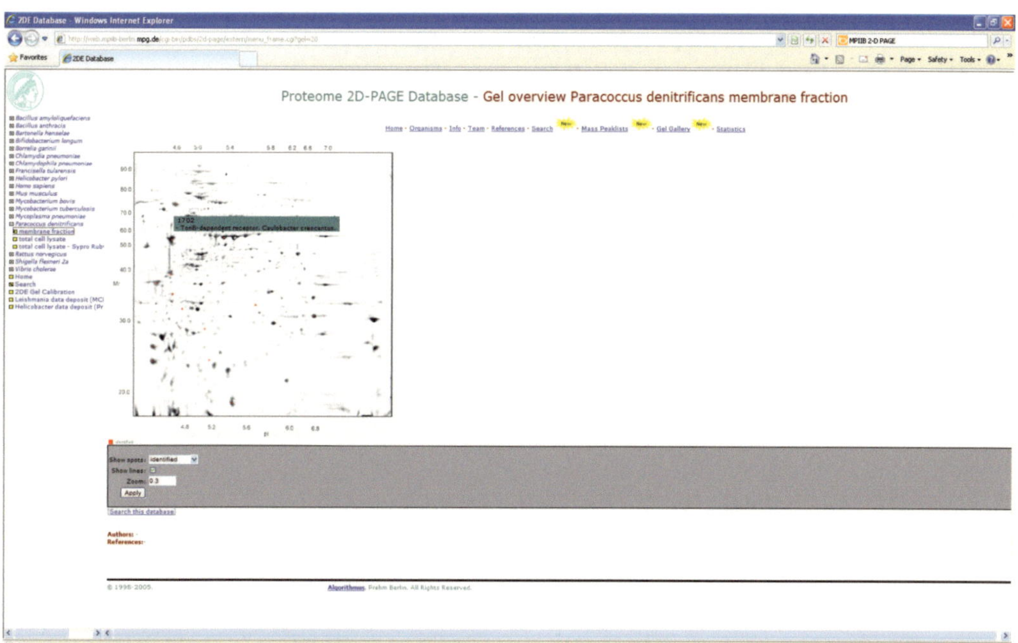

Fig. 6. The window of *P. denitrificans* dataset in the 2D-PAGE database: 2D-PAGE map of the membrane fraction.

Fig. 7. Sypro Ruby-stained total cell lysate map gel containing 640 identified proteins. The list of corresponding open reading frames identified by MS can be downloaded directly from this web page.

protein spot, the page with more detailed protein information is opened, including spot molecular weight, spot pI, UniProt accession number, gene locus, protein name, identification method, identification status and sequence coverage. If the visitor is interested in protein amount changes among different growth conditions, he can get detailed information in publicly accessible supplementary tables related to each published project.

Originally, only 8 proteins were identified by peptide mass fingerprinting from 49 analyzed spots within the first Coomassie-stained total cell lysate map and in membrane fraction map due to unknown sequence of *P. denitrificans* genome (Bouchal *et al.* 2004). Subsequent genome sequencing at Joint Genome Institute (http://www.jgi.doe.gov) resulted in public release of a *P. denitrificans* complete genome sequence in 2006. Thereafter the translated sequence data were transformed in the file applicable for MASCOT searches. Remaining 41 proteins were successfully identified using available MS spectra and the database gels were updated by new identifications. This substantial update underlines the significance of genome information on the effectivity of protein identification in proteomics.

5. *P. denitrificans* -omics projects facilitated *via* web accessible 2D-PAGE database

Because of the current progress in -omics methods and applications, it is expected that other laboratories interested in *P. denitrificans* biology will implement proteomic approach into their method toolboxes. The *P. denitrificans* web accessible dataset can help them with formulating their hypotheses, planning their experiments, orientation in their own proteome maps as well as facilitating communication among laboratories.

The perspective results coming from future studies on *P. denitrificans* are likely to be important from the bioenergetic point of view, providing a basis for understanding the well-known nutritional versatility of this bacterium. Since the rate of metabolic processes, in many cases depending on the substrate used, is often related to levels of involved proteins, identification of pivotal metabolic enzymes is one of the first tasks of proteomic research. From this point of view, the first important finding among the published results is the identification of proteins involved in denitrification. Nitrate reductase β-subunit, nitrite reductase and nitrous oxide reductase are the key enzymes of denitrification pathway and all of them were detected using proteomic approach in Sypro Ruby total cell lysate proteome maps as spots 5710 (nitrate reductase β-subunit), 1701 (nitrite reductase) and 2701 (nitrous oxide reductase).

Since the synthesis of denitrification enzymes is regulated by FNR-type transcription regulators (FnrP, NNR, NarR), looking for spot position of these regulatory proteins in 2-D maps is a next logical step of the expert user. Although their level is probably too low for detection on 2-D PAGE gels, an *UspA* gene product (gene 1849) as a direct neighbour of *fnrP* in the genome was detected in spot 3211 (Sypro Ruby 2-D map). Its expression profile was similar to *fnrP* gene product, suggesting their co-expression in single operone (Bouchal *et al.* 2010). Terminal oxidases (*aa*$_3$, *cbb*$_3$ and *ba*$_3$ types in *P. denitrificans*, Fig. 4) are very hydrophobic proteins as most of their subunits contain more than one transmembrane domain (information obtained using PSORT algorithm, http://psort.nibb.ac.jp) and their identification on 2-D PAGE gels cannot be expected - for review, see (Santoni *et al.* 2000). On the other hand, we identified α, β and ε-subunits of F_0F_1-ATPase (spot 2103 in Coomassie-stained total cell lysate map and spots 0011, 0120, 0503, 1204, 1319, 5529 in a map of membrane fraction) being in many cases downregulated under anaerobic growth conditions as a probable result of general slowing down of the energetic metabolism. Among the proteins induced with azide (Bouchal *et al.* 2004; Bouchal *et al.* 2011) Fe/Mn superoxide dismutase (spot 6106 in Coomassie-stained total cell lysate map and spot 5001 in a map of membrane fraction) was identified, indicating generation of an increased amount of reactive oxygen species, possibly as a result of the increased degree of reduction of respiratory components. This was independently confirmed at transcript and enzyme activity level (Bouchal *et al.* 2011). Furthermore, synthesis of Fe/Mn superoxide dismutase is independent of FNR-type regulators (Bouchal *et al.* 2010). The only protein in membrane fraction induced synergically by nitrate and azide (spot 1702 in membrane fraction map) was TonB dependent receptor, a protein involved in iron transport. His ORF is located very close to preudoazurin gene, an alternative electron-transporter in a denitrification pathway. We also found glyceraldehyde-3-phosphate dehydrogenase (spot 7402 in Coommassie-stained total cell lysate map) non-affected with different growth conditions, so it could serve as an internal standard for gene expression comparison.

Our subsequent comprehensive study focused on FNR-type transcription regulators (Bouchal *et al.* 2010) revealed four significant protein clusters according to correlation of their levels under aerobic, semiaerobic and semiaerobic with nitrate growth conditions (see Fig. 8, spot numbers correspond to Sypro Ruby-stained proteome map): (i) The first cluster contains proteins involved in the FnrP regulon. It involves nitrous oxide reductase (spot 2701), UspA protein (spot 3211), and two OmpW proteins (spots 4107 and 5105) as well as two spots 501 and 8701 identified as unknown proteins. The direct regulation of nitrous oxide reductase, UspA and OmpW proteins by FnrP is a new finding from the mentioned study. (ii) Second cluster involves proteins regulated *via* additional regulators, including

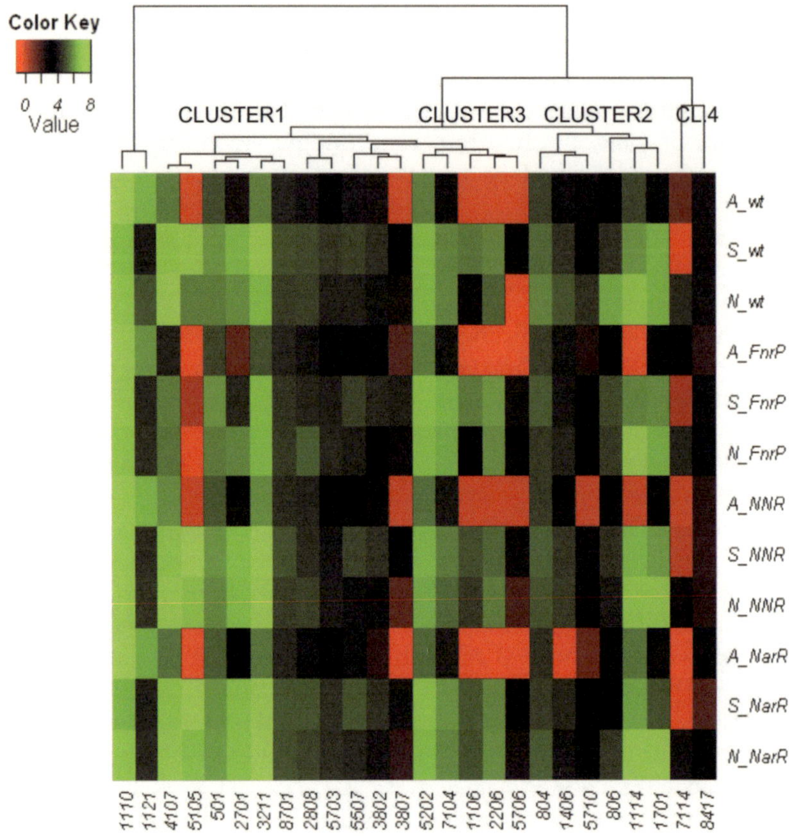

Fig. 8. The Spearman correlated clustering of the regulation profiles of proteins differentially regulated in response to mutation in FnrP, NNR and NarR transcription regulators and/or to oxygen and nitrate. See (Bouchal *et al.* 2010) for details.

proteins involved in NNR and NarR regulons. This cluster contains two TonB dependent receptors (spots 804, 806 and 1814), nitrate reductase β-subunit (spot 5710), a TenA-type transcription regulator (spot 1114), nitrite reductase (spot 1701) and an unknown protein with an alpha/beta hydrolase fold (spot 1406). The clustering of the TenA transcription regulator with nitrite reductase might well be indicative for the involvement of such an additional regulator. This clustering also indicates that ranking of the above mentioned proteins specifically under the NNR or the NarR regulon is less straightforward, probably since both regulators are activated by the reduction products of a common substrate, nitrate. (iii) The third cluster involves proteins whose amount is affected by the growth condition rather than by mutations in the FNR-type proteins. As such, these proteins may be part of a more global regulatory switch. This cluster contains SSU ribosomal protein S305 / σ54 modulation protein (spot 5202) and two SDR proteins (spots 2206 and 7104). (iv) The fourth cluster contains only the proteins specifically upregulated in cells grown semiaerobically in the presence of nitrate: one uncharacterized protein (spot 7114) and an ABC-type transporter of unknown function (spot 8417) (Bouchal *et al.* 2010).

6. Perspectives

In the upcoming time, development of the *P. denitrificans* 2-D PAGE dataset can continue in the following ways. (1) The identification of gel proteins spots not yet analyzed *via* mass spectrometry will continue, followed by immediate updates of the dataset. Resolution at the protein species level as obtained by 2D-PAGE-MS methods has in contrast to the bottom-up LC-MS approach the advantage to consider protein speciation (Jungblut *et al.* 2008) and will allow the analysis of protein species-specific regulation as already described within the phosphoproteome of *Helicobacter pylori* infected human stomach adenocarcinoma AGS cell line (Holland *et al.* 2011). (2) Implementation of new perspective analytical methods is in progress. 2DLC-MS/MS-based approaches with stable isotope labelling (iTRAQ, SILAC), or with label-free quantification can serve as a method complementary to 2D-PAGE-MS. These approaches are also helpful when identification of integral membrane proteins with non-membrane domains is required (Wu *et al.* 2003; Bouchal *et al.* 2009). (3) In the case of the most hydrophobic or low abundant proteins, non-proteomic approaches like qRT-PCR and cDNA chips are available for the study of gene expression. (4) Further progress of *P. denitrificans* genome information and annotation is expected. Direct accession from the *P. denitrificans* dataset in the 2D-PAGE database into *P. denitrificans* genome database would underline the integrity of genomic and proteomic data and facilitate finding the relations between protein level, gene location and gene function.

The data collection in the similar format raises the possibility of data comparison between different proteomes. During reading the communications about various microbial proteomic studies, one can be surprised how many identical (or very relative) proteins have been identified in different organisms, while other physiologically important proteins may be underrepresented in 2-D gels. It is more interesting if the number of theoretical ORFs in genomes is taken into account. It is obvious that protein hydrophobicity and solubility together with copy number plays an important role in these observations (Wilkins *et al.* 1998; Santoni *et al.* 2000; Jungblut *et al.* 2010). Using an integrated "Proteome 2D-PAGE Database" (http://www.mpiib-berlin.mpg.de/2D-PAGE/) covering a number of bacterial proteomes, it is easier to predict whether the protein of one's interest will, or will not be identified using proteomics approach and can be quantified this way. Such a tool is very useful for people making a choice of the best method for a screening of the gene(s) expression, or the protein(s) synthesis in *Paracoccus denitrificans* focused projects.

7. Conclusions

We feel that -omics approach is a powerful tool for a study of thousands of cellular genes and proteins in *P. denitrificans* and their variability between different growth conditions. Keeping in mind the principles of proteomics, it can be viewed as a screening tool able to reveal the specific changes, the detection of which would require significantly larger amount of work using classical approaches. With regard to nature of various environmental effects, „omics" approaches itself cannot provide final evidences for their mechanisms. However, the hypotheses obtained using this toolbox can provide a firm bases for targeted functional studies using integrated modern biochemical, bioinformatic and molecular-biological methodologies. We hope that our dataset within the 2D-PAGE database will be useful in designing such integrated *P. denitrificans* projects and in facilitating the international interlaboratory cooperation.

8. Acknowledgements

The authors would like to thank Dr. Rob van Spanning, Vrije Universiteit Amsterdam, the Netherlands, for providing the bacterial strains and Dr. Eva Budinská, Masaryk University Brno, Czech Republic & Swiss Institute for Bioinformatics, Lausanne, Switzerland, for statistical analyses. The experiments and chapter publishing costs were covered by Czech Ministry of Education (project No. MSM0021622413). P.B. was partly supported by European Regional Development Fund and the State Budget of the Czech Republic (RECAMO; CZ 1.05/2.1.00/03.0101) and by Czech Ministry of Health (MZ0MOU2005), Z.Z. was supported by the project "CEITEC - Central European Institute of Technology" (CZ.1.05/1.1.00/02.0068) from European Regional Development Fund.

9. References

Bouchal, P. and I. Kucera (2002) Two-dimensional electrophoresis in proteomics: Principles and applications. *Chemicke Listy*. 97, 1, 29-36, ISSN 0009-2770

Bouchal, P. and I. Kucera (2004) Examination of membrane protein expression in Paracoccus denitrificans by two-dimensional gel electrophoresis. *Journal of Basic Microbiology*. 44, 1, 17-22, ISSN 0233-111X

Bouchal, P., P. Precechtelova, Z. Zdrahal and I. Kucera (2004) Protein composition of *Paracoccus denitrificans* cells grown on various electron acceptors and in the presence of azide. *Proteomics*. 4, 9, 2662-2671, ISSN 1615-9853

Bouchal, P., T. Roumeliotis, R. Hrstka, R. Nenutil, B. Vojtesek and S. D. Garbis (2009) Biomarker discovery in low-grade breast cancer using isobaric stable isotope tags and two-dimensional liquid chromatography-tandem mass spectrometry (iTRAQ-2DLC-MS/MS) based quantitative proteomic analysis. *Journal of Proteome Research*. 8, 1, 362-373, ISSN 1535-3893 (Print)

Bouchal, P., I. Struharova, E. Budinska, O. Sedo, T. Vyhlidalova, Z. Zdrahal, R. van Spanning and I. Kucera (2010) Unraveling an FNR based regulatory circuit in *Paracoccus denitrificans* using a proteomics-based approach. *Biochimica Et Biophysica Acta-Proteins and Proteomics*. 1804, 6, 1350-1358, ISSN 1570-9639

Bouchal, P., T. Vyhlidalova, I. Struharova, Z. Zdrahal and I. Kucera (2011) Fe/Mn superoxide dismutase-encoding gene in *Paracoccus denitrificans* is induced by azide and expressed independently of the FNR-type regulators. *Folia Microbiologica*. 56, 1, 13-17, ISSN 1874-9356 (Electronic) 0015-5632 (Linking)

Burnell, J. N., P. John and F. R. Whatley (1975) The reversibility of active sulphate transport in membrane vesicles of Paracoccus denitrificans. *Biochemical Journal*. 150, 3, 527-536, ISSN 0264-6021 (Print) 0264-6021 (Linking)

Havlis, J., H. Thomas, M. Sebela and A. Shevchenko (2003) Fast-response proteomics by accelerated in-gel digestion of proteins. *Analytical Chemistry*. 75, 6, 1300-1316, ISSN 0003-2700 (Print)

Holland, C., M. Schmid, U. Zimny-Arndt, J. Rohloff, R. Stein, P. R. Jungblut and T. F. Meyer (2011). Quantitative phosphoproteomics reveals link between *Helicobacter pylori* infection and RNA splicing modulation in host cells. *Proteomics*. 11, 14, 2798-2811, ISSN 1615-9861 (Electronic) 1615-9853 (Linking)

Jungblut, P. R., H. G. Holzhutter, R. Apweiler and H. Schluter (2008) The speciation of the proteome. *Chemistry Central Journal.* 2, 16, ISSN 1752-153X (Electronic) 1752-153X (Linking)

Jungblut, P. R., F. Schiele, U. Zimny-Arndt, R. Ackermann, M. Schmid, S. Lange, R. Stein and K. P. Pleissner (2010) Helicobacter pylori proteomics by 2-DE/MS, 1-DE-LC/MS and functional data mining. *Proteomics.* 10, 2, 182-193, ISSN 1615-9861 (Electronic) 1615-9853 (Linking)

Koutny, M. and I. Kucera (1999) Kinetic analysis of substrate inhibition in nitric oxide reductase of Paracoccus denitrificans. *Biochemical and Biophysycal Research Communications.* 262, 2, 562-564, ISSN 0006-291X (Print)

Laemmli, U. K. (1970) Cleavage of structural proteins during the assembly of the head of bacteriophage T4. *Nature.* 227, 5259, 680-685, ISSN 0028-0836 (Print) 0028-0836 (Linking)

Mollenkopf, H. J., P. R. Jungblut, B. Raupach, J. Mattow, S. Lamer, U. Zimny-Arndt, U. E. Schaible and S. H. Kaufmann (1999) A dynamic two-dimensional polyacrylamide gel electrophoresis database: the mycobacterial proteome via Internet. *Electrophoresis.* 20, 11, 2172-2180, ISSN 0173-0835 (Print) 0173-0835 (Linking)

Planeta, J., P. Karasek and J. Vejrosta (2003) Development of packed capillary columns using carbon dioxide slurries. *Journal of Separation Science.* 26, 6-7, 525-530, ISSN 1615-9306 (Print), 1615-9314 (Online)

R_Development_Core_Team (2008) A language and environment for statistical computing, R Foundation for Statistical Computing Vienna, Austria

Rabilloud, T. (1992) A comparison between low background silver diammine and silver nitrate protein stains. *Electrophoresis.* 13, 7, 429-439, ISSN 0173-0835 (print), 1522-2683 (online)

Santoni, V., M. Molloy and T. Rabilloud (2000) Membrane proteins and proteomics: un amour impossible? *Electrophoresis.* 21, 6, 1054-1070, ISSN 0173-0835 (Print) 0173-0835 (Linking)

Tusher, V. G., R. Tibshirani and G. Chu (2001) Significance analysis of microarrays applied to the ionizing radiation response. *Proceedings of National Academy of Sciences U S A.* 98, 9, 5116-5121, ISSN 0027-8424 (Print)

Van Spanning, R. J., A. P. De Boer, W. N. Reijnders, S. Spiro, H. V. Westerhoff, A. H. Stouthamer and J. Van der Oost (1995) Nitrite and nitric oxide reduction in *Paracoccus denitrificans* is under the control of NNR, a regulatory protein that belongs to the FNR family of transcriptional activators. *FEBS Letters.* 360, 2, 151-154, ISSN 0014-5793 (Print)

Van Spanning, R. J. M., A. P. N. De Boer, W. N. M. Reijnders, H. V. Westerhoof, A. H. Stouthamer and J. van der Oost (1997) FnrP and NNR of *Paracoccus denitrificans* are both members of the FNR family of transcriptional activators but have distinct roles in respiratory adaptation in response to oxygen limitation. *Molecular Microbiology.* 12, 5, 893-907, ISSN 0950-382X (Print), 1365-2958 (Online)

Veldman, R., W. N. M. Reijnders and R. J. M. van Spanning (2006) Specificity of FNR-type regulators in *Paracoccus denitrificans. Biochemical Society Transactions.* 34, 1, 94-96, ISSN 0300-5127 (Print), 1470-8752 (Electronic)

Wilkins, M. R., E. Gasteiger, L. Tonella, K. Ou, M. Tyler, J. C. Sanchez, A. A. Gooley, B. J. Walsh, A. Bairoch, R. D. Appel, K. L. Williams and D. F. Hochstrasser (1998)

Protein identification with N and C-terminal sequence tags in proteome projects. *Journal of Molecular Biology*. 278, 3, 599-608, ISSN 0022-2836 (Print) 0022-2836 (Linking)

Wood, N. J., T. Alizadeh, S. Bennett, J. Pearce, S. J. Ferguson, D. J. Richardson and J. W. Moir (2001) Maximal expression of membrane-bound nitrate reductase in *Paracoccus* is induced by nitrate via a third FNR-like regulator named NarR. *Journal of Bacteriology*. 183, 12, 3606-3613, ISSN 0021-9093 (Print), 1098-5530 (Online)

Wu, C. C., M. J. MacCoss, K. E. Howell and J. R. Yates, 3rd (2003) A method for the comprehensive proteomic analysis of membrane proteins. *Nature Biotechnology*. 21, 5, 532-518, ISSN 1087-0156 (Print) 1087-0156 (Linking)

Zumft, W. G. (1997) Cell biology and molecular basis of denitrification. *Microbiology and Molecular Biology Reviews*. 61, 533-616, ISSN 1092-2172 (Print), 1098-5557 (Online)

Part 4

Subproteomes Analyses

9

Targeted High-Throughput Glycoproteomics for Glyco-Biomarker Discovery

Eunju Choi[1,2] and Michelle M. Hill[1]
[1]*The University of Queensland Diamantina Institute, Brisbane,*
[2]*The University of Queensland, School of Veterinary Science, Brisbane*
Australia

1. Introduction

Biomarker discovery has become a major research area in proteomics as protein markers are more readily developed into clinical diagnostic tests than nucleic acid biomarkers. This is reflected by the fact that all United States Food and Drug Administration (US FDA)-approved biomarkers currently available for clinical use are protein molecules (Srivastava, Verma, and Gopal-Srivastava 2005). Proteomic technologies for the global study of proteins have evolved in the past decade, in response to the growing demand for body fluid biomarker development (Anderson and Hunter 2006; Wang, Whiteaker, and Paulovich 2009). While mass spectrometry technology is improving in sensitivity and speed, several technical challenges in protein biomarker discovery still requires optimization. These include maximizing sample throughput to process adequate number of samples, reaching high sensitivity, specificity and reproducibility required for FDA approval, and managing the costs for biomarker discovery and assay development. This chapter will discuss the application of a targeted proteomics approach using lectins as affinity reagent throughout the biomarker discovery pipeline, and automation with magnetic beads to increase throughput.

2. Biomarkers

Biomarkers are biological molecules that correlate with a disease condition or phenotype. The search for cancer biomarkers has increased as the traditional tumor node metastases (TNM) system, a morphological pathology-based system used to determine the treatment strategy and prognosis in cancer patients, cannot correlate cancer subtypes with clinical outcomes (Ludwig and Weinstein 2005). Many studies using gene expression profiling have been published in the past decade contributing to a detailed molecular classification of each tumor subtype (Srivastava and Gopal-Srivastava 2002). Genomic profiling of tumor samples allowed the access to individualized genomic data to determine the appropriate treatment method or prognosis. For example, nonsmall cell lung cancer patients with mutated epidermal growth factor receptor (EGFR) will be able to receive an inhibitor of the EGFR tyrosine kinase activity called gefitinib (Belda-Iniesta, de Castro, and Perona 2011). The availability of specific non-invasive biomarkers will facilitate this type of tailored or personalized medicine to improve therapy and patient outcomes.

2.1 Types of biomarkers

Biomarkers can be divided into types based on clinical significance; including predictive, detection, diagnostic and prognostic markers (Mishra and Verma 2010). Predictive markers or response markers are used to assess the response of a specific drug to allow selection of appropriate treatment regimes for each patient. For example, in breast cancer patients, Her2/Neu overexpression will lead to treatment using Herceptin®, whereas for other types of breast cancer, tamoxifen provides the best patient outcomes (Hudis 2007). Thus, Her-2/Neu is a predictive cancer biomarker for some breast cancer therapies (Roses et al. 2009). Likewise, drugs such as INGN 201 (ADVEXIN®), which targets abnormal p53 tumor suppressor function, can be administered as monotherapy or in combination with radiation and/or chemotherapeutic agents in cancers showing abnormal p53 function (Gabrilovich 2006). Pharmacodynamic markers are used to select the appropriate dose of chemotherapeutic drugs. These markers help in optimizing cancer drug doses to minimize cytotoxicity and are often used in clinical trials. Mitogen-activated protein kinase (MAPK), Akt, or p27 which are downstream receptor-dependent molecules of phosphorylated EGFR are pharmacodynamic biomarkers for certain EGFR tyrosine kinase inhibitors (Albanell, Rojo, and Baselga 2001). Diagnostic markers can be used for early detection, determination of stage, tissue or relapse (Verma and Manne 2006). For example, the presence of bladder tumor antigen (BTA) and nuclear matrix protein-22 (NMP-22) in urine indicates the presence of bladder cancer (Lau et al. 2009) and serum alpha-fetoprotein is useful to diagnose nonseminomatous testicular cancer (Sturgeon et al. 2008). Prognostic biomarkers are used to discriminate benign from malignant tumors. For example, human papillomavirus (HPV) associated in oral cancer has a better survival time compared to other types of oral cancer (Mishra et al. 2006). Commercially available tests based on the genetic expression of the virus can be used to determine the prognosis. Some biomarkers can have overlapping uses, i.e. carcinoembryonic antigen (CEA) is used as a prognostic and diagnostic marker and so can be used in postoperative surveillance and monitoring of the effectiveness of therapy in advanced colorectal cancer (Sturgeon et al. 2008).

Biomarkers can be based on any biomolecule including DNA, RNA, protein, and carbohydrate markers (Mishra and Verma 2010). Single nucleotide polymorphisms (SNP), loss of heterozygosity, copy number variants, chromosomal aberrations such as microsatellite instability and epigenetic modifications, and mutations in oncogenes or tumor suppressor genes are all examples of DNA markers (Ludwig and Weinstein 2005). RNA markers are usually identified from microarray analysis, and can be validated using qRT-PCR (Gray and Collins 2000). The potential of microRNAs (miRNA) or small non-coding RNAs for use as cancer biomarkers has also been documented (Bartels and Tsongalis 2009). DNA and RNA markers have improved the molecular characterization of specific tumors and their subtypes, but the practical usefulness in the clinical setting may be limited as the tests involve intensive processing, and are far from being noninvasive, simple and cost effective. Protein biomarkers are clinically useful because cancer cells secrete or shed proteins and peptides into body fluids, allowing minimally invasive tests. Hence mass spectrometry based proteomics techniques have evolved with a purpose driven aim to discover novel protein biomarkers.

2.2 Biomarker discovery

Ideal biomarker tests should be noninvasive, cheap, simple to perform, informative and accurate (Boja et al. 2011; Negm, Verma, and Srivastava 2002; Srivastava and Gopal-

Srivastava 2002). The process of developing such a test is a difficult and uncertain task, as reflected by the declining number of newly approved biomarker tests by the FDA. However, despite this, there are a growing number of articles published on potential biomarker candidates (Anderson and Anderson 2002; Polanski and Anderson 2007; Rifai, Gillette, and Carr 2006). Depending on the purpose of the biomarker and its application in clinics, the criteria and developmental approach for each biomarker varies. The conventional biomarker discovery pipeline involves five stages. Clearly defined issues should be addressed at each stage to guide the process through to success (Fig. 1) (Surinova et al. 2011; Pepe et al. 2001).

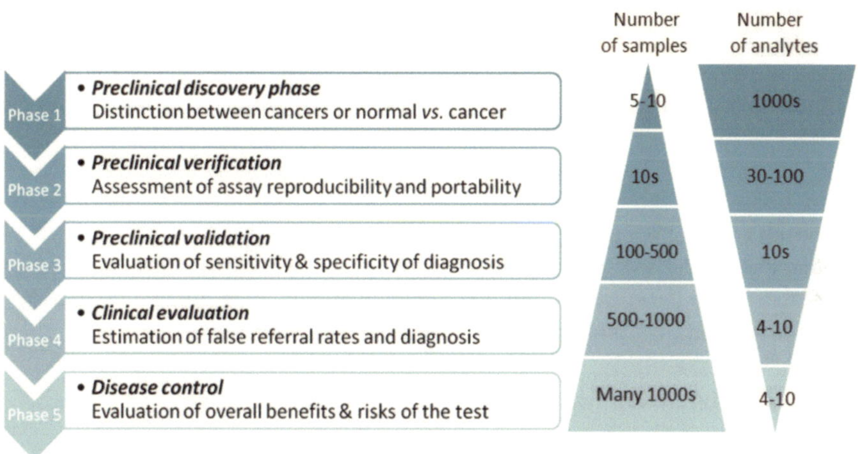

Fig. 1. Biomarker discovery workflow and study objectives for each phase. Modified from Pepe et al. (Pepe et al. 2001)

Phase 1 - Preclinical Discovery phase. Phase 1 is dedicated to hypothesis driven identification of candidate biomarkers, ranking and/or finding suitable combinations of potential biomarkers. The clinical question is defined and a small number of samples are obtained and analyzed to generate a list of candidates with their fold changes (Pepe et al. 2001).

Phase 2 - Preclinical verification. Phase 2 evaluates the (ranked) list of potential biomarkers generated in phase 1 using clinical samples from cases with known diagnosis. The end point of the assay may be mean concentration of candidate protein(s) or a unique signature associated with either one of the groups (Alonzo, Pepe, and Moskowitz 2002). The reproducibility, dynamic range and limit of detection (sensitivity) are determined in a relatively small cohort of patients, but with more patients than phase 1 (Rifai, Gillette, and Carr 2006). Another aim of the verification phase is to determine the sample size required for the Preclinical validation phase, to achieve statistical significance.

Phase 3 - Preclinical validation. The third phase is still within the scope of preclinical assessment but the aim is to generate a disease signature to determine whether the study objective can be met by the platform. The control and patient groups are designed retrospectively and the numbers used depend on the sensitivity and specificity of the biomarker determined in the previous phase, and the prevalence of the cancer in the population. The results are evaluated for analytical performance including test accuracy and precision, and clinical performance (Gutman and Kessler 2006), which must meet single-digit measurement coefficient-of-variation values (CVs) from measurement of thousands of patient samples. If the performance of the optimized assay meets the clinical objective, the process proceeds to the next phase, clinical evaluation.

Phase 4 - Clinical evaluation. Phase 4 is the development of a clinical assay and clinical evaluation of the biomarker as an in vitro diagnostic test. This phase is prospective and involves new control subjects and patients who are yet to be diagnosed (Manolio, Bailey-Wilson, and Collins 2006). The patient group sizes increase again based on the results from phase 3. The aim of phase 4 is to fulfil the clinical requirements and determine the true positive and false positive rates.

Phase 5 - Disease control. The last phase aims to determine the effect of the biomarker on disease management in the target population. Therefore, the biomarker proceeds into phase 5 when it is approved and accepted for clinical use. Phase 5 consists of the largest sample size and thus takes many years to complete. Data pertaining to cost of the test, as well as the consequences from the use of the biomarker are determined.

Biomarker development has had limited progress due to the lack of effective technology, established guidelines for designing clinical sample groups in each phase, standardized procedures for the development of the biomarker pipeline and quality assessment of the studies published (Mischak et al. 2007; Surinova et al. 2011). Therefore, by addressing the study objective clearly and by applying considerations for each phase, biomarker research should lead to more translatable candidates in the clinical context.

3. Proteomics for biomarker discovery

As described above, the road to discover biomarkers is a long and uncertain path consisting of different stages and multiple validation steps. The decisions made especially in the first few phases on the ranking of candidates or the best combination of candidates to maximize the sensitivity and specificity have enormous effects on the outcome of a successful biomarker assay. Consistency in the proteomics techniques and sample type used for each phase is crucial to successful biomarker discovery and validation.

3.1 Choice of sample type

The choice of sample type may be determined by availability, as well as complexity of the sample type for the available technology. Although the final preferred outcome are body fluid (commonly blood) tests, plasma or serum as a sample for proteomics is technically challenging due to dilution of potential biomarkers and the presence of high abundance proteins masking the lower abundance disease-associated proteins. Estimates suggest that there are more than 10^6 proteins in the blood proteome while one protein (albumin) accounts for more than half of all blood proteins (Zhang, Faca, and Hanash 2011). Approximately 22 proteins, including globulins, transferrins and fibrinogen make up 99% of the total blood proteins. Additionally, the concentration of a blood protein can range from less than 1-5 pg/ml to more than 55 billion pg/ml, stretching across seven logs (Zhang, Faca, and Hanash 2011).

Immunodepletion columns have been developed to remove the top 6, 7, 12, 14, or 20 proteins from plasma/serum, prior to proteome profiling (Smith et al. 2011; Gong et al. 2006; Tu et al. 2010). However, this procedure may also deplete potential proteins of interest that are bound to albumin in the blood stream, as well as low abundance proteins due to non-specific binding (Gong et al. 2006). Due to these technical difficulties, many studies choose to use tissue samples in the discovery phase; however, it is difficult to predict which proteins will be easily detected in the blood as data derived from tissue is not always translatable to blood (Abbott and Pierce 2010). Therefore, direct analysis of plasma or serum rather than tissue may be useful in the initial discovery phase (Rifai, Gillette, and Carr 2006; Kulasingam and Diamandis 2008).

3.2 Choice of technology

Ideally, similar or compatible techniques are used throughout the biomarker discovery and validation pipeline. However, no single technique can fulfill the requirements of all 5 phases with sufficient throughput, sensitivity and accuracy. Phase 1 requires the measurement of thousands of analytes in few samples, while phases 2-4 require the (simultaneous) measurements of fewer analytes in increasing number of samples. Furthermore, clinical assays (phase 4-5) ideally requires minimal sample handling.

Current proteomic profiling methods used in the discovery phase are not suitable for later phases since techniques such as two-dimensional difference gel electrophoresis (2D-DIGE) and multidimensional protein identification technology (MuDPIT) can only analyze one sample at a time, and require days of processing. Current technologies for preclinical and clinical phases such as radioimmunoassay (RIA), enzyme-linked immunosorbent assay (ELISA) and multiplex fluorescent detection technology are antibody-based assays requiring identified target, and hence not applicable to the discovery phase. The development and use of Selected Reaction Monitoring mass spectrometry (SRM-MS) as pre-clinical and potentially clinical assays not only provide a link between discovery, validation and clinical techniques, it also avoids the significant cost outlay for antibody development. Hence SRM-MS technology is fast becoming the method of choice for pre-clinical phases, and is set to make it into the clinical arena.

3.3 Improving throughput

Due to the high cost and low sample throughput in proteomics technology, biomarker discovery workflows have commonly suffered from the lack of sufficient technical and biological replicates. To address these short-comings, significant effort has been spent on sample preparation and separation using automation on robotic liquid handler, and the introduction of nanomaterial for nanoproteomics (Ray et al. 2011). Increased throughput in mass spectrometry can be achieved by means of multiplexing samples (Boersema et al. 2009; Chen et al. 2007) and/or shortening bioinformatic analysis time after the generation of mass spectrometry data (Martens 2011a, 2011b).

3.4 Targeted proteomics

Discovery proteomics workflows generally require multiple steps of separation due to high sample complexity. One strategy to reduce extensive separation steps is to enrich for a subset of proteins that are disease-relevant. In this chapter, we focus on the potential of targeted glycoproteomics as an all-encompassing technology for the phases of (glyco-) biomarker discovery.

4. Glycoproteomics

Glycoproteomics, an area of proteomics with biological and clinical significance, is an emerging field in biomarker research (Pan et al. 2011; Meany and Chan 2011).

Glycoproteins are a group of proteins in which one or more glycans (sugars) are covalently bonded to the protein through a process called glycosylation. There are two main types of protein glycosylation: (i) N-linked glycosylation whereby the glycan is attached to the amide nitrogen of asparagine in a consensus Asparagine-X-Serine/Threonine (Asp-X-Ser/Thr) sequence, where X can be any amino acid except proline and (ii) O-linked glycosylation in which the glycan is attached to the hydroxyl oxygen of serine or threonine in the protein. Glycosylation is the most abundant posttranslational modification and the most structurally diverse. There are at least 14 different monosaccharides and 8 different amino acids involved in this process with at least 41 different chemical bonds in glycan-protein linkage.

Glycoproteins are important targets in the search for biomarkers for the following reasons: (i) more than 50% of secreted proteins are glycoproteins, (ii) glycosylation changes in tissues, blood and serum from patients with disease has been implicated in pathogenesis, (iii) changes in glycosylation can be more distinctive than changes in protein expression, as specific glycan structures are generally not present normally, but increase in disease states, (iv) changes in glycosylation occur in many proteins including abundant proteins, thus increasing the likelihood of early detection, (v) the glycosylated form of a particular protein site is generally stable for a given cell type and physiological state, and (vi) as one of the important functions of glycans is in cell-cell interactions and consequently the control of cell function, alterations of protein glycosylation can be diagnostic for a disease (Pan et al. 2011; Packer et al. 2008). Altered glycosylation can be seen in diseases as hypo, hyper or newly glycosylated sites, and/or altered carbohydrate moieties (Pan et al. 2011).

Although advances in technologies used in glycoprotein research has been slow due to the complicated nature and vast variety of changes in glycosylation, advances in proteomic technologies have facilitated glycoproteomics research. An excellent example of a glyco-biomarker is alpha-fetoprotein (AFP), a marker for hepatocellular carcinoma (HCC) (Sturgeon et al. 2010). The specificity for AFP in HCC is low, limiting the use in the clinic (Meany, Sokoll, and Chan 2009), however, recent studies have shown that the fucosylated form of AFP which is highly reactive with the *Lens culinaris* agglutinin, also known as AFP-L3, improves the specificity (Masuda and Miyoshi 2011), demonstrating the utility of glyco-biomarkers.

4.1 Glycoproteomic approaches for biomarker discovery

A typical glycoproteomics pipeline consists of glycoprotein enrichment techniques, followed by multidimensional chromatographic separation, and mass spectrometry with bioinformatic data analysis. Glycoproteomics approaches can be divided into glycoprotein-based and glycopeptide-based methods (Fig. 2). Glycoprotein-based enrichment methods, also known as the top-down workflow, enrich for the glycoproteins prior to proteolytic digestion with enzymes such as trypsin. Glycan cleavage is performed before or after proteolytic digestion. In glycopeptide enrichment methods, proteolytic digestion is performed before enrichment. This is also known as the bottom-up workflow. The bottom-up workflow is more popular as it provides detailed information of a glycoprotein profile, and also specific mapping of glycosylation sites. However, the bottom-up workflow can result in very low sample throughput, and current technology is not capable of determining detailed glycan structure of glycoproteins in one analysis (Pan et al. 2011). On the other hand, the top-down workflow may not accurately provide mapping of glycosylation sites,

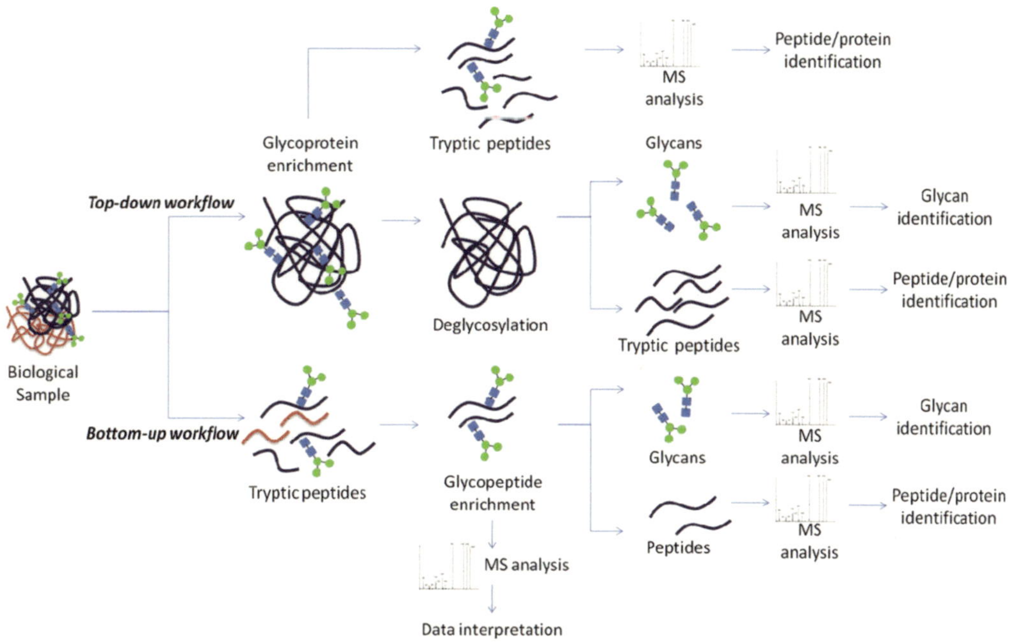

Fig. 2. Glycoproteomic approaches for glycan, deglycosylated and intact glycopeptide analysis. In the top-down workflow, glycoprotein enrichment is performed which may or may not follow deglycosylation. In the bottom-up workflow, proteins are digested then glycopeptides are enriched for further analysis.

although it results in greater glycoprotein sequence coverage. Therefore, the technique used will depend on the specific research question asked.

4.2 Glycoproteome enrichment techniques

Several techniques have been used for enrichment of glycans, glycopeptides and glycoproteins (Tousi, Hancock, and Hincapie 2011; Rakus and Mahal 2011; Pan et al. 2011), including hydrazide chemistry-based solid phase extraction methods, boronic acid-based solid phase extraction, size exclusion chromatography, hydrophilic interaction liquid chromatography (HILIC), activated graphitized carbon and lectin affinity based methods (Table 1). This chapter will discuss the potential of lectins as a universal enrichment tool in all phases of the glyco-biomarker discovery workflow. Lectins are naturally occurring sugar binding proteins which are highly specific for their sugar moieties. Their abilities to recognize and bind to specific glycans make them ideal for glycan structure specific glycoprotein enrichment. Lectins have been used in biological research as an affinity reagent for the past few decades, with applications such as lectin histochemistry (Brooks et al. 1996; Carter and Brooks 2006), lectin blotting (Welinder et al. 2009), lectin-affinity chromatography in combination with mass spectrometry (Abbott and Pierce 2010; Yang et al. 2006; Zhao et al. 2006; Xu et al. 2007; Qiu et al. 2008; Jung, Cho, and Regnier 2009) and lectin microarray (Gupta, Surolia, and Sampathkumar 2010; Katrlik et al. 2010) to examine the glycoproteome of serum and plasma.

Table 1. Glycoproteome enrichment techniques

Technique	Chemistry	Advantage	Disadvantage
Hydrazide chemistry (Zhang et al. 2003; Sun et al. 2007)	Sodium periodate oxidizes cis-diols of the glycan to aldehydes which are then coupled to hydrazide resin	· High-coupling efficiency of all types of glycans · Reduction of non-specific binding from complex mixtures	· Loss of structural information of glycans by oxidative coupling · Glycopeptide identification only for N-linked glycans using PNGase F
Boronic acid-based solid phase extraction (Sparbier, Wenzel, and Kostrzewa 2006)	Formation of diesters with cis-diols of glycans and borate under high pH	· High-coupling efficiency of all glycans	· Not specific to glycans · Cannot enrich for glycans that terminate in N-acetylgalactosamine
Size exclusion chromatography (Alvarez-Manilla et al. 2006)	Size exclusion to separate high mass glycopeptides from non-glycopeptides	· Simple, nonselective, and rapid	· Not glycan specific · Glycopeptide identification only for N-linked glycans owing to PNGase F
Hydrophilic interaction liquid chromatography (HILIC) (Hagglund et al. 2004)	Interaction of hydrophilic glycans and stationary phase through dipole-dipole interaction, electrostatic interaction and hydrogen binding	· Selective separation of glycopeptides · Allows direct MS analysis by desalting and concentrating the sample · Neutral and sialylated N-glycopeptides can be separated on the basis of hydrophilic interaction	· Not glycan specific · Favors N- over O-glycoproteome · Resolution of structural isomers is generally not observed
Activated graphitized carbon (de Leoz et al. 2011)	Retention of hydrophilic compounds by polar, hydrophobic, & ionic binding	· Retain glycoconjugates and separate structural isomers	· Nonspecific separation of hydrophilic species of glycans, peptides and phosphopeptides
Lectins (Durham and Regnier 2006; Yang and Hancock 2004)	Affinity to specific glycan moiety	· Enriches in glycan structure dependent manner · Can be coupled to various solid support systems for such as affinity chromatography, lectin array, etc.	· Enriches only for a particular subset of glycoconjugates · Relatively weak binding of glycans · Binding specificity not well known · Nonspecific binding

5. Use of lectins in glyco-biomarker discovery

The potential of a lectin-enrichment step to be coupled to different downstream assay techniques is attractive in glyco-biomarker discovery as it reduces the potential variation introduced by the change of enrichment methods going from one phase to another (Fig. 3). For example, in the discovery workflow of phase 1, lectin-enrichment can be followed by glycoprotein or glycopeptide separation and identification by tandem mass spectrometry (MS/MS), to measure hundreds of analytes. In the preclinical stages (phases 2 and 3), lectin affinity isolation may be coupled to SRM-MS for targeted quantification of a reduced number of candidates. Although SRM-MS assays may have the desired sensitivity and reproducibility, routine use in clinical pathology laboratories will need additional technology optimization. Lectin affinity can also be incorporated into other preclinical verification technology such as multiplexed immunoassay incorporating fluorescence-labeled microspheres with specific antibodies (Li et al. 2011), multiplexed protein analysis using antibody-conjugated microbead arrays (Theilacker et al. 2011), and multiplex proteins assays using magnetic nanotag sensing (Osterfeld et al. 2008). For clinical phases 3-5, existing antibodies may be used or antibodies may be developed for use in lectin microarrays or lectin-immunosorbent assays.

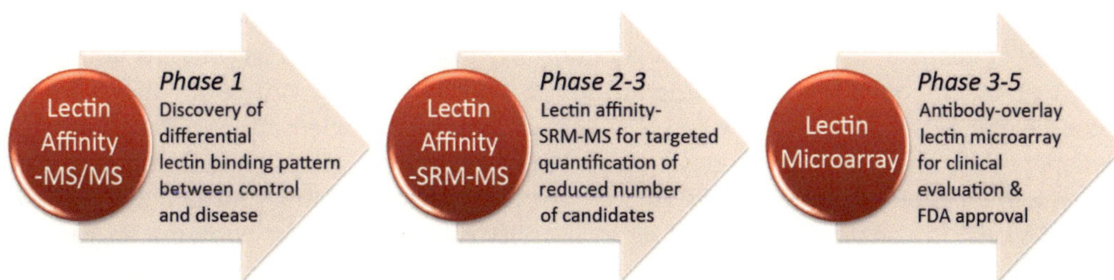

Fig. 3. Biomarker discovery pipeline using lectins.

5.1 Lectin affinity chromatography for glyco-biomarker discovery

Lectin affinity chromatography is a technique that employs one or more lectins to enrich for structurally similar subset(s) of glycoproteins or glycopeptides (Jung, Cho, and Regnier 2009; Durham and Regnier 2006; Yang et al. 2006). By coupling this technique to mass spectrometry analysis, bound and unbound fractions can be analysed to identify proteins in the two fractions. Lectin affinity chromatography can be performed in different formats including tubes, packed columns, microfluidic channels and high pressure liquid chromatography (HPLC) (Mechref, Madera, and Novotny 2008). Different types of support matrices can be used to immobilize the lectins, such as sepharose/agarose beads (Kobata and Endo 1992; Mechref, Madera, and Novotny 2008), magnetic beads (Lin et al. 2008), silica or styrene-divinylbenzene co-polymers coated with a cross-linked polyhydroxylated polymer (POROS) (Tousi, Hancock, and Hincapie 2011). Commonly used lectins include mannose and glucose binding concanavalin A (ConA) and N-acetylglucosamine binding wheat germ agglutinin (WGA) for their broad binding specificities and affinity to most N-linked glycans in biological material. For O-linked glycans, jacalin (JAC) is added to these two lectins for a global range of glycoprotein enrichment. For more specific enrichment, sialic acid and/or fucose binding lectins can be used, such as Sambucus nigra agglutinin

(SNA) and Maackia amurensis agglutinin (MAA) for sialic acid and Aleuria aurantia lectin (AAL) for fucose. A wide range of different sample types have been used including soluble and membrane derived glycoconjugates from serum/plasma, cell lysates and tissue homogenates. Elution of bound glycoproteins/peptides is commonly achieved using competitive sugar of relatively low concentrations (5-100 mM) (West and Goldring 1996) or low pH such as acidic solutions (Green, Brodbeck, and Baenziger 1987).

Lectin affinity chromatography can be incorporated into top down or bottom up proteomics workflows, where the glycoproteins or the glycopeptides are identified by LC-MS/MS, respectively. Top down workflows identify lectin-reactive glycoproteins primarily by the non-glycosylated peptides in the isolated glycoproteins. The advantages are high sensitivity and ease of use, but the top down approach does not identify the actual glycopeptide(s) that bound to the lectins. Bottom up workflows directly identify the captured glycopeptides, but is technically more challenging due to the lower amount of targets. Top down and bottom up approaches generate complementary data and have both been successfully applied in glyco-biomarker discovery (see 5.1.2).

Modified versions of lectin affinity chromatography has been reported including Serial Lectin Affinity Chromatography (S-LAC) which uses a series of sequential lectin affinity steps (Durham and Regnier 2006) or Multi-lectin Affinity Chromatography (M-LAC) which combines 3 or more different lectins for one-step isolation (Yang and Hancock 2004; Ahn et al. 2010; Na et al. 2009). Both methods can be incorporated into the top down and bottom up workflow. However, the bottom up workflow is preferred for S-LAC as proteins with more than 1 glycosylation site with binding affinity to both lectin, may not be identified by the second lectin. S-LAC using ConA and JAC was shown to be efficient for enriching O-linked glycopeptides, since ConA removes most N-linked glycopeptides containing mannose which will facilitate the binding of O-linked glycopeptides to Jacalin (Durham and Regnier 2006). M-LAC is also an effective system to simplify complex samples allowing enrichment of approximately 50% of the plasma proteome in one-step (Dayarathna, Hancock, and Hincapie 2008). The bound fraction of M-LAC using ConA, WGA and JAC has been used by Zeng and others for the initial identification of candidate biomarkers in serum from breast cancer patients (Zeng et al. 2011). M-LAC was coupled with 1D SDS-PAGE, isoelectric focusing and lectin-overlay antibody microarray to identify several glycoproteins such as alpha-1B-glycoprotein and complement C3 as potential candidates (Zeng et al. 2011). Kullolli et al. further developed M-LAC into a high performance multi-lectin affinity chromatography (HP-MLAC), involving targeted albumin and immunoglobulin depletion in-line with glycoprotein affinity isolation using M-LAC (Kullolli, Hancock, and Hincapie 2010). This method has shown reproducibility and consistency of the bound and unbound fraction over 200 runs which promises to provide quality plasma glycoproteome data for clinical proteomics.

5.1.1 Technical aspects of lectin affinity enrichment

Although widely used, significant binding of non-glycosylated proteins during lectin affinity enrichment has been reported (Lee et al. 2009). Potential causes of the non-specific binding include the presence of protein complexes and prolonged incubation leading to non-specific binding to support beads. To optimize binding conditions, we investigated glycoprotein capture using Concanavalin A (ConA)-magnetic beads with a range of mild to stringent binding buffers, using a short incubation time of 30 minutes (Loo, Jones, and Hill 2010). In order to disrupt protein-protein complexes which may result in binding of non-glycosylated proteins to lectin beads, we included a reducing agent (1 mM DTT) and a strong detergent (0.2% SDS) in the binding and washing steps. Although this resulted in ~20% loss of protein binding compared to previous lectin-affinity buffer (Yang et al. 2006), we still observed strong affinity between lectin and their cognate glycans (Loo, Jones, and Hill 2010). Using the most stringent buffer condition, we have shown reproducibility of lectin-glycoprotein binding, confirming this buffer condition helps to avoid non-specific binding of lectins while enriching for glycoproteins with the highest affinity to the individual lectins (Loo, Jones, and Hill 2010).

5.1.2 Application of lectin affinity enrichment in biomarker discovery

Top down workflows that incorporate lectin affinity chromatography have been used to identify potential biomarkers in diseases including psoriasis (Plavina et al. 2007), hepatocellular carcinoma (Na et al. 2009), diabetic nephropathy (Ahn et al. 2010) and bladder cancer (Yang et al. 2011). Plavina et al. depleted the two most abundant plasma proteins, albumin and immunoglobulin, and performed M-LAC consisting of ConA, WGA and JAC to identify numerous tissue leakage proteins present in plasma at low ng/mL concentrations, such as galectin-binding protein 3, which was subsequently verified by ELISA (Plavina et al. 2007). Na et al. used M-LAC consisting of ConA, WGA, JAC, SNA, and AAL and 2D-DIGE with liver tissue samples to identify human plasma carboxylesterase 1 as a potential biomarker for hepatocellular carcinoma (Na et al. 2009). Ahn et al. used M-LAC to capture plasma glycoproteins and found 13 up-regulated and 14 down-regulated glycoproteins in diabetic nephropathy (Ahn et al. 2010). Yang et al. used ConA and WGA for dual-lectin affinity chromatography to enrich for glycoproteins in urine to identify biomarker candidates for bladder cancer and identified 265 glycoproteins with higher abundance in the cancer group compared to the control group (Yang et al. 2011). While there was an overlap of the proteins identified, 240 glycoproteins were uniquely identified by each of the methods. Furthermore, lectin affinity chromatography of glycoproteins has been used for a cell cycle study which combined MAA-affinity chromatography of glycoproteins from cell lysates of the cervical cancer cell line, HeLa cells, and periodate labeling of membrane proteins of intact cells coupled to hydrazide chemistry, to identify distinct expression patterns during the cell cycle which demonstrated a 4-fold change in membrane protein expression during different cell cycles (McDonald et al. 2009).

Bottom up lectin-affinity has also been successfully applied in glyco-biomarker discovery. For example, Drake et al. utilized immunoaffinity depletion and subsequent M-LAC with SNA and AAL to identify 122 human plasma glycoproteins with 247 unique glycosites (Drake et al. 2011). Alvarez-Manilla et al. used ConA-sepharose to identify 18 glycoproteins unique to mouse embryonic stem cells and 45 proteins exclusively found in cells of differentiated embryoid bodies (Alvarez-Manilla et al. 2010). Furthermore, the bottom up method coupled with filter-aided sample preparation (FASP) was shown to detect 6367 N-glycosites on 2352 proteins which accounts for 74% of known mouse N-glycosites and 5753 unique sites in four mouse tissues and blood plasma, demonstrating the ability of lectin affinity chromatography techniques to enrich for glycopeptides (Zielinska et al. 2010).

5.2 Lectin magnetic bead array for high-throughput glyco-biomarker discovery and preclinical verification

Differential binding to a panel of lectins (a lectin signature) can be used as disease biomarker. This is the principle behind lectin microarrays (see section 5.3) for known target proteins, however, there is a lack of high-throughput methodology for de novo discovery of lectin signatures for potential glyco-biomarkers. To this end, we introduced the concept of a high-throughput lectin-magnetic bead array (LeMBA), consisting of a panel of individual lectin-magnetic beads arrayed in a microplate (Loo, Jones, and Hill 2010). The use of magnetic beads allows liquid handler-assisted automation to increase the throughput while assessing individual lectin-binding sub-glycoproteomes. Direct coupling to LC-MS/MS for glyco-protein (top down) or glyco-peptide (bottom up) analysis enables the simultaneous identification of glyco-biomarker and its lectin signature.

While most (glyco-)biomarker discovery workflows focus on low abundance proteins in the serum/plasma, LeMBA-MS screens for specific glycan structure changes by determining the lectin signatures of the glyco-proteome. Hence, instead of identifying new, low abundance proteins secreted or leaked by the diseased cells, the LeMBA approach focuses on alteration in the glycosylation structure of medium- to high-abundance secreted proteins. Since altered glycosylation of secreted and/or cell surface proteins reflects cell function and hence disease progression (Pan et al. 2011; Packer et al. 2008), this approach is likely to discover disease-relevant glyco-biomarkers. Previous studies aimed to find glyco-biomarkers have identified high abundance proteins in the blood as potential biomarker candidates, such as haptoglobin (Yoon et al. 2010; Fujimura et al. 2008), hemopexin (Comunale et al. 2009), transferrin (Zeng et al. 2011; Bones et al. 2010) and alpha-1B-glycoprotein (Zeng et al. 2011).

LeMBA results will be trading low abundance for high specificity as glycosylation changes detected by multiple lectins will be unique for the altered glycan structure. This approach also holds promise for early diagnostic biomarkers since detection of low abundance early diagnostic markers is extremely difficult to achieve with any throughput using the current detection systems and workflows. If glycosylation changes are identified in early stages of diseases in medium to high abundance proteins, these changes can be developed into biomarkers with reasonable sensitivity and specificity as the proteins carrying the altered glycan will be easy to detect.

Taken together, it is expected that candidate biomarkers resulting from LeMBA-MS screen will increase the sensitivity and specificity of glyco-biomarker, owing to the ability of lectin signatures to identify overall and subtle changes. For biomarker discovery phase 2, combinations of lectin signatures that show the biggest changes between normal and disease will result in a panel of potential biomarker candidates that can be verified using LeMBA coupled to SRM-MS for verification and antibody-overlay lectin microarrays for further validation (Boja and Rodriguez 2011).

5.3 Lectin microarray as high-throughput glyco-biomarker validation assay

Since their introduction in 2005, lectin microarrays have emerged as a new technology that utilizes lectins as a glyco-profiling tool. A typical microarray contains 6 to 43 lectins immobilized on a solid surface and binding of glycoproteins to lectins is, in most cases, detected by standard fluorescence microarray scanners (Gemeiner et al. 2009). Lectin microarrays are a rapid, sensitive and high-throughput screening tool, highly suitable for all phases of glyco-biomarker discovery, depending on the type used.

5.3.1 Types of lectin microarrays and their use in biomarker discovery

Generally, there are two types of lectin microarrays: the direct assay and reverse-phase dot-blot lectin array (Gemeiner et al. 2009; Gupta, Surolia, and Sampathkumar 2010). The direct assay format immobilizes lectins on a solid surface and applies prelabeled sample over the surface. On the other hand, reverse-phase dot-blot lectin array immobilizes glycoproteins on a solid surface and applies prelabeled lectins. These two types have been used for biomarker discovery phase 1 for pancreatic cancer (Li et al. 2009; Patwa et al. 2006; Liu et al. 2010), glioblastoma (He et al. 2010), HCC (Zhao et al. 2007) and colorectal cancer (Qiu et al. 2008) to investigate differential glycosylation between control and disease.

The direct assay can also be modified into a sandwich assay called the antibody-overlay lectin microarray (ALM) or lectin-overlay antibody microarray (LAM). In ALM, lectins are immobilized on a solid surface; glycoproteins are added, followed by a biotinylated antibody overlay that binds to the protein. Then, streptavidin with a fluorophore attached is added, and the fluorescence is detected. The difference between ALM and LAM is that in LAM, the antibody is attached to a solid surface and biotinylated lectins are overlaid to bind to the glycan structure (Fig. 4). These types of lectin microarrays may be used for biomarker discovery phase 3 and higher and can be developed into clinical assays with a condition that they are reproducible with less than 10% CV (Fung 2010).

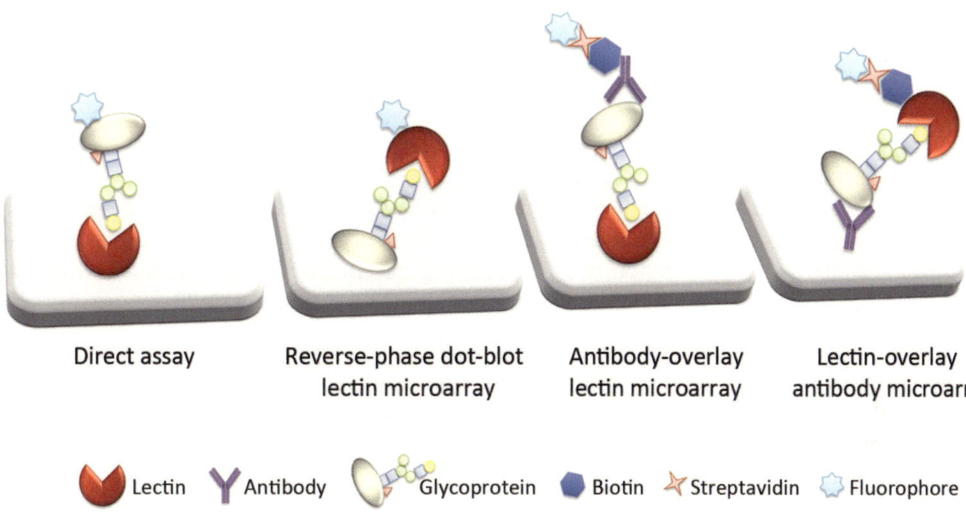

Fig. 4. Different types of lectin microarrays.

5.3.2 Technological aspects of lectin microarrays for phase 3 and above biomarker assay development

Preserving the carbohydrate recognition domain (CRD) is important for the reproducibility of the assay for assays with immobilized lectins. Popular methods of lectin immobilization include adsorption on nitrocellulose, attachment of amine functional group of protein backbone of lectins to a solid surface through epoxy- or N-hydroxysuccinimidyl-derived ester coated glass slides (Kuno et al. 2005) and use of self-assembled monolayers of thiols on gold-coated surfaces (Zheng, Peelen, and Smith 2005). Other methods include biotinylated lectin-neutravidin bridging (Angeloni et al. 2005), DNA-driven immobilization of lectins on polystyrene latex particles (Fromell et al. 2005), and binding to hydrogel based surfaces (Koshi et al. 2006). Unfortunately, no method can control for the optimal orientation of the CRD of lectins, to maximize the lectin binding ability and for the reproducibility of the assay. Techniques such as covalent bonding of lectins by carbenes have shown to immobilize the lectins but failed to preserve the carbohydrate binding activity (Angeloni et al. 2005) indicating the importance of preserving the CRD of lectins when lectin arrays are generated. The lack of control for lectin immobilization may lead to increased variation of assays. The variations of spotting have been reported to be 10-20% (Kuno et al. 2005) and the variation of a reverse-phase dot-blot assay, 10% (Patwa et al. 2006), which may be too high to qualify for FDA approval. To preserve the CRD, it has been suggested that glycans of glycosylated lectins may be used as an anchor point for attachment, followed by anchoring to hydroxylamine or hydrazine containing solid surface, which would preserve the CRD of the lectin (Gupta, Surolia, and Sampathkumar 2010). Of course, not all lectins are glycosylated, but this may help lower the variation of a biomarker assay. Additionally, the LAM type may be more suitable for phase 3 and above biomarker assays to avoid this issue.

As in most protein arrays, binding is, in most cases, detected by fluorescence (Pilobello and Mahal 2007; Gemeiner et al. 2009) using fluophores such as Cy3/Cy5, Alexa Fluor 555, and phycoerythrin. A number of different technologies have been introduced to increase the sensitivity of detection and salvage weak lectin-glycan bonds. Kuno et al. have introduced the use of evanescent-field fluorescence which allows *in situ* detection without a washing step to wash away any unbound material (Kuno et al. 2005). However, this technique requires a specialized evanescent-field fluorescence scanner. Other methods proposed include a modified fluorescence resonance energy transfer (FRET) method which demonstrated that a biomolecular fluorescence quenching and recovery (BFQR) technique can be used together with a supramolecular hydrogel matrix for the selective recognition of lectin-glycan bonds in reverse-phase dot-blot assays (Koshi et al. 2006). The use of tyramide signal amplification (TSA), which is a horseradish peroxidase (HRP)-mediated signal amplification method for ALM, has also shown to enhance signaling and therefore, increase the sensitivity of ALM over 100 times and allowed the detection of weak lectin-glycan interactions as demonstrated with as low as 20 ng of prostate specific antigen from seminal fluid (Meany et al. 2011).

6. Conclusions

There is no doubt that advancement in proteomics has and will contribute to protein biomarker discovery. Especially, technological advancement has enabled glyco-biomarker research. Medium to high abundance blood glycoproteins with disease-specific glycosylation structures are attractive as glyco-biomarkers, with potential for development of robust clinical assays compared to low abundance blood proteins. However, there is still a general lack of high-throughput glycoproteomics platforms to facilitate the discovery and validation of candidate glyco-biomarkers. The technologies and sample types used in the phases of glyco-biomarker discovery are critical to the final outcome, that is, development of a clinical assay.

In this chapter, we highlight the potential of lectins as a unifying glycan affinity tool for glyco-biomarker discovery. Lectin-based glycoprotein enrichment methods such as lectin affinity chromatography and high-throughput LeMBA can be coupled with LC-MS/MS to generate candidate biomarkers (phase 1 biomarker discovery). After the discovery of potential biomarkers, lectin affinity techniques such as LeMBA can be coupled to SRM-MS for high-throughput verification of a large number of patient samples. Finally, for phase 3 and onwards, ALM or LAM type lectin microarrays or lectin-coupled immunosorbent assays can be used for further validation of the biomarker assay to ensure high clinical and analytical performance. Having a unifying affinity reagent will improve the consistency and, therefore, success rate of transfer between the phases of biomarker discovery.

Combined with the appropriate bioinformatics tools, such as the recently developed serum glycopeptide SRM atlas (Schiess, Wollscheid, and Aebersold 2009) and glycan databases (reviewed in Frank and Schloissnig 2010), glyco-biomarker discovery and validation will surely contribute to biomarker research.

7. Acknowledgements

MH is supported by Career Development Award No. 569512 from the National Health and Medical Research Council of Australia. EC is supported by the University of Queensland International Research Tuition Awards and the University of Queensland Research Scholarship. Development of LeMBA was supported by an Australian Animal Cancer Foundation grant and a University of Queensland Collaboration and Industry Engagement Fund. High-throughput proteomics sample preparation station for the University of Queensland Diamantina Institute was supported by a Ramaciotti Foundations Equipment Gift.

8. References

Abbott, K. L., and J. M. Pierce. 2010. "Lectin-based glycoproteomic techniques for the enrichment and identification of potential biomarkers." *Methods Enzymol* no. 480:461-76. doi: S0076-6879(10)80020-5 [pii] 10.1016/S0076-6879(10)80020-5.

Ahn, J. M., B. G. Kim, M. H. Yu, I. K. Lee, and J. Y. Cho. 2010. "Identification of diabetic nephropathy-selective proteins in human plasma by multi-lectin affinity chromatography and LC-MS/MS." *Proteomics. Clinical applications* no. 4 (6-7):644-53. doi: 10.1002/prca.200900196.

Albanell, J., F. Rojo, and J. Baselga. 2001. "Pharmacodynamic studies with the epidermal growth factor receptor tyrosine kinase inhibitor ZD1839." *Semin Oncol* no. 28 (5 Suppl 16):56-66. doi: asonc02805n0056 [pii].

Alhadeff, J. A., and R. T. Holzinger. 1982. "Sialyltransferase, sialic acid and sialoglycoconjugates in metastatic tumor and human liver tissue." *The International journal of biochemistry* no. 14 (2):119-26.

Alonzo, T. A., M. S. Pepe, and C. S. Moskowitz. 2002. "Sample size calculations for comparative studies of medical tests for detecting presence of disease." *Stat Med* no. 21 (6):835-52. doi: 10.1002/sim.1058 [pii].

Alvarez-Manilla, G., J. Atwood, 3rd, Y. Guo, N. L. Warren, R. Orlando, and M. Pierce. 2006. "Tools for glycoproteomic analysis: size exclusion chromatography facilitates identification of tryptic glycopeptides with N-linked glycosylation sites." *J Proteome Res* no. 5 (3):701-8. doi: 10.1021/pr050275j.

Alvarez-Manilla, G., N. L. Warren, J. Atwood, 3rd, R. Orlando, S. Dalton, and M. Pierce. 2010. "Glycoproteomic analysis of embryonic stem cells: identification of potential glycobiomarkers using lectin affinity chromatography of glycopeptides." *Journal of Proteome Research* no. 9 (5):2062-75. doi: 10.1021/pr8007489.

Anderson, L., and C. L. Hunter. 2006. "Quantitative mass spectrometric multiple reaction monitoring assays for major plasma proteins." *Mol Cell Proteomics* no. 5 (4):573-88. doi: M500331-MCP200 [pii] 10.1074/mcp.M500331-MCP200.

Anderson, N. L., and N. G. Anderson. 2002. "The human plasma proteome: history, character, and diagnostic prospects." *Mol Cell Proteomics* no. 1 (11):845-67.

Angeloni, S., J.L. Ridet, N. Kusy, H. Gao, F. Crevoisier, S. Guinchard, S. Kochhar, H. Sigrist, and N. Sprenger. 2005. "Glycoprofiling with micro-arrays of glycoconjugates and lectins." *Glycobiology* no. 15 (1):31-41. doi: 10.1093/glycob/cwh143.

Bartels, C. L., and G. J. Tsongalis. 2009. "MicroRNAs: novel biomarkers for human cancer." *Clin Chem* no. 55 (4):623-31. doi: clinchem.2008.112805 [pii] 10.1373/clinchem.2008.112805.

Belda-Iniesta, C., J. de Castro, and R. Perona. 2011. "Translational proteomics: what can you do for true patients?" *Journal of proteome research* no. 10 (1):101-4. doi: 10.1021/pr100853a.

Boersema, P. J., R. Raijmakers, S. Lemeer, S. Mohammed, and A. J. Heck. 2009. "Multiplex peptide stable isotope dimethyl labeling for quantitative proteomics." *Nature protocols* no. 4 (4):484-94. doi: 10.1038/nprot.2009.21.

Boja, E., T. Hiltke, R. Rivers, C. Kinsinger, A. Rahbar, M. Mesri, and H. Rodriguez. 2011. "Evolution of clinical proteomics and its role in medicine." *J Proteome Res* no. 10 (1):66-84. doi: 10.1021/pr100532g.

Boja, E. S., and H. Rodriguez. 2011. "The path to clinical proteomics research: integration of proteomics, genomics, clinical laboratory and regulatory science." *The Korean journal of laboratory medicine* no. 31 (2):61-71. doi: 10.3343/kjlm.2011.31.2.61.

Bones, J., S. Mittermayr, N. O'Donoghue, A. Guttman, and P. M. Rudd. 2010. "Ultra performance liquid chromatographic profiling of serum N-glycans for fast and efficient identification of cancer associated alterations in glycosylation." *Analytical chemistry* no. 82 (24):10208-15. doi: 10.1021/ac102860w.

Brooks, S. A., M. Lymboura, U. Schumacher, and A. J. Leathem. 1996. "Histochemistry to detect Helix pomatia lectin binding in breast cancer: methodology makes a difference." *J Histochem Cytochem* no. 44 (5):519-24.

Carter, T. M., and S. A. Brooks. 2006. "Detection of aberrant glycosylation in breast cancer using lectin histochemistry." *Methods Mol Med* no. 120:201-16.

Chen, X., L. Sun, Y. Yu, Y. Xue, and P. Yang. 2007. "Amino acid-coded tagging approaches in quantitative proteomics." *Expert review of proteomics* no. 4 (1):25-37. doi: 10.1586/14789450.4.1.25.

Comunale, M. A., M. Wang, J. Hafner, J. Krakover, L. Rodemich, B. Kopenhaver, R. E. Long, O. Junaidi, A. M. Bisceglie, T. M. Block, and A. S. Mehta. 2009. "Identification and development of fucosylated glycoproteins as biomarkers of primary hepatocellular carcinoma." *J Proteome Res* no. 8 (2):595-602. doi: 10.1021/pr800752c [pii].

Dayarathna, M. K., W. S. Hancock, and M. Hincapie. 2008. "A two step fractionation approach for plasma proteomics using immunodepletion of abundant proteins and multi-lectin affinity chromatography: Application to the analysis of obesity, diabetes, and hypertension diseases." *J Sep Sci* no. 31 (6-7):1156-66. doi: 10.1002/jssc.200700271.

de Leoz, M. L., L. J. Young, H. J. An, S. R. Kronewitter, J. Kim, S. Miyamoto, A. D. Borowsky, H. K. Chew, and C. B. Lebrilla. 2011. "High-mannose glycans are elevated during breast cancer progression." *Mol Cell Proteomics* no. 10 (1):M110 002717. doi: M110.002717 [pii] 10.1074/mcp.M110.002717.

Drake, P. M., B. Schilling, R. K. Niles, M. Braten, E. Johansen, H. C. Liu, M. Lerch, D. J. Sorensen, B. S. Li, S. Allen, S. C. Hall, H. E. Witkowska, F. E. Regnier, B. W. Gibson, and S. J. Fisher. 2011. "A lectin affinity workflow targeting glycosite-specific, cancer-related carbohydrate structures in trypsin-digested human plasma." *Analytical biochemistry* no. 408 (1):71-85. doi: 10.1016/j.ab.2010.08.010.

Durham, M., and F. E. Regnier. 2006. "Targeted glycoproteomics: serial lectin affinity chromatography in the selection of O-glycosylation sites on proteins from the human blood proteome." *J Chromatogr A* no. 1132 (1-2):165-73. doi: S0021-9673(06)01467-1 [pii] 10.1016/j.chroma.2006.07.070.

Frank, M., and S. Schloissnig. 2010. "Bioinformatics and molecular modeling in glycobiology." *Cellular and molecular life sciences : CMLS* no. 67 (16):2749-72. doi: 10.1007/s00018-010-0352-4.

Fromell, K., M. Andersson, K. Elihn, and K. D. Caldwell. 2005. "Nanoparticle decorated surfaces with potential use in glycosylation analysis." *Colloids and surfaces. B, Biointerfaces* no. 46 (2):84-91. doi: 10.1016/j.colsurfb.2005.06.017.

Fujimura, T., Y. Shinohara, B. Tissot, P. C. Pang, M. Kurogochi, S. Saito, Y. Arai, M. Sadilek, K. Murayama, A. Dell, S. Nishimura, and S. I. Hakomori. 2008. "Glycosylation status of haptoglobin in sera of patients with prostate cancer vs. benign prostate disease or normal subjects." *International journal of cancer. Journal international du cancer* no. 122 (1):39-49. doi: 10.1002/ijc.22958.

Fung, E. T. 2010. "A recipe for proteomics diagnostic test development: the OVA1 test, from biomarker discovery to FDA clearance." *Clinical chemistry* no. 56 (2):327-9. doi: 10.1373/clinchem.2009.140855.

Gabrilovich, D. I. 2006. "INGN 201 (Advexin): adenoviral p53 gene therapy for cancer." *Expert Opin Biol Ther* no. 6 (8):823-32. doi: 10.1517/14712598.6.8.823.

Gemeiner, P., D. Mislovicova, J. Tkac, J. Svitel, V. Patoprsty, E. Hrabarova, G. Kogan, and T. Kozar. 2009. "Lectinomics II. A highway to biomedical/clinical diagnostics." *Biotechnol Adv* no. 27 (1):1-15. doi: S0734-9750(08)00081-5 [pii] 10.1016/j.biotechadv.2008.07.003.

Gong, Y., X. Li, B. Yang, W. Ying, D. Li, Y. Zhang, S. Dai, Y. Cai, J. Wang, F. He, and X. Qian. 2006. "Different immunoaffinity fractionation strategies to characterize the human plasma proteome." *Journal of proteome research* no. 5 (6):1379-87. doi: 10.1021/pr0600024.

Gray, J. W., and C. Collins. 2000. "Genome changes and gene expression in human solid tumors." *Carcinogenesis* no. 21 (3):443-52.

Green, E. D., R. M. Brodbeck, and J. U. Baenziger. 1987. "Lectin affinity high-performance liquid chromatography: interactions of N-glycanase-released oligosaccharides with leukoagglutinating phytohemagglutinin, concanavalin A, Datura stramonium agglutinin, and Vicia villosa agglutinin." *Analytical biochemistry* no. 167 (1):62-75.

Gupta, G., A. Surolia, and S. G. Sampathkumar. 2010. "Lectin microarrays for glycomic analysis." *OMICS* no. 14 (4):419-36. doi: 10.1089/omi.2009.0150.

Gutman, S., and L. G. Kessler. 2006. "The US Food and Drug Administration perspective on cancer biomarker development." *Nat Rev Cancer* no. 6 (7):565-71. doi: nrc1911 [pii] 10.1038/nrc1911.

Hagglund, P., J. Bunkenborg, F. Elortza, O. N. Jensen, and P. Roepstorff. 2004. "A new strategy for identification of N-glycosylated proteins and unambiguous assignment of their glycosylation sites using HILIC enrichment and partial deglycosylation." *J Proteome Res* no. 3 (3):556-66.

He, J., Y. Liu, X. Xie, T. Zhu, M. Soules, F. DiMeco, A. L. Vescovi, X. Fan, and D. M. Lubman. 2010. "Identification of cell surface glycoprotein markers for glioblastoma-derived stem-like cells using a lectin microarray and LC-MS/MS approach." *Journal of proteome research* no. 9 (5):2565-72. doi: 10.1021/pr100012p.

Hirabayashi, J. 2008. "Concept, strategy and realization of lectin-based glycan profiling." *Journal of biochemistry* no. 144 (2):139-47. doi: 10.1093/jb/mvn043.

Hudis, C. A. 2007. "Trastuzumab--mechanism of action and use in clinical practice." *N Engl J Med* no. 357 (1):39-51. doi: 357/1/39 [pii] 10.1056/NEJMra043186.

Jung, K., W. Cho, and F. E. Regnier. 2009. "Glycoproteomics of plasma based on narrow selectivity lectin affinity chromatography." *J Proteome Res* no. 8 (2):643-50. doi: 10.1021/pr8007495 10.1021/pr8007495 [pii].

Katrlik, J., J. Svitel, P. Gemeiner, T. Kozar, and J. Tkac. 2010. "Glycan and lectin microarrays for glycomics and medicinal applications." *Med Res Rev* no. 30 (2):394-418. doi: 10.1002/med.20195.

Kobata, A., and T. Endo. 1992. "Immobilized Lectin Columns - Useful Tools for the Fractionation and Structural-Analysis of Oligosaccharides." *Journal of Chromatography* no. 597 (1-2):111-122.

Koshi, Y., E. Nakata, H. Yamane, and I. Hamachi. 2006. "A fluorescent lectin array using supramolecular hydrogel for simple detection and pattern profiling for various glycoconjugates." *Journal of the American Chemical Society* no. 128 (32):10413-22. doi: 10.1021/ja0613963.

Kulasingam, V., and E. P. Diamandis. 2008. "Strategies for discovering novel cancer biomarkers through utilization of emerging technologies." *Nature clinical practice. Oncology* no. 5 (10):588-99. doi: 10.1038/ncponc1187.

Kullolli, M., W. S. Hancock, and M. Hincapie. 2010. "Automated platform for fractionation of human plasma glycoproteome in clinical proteomics." *Anal Chem* no. 82 (1):115-20. doi: 10.1021/ac9013308.

Kuno, A., N. Uchiyama, S. Koseki-Kuno, Y. Ebe, S. Takashima, M. Yamada, and J. Hirabayashi. 2005. "Evanescent-field fluorescence-assisted lectin microarray: a new strategy for glycan profiling." *Nature methods* no. 2 (11):851-6. doi: 10.1038/nmeth803.

Lau, P., J. L. Chin, S. Pautler, H. Razvi, and J. I. Izawa. 2009. "NMP22 is predictive of recurrence in high-risk superficial bladder cancer patients." *Can Urol Assoc J* no. 3 (6):454-8.

Lee, A., M. Nakano, M. Hincapie, D. Kolarich, M. S. Baker, W. S. Hancock, and N. H. Packer. 2010. "The lectin riddle: glycoproteins fractionated from complex mixtures have similar glycomic profiles." *Omics : a journal of integrative biology* no. 14 (4):487-99. doi: 10.1089/omi.2010.0075.

Li, C., D. M. Simeone, D. E. Brenner, M. A. Anderson, K. A. Shedden, M. T. Ruffin, and D. M. Lubman. 2009. "Pancreatic cancer serum detection using a lectin/glyco-antibody array method." *J Proteome Res* no. 8 (2):483-92. doi: 10.1021/pr8007013 10.1021/pr8007013 [pii].

Li, C., E. Zolotarevsky, I. Thompson, M. A. Anderson, D. M. Simeone, J. M. Casper, M. C. Mullenix, and D. M. Lubman. 2011. "A multiplexed bead assay for profiling glycosylation patterns on serum protein biomarkers of pancreatic cancer." *Electrophoresis*. doi: 10.1002/elps.201000693.

Lin, S. H., Y. C. Lee, G. Block, H. Chen, E. Folch-Puy, R. Foronjy, R. Jalili, C. B. Jendresen, M. Kimura, E. Kraft, S. Lindemose, J. Lu, T. McLain, L. Nutt, S. Ramon-Garcia, J. Smith, A. Spivak, M. L. Wang, and M. Zanic. 2008. "One-step isolation of plasma membrane proteins using magnetic beads with immobilized concanavalin A." *Protein Expression and Purification* no. 62 (2):223-229. doi: 10.1016/j.pep.2008.08.003.

Liu, Y., J. He, C. Li, R. Benitez, S. Fu, J. Marrero, and D. M. Lubman. 2010. "Identification and confirmation of biomarkers using an integrated platform for quantitative analysis of glycoproteins and their glycosylations." *Journal of proteome research* no. 9 (2):798-805. doi: 10.1021/pr900715p.

Loo, D., A. Jones, and M. M. Hill. 2010. "Lectin magnetic bead array for biomarker discovery." *J Proteome Res*. doi: 10.1021/pr100472z.

Ludwig, J. A., and J. N. Weinstein. 2005. "Biomarkers in cancer staging, prognosis and treatment selection." *Nat Rev Cancer* no. 5 (11):845-56. doi: nrc1739 [pii] 10.1038/nrc1739.

Manolio, T. A., J. E. Bailey-Wilson, and F. S. Collins. 2006. "Genes, environment and the value of prospective cohort studies." *Nat Rev Genet* no. 7 (10):812-20. doi: nrg1919 [pii] 10.1038/nrg1919.

Martens, L. 2011a. "Bioinformatics challenges in mass spectrometry-driven proteomics." *Methods in molecular biology* no. 753:359-71. doi: 10.1007/978-1-61779-148-2_24.

Martens, L. 2011b. "Data management in mass spectrometry-based proteomics." *Methods in molecular biology* no. 728:321-32. doi: 10.1007/978-1-61779-068-3_21.

Masuda, Tomomi, and Eiji Miyoshi. 2011. "Cancer biomarkers for hepatocellular carcinomas: from traditional markers to recent topics." *Clinical Chemistry and Laboratory Medicine* no. 49 (6):959-966. doi: 10.1515/cclm.2011.152.

McDonald, C. A., J. Y. Yang, V. Marathe, T. Y. Yen and Macher, B. A. 2009. "Combining Results from Lectin Affinity Chromatography and Glycocapture Approaches Substantially Improves the Coverage of the Glycoproteome." *Molecular & Cellular Proteomics* no. 8 (2):287-301. doi: 10.1074/mcp.M800272-MCP200.

Meany, D. L., L. Hackler, Jr., H. Zhang, and D. W. Chan. 2011. "Tyramide signal amplification for antibody-overlay lectin microarray: a strategy to improve the sensitivity of targeted glycan profiling." *Journal of proteome research* no. 10 (3):1425-31. doi: 10.1021/pr1010873.

Meany, D. L., L. J. Sokoll, and D. W. Chan. 2009. "Early Detection of Cancer: Immunoassays for Plasma Tumor Markers." *Expert opinion on medical diagnostics* no. 3 (6):597-605. doi: 10.1517/17530050903266830.

Meany, Danni, and Daniel Chan. 2011. "Aberrant glycosylation associated with enzymes as cancer biomarkers." *Clinical Proteomics* no. 8 (1):7.

Mechref, Y., M. Madera, and M. V. Novotny. 2008. "Glycoprotein enrichment through lectin affinity techniques." *Methods Mol Biol* no. 424:373-96. doi: 10.1007/978-1-60327-064-9_29.

Mischak, H., R. Apweiler, R. E. Banks, M. Conaway, J. Coon, A. Dominiczak, J. H. H. Ehrich, D. Fliser, M. Girolami, H. Hermjakob, D. Hochstrasser, J. Jankowski, B. A. Julian, W. Kolch, Z. A. Massy, C. Neusuess, J. Novak, K. Peter, K. Rossing, J. Schanstra, O. J. Semmes, D. Theodorescu, V. Thongboonkerd, E. M. Weissinger, J. E. Van Eyk, and T. Yamamoto. 2007. "Clinical proteomics: A need to define the field and to begin to set adequate standards." *Proteomics Clinical Applications* no. 1 (2):148-156. doi: DOI 10.1002/prca.200600771.

Mishra, A., A. C. Bharti, P. Varghese, D. Saluja, and B. C. Das. 2006. "Differential expression and activation of NF-kappaB family proteins during oral carcinogenesis: Role of high risk human papillomavirus infection." *Int J Cancer* no. 119 (12):2840-50. doi: 10.1002/ijc.22262.

Mishra, Alok, and Mukesh Verma. 2010. "Cancer Biomarkers: Are We Ready for the Prime Time?" *Cancers* no. 2 (1):190-208.

Na, K., E. Y. Lee, H. J. Lee, K. Y. Kim, H. Lee, S. K. Jeong, A. S. Jeong, S. Y. Cho, S. A. Kim, S. Y. Song, K. S. Kim, S. W. Cho, H. Kim, and Y. K. Paik. 2009. "Human plasma carboxylesterase 1, a novel serologic biomarker candidate for hepatocellular carcinoma." *Proteomics* no. 9 (16):3989-99. doi: 10.1002/pmic.200900105.

Negm, R. S., M. Verma, and S. Srivastava. 2002. "The promise of biomarkers in cancer screening and detection." *Trends Mol Med* no. 8 (6):288-93. doi: S1471491402023535.

Osterfeld, Sebastian J., Heng Yu, Richard S. Gaster, Stefano Caramuta, Liang Xu, Shu-Jen Han, Drew A. Hall, Robert J. Wilson, Shouheng Sun, Robert L. White, Ronald W. Davis, Nader Pourmand, and Shan X. Wang. 2008. "Multiplex protein assays based on real-time magnetic nanotag sensing." *Proceedings of the National Academy of Sciences* no. 105 (52):20637-20640. doi: 10.1073/pnas.0810822105.

Packer, N. H., C. W. von der Lieth, K. F. Aoki-Kinoshita, C. B. Lebrilla, J. C. Paulson, R. Raman, P. Rudd, R. Sasisekharan, N. Taniguchi, and W. S. York. 2008. "Frontiers in glycomics: bioinformatics and biomarkers in disease. An NIH white paper prepared from discussions by the focus groups at a workshop on the NIH campus, Bethesda MD (September 11-13, 2006)." *Proteomics* no. 8 (1):8-20. doi: 10.1002/pmic.200700917.

Pan, S., R. Chen, R. Aebersold, and T. A. Brentnall. 2011. "Mass spectrometry based glycoproteomics--from a proteomics perspective." *Mol Cell Proteomics* no. 10 (1):R110 003251. doi: R110.003251 [pii] 10.1074/mcp.R110.003251.

Patwa, Tasneem H., Jia Zhao, Michelle A. Anderson, Diane M. Simeone, and David M. Lubman. 2006. "Screening of Glycosylation Patterns in Serum Using Natural Glycoprotein Microarrays and Multi-Lectin Fluorescence Detection." *Analytical chemistry* no. 78 (18):6411-6421. doi: 10.1021/ac060726z.

Pepe, M. S., R. Etzioni, Z. Feng, J. D. Potter, M. L. Thompson, M. Thornquist, M. Winget, and Y. Yasui. 2001. "Phases of biomarker development for early detection of cancer." *J Natl Cancer Inst* no. 93 (14):1054-61.

Pilobello, K. T., and L. K. Mahal. 2007. "Lectin microarrays for glycoprotein analysis." *Methods in molecular biology* no. 385:193-203.

Plavina, T., E. Wakshull, W. S. Hancock, and M. Hincapie. 2007. "Combination of abundant protein depletion and multi-lectin affinity chromatography (M-LAC) for plasma protein biomarker discovery." *Journal of proteome research* no. 6 (2):662-71. doi: 10.1021/pr060413k.

Polanski, M., and N. L. Anderson. 2007. "A list of candidate cancer biomarkers for targeted proteomics." *Biomark Insights* no. 1:1-48.

Qiu, Y., T. H. Patwa, L. Xu, K. Shedden, D. E. Misek, M. Tuck, G. Jin, M. T. Ruffin, D. K. Turgeon, S. Synal, R. Bresalier, N. Marcon, D. E. Brenner, and D. M. Lubman. 2008. "Plasma glycoprotein profiling for colorectal cancer biomarker identification by lectin glycoarray and lectin blot." *J Proteome Res* no. 7 (4):1693-703. doi: 10.1021/pr700706s.

Rakus, J. F., and L. K. Mahal. 2011. "New technologies for glycomic analysis: toward a systematic understanding of the glycome." *Annual review of analytical chemistry* no. 4 (1):367-92. doi: 10.1146/annurev-anchem-061010-113951.

Ray, S., P. J. Reddy, S. Choudhary, D. Raghu, and S. Srivastava. 2011. "Emerging nanoproteomics approaches for disease biomarker detection: A current perspective." *Journal of proteomics*. doi: 10.1016/j.jprot.2011.04.027.

Rifai, N., M. A. Gillette, and S. A. Carr. 2006. "Protein biomarker discovery and validation: the long and uncertain path to clinical utility." *Nat Biotechnol* no. 24 (8):971-83. doi: nbt1235 [pii] 10.1038/nbt1235.

Roses, R. E., E. C. Paulson, A. Sharma, J. E. Schueller, H. Nisenbaum, S. Weinstein, K. R. Fox, P. J. Zhang, and B. J. Czerniecki. 2009. "HER-2/neu overexpression as a predictor for the transition from in situ to invasive breast cancer." *Cancer Epidemiol Biomarkers Prev* no. 18 (5):1386-9. doi: 1055-9965.EPI-08-1101 [pii] 10.1158/1055-9965.EPI-08-1101.

Schiess, Ralph, Bernd Wollscheid, and Ruedi Aebersold. 2009. "Targeted proteomic strategy for clinical biomarker discovery." *Molecular Oncology* no. 3 (1):33-44. doi: 10.1016/j.molonc.2008.12.001.

Shamberger, R. J. 1984. "Serum sialic acid in normals and in cancer patients." *Journal of clinical chemistry and clinical biochemistry. Zeitschrift fur klinische Chemie und klinische Biochemie* no. 22 (10):647-51.

Silver, H. K., K. A. Karim, E. L. Archibald, and F. A. Salinas. 1979. "Serum sialic acid and sialyltransferase as monitors of tumor burden in malignant melanoma patients." *Cancer research* no. 39 (12):5036-42.

Smith, M. P., S. L. Wood, A. Zougman, J. T. Ho, J. Peng, D. Jackson, D. A. Cairns, A. J. Lewington, P. J. Selby, and R. E. Banks. 2011. "A systematic analysis of the effects of increasing degrees of serum immunodepletion in terms of depth of coverage and other key aspects in top-down and bottom-up proteomic analyses." *Proteomics* no. 11 (11):2222-35. doi: 10.1002/pmic.201100005.

Sparbier, K., T. Wenzel, and M. Kostrzewa. 2006. "Exploring the binding profiles of ConA, boronic acid and WGA by MALDI-TOF/TOF MS and magnetic particles." *J Chromatogr B Analyt Technol Biomed Life Sci* no. 840 (1):29-36. doi: S1570-0232(06)00526-5 [pii] 10.1016/j.jchromb.2006.06.028.

Srivastava, S., and R. Gopal-Srivastava. 2002. "Biomarkers in cancer screening: a public health perspective." *J Nutr* no. 132 (8 Suppl):2471S-2475S.

Srivastava, S., M. Verma, and R. Gopal-Srivastava. 2005. "Proteomic maps of the cancer-associated infectious agents." *J Proteome Res* no. 4 (4):1171-80. doi: 10.1021/pr050017m.

Sturgeon, C. M., M. J. Duffy, B. R. Hofmann, R. Lamerz, H. A. Fritsche, K. Gaarenstroom, J. Bonfrer, T. H. Ecke, H. B. Grossman, P. Hayes, R. T. Hoffmann, S. P. Lerner, F. Lohe, J. Louhimo, I. Sawczuk, K. Taketa, and E. P. Diamandis. 2010. "National Academy of Clinical Biochemistry Laboratory Medicine Practice Guidelines for use of tumor markers in liver, bladder, cervical, and gastric cancers." *Clin Chem* no. 56 (6):e1-48. doi: clinchem.2009.133124 [pii] 10.1373/clinchem.2009.133124.

Sturgeon, C. M., M. J. Duffy, U. H. Stenman, H. Lilja, N. Brunner, D. W. Chan, R. Babaian, R. C. Bast, Jr., B. Dowell, F. J. Esteva, C. Haglund, N. Harbeck, D. F. Hayes, M. Holten-Andersen, G. G. Klee, R. Lamerz, L. H. Looijenga, R. Molina, H. J. Nielsen, H. Rittenhouse, A. Semjonow, M. Shih Ie, P. Sibley, G. Soletormos, C. Stephan, L. Sokoll, B. R. Hoffman, and E. P. Diamandis. 2008. "National Academy of Clinical Biochemistry laboratory medicine practice guidelines for use of tumor markers in testicular, prostate, colorectal, breast, and ovarian cancers." *Clin Chem* no. 54 (12):e11-79. doi: 54/12/e11 [pii] 10.1373/clinchem.2008.105601.

Sun, B., J. A. Ranish, A. G. Utleg, J. T. White, X. Yan, B. Lin, and L. Hood. 2007. "Shotgun glycopeptide capture approach coupled with mass spectrometry for comprehensive glycoproteomics." *Mol Cell Proteomics* no. 6 (1):141-9. doi: T600046-MCP200 [pii] 10.1074/mcp.T600046-MCP200.

Surinova, S., R. Schiess, R. Huttenhain, F. Cerciello, B. Wollscheid, and R. Aebersold. 2011. "On the development of plasma protein biomarkers." *J Proteome Res* no. 10 (1):5-16. doi: 10.1021/pr1008515.

Theilacker, Nora, Eric E. Roller, Kristopher D. Barbee, Matthias Franzreb, and Xiaohua Huang. 2011. "Multiplexed protein analysis using encoded antibody-conjugated microbeads." *Journal of The Royal Society Interface* no. 8 (61):1104-1113. doi: 10.1098/rsif.2010.0594.

Tousi, Fateme, William S. Hancock, and Marina Hincapie. 2011. "Technologies and strategies for glycoproteomics and glycomics and their application to clinical biomarker research." *Analytical Methods* no. 3 (1):20-32.

Tu, C., P. A. Rudnick, M. Y. Martinez, K. L. Cheek, S. E. Stein, R. J. Slebos, and D. C. Liebler. 2010. "Depletion of abundant plasma proteins and limitations of plasma proteomics." *Journal of proteome research* no. 9 (10):4982-91. doi: 10.1021/pr100646w.

Verma, M., and U. Manne. 2006. "Genetic and epigenetic biomarkers in cancer diagnosis and identifying high risk populations." *Crit Rev Oncol Hematol* no. 60 (1):9-18. doi: S1040-8428(06)00087-4 [pii] 10.1016/j.critrevonc.2006.04.002.

Wang, P., J. R. Whiteaker, and A. G. Paulovich. 2009. "The evolving role of mass spectrometry in cancer biomarker discovery." *Cancer Biol Ther* no. 8 (12):1083-94. doi: 8634 [pii].

Welinder, C., B. Jansson, M. Ferno, H. Olson, and B. Baldetorp. 2009. "Expression of Helix pomatia lectin binding glycoproteins in women with breast cancer in relationship to their blood group phenotypes." *J Proteome Res* no. 8 (2):782-7. doi: 10.1021/pr800444b.

West, I., and O. Goldring. 1996. "Lectin affinity chromatography." *Methods in molecular biology* no. 59:177-85. doi: 10.1385/0-89603-336-8:177.

Xu, Z., X. Zhou, H. Lu, N. Wu, H. Zhao, L. Zhang, W. Zhang, Y. L. Liang, L. Wang, Y. Liu, P. Yang, and X. Zha. 2007. "Comparative glycoproteomics based on lectins affinity capture of N-linked glycoproteins from human Chang liver cells and MHCC97-H cells." *Proteomics* no. 7 (14):2358-70. doi: 10.1002/pmic.200600041.

Yang, N., S. Feng, K. Shedden, X. L. Xie, Y. S. Liu, C. J. Rosser, D. M. Lubman, and S. Goodison. 2011. "Urinary Glycoprotein Biomarker Discovery for Bladder Cancer Detection Using LC/MS-MS and Label-Free Quantification." *Clinical Cancer Research* no. 17 (10):3349-3359. doi: 10.1158/1078-0432.CCR-10-3121.

Yang, Z., and W. S. Hancock. 2004. "Approach to the comprehensive analysis of glycoproteins isolated from human serum using a multi-lectin affinity column." *J Chromatogr A* no. 1053 (1-2):79-88.

Yang, Z., and W. S. Hancock. 2004. "Approach to the comprehensive analysis of glycoproteins isolated from human serum using a multi-lectin affinity column." *J Chromatogr A* no. 1053 (1-2):79-88.

Yang, Z., L. E. Harris, D. E. Palmer-Toy, and W. S. Hancock. 2006. "Multilectin affinity chromatography for characterization of multiple glycoprotein biomarker candidates in serum from breast cancer patients." *Clin Chem* no. 52 (10):1897-905. doi: clinchem.2005.065862 [pii] 10.1373/clinchem.2005.065862.

Yoon, S. J., S. Y. Park, P. C. Pang, J. Gallagher, J. E. Gottesman, A. Dell, J. H. Kim, and S. I. Hakomori. 2010. "N-glycosylation status of beta-haptoglobin in sera of patients with prostate cancer vs. benign prostate diseases." *Int J Oncol* no. 36 (1):193-203.

Zeng, Z., M. Hincapie, S. J. Pitteri, S. Hanash, J. Schalkwijk, J. M. Hogan, H. Wang, and W. S. Hancock. 2011. "A proteomics platform combining depletion, multi-lectin affinity chromatography (M-LAC), and isoelectric focusing to study the breast cancer proteome." *Analytical chemistry* no. 83 (12):4845-54. doi: 10.1021/ac2002802.

Zhang, H., X. J. Li, D. B. Martin, and R. Aebersold. 2003. "Identification and quantification of N-linked glycoproteins using hydrazide chemistry, stable isotope labeling and mass spectrometry." *Nature biotechnology* no. 21 (6):660-6. doi: 10.1038/nbt827.

Zhang, Q., V. Faca, and S. Hanash. 2011. "Mining the plasma proteome for disease applications across seven logs of protein abundance." *J Proteome Res* no. 10 (1):46-50. doi: 10.1021/pr101052y.

Zhao, J., T. H. Patwa, W. Qiu, K. Shedden, R. Hinderer, D. E. Misek, M. A. Anderson, D. M. Simeone, and D. M. Lubman. 2007. "Glycoprotein microarrays with multi-lectin detection: unique lectin binding patterns as a tool for classifying normal, chronic pancreatitis and pancreatic cancer sera." *Journal of proteome research* no. 6 (5):1864-74. doi: 10.1021/pr070062p.

Zhao, J., D. M. Simeone, D. Heidt, M. A. Anderson, and D. M. Lubman. 2006. "Comparative serum glycoproteomics using lectin selected sialic acid glycoproteins with mass spectrometric analysis: application to pancreatic cancer serum." *J Proteome Res* no. 5 (7):1792-802. doi: 10.1021/pr060034r.

Zheng, T., D. Peelen, and L. M. Smith. 2005. "Lectin arrays for profiling cell surface carbohydrate expression." *Journal of the American Chemical Society* no. 127 (28):9982-3. doi: 10.1021/ja0505550.

Zielinska, D. F., F. Gnad, J. R. Wisniewski, and M. Mann. 2010. "Precision mapping of an in vivo N-glycoproteome reveals rigid topological and sequence constraints." *Cell* no. 141 (5):897-907. doi: 10.1016/j.cell.2010.04.012.

10

Recent Advances in Glycosylation Modifications in the Context of Therapeutic Glycoproteins

Xiaotian Zhong* and Will Somers
Pfizer Global BioTherapeutics Technologies,
Cambridge, MA
USA

1. Introduction

Glycosylation is one of the most complex post-translation modifications, commonly found in many cell surface and secreted eukaryotic proteins. 1-2% of the human transcriptome encodes proteins that link to glycosylation. Many protein-based biotherapeutics approved or in clinical trials are glycoproteins. The oligosaccharides covalently attached to therapeutic glycoproteins pose biological benefits as well as manufacturing challenges. The present chapter reviews the structure and function of glycosylation, glycoform patterns observed for the biotherapeutic proteins produced by various host systems, and analytic methods for the characterization of glycoforms. Recent advances in utilizing glycosylation as a strategy to improve biotherapeutics properties are also discussed.

2. Glycosylation as a major post-translational modification

Glycosylation has been studied intensively for the past two decades as the most common covalent protein modification in eukaryotic cells (Varki 2009). Sophisticated oligosaccharide analysis has revealed a remarkable complexity and diversity of this post-translational modification. About 1-2% of the human transcriptome (about 250-500 glycogenes) has been predicted to encode proteins that are involved in glycosylation processing (Campbell and Yarema 2005). Majority of proteins synthesized in the endoplasmic reticulum (ER) such as cell surface and extracellular eukaryotic proteins are glycoproteins. It has been estimated that more than 50% of proteins in human are glycosylated (Apweiler et al. 1999; Wong 2005).

Glycoproteins can be classified into four groups: N-linked, O-linked, glycosaminoglycans, and glycosylphosphatidylinositol-anchored proteins (Table 1). This chapter focuses only on N- and O-linked glycosylation. N-linked glycosylation is through the side chain amide nitrogen of a specific asparagine residue, while O-linked glycosylation is through the oxygen atom in the side chain of serine or threonine residues. The N-linked modification takes place in both ER and Golgi, while the O-linked glycosylation in higher eukaryotes occurs exclusively in the Golgi.

* Corresponding Author

Type	Consensus site	Sugar structures	Synthesis Location
(1) N-linked	N-X-T/S	High mannose Complex-type Hybrid-type	ER, Golgi
(2) O-linked	Ser/Thr	Mucin-type O-linked fucose O-linked glucose O-linked GlcNAc	Golgi
(3) Glycosaminoglycans	Asn/Ser/Thr	Glycosaminoglycans	ER, Golgi
(4) Glycosylphosphatidylinositol	phosphatidylinositol / phosphoethanolamine to protein carboxyl terminus	glycosylphosphatidylinositol	ER, Golgi

Table 1. Glycoproteins Categories

Comparing to other major molecular constituents of cells such as nucleic acids and proteins, the biological importance of glycans or carbohydrates in the post-translational modification has been much later appreciated (Varki 2009). There is no single theory explaining why cells go through such complex and highly conserved biosynthetic machineries. Though not all answers are known, it is now clear that glycosylation plays many key biological functions such as protein folding, stability, intracellular and inter-cellular trafficking, cell-cell and cell-matrix interaction (Varki 1993; Varki 2009).

It is therefore not surprising that congenital disorders with serious medical consequences have been identified linked to the defects in a number of genes in glycosylation pathway (Freeze 2006). Over 40 such disorders have been reported to be associated with glycogene mutations, and many more to be discovered. In addition, glycosylation profiles of specific proteins change as certain diseases progress, such as cancers and rheumatoid arthritis, and have been regarded as disease and diagnostic markers.

This chapter focuses on the biological structures and physiological roles of glycosylation modification in the context of biotherapeutics. Glycosylation differences in proteins produced by various host systems, and the potential impacts on biotherapeutics safety and side effects, are described. Various analytical characterization methods for glycoforms are also described. Lastly, several therapeutic examples with glycoengineering application are illustrated and discussed.

2.1 Structure and biosynthesis

N-linked glycosylation occurs in the sequon of Asn-X-Ser/Thr where X can be any amino acid except proline and aspartic acid (Helenius and Aebi 2004; Kornfeld and Kornfeld 1985). Glycosylation at Asn-Ala-Cys has also been reported (Stenflo and Fernlund 1982). Glycosylation efficiency of these Threonine, Serine, and Cysteine containing sequon is very different with an order of Thr>Ser>Cys (Bause and Legler 1981). N-linked oligosaccharides are added to proteins *en bloc* in the lumen of ER as pre-synthesized core units of 14 saccharides ($Glc_3Man_9GlcNAc_2$) in virtually all eukaryotes. This core glycan is the product of a biosynthesis pathway in which monosaccharides are added to a lipid carrier (dolichol-pryophosphate) on both sides of the ER membrane by monosaccharyltransferases in the membrane. The sugar moiety is translocated from cytosolic side to the luminal side of the ER by an ATP-independent bidirectional flippase (Hirschberg and Snider 1987). The oligosaccharyltransferase then scans the emerging polypeptide from translocon complex for glycosylation sequon and adds the core glycan unit to the side chain nitrogen of the Asn residue by N-glycosidic bond. The oligosaccharides are added to the sequon when it is only 12-14 residues into the ER lumen, as the active site of the oligosaccharytransferase is no further than 5nm away from the exit of the protein translocon (Nilsson and von Heijne 1993).

After the core glycan is added to the growing nascent polypeptide chain, the oligosaccharide portion is modified by a series of glycosidases and glycosyl transferases. Various complex, hybrid, and high mannose types of N-linked oligosaccharides are generated. Glucosidase I and II located in the ER remove all three glucose residues from the core unit to produce a $Man_9GlcNAc_2$ high mannose structure. Hybrid and complex oligosaccharides can be produced from high mannose structures, from which α-mannosidases in the ER and the Golgi remove 4-6 mannoses. Then Golgi-bound glycosyl transferases add GlcNAc as well as galactoses and sialic acids to produce complex types of oligosaccharides. These modifications reflect a spectrum of functions related to glycoprotein folding, quality control, sorting, degradation, and secretion.

O-linked glycosylation normally takes place in the Golgi, most commonly initiated with a transfer of N-acetylgalactosamine (GalNAc) to a serine or threonine residue by an N-acetyl galactosaminyltransferase (Van den Steen et al. 1998). After the addition of the first GalNAc, a number of glycosyltransferases and enzymes in the Golgi can elongate the core structure and modify it with sialylation, fucosylation, sulphatation, methylation or acetylation(Van den Steen et al. 1998). O-linked glycosylation site is not readily predicted, any serine or threonine residue is a potential site and O-linked sugars are frequently clustered in short regions of peptide chain that contain repeating units of Serine, Threonine, and Proline. There are various types of O-linked sugars, including mucin-type O-glycans commonly found in many secreted and membrane-bound glycoproteins in higher eukaryotes, O-linked fucose and O-linked glucose found in the epidermal growth factor domains of different proteins, and O-linked GlcNAc on cytosolic and nuclear proteins. Yeast's O-linked oligomannose glycans take place in the ER utilizing dolichol-phosphate-mannose instead of a sugar nucleotide, which is similar to N-linked glycosylation occurred co-translationally (van den Steen et al, 1998).

2.2 Physiological function and roles

Protein folding and conformation stabilization function of N-linked glycans were first suggested by the early studies with tunicamycin, a glycosylation inhibitor (Olden et al. 1982). The sequential processing by glucosidases, mannosidases, and glycotranferases, of the core unit of 14 saccharides, provides recognition tags for lectins mediated folding pathway (Helenius and Aebi 2004). The content of oligosaccharides can regulate protein half-life. Large amount of sialic acids can increase plasma half-life while exposure of galactose and mannose can decrease half-life (Walsh and Jefferis 2006). N-glycans also play a critical role in intracellular trafficking with a well understood example of mannose-6-phosphate of lysosomal enzymes (Kornfeld and Mellman 1989). For Antibodies, oligosaccharide moieties covalently attached at the highly conserved Asn297 at the CH2 domain of the Fc (crystallizable fragment) region, is critical to the activation of downstream effector mechanisms (Jefferis 2009; Natsume et al. 2009). Completely aglycosylated or deglycosylated IgGs do not bind effector receptors such as FcγRI, FcγRII, and FcγRIII (Leader et al. 1991; Leatherbarrow et al. 1985; Walker et al. 1989). Sialylated IgGs have a lower affinity to FcγRIIIA than non-sialylated IgGs, consequently a lower antibody-dependent cellular cytotoxicity (ADCC) activity (Kaneko et al. 2006; Scallon et al. 2007). Removal of terminal galactose residues from Fc glycans reduces complement-dependent cytotoxicity (CDC) activity (Boyd et al. 1995; Kumpel et al. 1995). Absence of a core α-1, 6 linked fucose from Fc glycans improves *in vitro* ADCC activity (Niwa et al. 2004; Shields et al. 2002).

O-linked glycosylation plays a role in maintaining secondary, tertiary, and quaternary structures of fully folded proteins. The examples are mucins and related molecules, in which peptide regions with O-linked sugar attachments assume a "bottle brush"-like structure (Carraway and Hull 1991; Gowda and Davidson 1994). Like N-glycans, O-glycans can modulate aggregation, maintain protein stability, confer protease and heat resistance. An example of O-linked sugars hindering protease cleavage is the modification at the hinge regions of IgA_1 and IgD (Field et al. 1994; Van den Steen et al. 1998). O-linked glycosylation is important for the expression and processing of particular proteins such as glycophorin A (Remaley et al. 1991) and IGF-II (Daughaday et al. 1993). They are also crucial for some glycoprotein-protein interaction, such as the interaction between P-selectin glycoprotein ligand-1 (PSGL-1) and P Selectin. Some O-linked oligosaccharides of PSGL-1 have a terminal sialyl-Lewis-x structure, which is important for its P-selectin receptor function (Hooper et al. 1996).

3. Glycoproteins as biotherapeutics

More than one-third of approved biotherapeutics and many in clinical trials are glycoproteins (Walsh and Jefferis 2006). The presence and nature of the oligosaccharides clearly affect these protein drugs' folding, stability, trafficking, immunogenicity as well as their primary activities.

3.1 Antibodies and Fc-fusion proteins

Therapeutic recombinant antibodies and fusion proteins of Fc region of immunoglobulin G_1 (IgG_1) represent a major class of biotherapeutics. An individual antibody molecule contains two light and two heavy polypeptide chains, forming two identical Fab (antigen-binding fragment) regions with a specific antigen-binding site, and a homodimeric IgG-Fc region. This Fc region is critical for phagocytosis, ADCC activity, CDC activity, and FcRn binding for recycling. As discussed above, the N-glycans attached to Asn297 in Fc region are critical to the activation of downstream effector mechanisms, while not affecting FcRn binding for catabolic half-life.

Besides the presence of core glycans at the Fc regions, about 30% of polyclonal human IgG molecules contain N-linked oligosaccharides within the IgG-Fab region (Jefferis 2009). The N-linked sites can be at the variable regions of either heavy chains or light chains or both. The licensed antibody therapeutics cetuximab has an N-linked glycan at Asn88 of the heavy chain variable region, and an unoccupied N-linked motif at Asn41 of the light chain variable region (Qian et al. 2007). Fab oligosaccharide is heterogeneous complex diantenary and hybrid oligosaccharides with sialic acids and galactoses, which are very different from the Fc oligosaccharides with predominantly fucosylated non-galactosylated diantennary oligosaccharides. The difference may be due to the inaccessibility of Fc N-glycan for further modification, as the N-glycans at the Fc regions are integral to the IgG structure and has a defined confirmation (Jefferies, 2009). Many Fc-fusion therapeutics proteins, such as TNFRII-Fc, CD2-Fc, and CTLA4-Fc, contain glycosylation modifications in the fusion portions, in addition to their Fc glycans. The contents of these glycosylations are very similar to those of Fab oligosaccharides.

3.2 Non-immunoproteins

Many non-immunoproteins such as growth factors, cytokines, hormones, and therapeutic enzymes, are glycoproteins. Growth factors such as erythropoietin (EPO) have three N-linked and one O-linked sugar side chains. Removal of either two (Asn38 and Asn83) or all three sites results in poor product secretion (Egrie 1993). Cytokines such as interferon(IFN)-β and IFN-γ are glycoproteins (Pestka et al. 1987). Although glycosylation is not essential for INFs protein efficacy or safety, lack of glycosylation decreases their biological activity and circulatory half-life. Oligosaccharide structures of follicle-stimulating hormone heterodimer play an important role in its biosynthesis, secretion, metabolic fate, and functional potency (Ulloa-Aguirre et al. 1999). The glycans at each subunit seem to exhibit distinct roles, with those in α subunit critical for dimer assembly, signal transduction, and secretion, and those in β subunit more crucial for circulation clearance. In addition, many therapeutics enzymes such as recombinant human glucocerebrosidase for Gaucher disease (Van Patten et al. 2007) are glycoproteins and N-glycosylation is important for its targeting and functional activities.

3.3 Effects of glycosylation on therapeutic efficacy of glycoproteins

In comparison to small-molecule drugs, therapeutic proteins display a number of favorable therapeutic properties, such as higher target specificity, good pharmacological potencies, and lower side effects, but they also possess intrinsic limitations like poor physicochemical and pharmacological properties. Glycosylation of therapeutic glycoproteins can improve therapeutic efficacy through its positive impact on protein pharmacodynamics (PD) and pharmacokinetics (PK).

Pharmacodynamics refers to the potency of therapeutic proteins as enzymatic rates and receptor binding affinities. Pharmacokinetics exams the time dependency of drug action, which is influenced by drug absorption, distribution, excretion, initial response times and duration of effects. The parameters include circulatory half-life, volumes of distribution, clearance rates, and total bioavailability. Protein drugs' PK/PD are typically affected by adverse local adsorption in subcutaneous administration due to variable protein hydropathy surface, and by rapid elimination from body in intravenous administration, via proteolytic, renal, hepatic, and receptor mediated clearance mechanisms (Mahmood and Green 2005; Tang et al. 2004).

Glycosylation has multiple impacts on PK/PD properties of therapeutics glycoproteins. First, glycosylation can shield non-specific proteolytic degradation, as discussed above. Second, sialic acids at the terminus of glycan chains carry negative charge, which reduces renal clearance most likely due to repulsion from negatively charged polysaccharides on membranes in the glomerular filter (Chang et al. 1975; Venkatachalam and Rennke 1978). Third, size of glycans can increase protein molecular weight and hydrodynamic radius of glycoprotein and therefore reduce glomerular filtration. Fourth, terminal sialic acids of glycan branches prevent the exposure of galactose, N-acetyl-glycosamine, or mannose that interacts with hepatic asialoglycoprotein receptor as well as other mammalian lectin-like receptors to be removed from circulation.

4. Glycosylation in various cell production systems

Glycosylation patterns of biotherapeutics are highly variable based on the production systems (Table 2) and their culture processes. Mammalian cells such as Chinese Hamster Ovary cells (CHO) and mouse myeloma cells (NS0, SP2/0) are the most commonly used systems. Alternative cell production systems are being developed and explored.

Host systems	Similarity to human glycans	Abnormal sugars
CHO	High	trace amount of α-Gal, NGNA
NS0/SP2/0	High	small amount of α-Gal, NGNA
Yeast	Low	high mannose
Plant	Low	bisecting β1,2 xylose, α1,3 fucose
Transgenic animals	Low	high mannose and NGNA

Table 2. Glycans comparison in various production systems

The glycoforms of CHO-produced IgGs are close to human IgGs, but having very little glycoform with the third N-acetylglucosamine bisecting arm, which makes up about 10% of human IgG glycoforms, and also very low amount of terminal N-acetyl neuraminic acid is generated. The glycosylation in mouse-derived cells such as NS0 and SP2/0 shows even more difference from human glycoforms. They produce small amounts of glycoforms with additional α-1,3-galactose (α-Gal) and a different predominant sialic acid, N-Glycolyl neuraminic acid (NGNA). NGNA is reported to be immunogenic in human (Sheeley et al. 1997), and in certain patient populations, α-Gal is associated with IgE-mediated anaphylactic responses, with the best known example of cetuximab (Chung et al. 2008). Detection of both α-Gal and NGNA in CHO-derived glycans is also reported, but only in trace amount (Hamilton and Gerngross 2007; van Bueren et al. 2011).

Yeast, insect cells, plants, and transgenic animals, are the alternative systems to the current mammalian hosts. They are being actively explored for biotherapeutics production because of their lower manufacture cost. However, restricted abilities to generate human-like glycoforms are their major limitations, as different glycosylation machinery yields immunogenic recombinant glycoproteins. For instance, complex type N-glycans are very different in plants and mammals. Plant N-glycans contain a bisecting β1,2 xylose in place of β mannose core, an α1,3 fucose instead of an α1,6 fucose, and are highly heterogeneous (Gomord et al. 2005), and allergenic. Glycans from yeast (Hamilton et al. 2006) and insect (Shi and Jarvis 2007) have a high mannose content, resulting a quick clearance through binding to macrophage mannose receptor in the liver. IgGs produced in the milk of transgenic goats contain 50% NGNA and a higher level of mannose (Edmunds et al. 1998). Tremendous efforts have focused on "humanization" of the glycosylation pathways in these alternative systems to improve product consistency and pharmacokinetics, while decreasing the potential immunogenicity for product antibody response.

5. Analytic characterization of glycoforms

Various glycosylation analysis approaches (Table 3) have been developed and utilized for glycoform characterization. Glycans can be enzymatically or chemically released from glycoproteins, prior to electrophoretic, chromatographic or mass spectrometric analysis. Glycoproteins can also be treated with endoproteinases, followed by glycosylation analysis at the glycopeptides level.

Methods	Principles	Major advantages and shortcoming
(1) Electrophoresis		
SDS-PAGE	Size	General equipment, cheap, fast High-throughput possible,
IEF	Charge	Limited resolution
(2) Chromatography (HPLC)	Polarity	High resolution Separation of major glycoforms Risk of hydrolysis
(3) Mass spectrometry	Mass	Detailed information, fast High resolution, precise Expensive equipment Trained personnel

Table 3. Glyco-analytical methods

5.1 Electrophoresis

Sodium Docedyl sulfate-Polyacrylamide Gel Electrophoresis (SDS-PAGE) and IsoElectrofocusing electrophoresis (IEF) are two methods that are routinely used for gross glycoprotein characterization. SDS-PAGE is for separation of mass variant due to the 2kDa mass addition of a single N-glycan. When treated with glycanase such as PNGase F and Endo-H, a migration shift can be detected. IEF is for separation of charge variants. The sialic acid content of glycans can increase negative charge of glycoproteins, while the PNGase F treatment generates a negatively charged aspartic acid instead of the neutral N-glycan linked asparagines.

5.2 Liquid chromatography

Normal phase high-performance liquid chromatography (NP-HPLC) is one of the most commonly used analytical methods to analyze oligosaccharides after enzymatic release and fluorescent labeling. The glycans can be accurately quantified and detected in sub-picomolar levels (Guile et al. 1996). Different peaks in an NP-HPLC chromatogram can be isolated and submitted to off-line analysis by mass spectrometry or to sequential digestion with selective exoglycosidases (neuraminidase, α-galactosidase, β-galactosidase, β-hexoaminidase, α-fucosidase, α-mannosidase, β-mannosidase) for further biochemical confirmation. NP-HPLC can also be used for routine IgGs glycan finger printing for IgGs expressed in different cell lines.

5.3 Mass spectrometry

Mass spectrometry is a fast and powerful method to differentiate and estimate the relative proportion of different glycoforms. Glycans and glycopeptides are traditionally ionized by fast atom bombardment and laser desorption. In the past two decades, softer ionization techniques such as Electrospray Ionization-Time-of-Flight (ESI-TOF) and Matrix-assisted laser desorption ionization (MALDI) provide a much higher sensitivity and precision. It allows measuring intact glycoproteins and investigating non-symmetry of N-linked biantennary oligosaccharides between two heavy chains on intact antibodies (Beck et al. 2008).

6. Glycoengineering to improve protein therapeutics

It is obvious that selectively producing a certain type of glycoforms of biotherapeutics protein could be advantageous in terms of efficacy and safety. Residue screening with site-directed mutagenesis is widely used to introduce or eliminate N-glycosylation sites (Zhong et al. 2009). Though there is no "one-size-for-all" principle and guideline, the process has been aided by knowledge of the known structure and function of the target protein so that the changes can retain *in vitro* biological activity, stability, and high sugar occupancy rate. Cell line engineering to knock-out and knock-in glycogenes is another approach to enrich desired glycoforms. It is also possible to use in vitro glycoenzymes to modify glycoform profiles. The following are a few specific examples.

6.1 Half-life extension

One well known glycoengineering application is altering pharmacokinetic property of therapeutic proteins. Introducing new N-linked glycosylation site into target proteins to increase sialic acid containing carbohydrates can increase *in vivo* activity due to a longer half-life. This technology has been successfully applied to produce a hyperglycosylated analogue of recombinant human erythropoietin (Elliott et al. 2003). This glycoengineered protein contains two additional N-linked carbohydrates, which result in a threefold increase in serum half-life and a less frequent dosing for anemic patients (Sinclair and Elliott 2005). Sialic acid containing carbohydrates are highly hydrophilic and therefore increase protein solubility by shielding hydrophobic residues. Similar approach has been applied to a number of therapeutic proteins, including human growth hormone (Flintegaard et al. 2010), follicle stimulating hormone(Perlman et al. 2003), Leptin and Mpl ligand (Elliott et al. 2003). In case of human growth hormone, the terminal half-life in rats for the sialylated protein with three additional N-linked glycans was prolonged by 24-fold compared with that of wild type protein (Flintegaard et al. 2010). The correlation between half-life optimization and N-linked carbohydrate addition remain unclear.

6.2 Glycoengineered antibody for ADCC modulation

N-glycans in the Fc-region of IgG_1 play a critical role in ADCC activity. Absence of a core α-1, 6 linked fucose improves binding to FcγRIII and *in vitro* ADCC activity (Niwa et al. 2004; Shields et al. 2002). Addition of bisecting GlcNAc, which also results in the removal of core fucose, significantly enhances ADCC activity (Davies et al. 2001; Shinkawa et al. 2003; Umana et al. 1999). ADCC enhancement has also been shown for non-fucosylated IgG4 (Niwa et al. 2005), for Fc fusion proteins (TNFRII-Fc) (Shoji-Hosaka et al. 2006), for single chain-Fc and bispecific antibodies (Natsume et al. 2006). Several glycoengineered antibodies such as anti-GD3 (BioWa), anti-CD20 (Glycart-Roche), and anti-IL5R (BioWa/Medimmune) are currently being investigated in clinical trials.

Besides defucosylation, sialylation is also utilized for antibody and Fc engineering. Sialylated IgGs have been found to possess a lower ADCC activity than non-sialylated IgGs (Kaneko et al. 2006; Scallon et al. 2007). Overexpressing gal and sialic transferases in CHO results in sialylation increase of ≥ 90% of available glycan branches in Fc-fusion proteins (Weikert et al. 1999).

6.3 Mannose for target delivery

Engineered glycosylation has been employed for targeted delivery to disease affected tissues. One well established example is the treatment of lysosomal storage diseases. Recombinant human enzymes such as glucocerebrosidase can be digested with exoglycosidases to expose mannose or mannose-6-phosphate that can efficiently target the enzymes into the lysosomes of macrophages. The high mannose modified enzymes can also be produced by a glycosylation mutant such as *Lec1* mutant (Van Patten et al. 2007), or by treatment of chemical inhibitors (Zhou et al. 2008). Targeting the protein drugs to the desired site by glycoengineering have significantly increased therapeutic efficacy of a number of replacement enzymes, including α-glucosidase, α-galatosidase, and α-L-iduronidase (Sola and Griebenow 2010).

7. Conclusions and future directions

Glycosylation modification offers both an opportunity and a challenge to biotherapeutics glycoproteins. Complexity and heterogeneity of oligosaccharides present a considerable challenge to the biopharmaceutical industry to manufacture biotherapeutics with a reproducible and consistent glycoform profile. Meanwhile, a better understanding of the structure and function of glycosylation modification to glycoproteins can better facilitate the development of next-generation of biotherapeutics with optimized glycoforms and therapeutic utilities. Further humanization of glycosylation machinery in non-mammalian expression systems may represent a trend in lowering the manufacture cost for biotherapeutics such as antibodies and Fc-fusion proteins. With a full development of glycoanalytical techniques, an improved knowledge on glycoprotein activity *in vivo* will certainly help design a safer and more efficacious biotherapeutics drugs.

8. Acknowledgement

We would like to thank Ronald Kriz for critical reading on the manuscript. This book chapter is dedicated to the centenary of the late Prof. Haoran Jian (1911-2011) (by X.Z.).

9. References

Apweiler R, Hermjakob H, Sharon N. 1999. On the frequency of protein glycosylation, as deduced from analysis of the SWISS-PROT database. Biochim Biophys Acta 1473(1):4-8.

Bause E, Legler G. 1981. The role of the hydroxy amino acid in the triplet sequence Asn-Xaa-Thr(Ser) for the N-glycosylation step during glycoprotein biosynthesis. Biochem J 195(3):639-44.

Beck A, Wagner-Rousset E, Bussat MC, Lokteff M, Klinguer-Hamour C, Haeuw JF, Goetsch L, Wurch T, Van Dorsselaer A, Corvaia N. 2008. Trends in glycosylation, glycoanalysis and glycoengineering of therapeutic antibodies and Fc-fusion proteins. Curr Pharm Biotechnol 9(6):482-501.

Boyd PN, Lines AC, Patel AK. 1995. The effect of the removal of sialic acid, galactose and total carbohydrate on the functional activity of Campath-1H. Mol Immunol 32(17-18):1311-8.

Campbell CT, Yarema KJ. 2005. Large-scale approaches for glycobiology. Genome Biol 6(11):236.

Carraway KL, Hull SR. 1991. Cell surface mucin-type glycoproteins and mucin-like domains. Glycobiology 1(2):131-8.

Chang RS, Robertson CR, Deen WM, Brenner BM. 1975. Permselectivity of the glomerular capillary wall to macromolecules. I. Theoretical considerations. Biophys J 15(9):861-86.

Chung CH, Mirakhur B, Chan E, Le QT, Berlin J, Morse M, Murphy BA, Satinover SM, Hosen J, Mauro D and others. 2008. Cetuximab-induced anaphylaxis and IgE specific for galactose-alpha-1,3-galactose. N Engl J Med 358(11):1109-17.

Daughaday WH, Trivedi B, Baxter RC. 1993. Serum "big insulin-like growth factor II" from patients with tumor hypoglycemia lacks normal E-domain O-linked glycosylation, a possible determinant of normal propeptide processing. Proc Natl Acad Sci U S A 90(12):5823-7.

Davies J, Jiang L, Pan LZ, LaBarre MJ, Anderson D, Reff M. 2001. Expression of GnTIII in a recombinant anti-CD20 CHO production cell line: Expression of antibodies with altered glycoforms leads to an increase in ADCC through higher affinity for FC gamma RIII. Biotechnol Bioeng 74(4):288-94.

Edmunds T, Van Patten SM, Pollock J, Hanson E, Bernasconi R, Higgins E, Manavalan P, Ziomek C, Meade H, McPherson JM and others. 1998. Transgenically produced human antithrombin: structural and functional comparison to human plasma-derived antithrombin. Blood 91(12):4561-71.

Egrie J, Grant, J., Gillies, D., Aoki, K., & Strickland, T. 1993. The role of carbohydrate on the biological activity of erythropoietin. Glycoconj J 10:263.

Elliott S, Lorenzini T, Asher S, Aoki K, Brankow D, Buck L, Busse L, Chang D, Fuller J, Grant J and others. 2003. Enhancement of therapeutic protein in vivo activities through glycoengineering. Nat Biotechnol 21(4):414-21.

Field MC, Amatayakul-Chantler S, Rademacher TW, Rudd PM, Dwek RA. 1994. Structural analysis of the N-glycans from human immunoglobulin A1: comparison of normal human serum immunoglobulin A1 with that isolated from patients with rheumatoid arthritis. Biochem J 299 (Pt 1):261-75.

Flintegaard TV, Thygesen P, Rahbek-Nielsen H, Levery SB, Kristensen C, Clausen H, Bolt G. 2010. N-glycosylation increases the circulatory half-life of human growth hormone. Endocrinology 151(11):5326-36.

Freeze HH. 2006. Genetic defects in the human glycome. Nat Rev Genet 7(7):537-51.

Gomord V, Chamberlain P, Jefferis R, Faye L. 2005. Biopharmaceutical production in plants: problems, solutions and opportunities. Trends Biotechnol 23(11):559-65.

Gowda DC, Davidson EA. 1994. Isolation and characterization of novel mucin-like glycoproteins from cobra venom. J Biol Chem 269(31):20031-9.

Guile GR, Rudd PM, Wing DR, Prime SB, Dwek RA. 1996. A rapid high-resolution high-performance liquid chromatographic method for separating glycan mixtures and analyzing oligosaccharide profiles. Anal Biochem 240(2):210-26.

Hamilton SR, Davidson RC, Sethuraman N, Nett JH, Jiang Y, Rios S, Bobrowicz P, Stadheim TA, Li H, Choi BK and others. 2006. Humanization of yeast to produce complex terminally sialylated glycoproteins. Science 313(5792):1441-3.

Hamilton SR, Gerngross TU. 2007. Glycosylation engineering in yeast: the advent of fully humanized yeast. Curr Opin Biotechnol 18(5):387-92.

Helenius A, Aebi M. 2004. Roles of N-Linked Glycans in the Endoplasmic Reticulum. Annu Rev Biochem 73:1019-1049.

Hirschberg CB, Snider MD. 1987. Topography of glycosylation in the rough endoplasmic reticulum and Golgi apparatus. Annu Rev Biochem 56:63-87.

Hooper LV, Manzella SM, Baenziger JU. 1996. From legumes to leukocytes: biological roles for sulfated carbohydrates. Faseb J 10(10):1137-46.

Jefferis R. 2009. Glycosylation as a strategy to improve antibody-based therapeutics. Nat Rev Drug Discov 8(3):226-34.

Kaneko Y, Nimmerjahn F, Ravetch JV. 2006. Anti-inflammatory activity of immunoglobulin G resulting from Fc sialylation. Science 313(5787):670-3.

Kornfeld R, Kornfeld S. 1985. Assembly of asparagine-linked oligosaccharides. Annu Rev Biochem 54:631-64.

Kornfeld S, Mellman I. 1989. The biogenesis of lysosomes. Annu Rev Cell Biol 5:483-525.

Kumpel BM, Wang Y, Griffiths HL, Hadley AG, Rook GA. 1995. The biological activity of human monoclonal IgG anti-D is reduced by beta-galactosidase treatment. Hum Antibodies Hybridomas 6(3):82-8.

Leader KA, Kumpel BM, Hadley AG, Bradley BA. 1991. Functional interactions of aglycosylated monoclonal anti-D with Fc gamma RI+ and Fc gamma RIII+ cells. Immunology 72(4):481-5.

Leatherbarrow RJ, Rademacher TW, Dwek RA, Woof JM, Clark A, Burton DR, Richardson N, Feinstein A. 1985. Effector functions of a monoclonal aglycosylated mouse IgG2a: binding and activation of complement component C1 and interaction with human monocyte Fc receptor. Mol Immunol 22(4):407-15.

Mahmood I, Green MD. 2005. Pharmacokinetic and pharmacodynamic considerations in the development of therapeutic proteins. Clin Pharmacokinet 44(4):331-47.

Natsume A, Niwa R, Satoh M. 2009. Improving effector functions of antibodies for cancer treatment: Enhancing ADCC and CDC. Drug Des Devel Ther 3:7-16.

Natsume A, Wakitani M, Yamane-Ohnuki N, Shoji-Hosaka E, Niwa R, Uchida K, Satoh M, Shitara K. 2006. Fucose removal from complex-type oligosaccharide enhances the antibody-dependent cellular cytotoxicity of single-gene-encoded bispecific antibody comprising of two single-chain antibodies linked to the antibody constant region. J Biochem 140(3):359-68.

Nilsson IM, von Heijne G. 1993. Determination of the distance between the oligosaccharyltransferase active site and the endoplasmic reticulum membrane. J Biol Chem 268(8):5798-801.

Niwa R, Natsume A, Uehara A, Wakitani M, Iida S, Uchida K, Satoh M, Shitara K. 2005. IgG subclass-independent improvement of antibody-dependent cellular cytotoxicity by fucose removal from Asn297-linked oligosaccharides. J Immunol Methods 306(1-2):151-60.

Niwa R, Shoji-Hosaka E, Sakurada M, Shinkawa T, Uchida K, Nakamura K, Matsushima K, Ueda R, Hanai N, Shitara K. 2004. Defucosylated chimeric anti-CC chemokine receptor 4 IgG1 with enhanced antibody-dependent cellular cytotoxicity shows potent therapeutic activity to T-cell leukemia and lymphoma. Cancer Res 64(6):2127-33.

Olden K, Parent JB, White SL. 1982. Carbohydrate moieties of glycoproteins. A re-evaluation of their function. Biochim Biophys Acta 650(4):209-32.

Perlman S, van den Hazel B, Christiansen J, Gram-Nielsen S, Jeppesen CB, Andersen KV, Halkier T, Okkels S, Schambye HT. 2003. Glycosylation of an N-terminal extension prolongs the half-life and increases the in vivo activity of follicle stimulating hormone. J Clin Endocrinol Metab 88(7):3227-35.

Pestka S, Langer JA, Zoon KC, Samuel CE. 1987. Interferons and their actions. Annu Rev Biochem 56:727-77.

Qian J, Liu T, Yang L, Daus A, Crowley R, Zhou Q. 2007. Structural characterization of N-linked oligosaccharides on monoclonal antibody cetuximab by the combination of orthogonal matrix-assisted laser desorption/ionization hybrid quadrupole-quadrupole time-of-flight tandem mass spectrometry and sequential enzymatic digestion. Anal Biochem 364(1):8-18.

Remaley AT, Ugorski M, Wu N, Litzky L, Burger SR, Moore JS, Fukuda M, Spitalnik SL. 1991. Expression of human glycophorin A in wild type and glycosylation-deficient Chinese hamster ovary cells. Role of N- and O-linked glycosylation in cell surface expression. J Biol Chem 266(35):24176-83.

Scallon BJ, Tam SH, McCarthy SG, Cai AN, Raju TS. 2007. Higher levels of sialylated Fc glycans in immunoglobulin G molecules can adversely impact functionality. Mol Immunol 44(7):1524-34.

Sheeley DM, Merrill BM, Taylor LC. 1997. Characterization of monoclonal antibody glycosylation: comparison of expression systems and identification of terminal alpha-linked galactose. Anal Biochem 247(1):102-10.

Shi X, Jarvis DL. 2007. Protein N-glycosylation in the baculovirus-insect cell system. Curr Drug Targets 8(10):1116-25.

Shields RL, Lai J, Keck R, O'Connell LY, Hong K, Meng YG, Weikert SH, Presta LG. 2002. Lack of fucose on human IgG1 N-linked oligosaccharide improves binding to human Fcgamma RIII and antibody-dependent cellular toxicity. J Biol Chem 277(30):26733-40.

Shinkawa T, Nakamura K, Yamane N, Shoji-Hosaka E, Kanda Y, Sakurada M, Uchida K, Anazawa H, Satoh M, Yamasaki M and others. 2003. The absence of fucose but not the presence of galactose or bisecting N-acetylglucosamine of human IgG1 complex-type oligosaccharides shows the critical role of enhancing antibody-dependent cellular cytotoxicity. J Biol Chem 278(5):3466-73.

Shoji-Hosaka E, Kobayashi Y, Wakitani M, Uchida K, Niwa R, Nakamura K, Shitara K. 2006. Enhanced Fc-dependent cellular cytotoxicity of Fc fusion proteins derived from TNF receptor II and LFA-3 by fucose removal from Asn-linked oligosaccharides. J Biochem 140(6):777-83.

Sinclair AM, Elliott S. 2005. Glycoengineering: the effect of glycosylation on the properties of therapeutic proteins. J Pharm Sci 94(8):1626-35.

Sola RJ, Griebenow K. 2010. Glycosylation of therapeutic proteins: an effective strategy to optimize efficacy. BioDrugs 24(1):9-21.

Stenflo J, Fernlund P. 1982. Amino acid sequence of the heavy chain of bovine protein C. J Biol Chem 257(20):12180-90.

Tang L, Persky AM, Hochhaus G, Meibohm B. 2004. Pharmacokinetic aspects of biotechnology products. J Pharm Sci 93(9):2184-204.

Ulloa-Aguirre A, Timossi C, Damian-Matsumura P, Dias JA. 1999. Role of glycosylation in function of follicle-stimulating hormone. Endocrine 11(3):205-15.

Umana P, Jean-Mairet J, Moudry R, Amstutz H, Bailey JE. 1999. Engineered glycoforms of an antineuroblastoma IgG1 with optimized antibody-dependent cellular cytotoxic activity. Nat Biotechnol 17(2):176-80.

van Bueren JJ, Rispens T, Verploegen S, van der Palen-Merkus T, Turinsky AL, Stapel S, Workman LJ, James H, van Berkel PH, van de Winkel JG and others. 2011. Anti-galactose-alpha-1,3-galactose IgE from allergic patients does not bind alpha-galactosylated glycans on intact therapeutic antibody Fc domains. Nat Biotechnol 29(7):574-6.

Van den Steen P, Rudd PM, Dwek RA, Opdenakker G. 1998. Concepts and principles of O-linked glycosylation. Crit Rev Biochem Mol Biol 33(3):151-208.

Van Patten SM, Hughes H, Huff MR, Piepenhagen PA, Waire J, Qiu H, Ganesa C, Reczek D, Ward PV, Kutzko JP and others. 2007. Effect of mannose chain length on targeting of glucocerebrosidase for enzyme replacement therapy of Gaucher disease. Glycobiology 17(5):467-78.

Varki A. 1993. Biological roles of oligosaccharides: all of the theories are correct. Glycobiology 3(2):97-130.

Varki A, Cummings, R.D., Esko, J.D., Freeze, H.H., Stanley, P., Bertozzi, C.R., Hart, G.W., Etzler, M.E. 2009. Essentials of Glycobiology, 2nd edition. Cold Spring Harbor (NY): Cold Spring Harbor Laboratory Press.

Venkatachalam MA, Rennke HG. 1978. The structural and molecular basis of glomerular filtration. Circ Res 43(3):337-47.

Walker MR, Lund J, Thompson KM, Jefferis R. 1989. Aglycosylation of human IgG1 and IgG3 monoclonal antibodies can eliminate recognition by human cells expressing Fc gamma RI and/or Fc gamma RII receptors. Biochem J 259(2):347-53.

Walsh G, Jefferis R. 2006. Post-translational modifications in the context of therapeutic proteins. Nat Biotechnol 24(10):1241-52.

Weikert S, Papac D, Briggs J, Cowfer D, Tom S, Gawlitzek M, Lofgren J, Mehta S, Chisholm V, Modi N and others. 1999. Engineering Chinese hamster ovary cells to maximize sialic acid content of recombinant glycoproteins. Nat Biotechnol 17(11):1116-21.

Wong CH. 2005. Protein glycosylation: new challenges and opportunities. J Org Chem 70(11):4219-25.

Zhong X, Pocas J, Liu Y, Wu PW, Mosyak L, Somers W, Kriz R. 2009. Swift residue-screening identifies key N-glycosylated asparagines sufficient for surface expression of neuroglycoprotein Lingo-1. FEBS Lett 583(6):1034-8.

Zhou Q, Shankara S, Roy A, Qiu H, Estes S, McVie-Wylie A, Culm-Merdek K, Park A, Pan C, Edmunds T. 2008. Development of a simple and rapid method for producing non-fucosylated oligomannose containing antibodies with increased effector function. Biotechnol Bioeng 99(3):652-65.

11

Detection of Protein Phosphorylation by Open-Sandwich Immunoassay

Yuki Ohmuro-Matsuyama[1], Masaki Inagaki[2] and Hiroshi Ueda[1]
[1]*The University of Tokyo*
[2]*Aichi Cancer Center Research Institute*
Japan

1. Introduction

Understanding the post-translational modifications of proteins represents a next major challenge in post-genomic era. Intricate cascades of phosphorylation reactions regulate cell proliferation, differentiation, migration and so on. For example, Olsen et al recently applied high-resolution mass spectrometry-based proteomics to phosphoproteome of human cell cycle, quantifying 6,027 proteins and 20,443 unique phosphorylation sites [Olsen, 2010]. Phosphorylation-dependent signals participate in several diseases, such as cancers, immune diseases, and Alzheimer's. The method to produce phosphorylation-site and phosphorylation state-specific antibodies was established in 1990's [Nishizawa, 1991; Yano, 1991; Nagata, 2001; Goto, 2007] after a discovery of phosphorylated cytoskeletal protein-specific antibodies [Sternberger, 1983]. Since then, investigators have successfully observed spatio-temporal dynamics of particular phosphorylation *in vitro* or *in situ* using the antibodies. Recently, these antibodies are also proven useful as a live cell imaging probe *in vivo* [Hayashi-Takanaka, 2009; Kimura, 2010; Hayashi-Takanaka, 2011], clinical diagnosis, and drug screening [Brumbaugh, 2011].
Here we show a couple of applications of phosphorylation-specific antibodies to proteomic studies, more specifically focusing on a novel immunoassay approach, which we call open-sandwich immunoassay.

2. Open-sandwich immunoassay

2.1 An innovative immunoassay: Open-sandwich immunoassay

The open-sandwich immunoassay (OS-IA) was developed in a rather fortuitous way as follows [Ueda, 1996; Ueda, 2002]. Previously, one of the authors had been interested in the regulation of the tyrosine kinase activity of epidermal growth factor (EGF) receptor by the ligands other than the natural one (EGF), as a prototype molecular biosensor [Ueda, 1992]. Since the EGF receptor was found activated by ligand-induced dimerization [Schlessinger, 1986], a possible way to realize such regulation was to replace the EGF-binding domain of EGF receptor with a pair of specific binding domains that dimerize upon addition of their ligand. Pairs of such binding domains are known to exist in nature. For example, FKBP12 and FRAP are known to associate in the presence of an antibiotic rapamycin [Rossi, 1997] and the two erythropoietin (Epo) binding domains of EPO receptor homodimerize upon binding

with EPO [Lacombe, 1999]. As an alternative of such domains, one of the authors was looking for antibody variable regions (V_H and V_L), wherein the association between V_H and V_L is strengthened by the antigen. In other words, at the beginning we had no intention to devise a novel immunoassay. We only wanted to determine the V_H/V_L interaction strength and the effect of an antigen on this interaction to devise such a hybrid receptor (Fig. 1A).
Theoretically, there might be a range of immunoglobulin variable regions that show (i) weak and (ii) strong V_H/V_L interaction irrespective of antigen binding, (iii) weak V_H/V_L interaction in the absence of an antigen, which is strengthened by antigen binding, and (iv) strong V_H/V_L interaction in the absence of an antigen, which is weakened by antigen binding. Before that time there were very few reports on the strength of the interaction between V_H and V_L fragments and its antigen dependency, we tried to measure the interaction between V_H and V_L of anti-hen egg lysozyme (HEL) antibody, HyHEL-10 using a surface plasmon resonance biosensor Biacore (Fig. 1B).

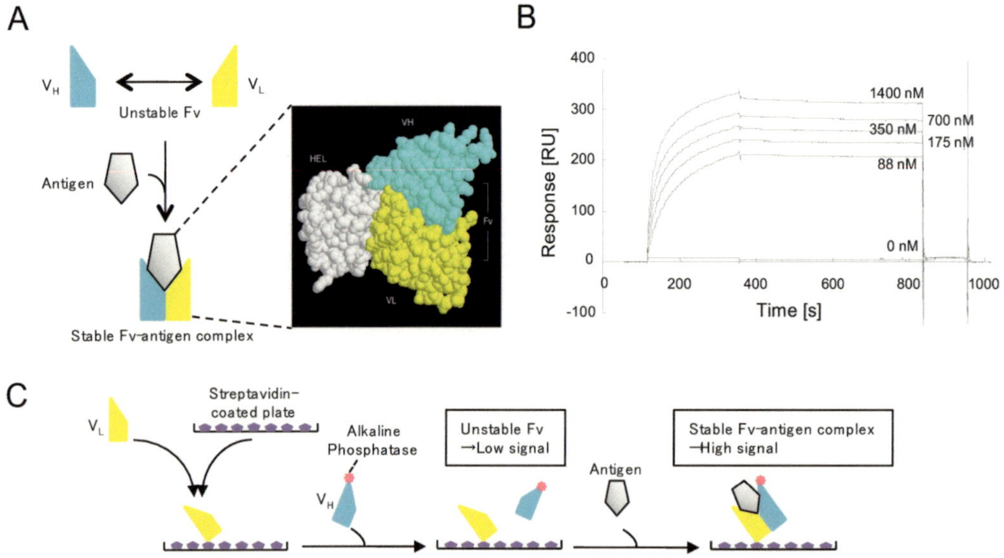

Fig. 1. Principle of open-sandwich immunoassay (OS-IA). (A) V_H/V_L/antigen ternary complex. While the intrinsic binding affinity of V_H and V_L is low, when they are added with antigen, that of V_H/V_L/antigen complex becomes high. (B) A measurement of V_H/V_L complex stability by Biacore. The association curve of V_H on immobilized V_L is significantly influenced by the co-existing antigen concentration as shown. (C) A basic procedure of OS-IA. V_L (or V_H) is immobilized, and labeled V_H (or V_L) is put together with a sample. To determine the affinity of V_H and V_L as a function of antigen concentration in a sample, the amount of labeled V_H (or V_L) bound onto the well is quantified.

When the binding of soluble V_H fragment to the immobilized V_L fragment was optically monitored, the interaction was calculated to be very weak in the absence of the antigen ($K_a < 10^5$/M), and markedly strengthened in the presence of the antigen ($K_a \sim 10^9$/M). The principal reason for this was found to be a remarkable reduction of dissociation rate k_{off} of the complex. This antigen-induced equilibrium shift was also observed in an ELISA, where the binding of phage particles displaying the V_H fragment to the V_L fragments immobilized on microplate wells was measured by horseradish peroxidase (HRP)-labeled anti-phage antibody (Fig. 1C). A reproducible standard curve could be drawn for the antigen HEL concentration in the sample (Fig. 2A). The results indicated such an assay could be utilized as a novel means to measure antigen concentration in a sample. We termed this type of assay an open-sandwich immunoassay (OS-IA) because the antigen to be measured was bound to two fragments of antibody, V_H and V_L like an open sandwich. Further studies using other many antibodies revealed that all four types of variable regions shown in the beginning of this paragraph have been identified, however, many anti-hapten antibodies exhibit the property categorize in (iii) described above which makes them suitable for OS-IA [Suzuki, 1999; Lim, 2007; Suzuki, 2007; Ihara, 2009; Islam, 2011].

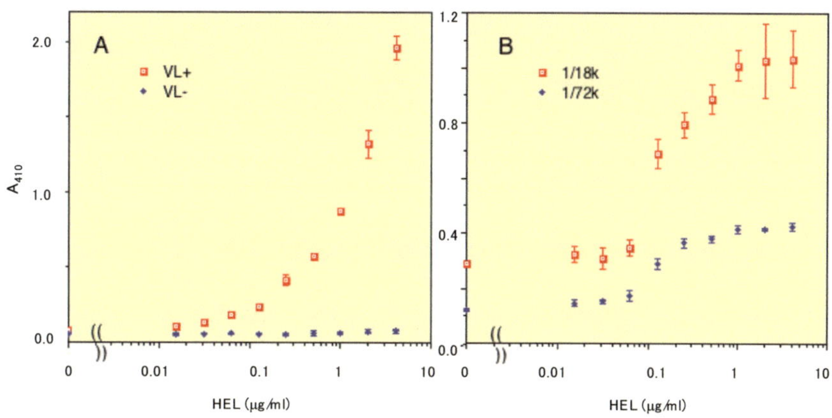

Fig. 2. Dose-response curves obtained with OS-IA and conventional sandwich immunoassay. (A) Association of phage-displayed V_H with immobilized V_L in the presence of hen egg lysozyme (HEL) was probed with labeled anti-phage antibody. A control without immobilized V_L is also shown. (B) Sandwich ELISA with HyHEL10 scFv and rabbit anti HEL serum at two dilutions as indicated. The signal was obtained with HRP-labeled anti rabbit IgG.

OS-IA has several advantages compared with conventional immunoassays. Sandwich immunoassay is one of conventional noncompetitive immunoassay with high sensitivity and a wide working range of more than three orders of magnitude. The principle of sandwich ELISA is to detect antigen in a sample captured first antibody immobilized on microplate by enzyme-linked second antibody. However, sandwich immunoassay has a fundamental limitation that the antigen to be measured must be large enough to have at least two epitopes to be captured; therefore, small monovalent antigens are not suitable for the assay. Another conventional immunoassay, competitive immunoassay is useful to measure monovalent antigens. The principle of assay is based on the competitive binding of labeled antigen and non-labeled antigen in sample, when captured by an antibody immobilized on a surface such as microplate wells. While competitive assay enables measurement of monovalent antigen, careful optimization of the reaction conditions is necessary to attain suitable sensitivity and working range, and a large amount of antigen is required. Furthermore, while the sensitivity is theoretically approximately 1/100 of K_d, the affinities of antibodies are limited by the surface area interacting antigens and paratopes of antibodies, hence, small antigens is generally undetectable with high sensitivity (Fig. 2B).

As a way to circumvent these limitations of conventional immunoassay, using OS-IA less than 10 ng/ml antigen is detected in a shorter time period than by a conventional sandwich assay, due to the omission of an incubation/washing cycle. In addition, OS-IA using HyHEL-10 resulted in better sensitivity than that obtained with the corresponding sandwich immunoassay. Also, the applications of OS-IA to small antigens could attain a similar or lower detection limit as well as wider working range than attained with the corresponding competitive assay. Why can OS-IA detect small antigens with a higher sensitivity and a wider working range? It may concerted with that Pellequer *et al.* categorized the changes in compactness of the V_H/V_L interface between bound and unbound antibodies on the size of the antigen and found that small antigens or haptens cause a closure of the interface, whereas lager protein antigens have little effect of the compactness of V_H/V_L interface [Pellequer, 1999]. This is also in accordance with previous observations that anti-hapten antibodies recognize their antigen between the V_H/V_L interfaces, whereas anti-protein antibodies do it on the upper surface of V_H/V_L dimers.

2.2 Application to homogeneous assays

In the field of healthcare, food safety and environmental monitoring, homogeneous assays are available for rapid and simple screening of components in samples. It is necessary for automation techniques for high-throughput screens. Here, we show applications of OS-IA to homogeneous assays.

First, we employed fluorescence resonance energy transfer (FRET), in which, one fluorophore as donor transfers its excited-state energy to another fluorophore as acceptor, resulting in emitting fluorescence of a different color. FRET generally occurs when the donor and acceptor are in approximate distance (10-100 Å) [Selvin, 1994]. It has been applicable to a homogeneous immunoassay by labeling antibody and antigen with donor and acceptor of fluorescence respectively. However, this is a competitive immunoassay with less sensitively than that of noncompetitive immunoassay, and requires a large amount of labeled antigen [Pradelles, 1994]. Furthermore, endogenous antigens cannot be detected in the assay.

We performed open-sandwich fluoroimmunoassay (OS-FIA) using fluorescein-labeled V_H and rhodamine-labeled V_L [Ueda, 1999] (Fig. 3A). A principle of OS-FIA is as follows; 1) Without antigen, the two fusion Fv fragments remain monomeric, so FRET between them is negligible. 2) The addition of antigen induced heterodimerization of the two chains, accompanied by the FRET. When the labeled fragments were added to the sample solution, the antigen concentration could be measured by monitoring V_H/V_L interaction with FRET. The each detection takes within only 2 minutes. However, a site-specific fluorolabeling for the assay is needed several laborious trials. Then we next use GFP variants fused V_H and V_L to obtain site-specific labeled probes [Arai, 2000] (Fig. 3B). The V_H and V_L fused to GFP variants are expressed in cytoplasm of mutant strains that have oxidized cytoplasmic environments to make proper S-S bonds in V_H or V_L. Using the purified V_H and V_L from *E. coli*. OS-FIA could be carried out without significant loss of sensitivity.

The second application makes use of bioluminescence resonance energy transfer (BRET) [Arai, 2001] (Fig. 3C). In BRET donor is a bioluminescence, and acceptor is a fluorescent protein. When donor and acceptor are in an approximate distance, luminescence of donor is transferred to acceptor, resulting in emitting fluorescence from acceptor. When V_H-Rluc and V_L-eYFP were mixed with a sample regent, an antigen in the sample dependent increase in BRET was measured. Compared with our comparable OS-FIA, the sensitivity is a 10-fold higher.

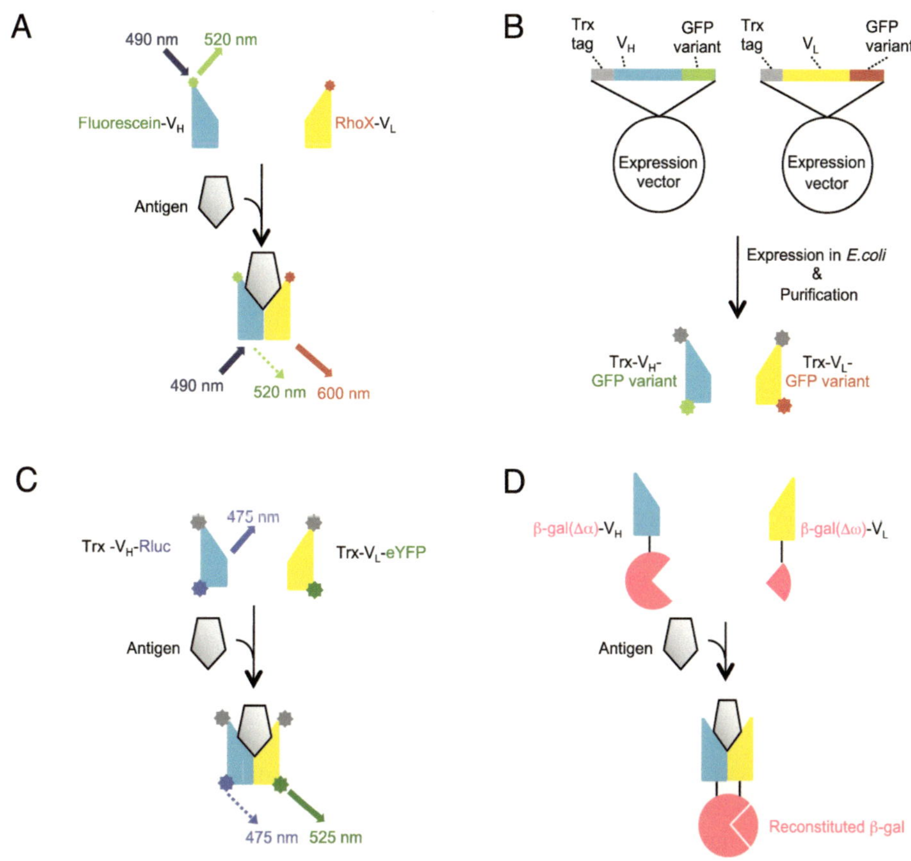

Fig. 3. Scheme of OS-IA applied to various homogeneous assays. (A) FRET-based OS-IA. Addition of antigen leads to decreased donor-derived (green) emission as well as increased acceptor-derived (red) emission. (B) A procedure to obtain site-specifically labeled Fv for OS-FIA. GFP variants are used as a label for the FRET-based assay. (C) BRET-based OS-IA. (D) Enzymatic complementation-based OS-IA. The two complementing fragments of β-galactosidase was used as a reporter for V_H / V_L association.

Thirdly, to obtain higher sensitivity we utilized β-galactosidase (β-gal) complementation [Yokozeki, 2002; Ueda, 2003] (Fig. 3D). Because of backgrounds of FRET due to relatively high protein concentrations of labeled V_H and V_L compared with dissociation constant, K_d of V_H/V_L interaction, we decided to employ β-gal complementation to reduce the amounts of proteins. A protein-protein interaction assay *in vivo* had been developed using enzymatic complementation between the two deleted mutants of β-gal fused to the respective interacting proteins. In our assay, an enzymatic complementation of β-gal was monitored between the antigen dependent interaction of V_H tethered to an N-terminal deletion mutant (Δα) of β-gal and V_L tethered to another deletion mutant (Δω) of β-gal. Without antigen, the almost deleted mutants of β-gal are detached from each other due to monomeric Fv, resulting in low enzymatic activity. The addition of antigen induced stabilization of the ternary complex, thus the mutants came close and reconstructed its enzyme activity. With the use of HyHEL10, the measurable concentration range was relatively broad from 0.1

ng/ml to more than 10 μg/ml. The lowest measurable concentration was almost 1000-fold less than those of OS-FIA and BRET based OS-IA, while an incubation for 40 minutes are needed to enzymatically amplify the signal in contrast to the cases of FRET and BRET, where the emitted light is readily measured.

Here we showed four examples of homogeneous OS-IA. Users are advised choose an appropriate format to consider both of their merits and demerits. For instance, the sensitivity of OS-IA based on enzyme complementation is much higher than that of OS-FIA, however, an enzyme complementation is generally irreversible and require rather long incubation time. OS-FIA needs the unconventional device in biological laboratories, while it has several merits; a high sensitivity and a facile standardization of signals. On the other hand, OS-FIA and OS-IA based on BRET are applicable for real-time imaging probes in live cells.

2.3 Split-Fv system

To rapidly evaluate and select antibody variable region (Fv) fragments that are suitable to OS-IA, we devised two phage-based systems. The first one is "split-Fv system". Phage display is a powerful method for screening functional antibody fragments retaining a high affinity to the antigen [Winter, 1994; Gao, 1999]. Although the method needs some technical training and some disadvantages compared with conventional monoclonal antibody technology exist, such as inability to display all the antibody fragments cloned from hybridoma cells for several reasons, it still has a power especially the use of recombinant antibody is essential. For the screening of the antigen binding ability of V_H/V_L by phage display, a simultaneous display of both fragments in close proximity on the same phage is necessary. On the other hand, to perform OS-ELISA, that measures V_H/V_L interaction rather than the antigen-antibody interaction, separate expression of V_H/V_L is necessary; for example, phage display of a V_H fragment and production of a soluble V_L fragment is desired.

To enable the facile switch of these two display formats, we adopted a filamentous phage p7-p9 display system to individually display formats V_H and V_L fragments as a functional Fv on the tip of the phage (Fig. 4A).

However, we put a modification into the reported system that an amber codon was placed between V_L and gene 7, which makes to enable/disenable display of V_L (tethered with His-myc tag) by changing the *sup* phenotype of host *E. coli*. When sup^+ strains are used for phage production, V_L-(His-myc)-p7 is expressed, resulting in displaying Fv on the phage. When sup^- strains are used, soluble V_L-(His-myc) along with a V_H-displaying phage is expressed. Thus, OS-IA can be performed in a plate in which V_L is immobilized by anti myc antibody or V_L-specific ligand protein L coated on the plate. Thorough the two-step selection the first selection of highest affinity binders to antigen give a higher possibility of spotting most suitable candidates to OS-IA [Aburatani, 2003].

While the split-Fv system was successfully used to select and clone many Fvs that are suitable for OS-IA, some other Fvs did not show positive antigen binding, or the level of secreted V_L fragment was too low to perform OS-IA, possibly due to limited stability of the isolated V_L domain. Compared with the scFv fragment that is known to have a high tendency to form multimers, the antibody Fab fragment is reported to stay monomeric, allowing selection for affinity in contrast to selection for avidity. Recently, we devised a Fab display system that can perform OS-IA [Dong, 2009] (Fig. 4B).

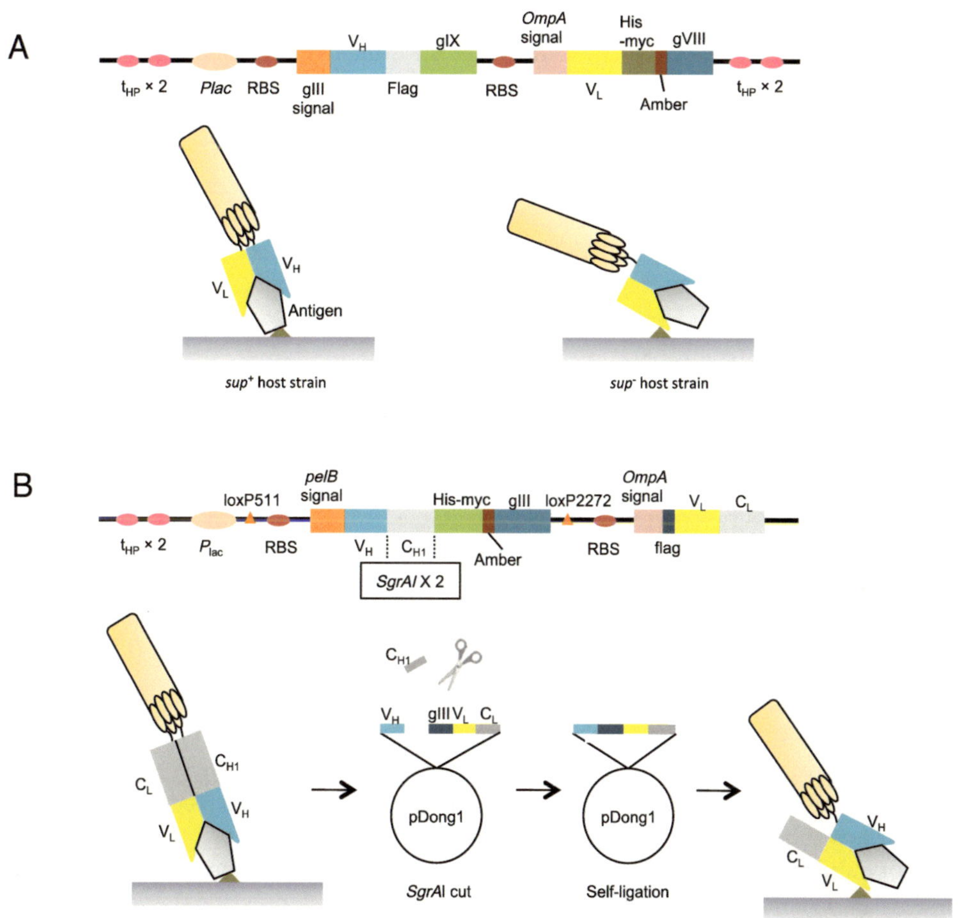

Fig. 4. Phage display systems to select antibodies for OS-IA. (A) Split Fv system. An amber codon was placed between V_L and gene 7, which makes to enable/disenable display of V_L (tethered with His-myc tag) by changing the *sup* phenotype of host *E. coli*. When *sup+* strains are used for phage production, V_L-(His-myc)-p7 is expressed, resulting in displaying Fv on the phage. When sup- strains are used, soluble V_L-(His-myc) along with a V_H-displaying phage is expressed. Thus, OS-IA can be performed in a plate in which V_L is immobilized by anti-myc antibody or V_L-specific ligand such as protein L coated on the plate. (B) pDong1 system. Human IgG_1 C_H1 and C_k domains were included in pDong1 to regard selecting human antibody using the system. An amber codon was placed between Fd gene and gene III, therefore, Fab fragment is displayed on the surface of phage using *sup+* host *E. coli* to check the affinity of Fab to antigen. On the other hand, two restriction sites for *SgrA*I are incorporated at the both ends of C_H1 gene to convert antigen-selected vectors for Fab display (Fab vector) to the vector for V_H display with simultaneous secretion of L chain (OS vector). After digestion with *SgrA*I and self-ligation, the culture supernatant of resultant phage-containing culture contains secreted L chain fused to flag tag and V_H-displaying phage. Thus, V_H-displaying phage is detected with enzyme-labeled anti-phage antibody on a plate in which L chain is immobilized by anti-flag or anti-L chain antibody coated on the plate to perform OS-IA.

A phagemid vector named pDong1 is shown in Fig 4B. pDong1 was designed to display Fab on a minor protein pIII of M13 phage. The genes for human IgG_1 C_H1 and C_k domains were included in pDong1 to regard selecting human antibody using the system. An amber codon was placed between Fd gene and gene III, therefore, Fab fragment is displayed on the surface of phage using sup^+ host *E. coli* to check the Fab affinity. On the other hand, a rare cutting restriction site, *SgrA*I is incorporated at both end of C_H1 gene to convert antigen-selected vectors for Fab display (Fab vector) to the vector for V_H display with simultaneous secretion of L chain (OS vector). After cutting by *SgrA*I the culture supernatant contains secreted L chain fused to flag tags and V_H-displaying phage. Thus, V_H-displaying phage is detected using HRP-labeled anti-phage antibody on a plate in which V_L is immobilized by anti flag or His or myc antibody or V_L-specific ligand protein L coated on the plate to perform OS-IA. In addition, pDong1 is encoding one each of two recombination sites, LoxP2272 and LoxP511, which were placed at the upstream and the downstream of Fd-FIII ORF, respectively, that enable increased diversity of phage library due to Fd/L chain exchange. It is useful to make a library with Cre recombinase.

These systems are very powerful to select suitable antibody fragments for OS-IA from natural and engineered libraries and allow detailed analysis on the molecular bases of variable V_H/V_L interaction strength and its antigen dependency [Masuda, 2006; Sasajima, 2006; Lim, 2007; Suzuki, 2007; Islam, 2011]

3. Application of anti-phosphotyrosine antibody to open-sandwich immunoassay

A most important and unique character of OS-IA is that it can noncompetitively detect small molecules including organic chemicals. Phosphate addition to a target amino acid affects very small change in molecular weight. As a way to rapidly assay protein phosphorylation events, here we show the use of OS-IA [Sasajima, 2006].

Phosphorylation of tyrosine is known to be very important in intracellular signal transduction events [Ullrich, 1990], and a number of good polyclonal [Ross, 1981] and monoclonal anti phosphotyrosine (PY) antibodies are reported [Frackelton, 1983; Glenney, 1988]. We decided to synthesize PY20 Fv gene based on the published amino acid sequence [Ruff-Jamison, 1991; Ruff-Jamison, 1993a], because PY20 Fv was reported to bind PY with high affinity (Kd= 1.55×10^{-7} M) [Ruff-Jamison, 1993b].

For OS-IA to detect phosphotyrosine we use the split Fv system. When the culture supernatant containing V_H-displaying phage and soluble V_L fragment was applied to the microplate wells immobilized with antigen conjugate (PY-BSA) for phage ELISA, a strong and specific binding of the phage to PY-BSA wells and a clear competitive inhibition by added PY during incubation were observed. The PY concentration that gave half-maximal inhibition was 10 μg/ml, indicating sufficient affinity (Fig. 5A). The V_H/V_L interaction strength and its PY-dependency of PY20 Fv were investigated using the same culture supernatant and microplate wells immobilized with Penta-His antibody. As a result, a PY-dependent increase with the maximum response of 30 % increment was observed (Fig. 5C). However, we reasoned the increase might not be sufficient when the assay is taken to FRET-based homogeneous format, where increased protein concentration and reduced dynamic range due to spectral overlap of the two fluorophores can limit its sensitivity. So we then improved the response of OS-ELISA by site-directed mutagenesis approach.

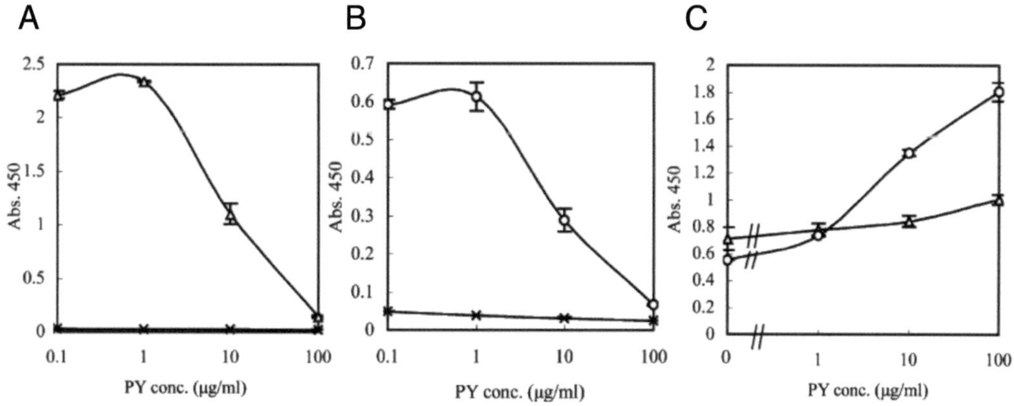

Fig. 5. Detection of phosphotyrosine with PY20. (A) Competitive split Fv-phage ELISA. Culture supernatant was mixed with twice the concentration of PY, and the bindings to immobilized PY-BSA (△) or BSA (×) were evaluated. (B) Competitive split Fv-phage ELISA with V_H(Q39R) mutant. (C) OS phage ELISA with the wild-type (△) and the mutant (O) culture supernatant.

We knew that a mutation in a V_H residue facing V_L interface (H39) can effectively modulate V_H/V_L interaction strength without affecting antigen-binding affinity [Masuda, 2006]. Since 39_H of PY20 is glutamine, probably making two hydrogen bonds with a corresponding V_L residue Gln(38_L), we introduced random mutation to this residue in order to get mutant(s) with lower V_H/V_L interaction strength in the absence of PY. After screening phage clones, a mutant showing higher response V_H(Q39R) was obtained. When the antigen binding activity of the mutant Fv and also its competition by PY were investigated by phage ELISA, significant binding to PY-BSA and its inhibition by PY similar to wild-type PY20 were observed (Fig. 5B). Similarly, when OS-IA was performed for this mutant, significantly higher PY-dependent signal increase of 200% with reduced background signal corresponding to V_H/V_L interaction was obtained. Surprisingly, the signal in the presence of PY was almost twice that of wild-type PY20, and the resulting sensitivity was higher than that with competitive assay (Fig. 5C).

To conduct FRET-based homogeneous assay, the gene for mutant V_H or the V_L was fused to eCFP or eYFP, respectively (Fig. 6A). V_H-eCFP was excited at 433 nm and the fluorescent spectra at 500-650 nm were recorded in the presence of several PY concentrations. The result shows a slight PY-concentration-dependent decrease in eCFP fluorescence around 475 nm and a significant increase in eYFP fluorescence peaking around 525 nm, resulting in increased eYFP/eCFP fluorescence ratio up to ~30 %. The result clearly showed that we could successfully detect PY in a homogeneous solution in a noncompetitive manner.

Next, when pp60 peptide encoding a physiologically tyrosine-phosphorylated protein c-src residue 521-533 containing pY527 was added, a clear increment in FRET was detected (Fig. 6B). Prior dephosphorylation of the peptide by calf intestinal alkaline phosphatase resulted in complete reversal of the spectrum compared to that of the control.

Because of its simple fluorescence ratiometric detection, this OS-FIA will be useful for diagnostics and facile *in situ* visualization of intracellular tyrosine phosphorylation, which includes Alzheimer, malignantly growing cells and immune abnormal cells and so on. In near future, in combination with an appropriate method will be proven to be a powerful

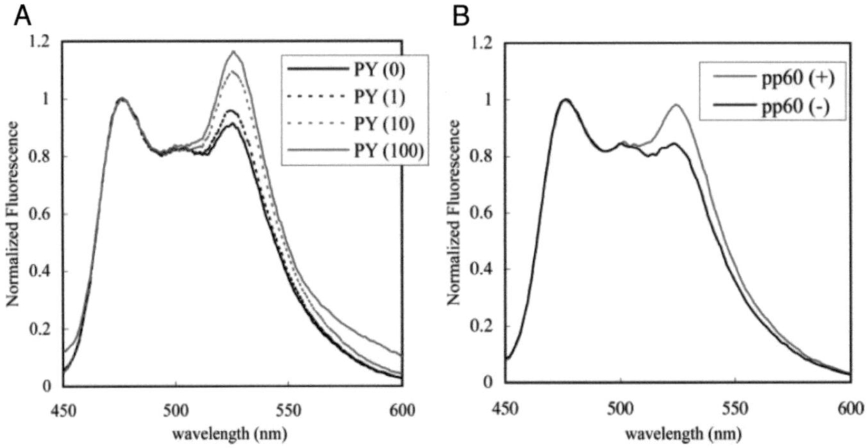

Fig. 6. FRET-based OS-IA for PY and peptide tyrosine phosphorylation (A) OS-FIA for PY. Fluorescence spectra of the probe (500 ng/ml) in the presence of 1 mg/ml BSA and indicated amounts of PY (in μg/ml). (B) OS-FIA with pp60 c-src peptide. The same spectra with (A) in the presence (gray line) or absence (black line) of the peptide. Phosphatase treatment of the peptide restored the signal (not shown).

tool to find indicators of several diseases in clinical specimens and to monitor intracellular imaging of protein tyrosine phosphorylation, as in the cases of other useful probes to monitor intracellular calcium concentration [Chung, 2009; Horikawa, 2010].

4. Application of vimentin phospho-specific antibody to open-sandwich immunoassay

4.1 Phosphorylation of intermediate filaments

Next, we used an antibody that recognizes site-specifically phosphorylated vimentin to detect more specific protein phosphorylation. The cytoskeletons are composed of three major groups distinguished by their diameter; microfilaments with 6 nm diameter, microtubules with 24 nm diameter, and intermediate filaments with its diameter between that of microfilaments and that of microtubules (10 nm). Vimentin is one of commonly observed intermediate filament. In 1980's, the mechanism of intermediate filament disassembly was considered to be its degradation by proteases, since intermediate filament has a stable structure with insolubility and chemical unreactivity. Furthermore, intermediate filaments are thought to play only a role to maintain cytoplasmic organization [Ishikawa, 1968; Lazarides, 1980b; Lazarides, 1980a; Lazarides, 1982]. On the other hand, it is known that vimentin exists in a phosphorylation form during mitosis [Celis, 1983; Bennett, 1988; Chou, 1990; Liao, 1997]. In 1987, it was shown that phosphorylation induces disassembly of the filaments *in vitro* [Inagaki, 1987; Inagaki, 1988; Inagaki, 1989; Inagaki, 1990]. The discovery of conformational dynamics of intermediate filaments leads to reconsider the roles in cell function. Recently, phosphorylation-dependent assembly/disassembly of intermediate filaments has been reported to be associated with cell cycle, cell migration and several diseases, while the entire roles are still obscure [Chou, 1990; Chou, 1991; Inagaki, 1994; Tsujimura, 1994a; Tsujimura, 1994b; Goto, 2000; Yasui, 2001; Eriksson, 2004; Yamaguchi, 2005; Izawa, 2006; Helfand, 2011].

Vimentin contains three substructures; head, rod, and tail domains. The head domain is phosphorylated by several kinases, namely, protein kinase A, protein kinase C, CaMKII, PAK, CDK1, Rho kinase, Aurora B, and Plk1. In regard to the phosphorylation sites, phosphorylation of Ser38, Ser55, Ser71, Ser72, Ser82 are necessary to divide normally into two daughter cells in M stage. When a mutant vimentin in which both Ser71 and Ser 72 are changed to Ala is expressed in T24 cells that do not express endogenous vimentin, incomplete cytokinesis is observed (ref.). We decided employ a rat monoclonal antibody TM71 that specifically binds phosphorylated Ser71 of human vimentin. While Ser71 is not phosphorylated during metaphase, the phosphorylation is observed at cleavage furrow, which is a distinguishable narrow part between the two dividing cells from anaphase to telophase.

4.2 OS-IA using TM71

First, antigen-binding affinity as well as specificity of recombinant TM71 Fab was examined. The cDNAs of TM71 V_H/V_L were cloned into pDong1 phagemid, and the Fab fragment was displayed on phages (Fig. 7A). The wells on a microplate were immobilized with BSA conjugated with phosphorylated or unphosphorylated antigen peptides containing native sequence around Ser71. As a result, specific binding of Fab-displaying phage to phosphorylated antigen was observed on the plate. Next, the V_H/V_L interaction strength and its dependency to phosphorylated antigen using the V_H-displaying phage and the secreted L chain in the culture supernatant (Fig. 7B). The V_H/V_L interaction was increased depending on the concentration of phosphorylated peptide, while it was not the case with unphosphorylated peptide. The detection limit was as low as 0.1 ng/ml.

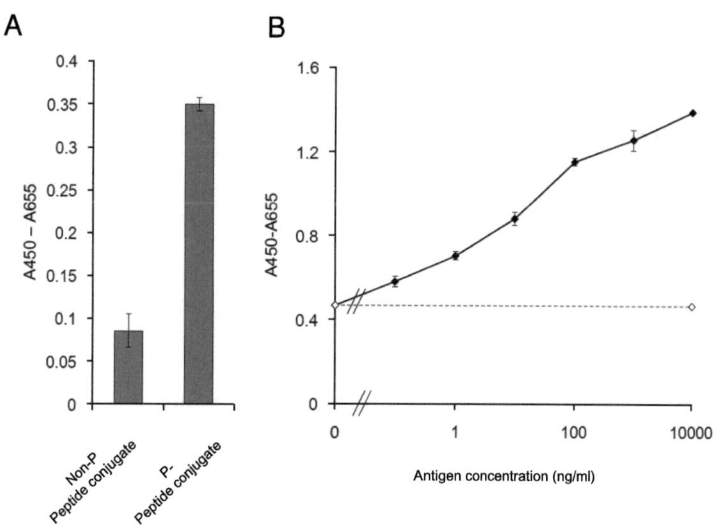

Fig. 7. Conventional phage ELISA and open-sandwich phage ELISA with anti-vimentin PSer71 (TM71). (A) Phage ELISA to detect immobilized antigens. (B) Open-sandwich phage ELISA. A solid line indicates the signal in the presence of phosphorylated antigen peptide. A dotted line indicates in the presence of non-phosphorylated peptide.

As described above, when H39 is Gln, the residue probably makes two hydrogen bonds with a corresponding V_L residue L38 (Gln), resulting a strong interaction between V_H and V_L

independent on antigen binding. Since H39 of TM71 is Gln, we randomized the DNA sequence for this residue, and obtained Q39R and Q39D mutants. When OS-IA was performed using these mutants, interaction strengths between V_H and V_L (backgrounds) were significantly decreased, resulting in higher signal/background ratios than that using wild type V_H (Fig. 7B).

Then we applied OS-IA to FRET using Q39R mutant V_H labeled with Rhodamine-X and V_L labeled with Alexa488. The result of the OS-FIA shows a slight phosphorylated antigen-concentration-dependent decrease in Alexa488 fluorescence and a significant increase in Rhodamine-X fluorescence, resulting in increased rhodamine-X/Alexa488 fluorescence ratio up to ~24 %. However, the increased FRET was not observed with unphosphorylated antigen. The result clearly showed that phosphorylated antigen specific detection was performed in a homogeneous solution in a noncompetitive manner.

An immunofluorescence staining of cells using TM71 shows Ser71 is phosphorylated only at cleavage furrow from anaphase to telophase (Fig. 8A). Finally, we try to detect endogenous phosphorylation in live cells. Human glioblastoma U251 cells were electroporated with both V_H(39R) labeled with Rhodamine-X and V_L labeled with Alexa48 by NEON™ Transfection System (Invitrogen, Carlsbad, CA). As a result, the fluorescence of the probe was observed at cleavage furrow, suggesting endogenous phosphorylated antigens were recognized between the V_H/V_L interface (Fig. 8B).

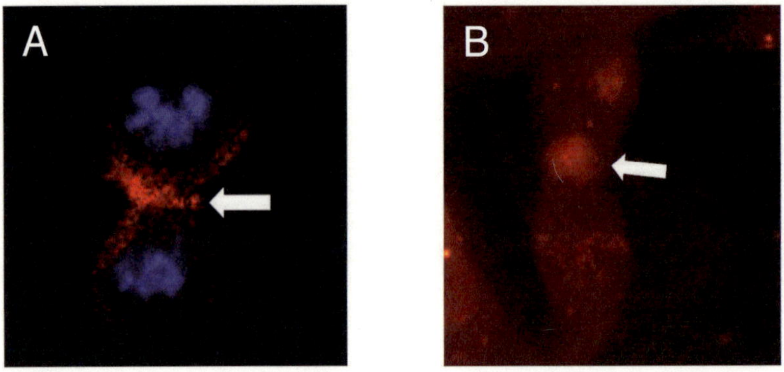

Fig. 8. Recognition of endogenous phosphorylated antigen using the probes based on OS-IA.(A) Immunofluorescence staining of U251 cells with TM71. A cleavage furrow is indicated by an arrow, where the signal from TM71 is observed (red). Nuclei are stained with DAPI (blue). (B) Recognition of endogenous phosphorylated antigen. The mixture of V_H labeled with rhodamine-X and V_L labeled with Alexa488 was electroporated into live U251 cells. Signal of rhodamine-X (red) was observed at a cleavage furrow (arrow).

5. Conclusion

In this chapter, we introduced OS-IA and its application in phosphoproteomics. First, we succeed in detecting general protein tyrosine phosphorylation by FRET approach. Second, the method was extended to site-specific detection of specific protein phosphorylation. Furthermore, it allowed observation of endogenous phosphorylation in a single live cell.

This method using OS-IA has the following merits. 1) Once the two appropriate antibody fragments, V_H and V_L are obtained in recombinant forms, many detection methods of protein-protein interaction, from heterogeneous assays such as ELISA to homogeneous assays such as FRET and PCA could be applied. In this sense, the application of OS-IA will be further expanded in future, including sensitive electrochemical detection [Sakata, 2009] and genetic reporters that allow high throughput screening on yeast [Gion, 2009]. Since most V_H fragments are known to be antigen-specific, future development of multiplexing with multiple reporters will be also possible. 2) Antigen specificity and binding affinity are adjustable by protein engineering with the help of powerful phage display systems. Especially, too strong V_H/V_L interaction in the absence of antigen can be effectively weakened to get larger signal in many OS-IAs. 3) Molecular weight of each antibody fragment is less than 13 kDa, which is small enough to allow fast diffusion in the cell and passing through the nuclear pores.

We are confident that this novel method will be exploited as a live cell imaging probe for phosphorylation and homogeneous assays for clinical diagnosis and drug screening.

6. Acknowledgment

We thank Yuichi Kitaoka for his experimental help with OS ELISA for phospho-vimentin.

7. References

Aburatani, T., Sakamoto, K., Masuda, K., Nishi, K., Ohkawa, H., Nagamune, T. & Ueda, H. (2003). A general method to select antibody fragments suitable for noncompetitive detection of monovalent antigens, *Anal Chem*, Vol. 75, No. 16, pp. 4057-4064, 0003-2700

Arai, R., Nakagawa, H., Tsumoto, K., Mahoney, W., Kumagai, I., Ueda, H. & Nagamune, T. (2001). Demonstration of a homogeneous noncompetitive immunoassay based on bioluminescence resonance energy transfer, *Anal Biochem*, Vol. 289, No. 1, pp. 77-81, 0003-2697

Arai, R., Ueda, H., Tsumoto, K., Mahoney, W. C., Kumagai, I. & Nagamune, T. (2000). Fluorolabeling of antibody variable domains with green fluorescent protein variants: application to an energy transfer-based homogeneous immunoassay, *Protein Eng*, Vol. 13, No. 5, pp. 369-376, 0269-2139

Bennett, G. S., Hollander, B. A. & Laskowska, D. (1988). Expression and phosphorylation of the mid-sized neurofilament protein NF-M during chick spinal cord neurogenesis, *J Neurosci Res*, Vol. 21, No. 2-4, pp. 376-390, 0360-4012

Brumbaugh, J. E., Morgan, S., Beck, J. C., Zantek, N., Kearney, S., Bendel, C. M. & Roberts, K. D. (2011). Blueberry muffin rash, hyperbilirubinemia, and hypoglycemia: a case of hemolytic disease of the fetus and newborn due to anti-Kp(a), *J. Perinatol.*, Vol. 31, No., pp. 373-376,

Celis, J. E., Larsen, P. M., Fey, S. J. & Celis, A. (1983). Phosphorylation of keratin and vimentin polypeptides in normal and transformed mitotic human epithelial amnion cells: behavior of keratin and vimentin filaments during mitosis, *J Cell Biol*, Vol. 97, No. 5 Pt 1, pp. 1429-1434, 0021-9525

Chou, Y. H., Bischoff, J. R., Beach, D. & Goldman, R. D. (1990). Intermediate filament reorganization during mitosis is mediated by p34cdc2 phosphorylation of vimentin, *Cell*, Vol. 62, No. 6, pp. 1063-1071, 0092-8674

Chou, Y. H., Ngai, K. L. & Goldman, R. (1991). The regulation of intermediate filament reorganization in mitosis. p34cdc2 phosphorylates vimentin at a unique N-terminal site, *J Biol Chem*, Vol. 266, No. 12, pp. 7325-7328, 0021-9258

Chung, A. S. & Chin, Y. E. (2009). Antibody array platform to monitor protein tyrosine phosphorylation in mammalian cells, *Methods Mol Biol*, Vol. 527, No., pp. 247-255, ix, 1064-3745

Dong, J., Ihara, M. & Ueda, H. (2009). Antibody Fab display system that can perform open-sandwich ELISA, *Anal Biochem*, Vol. 386, No. 1, pp. 36-44, 1096-0309

Eriksson, J. E., He, T., Trejo-Skalli, A. V., Harmala-Brasken, A. S., Hellman, J., Chou, Y. H. & Goldman, R. D. (2004). Specific in vivo phosphorylation sites determine the assembly dynamics of vimentin intermediate filaments, *J Cell Sci*, Vol. 117, No. Pt 6, pp. 919-932, 0021-9533

Frackelton, A. R., Jr., Ross, A. H. & Eisen, H. N. (1983). Characterization and use of monoclonal antibodies for isolation of phosphotyrosyl proteins from retrovirus-transformed cells and growth factor-stimulated cells, *Mol Cell Biol*, Vol. 3, No. 8, pp. 1343-1352, 0270-7306

Gao, C., Mao, S., Lo, C. H., Wirsching, P., Lerner, R. A. & Janda, K. D. (1999). Making artificial antibodies: a format for phage display of combinatorial heterodimeric arrays, *Proc Natl Acad Sci U S A*, Vol. 96, No. 11, pp. 6025-6030, 0027-8424

Gion, K., Sakurai, Y., Watari, A. & Inui, H. (2009). Designed recombinant transcription factor with antibody-variable regions, *Anal. Chem.*, Vol. 81, No., pp. 10162-10166,

Glenney, J. R., Jr., Zokas, L. & Kamps, M. P. (1988). Monoclonal antibodies to phosphotyrosine, *J Immunol Methods*, Vol. 109, No. 2, pp. 277-285, 0022-1759

Goto, H. & Inagaki, M. (2007). Production of a site- and phosphorylation state-specific antibody, *Nat Protoc*, Vol. 2, No. 10, pp. 2574-2581, 1750-2799

Goto, H., Kosako, H. & Inagaki, M. (2000). Regulation of intermediate filament organization during cytokinesis: possible roles of Rho-associated kinase, *Microsc Res Tech*, Vol. 49, No. 2, pp. 173-182, 1059-910X

Hayashi-Takanaka, Y., Yamagata, K., Nozaki, N. & Kimura, H. (2009). Visualizing histone modifications in living cells: spatiotemporal dynamics of H3 phosphorylation during interphase, *J Cell Biol*, Vol. 187, No. 6, pp. 781-790, 1540-8140

Hayashi-Takanaka, Y., Yamagata, K., Wakayama, T., Stasevich, T. J., Kainuma, T., Tsurimoto, T., Tachibana, M., Shinkai, Y., Kurumizaka, H., Nozaki, N. & Kimura, H. (2011). Tracking epigenetic histone modifications in single cells using Fab-based live endogenous modification labeling, *Nucleic Acids Res*, Vol., No., pp., 1362-4962

Helfand, B. T., Mendez, M. G., Murthy, S. N., Shumaker, D. K., Grin, B., Mahammad, S., Aebi, U., Wedig, T., Wu, Y. I., Hahn, K. M., Inagaki, M., Herrmann, H. & Goldman, R. D. (2011). Vimentin organization modulates the formation of lamellipodia, *Mol Biol Cell*, Vol. 22, No. 8, pp. 1274-1289, 1939-4586

Horikawa, K., Yamada, Y., Matsuda, T., Kobayashi, K., Hashimoto, M., Matsu-ura, T., Miyawaki, A., Michikawa, T., Mikoshiba, K. & Nagai, T. (2010). Spontaneous network activity visualized by ultrasensitive Ca(2+) indicators, yellow Cameleon-Nano, *Nat Methods*, Vol. 7, No. 9, pp. 729-732, 1548-7105

Ihara, M., Suzuki, T., Kobayashi, N., Goto, J. & Ueda, H. (2009). Open-sandwich enzyme immunoassay for one-step noncompetitive detection of corticosteroid 11-deoxycortisol, *Anal Chem*, Vol. 81, No. 20, pp. 8298-8304, 1520-6882

Inagaki, M., Gonda, Y., Ando, S., Kitamura, S., Nishi, Y. & Sato, C. (1989). Regulation of assembly-disassembly of intermediate filaments in vitro, *Cell Struct Funct*, Vol. 14, No. 3, pp. 279-286, 0386-7196

Inagaki, M., Gonda, Y., Matsuyama, M., Nishizawa, K., Nishi, Y. & Sato, C. (1988). Intermediate filament reconstitution in vitro. The role of phosphorylation on the assembly-disassembly of desmin, *J Biol Chem*, Vol. 263, No. 12, pp. 5970-5978, 0021-9258

Inagaki, M., Gonda, Y., Nishizawa, K., Kitamura, S., Sato, C., Ando, S., Tanabe, K., Kikuchi, K., Tsuiki, S. & Nishi, Y. (1990). Phosphorylation sites linked to glial filament disassembly in vitro locate in a non-alpha-helical head domain, *J Biol Chem*, Vol. 265, No. 8, pp. 4722-4729, 0021-9258

Inagaki, M., Nakamura, Y., Takeda, M., Nishimura, T. & Inagaki, N. (1994). Glial fibrillary acidic protein: dynamic property and regulation by phosphorylation, *Brain Pathol*, Vol. 4, No. 3, pp. 239-243, 1015-6305

Inagaki, M., Nishi, Y., Nishizawa, K., Matsuyama, M. & Sato, C. (1987). Site-specific phosphorylation induces disassembly of vimentin filaments in vitro, *Nature*, Vol. 328, No. 6131, pp. 649-652, 0028-0836

Ishikawa, H., Bischoff, R. & Holtzer, H. (1968). Mitosis and intermediate-sized filaments in developing skeletal muscle, *J Cell Biol*, Vol. 38, No. 3, pp. 538-555, 0021-9525

Islam, K. N., Ihara, M., Dong, J., Kasagi, N., Mori, T. & Ueda, H. (2011). Direct construction of an open-sandwich enzyme immunoassay for one-step noncompetitive detection of thyroid hormone T4, *Anal Chem*, Vol. 83, No. 3, pp. 1008-1014, 1520-6882

Izawa, I. & Inagaki, M. (2006). Regulatory mechanisms and functions of intermediate filaments: a study using site- and phosphorylation state-specific antibodies, *Cancer Sci*, Vol. 97, No. 3, pp. 167-174, 1347-9032

Kimura, H., Hayashi-Takanaka, Y. & Yamagata, K. (2010). Visualization of DNA methylation and histone modifications in living cells, *Curr Opin Cell Biol*, Vol. 22, No. 3, pp. 412-418, 1879-0410

Lacombe, C. & Mayeux, P. (1999). Erythropoietin (Epo) receptor and Epo mimetics, *Adv Nephrol Necker Hosp*, Vol. 29, No., pp. 177-189, 0084-5957

Lazarides, E. (1980a). Desmin and intermediate filaments in muscle cells, *Results Probl Cell Differ*, Vol. 11, No., pp. 124-131, 0080-1844

Lazarides, E. (1980b). Intermediate filaments as mechanical integrators of cellular space, *Nature*, Vol. 283, No. 5744, pp. 249-256, 0028-0836

Lazarides, E. (1982). Intermediate filaments: a chemically heterogeneous, developmentally regulated class of proteins, *Annu Rev Biochem*, Vol. 51, No., pp. 219-250, 0066-4154

Liao, J., Ku, N. O. & Omary, M. B. (1997). Stress, apoptosis, and mitosis induce phosphorylation of human keratin 8 at Ser-73 in tissues and cultured cells, *J Biol Chem*, Vol. 272, No. 28, pp. 17565-17573, 0021-9258

Lim, S. L., Ichinose, H., Shinoda, T. & Ueda, H. (2007). Noncompetitive detection of low molecular weight peptides by open sandwich immunoassay, *Anal Chem*, Vol. 79, No. 16, pp. 6193-6200, 0003-2700

Masuda, K., Sakamoto, K., Kojima, M., Aburatani, T., Ueda, T. & Ueda, H. (2006). The role of interface framework residues in determining antibody V(H)/V(L) interaction strength and antigen-binding affinity, *FEBS J*, Vol. 273, No. 10, pp. 2184-2194, 1742-464X

Nagata, K., Izawa, I. & Inagaki, M. (2001). A decade of site- and phosphorylation state-specific antibodies: recent advances in studies of spatiotemporal protein phosphorylation, *Genes Cells*, Vol. 6, No. 8, pp. 653-664, 1356-9597

Nishizawa, K., Yano, T., Shibata, M., Ando, S., Saga, S., Takahashi, T. & Inagaki, M. (1991). Specific localization of phosphointermediate filament protein in the constricted area of dividing cells, *J Biol Chem*, Vol. 266, No. 5, pp. 3074-3079, 0021-9258

Olsen, J. V., Vermeulen, M., Santamaria, A., Kumar, C., Miller, M. L., Jensen, L. J., Gnad, F., Cox, J., Jensen, T. S., Nigg, E. A., Brunak, S. & Mann, M. (2010). Quantitative phosphoproteomics reveals widespread full phosphorylation site occupancy during mitosis, *Science signaling*, Vol. 3, No., pp. ra3,

Pellequer, J. L., Chen, S., Roberts, V. A., Tainer, J. A. & Getzoff, E. D. (1999). Unraveling the effect of changes in conformation and compactness at the antibody V(L)-V(H) interface upon antigen binding, *J Mol Recognit*, Vol. 12, No. 4, pp. 267-275, 0952-3499

Pradelles, P., Grassi, J., Creminon, C., Boutten, B. & Mamas, S. (1994). Immunometric assay of low molecular weight haptens containing primary amino groups, *Anal Chem*, Vol. 66, No. 1, pp. 16-22, 0003-2700

Ross, A. H., Baltimore, D. & Eisen, H. N. (1981). Phosphotyrosine-containing proteins isolated by affinity chromatography with antibodies to a synthetic hapten, *Nature*, Vol. 294, No. 5842, pp. 654-656, 0028-0836

Rossi, F., Charlton, C. A. & Blau, H. M. (1997). Monitoring protein-protein interactions in intact eukaryotic cells by beta-galactosidase complementation, *Proc Natl Acad Sci U S A*, Vol. 94, No. 16, pp. 8405-8410, 0027-8424

Ruff-Jamison, S., Campos-Gonzalez, R. & Glenney, J. R., Jr. (1991). Heavy and light chain variable region sequences and antibody properties of anti-phosphotyrosine antibodies reveal both common and distinct features, *J Biol Chem*, Vol. 266, No. 10, pp. 6607-6613, 0021-9258

Ruff-Jamison, S. & Glenney, J. R., Jr. (1993a). Molecular modeling and site-directed mutagenesis of an anti-phosphotyrosine antibody predicts the combining site and allows the detection of higher affinity interactions, *Protein Eng*, Vol. 6, No. 6, pp. 661-668, 0269-2139

Ruff-Jamison, S. & Glenney, J. R., Jr. (1993b). Requirement for both H and L chain V regions, VH and VK joining amino acids, and the unique H chain D region for the high affinity binding of an anti-phosphotyrosine antibody, *J Immunol*, Vol. 150, No. 8 Pt 1, pp. 3389-3396, 0022-1767

Sakata, T., Ihara, M., Makino, I., Miyahara, Y. & Ueda, H. (2009). Open sandwich-based immuno-transistor for label-free and noncompetitive detection of low molecular weight antigen, *Anal Chem*, Vol. 81, No. 18, pp. 7532-7537, 1520-6882

Sasajima, Y., Aburatani, T., Sakamoto, K. & Ueda, H. (2006). Detection of protein tyrosine phosphorylation by open sandwich fluoroimmunoassay, *Biotechnol Prog*, Vol. 22, No. 4, pp. 968-973, 8756-7938

Schlessinger, J. (1986). Allosteric regulation of the epidermal growth factor receptor kinase, *J. Cell Biol.*, Vol. 103, No., pp. 2067-2072, 1540-8140

Selvin, P. R. & Hearst, J. E. (1994). Luminescence energy transfer using a terbium chelate: improvements on fluorescence energy transfer, *Proc Natl Acad Sci U S A*, Vol. 91, No. 21, pp. 10024-10028, 0027-8424

Sternberger, L. A. & Sternberger, N. H. (1983). Monoclonal antibodies distinguish phosphorylated and nonphosphorylated forms of neurofilaments in situ, *Proc Natl Acad Sci U S A*, Vol. 80, No. 19, pp. 6126-6130, 0027-8424

Suzuki, C., Ueda, H., Tsumoto, K., Mahoney, W. C., Kumagai, I. & Nagamune, T. (1999). Open sandwich ELISA with V(H)-/V(L)-alkaline phosphatase fusion proteins, *J Immunol Methods*, Vol. 224, No. 1-2, pp. 171-184, 0022-1759

Suzuki, T., Munakata, Y., Morita, K., Shinoda, T. & Ueda, H. (2007). Sensitive detection of estrogenic mycotoxin zearalenone by open sandwich immunoassay, *Anal Sci*, Vol. 23, No. 1, pp. 65-70, 0910-6340 (Print) 0910-6340 (Linking)

Tsujimura, K., Ogawara, M., Takeuchi, Y., Imajoh-Ohmi, S., Ha, M. H. & Inagaki, M. (1994a). Visualization and function of vimentin phosphorylation by cdc2 kinase during mitosis, *J Biol Chem*, Vol. 269, No. 49, pp. 31097-31106, 0021-9258

Tsujimura, K., Tanaka, J., Ando, S., Matsuoka, Y., Kusubata, M., Sugiura, H., Yamauchi, T. & Inagaki, M. (1994b). Identification of phosphorylation sites on glial fibrillary acidic protein for cdc2 kinase and Ca(2+)-calmodulin-dependent protein kinase II, *J Biochem*, Vol. 116, No. 2, pp. 426-434, 0021-924X

Ueda, H. (2002). Open sandwich immunoassay: a novel immunoassay approach based on the interchain interaction of an antibody variable region, *J Biosci Bioeng*, Vol. 94, No. 6, pp. 614-619, 1389-1723

Ueda, H., Kikuchi, M., Yagi, S. & Nishimura, H. (1992). Antigen responsive antibody-receptor kinase chimera, *Biotechnology (N Y)*, Vol. 10, No. 4, pp. 430-433, 0733-222X

Ueda, H., Kubota, K., Wang, Y., Tsumoto, K., Mahoney, W., Kumagai, I. & Nagamune, T. (1999). Homogeneous noncompetitive immunoassay based on the energy transfer between fluorolabeled antibody variable domains (open sandwich fluoroimmunoassay), *Biotechniques*, Vol. 27, No. 4, pp. 738-742, 0736-6205

Ueda, H., Tsumoto, K., Kubota, K., Suzuki, E., Nagamune, T., Nishimura, H., Schueler, P. A., Winter, G., Kumagai, I. & Mohoney, W. C. (1996). Open sandwich ELISA: a novel immunoassay based on the interchain interaction of antibody variable region, *Nat Biotechnol*, Vol. 14, No. 13, pp. 1714-1718, 1087-0156

Ueda, H., Yokozeki, T., Arai, R., Tsumoto, K., Kumagai, I. & Nagamune, T. (2003). An optimized homogeneous noncompetitive immunoassay based on the antigen-driven enzymatic complementation, *J. Immunol. Methods*, Vol. 279, No. 1-2, pp. 209-218,

Ullrich, A. & Schlessinger, J. (1990). Signal transduction by receptors with tyrosine kinase activity, *Cell*, Vol. 61, No. 2, pp. 203-212, 0092-8674

Winter, G., Griffiths, A. D., Hawkins, R. E. & Hoogenboom, H. R. (1994). Making antibodies by phage display technology, *Annu Rev Immunol*, Vol. 12, No., pp. 433-455, 0732-0582

Yamaguchi, T., Goto, H., Yokoyama, T., Sillje, H., Hanisch, A., Uldschmid, A., Takai, Y., Oguri, T., Nigg, E. A. & Inagaki, M. (2005). Phosphorylation by Cdk1 induces Plk1-mediated vimentin phosphorylation during mitosis, *J Cell Biol*, Vol. 171, No. 3, pp. 431-436, 0021-9525

Yano, T., Taura, C., Shibata, M., Hirono, Y., Ando, S., Kusubata, M., Takahashi, T. & Inagaki, M. (1991). A monoclonal antibody to the phosphorylated form of glial fibrillary acidic protein: application to a non-radioactive method for measuring protein kinase activities, *Biochem Biophys Res Commun*, Vol. 175, No. 3, pp. 1144-1151, 0006-291X

Yasui, Y., Goto, H., Matsui, S., Manser, E., Lim, L., Nagata, K. & Inagaki, M. (2001). Protein kinases required for segregation of vimentin filaments in mitotic process, *Oncogene*, Vol. 20, No. 23, pp. 2868-2876, 0950-9232

Yokozeki, T., Ueda, H., Arai, R., Mahoney, W. & Nagamune, T. (2002). A homogeneous noncompetitive immunoassay for the detection of small haptens, *Anal. Chem.*, Vol. 74, No. 11, pp. 2500-2504,

12

Phosphoproteomics: Detection, Identification and Importance of Protein Phosphorylation

Min Jia, Kah Wai Lin and Serhiy Souchelnytskyi
Karolinska Institutet, Stockholm
Sweden

1. Introduction

Reversible protein phosphorylation is one of the most important and well explored post-translational modifications. It is estimated that 30- 50% of the proteins are phosphorylated at some time point (Kalume, Molina, & Pandey, 2003). Phosphorylation is a major regulatory mechanism that controls many basic cellular processes. It may mediate a signal from the plasma membrane to the nucleus using a cascade of proteins, by which to regulate physiological and pathological processes such as cell growth, proliferation, differentiation and apoptosis (Blume-Jensen & Hunter, 2001; Hunter, 2000). Protein phosphorylation may result in alteration in protein- protein interactions, protein intracellular localization, and its activity (Blume-Jensen & Hunter, 2001; Kalume et al., 2003). Approximately 30% of drug discovery programs and R&D investment by the pharmaceutical industry target protein kinases. Knowledge of exactly when and where phosphorylation occurs and the consequences of this modification for the protein of interest can lead to an understanding of the detailed mechanism of the protein action, and ultimately to the discovery of new drug targets.

Protein phosphorylation is a fast and reversible process. It is catalyzed by kinases by attaching phosphate groups onto specific amino acids. Opposed to phosphorylation, dephosphorylation removes the phosphate groups from proteins by phosphatases. Dephosphorylation plays important role in balancing the protein phosphorylation status in signaling proteins. About 2-3% of the human genome encodes 518 distinct protein kinases (Manning et al., 2002). Four types of phosphorylation have been described based on the phosphorylation sites: (a) O-phosphorylation (serine, threonine and tyrosine), (b) N-phosphorylation (arginine, histidine and lysine), (c) S-phosphorylation (cysteine) and (d) acylphosphorylation (aspartic acid and glutamic acid) (Reinders & Sickmann, 2005). Currently, analytical methods have mainly been developed for O-phosphorylation, which is due to chemical stability of O-phosphorylation in acidic and in neutral milieu. Therefore, O-phosphorylation is the best studied among various types of phosphorylation (Reinders, 2002). In eukaryotic cells, phosphorylation occurs primarily on serine (pSer), threonine (pThr), and tyrosine (pTyr) residues, that is estimated to be in the ratio of 1800:200:1/pSer: pThr: pTyr (Kersten et al, 2006).

As aforementioned, phosphorylation is of importance for cell signaling and drug development. The lack of technologies to study all types of phosphorylation, differences in abundance and high dynamics make it difficult to have a comprehensive cover of all phosphorylation events in cells. This chapter summarizes strategies that have been developed to characterize the phosphoproteome. These strategies include identification of phosphoproteins and phosphopeptides, localization of the exact phosphorylation sites and quantitation of phosphorylation. In addition, the applications of phosphoproteomics in life science are discussed.

2. Phosphorylation

2.1 Detection of phosphoproteins

2.1.1 Radioactive labeling of proteins with ^{32}P isotope

Radioactive labeling of proteins with ^{32}P or ^{33}P is the oldest, but still one of the most sensitive approaches for detection of phosphorylation. Under the appropriate condition, the phosphoryl groups of ^{32}P or ^{33}P are enzymatic added to the proteins. The phosphorylated proteins are then detected by autoradiography. Therefore, radioactive labeling detects all types of phosphorylation, and is not specific to only one type of phosphorylation. The proteins can be labeled with $^{32}P/^{33}P$ isotopes in vitro and in vivo. For in vitro labeling, [γ-$^{32}P/^{33}P$]-ATP is used. It is fast and convenient process, that requires (semi)purified kinase and substrate (Springer, 1991). A kinase phosphorylates its substrate in a defined mixture of the kinase, substrate, buffer, ions, ATP and [γ $^{32}P/^{33}P$]-ATP. However, since the enzymatic reaction takes place in vitro, the major disadvantage is that it may not reflect the kinase activity under physiological conditions. This problem was overcome by introduction of the in vivo metabolic labeling (Wyttenbach & Tolkovsky, 2006). [$^{32}P/^{33}P$]Orthophosphate is used in in vivo labeling as a source of the isotope. The radioactive orthophosphates are incorporated during metabolic processes by kinases in cells. The significant advantage of in vivo labeling is that it provides a more accurate scenario of physiological enzymatic events, and reflects cellular responses as a consequence of treatments. The drawback of in vivo labeling also exists, e.g. it has been reported that in vivo labeling with doses of radioactivity may induce DNA fragmentation, DNA repair processes, subsequently may result in the cell cycle arrest and apoptosis. Another concern is that for in vivo labeling is usually used phosphate-free medium to culture cells. This medium may differ from the medium cells are cultured. Therefore, in vivo labeling experiments are often limited in time to 4- 8 hours. The third concern of radioactive labeling (in vitro and in vivo) is that only very small amount of radioactivity will be incorporated in proteins. This requires protocols for thorough removal of non-incorporated radioactivity from phosphorpoteins. The fourth concern of radioactive labeling is safety requirements. As the assays use radioactivity, corresponding safety rules have to be applied. Thus, it is very important to control quantity of the isotope and duration of labeling, take care of safety issues, and to minimize artificially-induced changes in phosphorylation.

2.1.2 Phospho-specific antibodies

In 1981, the first documented phospho-antibody was produced in rabbits immunized with benzonyl phosphonate conjugated to keyhole limpet hemocyanin (KLH) (Ignatoski, 2001). This antibody broadly recognized proteins containing phosphotyrosine. After that, there has been a rapid development in production of the phospho-antibodies. Nowadays, a large amount of phospho specific antibodies targeted to different amino acids (Ser, Thr, Tyr) at distinct sites in proteins have been produced, and widely used in the basic and clinic research (Ignatoski, 2001; Izaguirre, Aguirre, Ji, Aneskievich, & Haimovich, 1999). The availability of phospho-specific antibodies has opened the door for the improvement of detection of phosphorylation. The advantages of phospho-antibodies consist in 4 issues. The first one is that the antibodies can be used not only with extracted proteins and peptides, but also for intact cells or tissues. The second issue is that the specificity and sensitivity may be very high if an antibody has really good quality, and antibody can detect the epitope down to femtomole range. The third issue is the antibodies can be used to enrich and purify phosphorylated proteins and peptides. The fourth issue is that there are many antibodies very useful for phosphotyrosine detection, with good specificity and minimal reactivity to either unphosphorylated tyrosine or phosphorylated serine/threonine residues. The major decisive factor for selection of antibodies is their specificity in detection of a phosphoprotein. Therefore the quality of antibodies becomes the key concern on their applications.

2.1.3 Phosphoprotein staining

Detection of phosphoproteins by staining proteins separated in the acrylamide gels with phosphor-specific dye has been widely used for almost forty years (Green et al., 1973). Historically, several phospho-staining protocols were used. The cationic carbocyanine dye "Stains-All" stains phosphoproteins, but also highly acidic proteins, DNA and RNA (Green et al., 1973). This dye is not commonly used due to its low sensitivity that is one order magnitude less sensitivity than Coomassie staining and several orders less than ^{32}P radioactivity labeling. An alternative method involves the alkaline hydrolysis of phosphate esters of serine or threonine, precipitation of the released inorganic phosphate with calcium, formation of an insoluble phosphomolybdate complex and then visualization of the complex with a dye such as malachite green, methyl green or rhodamine B (Debruyne, 1983). The detection sensitivity of the staining method is still very poor, as a protein containing roughly 100 phosphoserine residues is detectable. Besides low sensitivity, phosphotyrosine is not to be detected as it cannot be hydrolyzed. Currently, ProQ Diamond has increasingly become the first choice of phosphoprotein dye (Steinberg et al., 2003). It is a fluorescent dye, and is suitable for the detection of phosphoserine-, phosphothreonine-, and phosphotyrosine-containing proteins directly in acrylamide gels. The sensitivity of ProQ Diamond staining has been improved significantly, and is down to 1- 16 ng. However, it is still considerably less sensitive than radioactive methods. The major advantages of ProQ Diamond are constituted of 1) it can be used in combination with a total protein stain, such as SYPRO Ruby protein gel stain, allowing protein phosphorylation levels and expression levels to be monitored in the same gel, 2) it is not dependent on kinase activity, 3) greater convenience and safety of handling, and 4) the stain also seems to be specific. However, in complex protein samples with thousands of protein species resolved by 2DE some nonphosphorylated, but rather abundant proteins may also be weakly stained (Stasyk et al., 2005).

2.1.4 Mass spectrometry (MS)

Mass spectrometry is one of the most modern techniques for detection of phosphorylation. Introduction of MS has significantly advanced the research in protein phosphorylation (Peters et al., 2004). It may be applied not only for detection of phosphorylation, but also identification of phosphorylation sites. Detection of phosphorylation by MS has been based on mass spectrum generated by trypsin-digested peptides. The mass shift of m/z 79.9 or neutral loss m/z 80 or 98 compared to its theoretical peptide mass has normally been considered as occurrence of phosphorylation. MS provides also a high speed and high sensitivity means for detection of phosphorylation. However, there are several inherent difficulties for the analysis of phospho-proteins. Firstly, signals from phosphopeptides are generally weaker as compared to non-phosphorylated peptides, as they are negatively charged and poorly ionized by MS performed in the positive mode. Secondly, it can be difficult to observe the signals from low-abundance phospho-proteins of interest in the high-background of abundant non-phosphorylated proteins. To overcome these drawbacks, enrichment of phophoproteins or phosphopeptides before MS is necessary to apply.

2.2 Isolation and enrichment of phosphorylated proteins and peptides

2.2.1 Immunoprecipitation

Phosphospecific antibodies are an efficient tool for enrichment of phosphorylated proteins (Rush et al., 2005). Antibodies specific to phosphorylated residues are used to immunoprecipitate full-length proteins and phosphopeptides. The most notable advantage of this approach is the sensitivity provided by antibodies, as we discussed in 2.1.2. Nowadays, a variety of commercial phsopho-specifc antibodies with high quality are available, especially antibodies for phosphotyrosine. The lack of high quality phosphoserine/ threonine antibodies impedes the characterization of serine or threonine phosphorylations.

2.2.2 Immobilized metal affinity chromatography (IMAC)

IMAC (Andersson & Porath, 1986) is the most frequently used technique for phosphopeptide and phosphoprotein enrichment, although it was originally introduced for purification of His-tagged proteins. It employs metal chelating compounds which are covalently bound to a chromatographic support for the coordination of metal ions. Phosphorylated peptides or proteins are bound to the IMAC stationary phase by electrostatic interactions of its negatively charged phosphate group with positively charged metal ions bound to the column material via nitriloacetic acid (NTA), iminodiacetic acid (IDA), and Tris (carboxymethyl) ethylenediamine (TED) linkers. Immobilized metal ions such as Ni^{2+}, Co^{2+}, or Mn^{2+} were initially shown to bind strongly to proteins with a high density of histidines. However, immobilized metal ions of Fe^{3+}, Ga^{3+}, and Al^{3+} have been demonstrated to show better binding with phosphopeptides. On the basis of measurements of ^{32}P or ^{33}P-radioactivity in whole cell extracts and in phosphoprotein samples after enrichment, IMAC-based techniques have been reported to recover up to 70–90% of total phosphoproteins (Dubrovska & Souchelnytskyi, 2005). IMAC procedures have become very popular rapidly due to its good compatibility with subsequent separation and detection techniques such as LC-ESI-MS/MS and MALDI MS. One of the major drawbacks of IMAC-based strategies is the nonspecific binding of peptides containing acidic amino acids, that is Glu and Asp, and the strong binding of multiply phosphorylated peptides. Nonspecific binding of acidic peptides can be diminished by esterification of carboxylic acids to methyl esters using HCl-saturated dried methanol (Ficarro et al., 2002). Reaction conditions have to be chosen carefully to avoid both incomplete esterification and side reactions because they increase sample complexity. Another disadvantage is that despite following a common binding-washing-eluting procedure, IMAC experimental conditions are very variable and care should be taken, as small variations in the experimental conditions (for example, pH, ionic strength, or organic composition of the solvents) could drastically affect the selectivity of the IMAC stationary phase.

2.2.3 Strong cation exchange chromatography (SCX)

Strong cation exchange chromatography has been used in the enrichment of phosphorylated peptides (Peng et al., 2003). This procedure is based on the fact that under acidic conditions (pH 2.7) phosphorylated peptides are single positively charged and amenable to further separation from nonphosphorylated peptides that usually have a net charge of 2+ at low pH. One of the main advantages of this method is that complex peptide mixtures can be analyzed directly, since it can be connected directly to LC-MS/MS for identification or sequencing (Villen & Gygi, 2008). However, this strategy does not have high specificity and the fractions enriched in phosphopeptides also contain a high percentage of contaminants. Therefore, it's very common to combine SCX with other enrichment methods, i.e IMAC and TiO2.

2.2.4 Titanium dioxide (TiO2)

A promising alternative to the use of IMAC for the enrichment of phosphorylated peptides was first described by Pinkse et al (Pinkse et al., 2004). The approach is based on the selective interaction of water-soluble phosphates with porous titanium dioxide microspheres via binding at the TiO2 surface. Phosphopeptides are trapped in a TiO2 precolumn under acidic conditions and desorbed under alkaline conditions. An increased specificity for phosphopeptides has been reported. Another advantage of this approach is that it can be easily coupled with a LC-MS/MS workflow (Ishihama et al., 2007; Marcantonio et al., 2008). Nevertheless, TiO2-based columns may retain nonphosphorylated acidic peptides. Peptide loading in 2, 5-dihydroxybenzoic acid (DHB) has been described to efficiently reduce the binding of nonphosphorylated peptides to TiO2 while retaining high binding affinity for phosphorylated peptides. This improved TiO2 procedure was found to be more selective than IMAC.

2.2.5 Chemical modification

Biotin tagging by β-elimination and Michael addition

A number of chemical modification strategies were developed in which the phosphate group has been replaced with a moiety that is chemically more stable than phosphate. One such method employs β-elimination of the phosphate from phosphothreonine or phosphoserine and results in the formation of dehydroaminobutyric acid or dehydroalanine, respectively. This product can be detected directly using tandem MS (Thompson et al., 2003). Alternatively, Michael addition is used to add a reactive thiol to dehydroaminobutyric acid or dehydroalanine to allow attachment of an affinity tag. Biotin is a widely used affinity tag and it permits purification of the chemically modified (previously phosphorylated) peptides (Meyer et al., 1991). This chemical modification is not applicable to phosphotyrosine residues and suffers from side reactions in which nonphosphorylated serine can be tagged.

2.3 Identification of phosphorylation sites

2.3.1 Two-dimensional (2D) phosphopeptide mapping

2D phosphopeptide mapping is a traditional biochemical method for identification of protein phosphorylation sites (Blaukat, 2004). After metabolically labeling cells with radioactive phosphate, the protein of interest is isolated by immunoprecipitation, subsequently subject to enzymatic digestion. The digested phosphopeptide is visualized by

2D phosphopeptide mapping (electrophoresis and thin- layer chromatography). To determine a phosphorylation site, labeled spots from the 2D phosphopeptide map are excised, and a combination of phosphoamino acid analysis and Edman sequencing is performed by monitoring the loss of radioactivity in each cycle. It should be noted that phosphorylation sites identified by 2D mapping need further validation (Nagahara et al., 1999). The most common way for confirmation is to mutate the phosphorylation sites and compare the phosphopeptide maps for the wild-type with those from mutant proteins. Although this method for phosphorylation identification is very useful, it still contains some limitations. It is time consuming process; care must be taken when label the cells with radioactivity; it only studies single protein, and can not apply to large scale identification of phosphorylation sites. Attempts were also made to combine 2D phosphopeptide mapping and MS analysis of recovered phosphopeptides by using 2D phosphopeptide mapping and HPLC purification before MS (Figure 1).

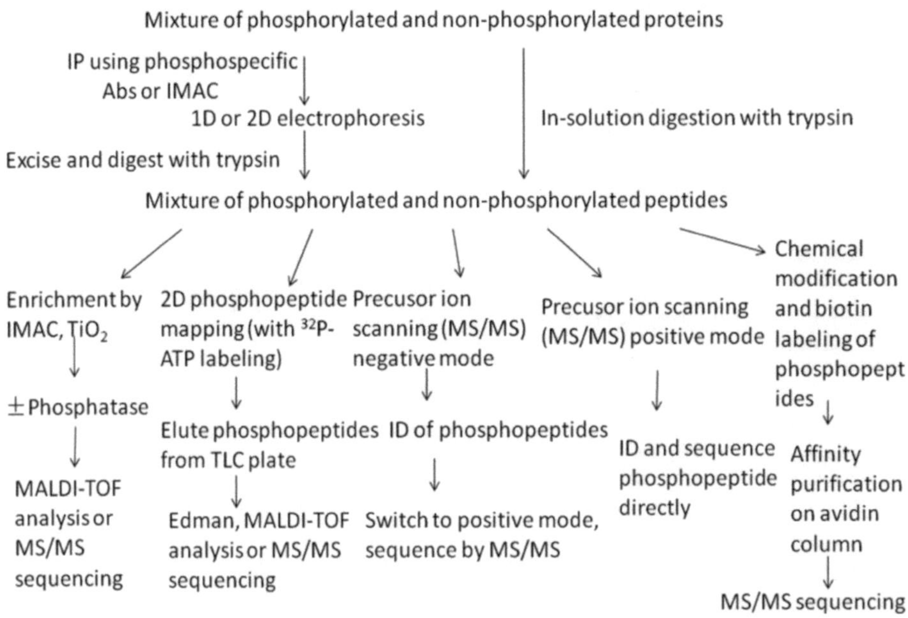

Fig. 1. An overview of techniques for enrichment and analysis of phosphorylated proteins or peptides using MS-based detection methods.

2.3.2 MS fragmentation

2.3.2.1 Collision-induced dissociation (CID)

PTMs on proteins often show greater susceptibility to cleavage by collision-induced fragmentation in the mass spectrometer than the peptide backbone. This characteristic may be used in different analytical strategies: 1) detection of the low mass 'signature' or 'marker' ions generated from the modification itself, 2) detection of the loss of the modification from the peptide precursor. Such targeted MS/MS analysis can enhance the specificity and sensitivity of phosphopeptide analysis, particularly for complex samples consisting of mixtures of phosphorylated and nonphosphorylated peptides. There are two most common precursor ion-scanning modes implemented on triple quadrupole mass spectrometers

(Figure 1). When phosphopeptides are fragmented by CID in the negative ion mode, a characteristic product ion (PO3) is generated giving rise to a peak at m/z 79 in the product spectrum (Collins et al., 2005). The detection of this marker ion has been used in various analytical setups. For example, a list of putative phosphopeptide ions can be generated by precursor ion scanning, a follow up analysis in positive ion mode is then performed to sequence these candidates by MS/MS using DDA (data-dependent acquisition) mode. Alternatively, detection of the precursor ion can be performed in positive ion mode, conduct MS/MS sequencing directly. Phosphotyrosine-containing peptides yield a characteristic immonium ion at m/z 216.043 from the loss of phosphotyrosine in the positive ion mode CID. Therefore, targeted monitoring of the precursor ion of 216.043 is useful for the detection of phosphotyrosine-containing peptides. This method was reported with good sensitivity, enabling the detection phosphotyrosine peptides from subpicomole amounts of gel-separated proteins (Steen et al., 2001).

2.3.2.2 Electron capture dissociation (ECD) and Electron transfer dissociation (ETD)

In tandem mass spectra of phosphopeptides generated by CID, limited or weak fragment ions spectra produce many false-negative as well as false-positive identifications especially for large, multiply charged and/or multiply phosphorylated peptides. Emerging alternative fragmentation techniques such as electron capture dissociation (ECD) and electron transfer dissociation (ETD) provide complementary sequence information for protein and peptide characterization, and are also applicable to the analysis of post-translational modifications (PTMs). These approaches induce more extensive cleavage along the peptide backbone and therefore provide excellent sequence tags, which retain labile PTMs (such as phosphorylation, glycosylation, acylation, ubiquitination and sumoylation) on backbone fragments. This feature enables direct and unambiguous assignment of the sites of modification. A further benefit is that these approaches are better suited for the analysis of large peptides, permitting the detection of multiple PTMs.

In ECD, multiply protonated ions capture low energy electrons and upon the following charge neutralisation, the resulting radical cations dissociate along the peptide backbone to produce a series of c and z type fragment ions while retaining the labile PTM group (Zubarev et al., 1998). Since the electron capture process requires low energy electrons (<10 eV) and long interaction times, the application of ECD was traditionally confined to instruments that employ static electromagnetic fields that avoid energizing or heating electrons, such as Fourier transform ion cyclotron resonance (FT-ICR) MS. Recently however, the addition of magnetic fields to ion traps have allowed for ECD in such electrodynamic trapping instruments (Baba et al., 2004) and the use of ECD in a digital ion trap mass spectrometer has also been reported (Ding & Brancia, 2006).

Electron transfer dissociation (ETD) is similar to ECD in that it also induces relatively non-selective cleavage of the N–Cα bond on a peptide's backbone producing c- and z-product ions, while maintaining phosphate groups and other potentially labile modifications (Syka et al., 2004). However, rather than involving the direct capture of an electron, ETD involves transfer of an electron to the multiply protonated precursor ion from a singly charged radical anion. The use of electron donors makes ETD amenable for use in quadrupole ion trap mass spectrometers which utilize rf fields for simultaneous storage and reaction of ions with positive and negative polarities. ETD fragmentation of phosphopeptides results in retention of phosphate groups in the sequence, allowing easier assignment of the exact site of modification. Moreover, these fragment ions are generated with good efficiency, making this a very promising approach for the analysis of phosphopeptides (Chi et al., 2007).

2.3.2.3 Photodissociation (PD)

Furthermore, ions in the gas-phase may be excited and subsequently dissociated by absorption of the photons. Photodissociation uses a laser that is directed through a window to irradiate the interior of the analyser. The mechanism of fragmentation by photodissociation involves the absorption by one or more photons. As each photon is absorbed, the ion increases its internal energy. The energy accumulates and finally it is sufficient to provoke dissociation resulting in gas-phase fragmentation of the ion. Ion activation may be achieved using infrared lasers (Brodbelt & Wilson, 2009). Due to its relatively low energy (-0.1 eV/photon), the absorption of multiple IR photons (tens to hundreds) are required for ion dissociation. Like CID, IRMPD is a "slow heating" method and allows for intramolecular energy redistribution over all of the vibrational degrees of freedom prior to the next photon absorption event (McLuckey & Goeringer, 1997). As a result, ergodic dissociation of low-energy pathways predominates and the resulting spectra are generally comparable with those obtained by CID. Photodissociation in the UV range has targeted common chromophores such as the amide bonds of a peptide using 193 and 157 nm light, as well as residue-specific chromophores such as aromatic amino acids using 220, 266, and 280 nm light (Reilly, 2009). Photodissociation has some advantages over the aforementioned methods. It is relatively selective, as only ions absorb the wavelength of the light used are activated. These techniques are most often used with ion trapping mass spectrometers.7

2.3.3 Site-specific microarrays

Site-specific microarray use oriented peptide libraries to map target specificity of kinases. This approach is based on kinase consensus sequences and phosphorylation prediction algorithm. It is thought that many phosphorylation sites tend to occur in accessible and flexible regions in three dimensional protein structures, suggesting that phosphorylation of linear peptide sequences in vitro should be similar to phosphorylation of the intact protein for the majority of sites. Data derived from peptide array experiments is consistent with known kinase consensus sequences, and is therefore a useful tool for studying phosphorylation. Peptide microarrays consist of synthetic peptide sequences deposited on to glass slides or attached to a derivatised surface, usually in triplicate, with phosphorylation site substitutions as controls. The peptides could map the entire sequence of a protein or correspond to a dataset of peptides that for example may have been identified from an in vivo sample, by MS. The in vitro phosphorylation reaction is performed in the presence of radiolabelled ATP, the array exposed to film and the image captured. Once the set of peptides have been synthesized, a large number of these microarrays can be made to screen many kinases relatively quickly (Kemp et al., 1975; Zetterqvist et al., 1976). The limitation of site-specific microarray is that, in vitro data is not sufficient on its own to definitively prove that a kinase may phosphorylate a given site in vivo. It is reasonable to use this peptide array technology as a first approach to screen for possible substrates (Diks et al., 2004; MacBeath & Schreiber, 2000), but further validation is required.

2.4 Quantification of protein phosphorylation

The field of phosphorylation quantitation by proteomics has made important advances over the last few years (Nita-Lazar et al., 2008). In general, quantitative phosphoproteomics can be performed as gel based and non-gel based (shotgun), e.g LC-MS/MS. Two-dimensional gel electrophoresis is a classical and powerful analytical method in proteomics that can separate complex mixtures of proteins based on charge (by isoelectric focusing) and apparent molecular mass (by sodium dodecyl sulfate polyacrylamide gel electrophoresis). In contrast to LC-MS/MS that analyzes digested peptides, 2DE delivers a map of intact proteins, which reflects changes in protein expression, isoforms or post-translational modifications. These changes can be confirmed by 1D or 2D Western blot analysis. Some forms of post-translational modification such as phosphorylation, glycosylation or limited proteolysis are easily located in 2DE gels as they appear as distinct spot trains along the horizontal and/or vertical axis. In 2DE, stoichiometry of phosphorylation can be readily determined by quantifying the spot intensity of each phosphorylated form. Furthermore, antibody-based approaches, using phosphorylation site-specific antibodies/Western blot analysis, ProQ Diamond staining and ^{32}P radioactive labeling are most frequently used approaches for gel-based phosphoprotein quantification (Agrawal & Thelen, 2006; Gorg et al., 2004).

Shotgun proteomics, where a peptide mixture from a sample is analyzed by LC-MS/MS without separating proteins on gels prior to the analysis, is a robust and high-throughput method that enables identification of thousands of proteins in a single analysis. There are many quantification methods in LC-MS/MS analysis, as summarized in the figure 2. Which method should be selected depends on the accuracy required, the sample source (from cultured cells or tissues) and the number of samples to be compared.

The easiest way is a label-free method based on the spectral counts of identified peptides. An abundant peptide is represented by a large LC peak eluted for a long time and has more chance of being analyzed by MS/MS. Thus, the number of observed spectra assigned to a particular peptide is a semi-quantitative measure of the abundance of the peptide. Although the accuracy of quantification using spectral counts is not high, it is convenient for analyzing large quantitative differences between samples. Another label-free method measures the intensity of MS chromatograms. A number of methods have been developed to quantify peptides/proteins from peak heights in shotgun proteomics using an internal control. Using high-resolution MS instruments, a peptide ion can be analyzed accurately in the low parts per million mass unit range, and it facilitates the peptide signal mapping across a few or multiple LC-MS measurements, using their mass to charge and retention time dimension. Thus, this method depends on the mass resolution, the mass precision and the consistency of the retention time to match the same peptides among different LC-MS analyses. It is essential to use a high-resolution MS, as well as a sensitive and reproducible nano-LC where the retention time of a particular peptide in crude extract behaves exactly the same.

Relative quantification based on differential stable isotope labeling is frequently used for quantitative phosphoproteomic analyses by MS. Although many techniques have been developed, only a few methods have been used in multiple laboratories. These include isotope-coded affinity tags, stable isotope labeling by amino acids in cell culture (SILAC) and the recently introduced chemical labeling by tandem mass tags, such as isobaric tag for relative and absolute quantitation (iTRAQ). SILAC and iTRAQ are currently the most frequently used techniques in quantitative MS-based phosphoproteomics. In SILAC, cell cultures to be compared are differentially labeled with amino acids containing stable isotopes, usually $^{13}C_6$-Lys and/or $^{13}C_6$-Arg, and normal amino acids. Lysates from differentially labeled cells are then mixed, digested with protease and analyzed by LC-MS/MS. As a result, differentially labeled peptides (light and heavy) with the same amino

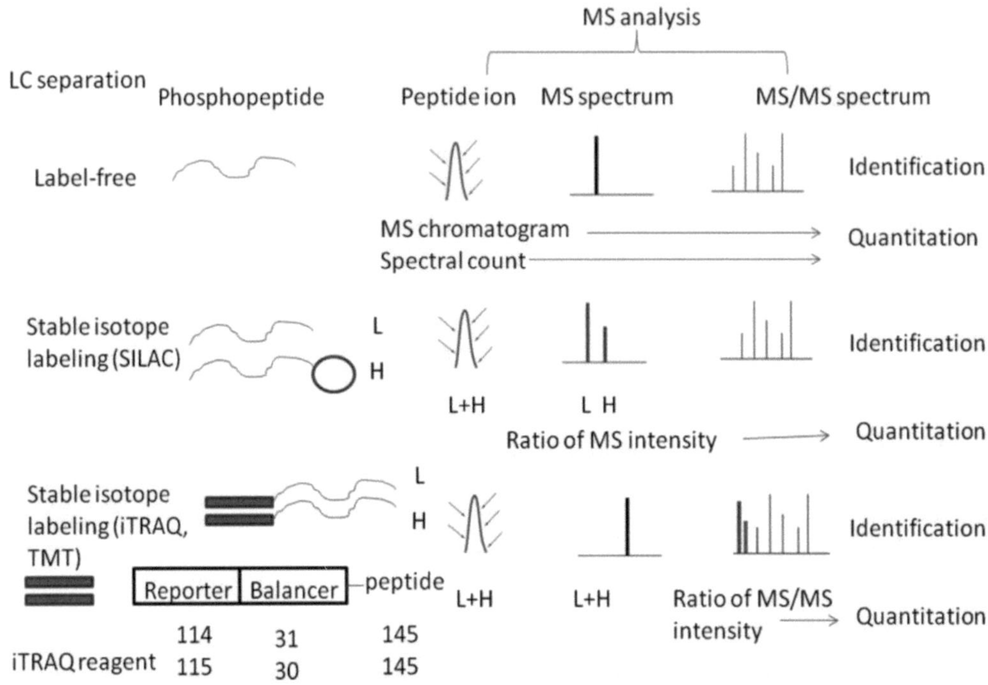

Fig. 2. Quantification methods for liquid chromatography-tandem mass spectrometry (LC-MS/MS) analysis. There are two major methods for quantification: label-free, labeling with stable isotopes. The two main non-labeling methods are based on the intensity of MS chromatograms and the spectral counts of identified peptides. Labeling methods are classified into two major groups: metabolic labeling and in vitro labeling. The representative of the metabolic labeling is SILAC. In SILAC, two cell cultures to be compared are differentially labeled with heavy amino acids containing stable isotopes (heavy) and normal amino acids (light). Lysates from differentially labeled cells are mixed, digested with protease and analyzed by LC-MS/MS. Differentially labeled peptides having the same amino acid sequence are detected in MS spectrum, and the relative abundance of the peptides can be compared by calculating their ratio. The representative of the in vitro labeling is performed using isobaric amine-specific tandem mass tags, such as iTRAQ. The iTRAQ reagent consists of reporter regions with 1 Da difference (molecular weight: 114, 115, 116…) and balance regions that adjust the molecular weight of the labeled parent ions (molecular weight: 31, 30, 29…). Each tag generates a unique reporter ion in the MS/MS spectra, and the relative abundance of the peptides can be compared by calculating their ratio.

acid sequence are detected in the MS spectrum, and the relative abundance of the peptides derived from different samples can be compared by calculating their ratio. Isobaric tagging for relative and absolute quantitation/tandem mass tags is a recently developed protein quantification method that uses isobaric amine-specific tandem mass tags and quantification in MS/MS instead of MS spectra. In MS spectra, the differentially labeled peptides possess the same mass by using the balance region in the tag and are represented in a combined single peak (Figure 2). However, each tag generates a unique reporter ion, and the intensities of the reporter ions in the MS/MS spectra are compared for protein quantification. iTRAQ can comparatively analyze four or eight different conditions in one experiment. This chemical labeling method is suitable for the phosphoproteomic analysis of tissue and clinical samples.

2.5 Application of phosphoproteomics

2.5.1 MS based applications

As described before, MS based phosphoprofiling can be performed at either gel or gel-free based level. Gel based phosphoproteome profiling indicates that intact phosphoproteins were isolated and enriched from samples, and subsequently subject to 1DE or 2DE. To explore changes in protein phosphorylation in MCF-7 cells overexpressing Samd2/3 in response to TGFβ1 stimulation, Stasyk et al (Stasyk et al., 2005) generated 2DE gels using ^{32}P labeled proteins, 32 proteins were identified with high confidence. One of the identified targets, transcription factor-II-I (TFII-I), was found in three phosphoprotein spots of similar molecular mass, suggesting at least three sites of phosphorylation of TFII-I. 2D phosphopeptide mapping of TFII-I identified Ser371 and Ser743 as two phosphorylation sites. Mutation of Ser 371 and Ser743 led to the abrogation of TGFβ1-dependent regulation of Cyclin D2, Cyclin D3, and E2F2 gene expression. Our lab previously developed a quantitative phosphoproteomics approach using phosphoprotein enrichment by Fe-IMAC followed by 2DE, which allows recovery of up to 90% of phosphoproteins, and can be applied to cultured cells and tissues (Dubrovska & Souchelnytskyi, 2005). We applied this approach to investigate the crosstalk of EGF and TGFβ signaling pathway in MCF-7 cell, and identified 47 convergent components of these two pathways. Systemic analysis identified MEK1 and CK1 as the primary common components of EGF and TGFβ signaling pathway in regulation of cell proliferation. Cell proliferation analysis showed that inhibition of MEK1 and CK1 can affect cell proliferation in the context of EGF and TGFβ treatment. And our experimental data also suggested that cross-talk between EGF and TGFβ may affect the responsiveness to Iressa in MCF-7 cell. Huber et al. performed two-dimensional differential gel electrophoresis (2D-DIGE) after purification of endosomes from EGF-treated mouse epithelial cells and identified 23 endosomal targets of EGF receptor signaling, such as R-Ras (Stasyk et al., 2007). Tang et al. performed 2D-DIGE analysis of phosphoprotein and plasma membrane fractions from brassinosteroid-treated Arabidopsis (Tang et al., 2008) and identified homologous protein kinases as key transducers of this steroid hormone signaling in plants (Tang et al., 2008). Thus, the combination of enrichment of phosphoproteins and 2D is a powerful proteomics approach for unraveling protein kinase-mediated signaling networks.

Gel-free (shotgun) proteomics, e.g LC-MS/MS analysis have been developed and successfully applied to quantify phosphopeptides from various cells and tissues. Ineffective erythropoiesis in human hematopoietic stem cells has been implicated in Hemoglobin E/beta-thalassemia. Ponnikorn et al. compared the phosphoprofiling of human hematopoietic cells between healthy donors and Hemoglobin E/beta-thalassemia patients (Ponnikorn et al., 2011). They enriched the phoshoproteins by IMAC, followed by LC-MS/MS for identification, and found 229 differentially phosphorylated proteins. To investigate the mechanisms of resistance to Her2 tyrosine kinase inhibitor lapatinib, Arteaga's group profiled the tyrosine phosphoproteome of sensitive and resistant cells using an immunoaffinityenrichment and mass spectrometry method (Rexer et al., 2011). Peptides containing phosphotyrosine were isolated directly from protease-digested cellular protein extracts with a phosphotyrosine-specific antibody and were identified by tandem mass spectrometry. They found increased phosphorylation of Src family kinases (SFKs) and putative Src substrates in several resistant cell lines. Treatment of these resistant cells with Src kinase inhibitors partially blocked PI3K-Akt signaling and restored lapatinib sensitivity. Further, SFK mRNA expression was upregulated in primary HER2+ tumors treated with lapatinib. Finally, they observed that the combination of lapatinib and the Src inhibitor AZD0530 was more effective than lapatinib alone at inhibiting pAkt and growth of established HER2-positive BT-474 xenografts in athymic mice. Ståhl et al. profiled phosphoproteome of ephrin and Eph signaling circuit. They combined SCX chromatography and TiO2 for enrichment of phosphopeptides followed by nano-LC and MS analysis (Stahl et al., 2011), and identified 1083 unique phosphorylated proteins. Out of these, 150 proteins were found only when ephrin B3 is expressed, whereas 66 proteins were found exclusively in U-1810 cells with silenced ephrin B3. Cantrell's group reported an unbiased analysis of the cytotoxic T lymphocyte (CTL) serine-threonine phosphoproteome by high-resolution mass spectrometry (Navarro et al., 2011). They used SILAC and IMAC based phosphopeptide enrichment, and identified approximately 2,000 phosphorylations in CTLs, of which approximately 450 were controlled by T cell antigen receptor (TCR) signaling. SILAC-based method also applied to study phosphorylation changes in EGF-stimulated HeLa cells (Olsen et al., 2006). After enrichment of phosphopeptides with SCX and TiO2, temporal profiles of 6600 unique phosphorylation sites on 2244 proteins were determined, including many known members of the EGF receptor signaling pathway. More recently, the cell cycle profiles of 20,443 phosphorylation sites in 6027 proteins have been determined and the site-specific stoichiometry of more than 5000 sites has been achieved by combining the results from corresponding non-phosphorylated peptides (Olsen et al., 2011). Comparative studies have revealed that different proteomic strategies are complementary to each other. For example, different phosphopeptide enrichment methods show distinct and partially overlapping preferences in phosphopeptide recovery. Bodenmiller and colleagues compared three different phosphopeptide enrichment approaches, phosphoramidate chemistry (PAC), IMAC and TiO2. They observed that among repeat isolates for each method pattern, overlap was ranging from an average of 80% for PAC (average of 6,643 features per run (FPR)), 76% for TiO2 (8,459 FPR), 74% for IMAC (9,312 FPR) (Bodenmiller et al., 2007). This suggested that no single method is sufficient for a comprehensive phosphoproteome analysis, and combination of different approaches for enrichment can improve the comprehensiveness of phosphoproteins. Furthermore, phosphoproteomic profiling of the ERK pathway using 2D-DIGE, label-free precursor ion scanning and SILAC identified surprisingly different subsets of ERK targets (Kosako & Nagano, 2011). Thus, a combination of various phosphoproteomic strategies, such as LC-MS/MS, 2DE and peptide (protein) microarrays, can increase the reliability and comprehensiveness of the data obtained.

2.5.2 Arrays-based application

Protein microarray technology offers the potential for profiling the proteome without employing separation techniques and evaluating protein biochemistry in a high-throughput and systematic manner. Synder's group have done large amount of work in protein microarray, including application in phosphoproteomics (Kafadar et al., 2003; Ptacek et al., 2005; Zhu et al., 2000). In 2000 his group screened 119 of the 122 yeast kinases with 17 different substrates (including the kinases themselves for monitoring autophosphorylation) on a prototype of protein microarray (Zhu et al., 2000). The substrates were immobilized onto nanowell protein chips and phosphorylation events were identified by adding ^{33}P-γ-

ATP and a specific yeast kinase and exposing the chip to a phosphoimager. They discovered that more than 60% of the kinases autophosphorylated themselves, and 94% of the tested kinases had at least one substrate in vitro, with 32 of them specifically phosphorylating one or two substrates. Twenty-seven kinases were found to phosphorylate poly (Tyr-Glu), which quadrupled the number of identified tyrosine kinases (seven) reported at that time. Moreover, these tyrosine kinases preferentially contain 3 conserved lysines and one conserved methionine near the catalytic region, indicating their potential roles in substrate selection. The same method was later used to identify Hrr25p as a kinase for the zinc-finger transcription factor Crz1, which turned out to negatively regulate Crz1 activity and nuclear localization by phosphorylation in vivo (Kafadar et al., 2003). His group later expanded the study to search for the substrates of 87 different *S. cerevisiae* kinases in a large set of more than 4400 full-length, functional yeast proteins with a yeast protein microarray containing 4400 yeast proteins (Ptacek et al., 2005). In this study they discovered about 4200 phosphorylation events affecting 1325 proteins and generated the first version of the phosphorylation network in yeast. In contrast to previous protein arrays that immobilize the probe, Paweletz et al developed reverse phase protein array, which immobilizes the whole repertoire of patient proteins that represent the state of individual tissue cell populations undergoing disease transitions (Paweletz et al., 2001). A high degree of sensitivity, precision and linearity was achieved, making it possible to quantify the phosphorylated status of signal proteins in human tissue cell subpopulations. Using this novel protein microarray they have analyzed the state of pro-survival checkpoint proteins at the transition stage from patient matched histologically normal prostate epithelium to prostate intraepithelial neoplasia (PIN) and then to invasive prostate cancer. Cancer progression was associated with increased phosphorylation of Akt ($P<0.04$), suppression of apoptosis pathways ($P<0.03$), as well as decreased phosphorylation of ERK ($P<0.01$). c-Src tyrosine kinase plays a critical role in signal transduction downstream of growth factor receptors, integrins and G protein-coupled receptors.

Amanchy et al. employed peptide microarrays approach and identified tyrosine phosphorylation sites in c-Src substrates (Amanchy et al., 2008). They designed custom peptide microarrays containing all possible tyrosine-containing peptides and their mutant counterparts containing a Tyr → Phe substitution from the identified substrates. In all, 624 WT or mutant (312 WT and 312 MUT) peptides from 14 proteins were spotted with each sequence being represented in triplicate, on to the glass slides. c-Src kinase assays were performed on the peptide microarrays and the arrays subsequently exposed to phosphorimager screen. From this analysis, 12 out of 14 proteins phosphorylation sites were identified.

3. Challenges, limitations, future directions and potential

The term 'phosphoproteomics' describes a subdiscipline of proteomics that is focused on deriving a comprehensive view of the extent and dynamics of protein phosphorylation. Phosphoproteomics greatly expands knowledge about the numbers and types of phosphoproteins, and promotes rapidly the analysis of entire phosphorylation-based signaling networks. The combination of quantitative methods and phosphoproteomics has generated powerful technologies for studying cellular signaling. However, there are still many challenges to the approach itself. Firstly, further improvements of the comprehensiveness are necessary. Ideally one could identify every single phosphorylation, independent of its concentration. However, currently most of the time people only look at most abundant ones. Lack of comprehensiveness impacts reproducibility. Data-dependent acquisition in MS is inherently irreproducible, so alternative ways of choosing ions for further fragmentation are needed. Additional complementary techniques are also needed, such as starting with proteases other than trypsin. Secondly, optimization of enrichment techniques for phosphopeptides and phosphoproteins also pose a significant challenge. As we discussed in 2.5.1, currently no single enrichment method can fully recovery of phosphoproteins and phosphopeptides. Combination of different enrichment methods will be an efficient way to this problem. Thirdly, the interpretation of quantitative phosphoproteomics studies is complicated because each differential phosphorylation event integrates both changes in protein expression and phosphorylation. Studies have been performed by parallel comparisons of protein expression and phosphorylation in *S. cerevisiae*, and it has been found that 25% of seemingly differential phosphopeptides now attributed to changes in protein expression (Wu et al., 2011). Hence, correct interpretation of comprehensive phosphorylation dynamics requires normalization by protein expression changes. In addition, despite the vast amount of quantitative phosphoproteomic data generated in recent studies, validation of these data has been quite limited. Furthermore, although large amount phosphorylation sites identified, most of these studies did not in-depth investigate the biological functions of the phosphorylation sites in signaling transduction.

4. References

Agrawal, G. K., Thelen, J. J. (2006). Large scale identification and quantitative profiling of phosphoproteins expressed during seed filling in oilseed rape. *Mol Cell Proteomics*, 5(11), 2044-2059.

Amanchy, R., Zhong, J., Molina, H., Chaerkady, R., Iwahori, A., Kalume, D. E. (2008). Identification of c-Src tyrosine kinase substrates using mass spectrometry and peptide microarrays. *J Proteome Res*, 7(9), 3900-3910.

Andersson, L., Porath, J. (1986). Isolation of phosphoproteins by immobilized metal (Fe3+) affinity chromatography. *Anal Biochem*, 154(1), 250-254.

Baba T, Hashimoto Y, Hasegawa H, Hirabayashi A, Waki I. (2004). Electron capture dissociation in a radio frequency ion trap. *Anal Chem* 76: 4263–4266.

Blaukat, A. (2004). Identification of G-protein-coupled receptor phosphorylation sites by 2D phosphopeptide mapping. *Methods Mol Biol*, 259, 283-297.

Blume-Jensen, P., Hunter, T. (2001). Oncogenic kinase signalling. *Nature*, 411(6835), 355-365.

Bodenmiller, B., Mueller, L. N., Mueller, M., Domon, B., Aebersold, R. (2007). Reproducible isolation of distinct, overlapping segments of the phosphoproteome. *Nat Methods*, 4(3), 231-237.

Brodbelt JS, Wilson JJ. (2009). Infrared multiphoton dissociation in quadrupole ion traps. *Mass Spectrom Rev* 28: 390–424.

Chi, A., Huttenhower, C., Geer, L. Y., Coon, J. J., Syka, J. E., Bai, D. L. (2007). Analysis of phosphorylation sites on proteins from Saccharomyces cerevisiae by electron transfer dissociation (ETD) mass spectrometry. *Proc Natl Acad Sci U S A*, 104(7), 2193-2198.

Collins, M. O., Yu, L., Coba, M. P., Husi, H., Campuzano, I., Blackstock, W. P. (2005). Proteomic analysis of in vivo phosphorylated synaptic proteins. *J Biol Chem*, 280(7), 5972-5982.

Debruyne, I. (1983). Staining of alkali-labile phosphoproteins and alkaline phosphatases on polyacrylamide gels. *Anal Biochem*, 133(1), 110-115.

Diks, S. H., Kok, K., O'Toole, T., Hommes, D. W., van Dijken, P., Joore, J. (2004). Kinome profiling for studying lipopolysaccharide signal transduction in human peripheral blood mononuclear cells. *J Biol Chem*, 279(47), 49206-49213.

Ding, L., Brancia, F. L. (2006). Electron capture dissociation in a digital ion trap mass spectrometer. *Anal Chem*, 78(6), 1995-2000.

Dubrovska, A., Souchelnytskyi, S. (2005). Efficient enrichment of intact phosphorylated proteins by modified immobilized metal-affinity chromatography. *Proteomics*, 5(18), 4678-4683.

Ficarro, S. B., McCleland, M. L., Stukenberg, P. T., Burke, D. J., Ross, M. M., Shabanowitz, J. (2002). Phosphoproteome analysis by mass spectrometry and its application to Saccharomyces cerevisiae. *Nat Biotechnol*, 20(3), 301-305.

Gorg, A., Weiss, W., Dunn, M. J. (2004). Current two-dimensional electrophoresis technology for proteomics. *Proteomics*, 4(12), 3665-3685.

Green, M. R., Pastewka, J. V., Peacock, A. C. (1973). Differential staining of phosphoproteins on polyacrylamide gels with a cationic carbocyanine dye. *Anal Biochem*, 56(1), 43-51.

Hunter, T. (2000). Signaling--2000 and beyond. *Cell*, 100(1), 113-127.

Ignatoski, K. M. (2001). Immunoprecipitation and western blotting of phosphotyrosine-containing proteins. *Methods Mol Biol*, 124, 39-48.

Ishihama, Y., Wei, F. Y., Aoshima, K., Sato, T., Kuromitsu, J., Oda, Y. (2007). Enhancement of the efficiency of phosphoproteomic identification by removing phosphates after phosphopeptide enrichment. *J Proteome Res*, 6(3), 1139-1144.

Izaguirre, G., Aguirre, L., Ji, P., Aneskievich, B., Haimovich, B. (1999). Tyrosine phosphorylation of alpha-actinin in activated platelets. *J Biol Chem*, 274(52), 37012-37020.

Kafadar, K. A., Zhu, H., Snyder, M., Cyert, M. S. (2003). Negative regulation of calcineurin signaling by Hrr25p, a yeast homolog of casein kinase I. *Genes Dev*, 17(21), 2698-2708.

Kalume, D. E., Molina, H., Pandey, A. (2003). Tackling the phosphoproteome: tools and strategies. *Curr Opin Chem Biol*, 7(1), 64-69.

Kemp, B. E., Bylund, D. B., Huang, T. S., Krebs, E. G. (1975). Substrate specificity of the cyclic AMP-dependent protein kinase. *Proc Natl Acad Sci U S A*, 72(9), 3448-3452.

Kersten, B., Agrawal, G. K., Iwahashi, H., Rakwal, R. (2006). Plant phosphoproteomics: a long road ahead. *Proteomics*, 6(20), 5517-5528.

Kosako, H., Nagano, K. (2011). Quantitative phosphoproteomics strategies for understanding protein kinase-mediated signal transduction pathways. *Expert Rev Proteomics*, 8(1), 81-94.

MacBeath, G., Schreiber, S. L. (2000). Printing proteins as microarrays for high-throughput function determination. *Science*, 289(5485), 1760-1763.

Manning, G., Whyte, D. B., Martinez, R., Hunter, T., Sudarsanam, S. (2002). The protein kinase complement of the human genome. *Science*, 298(5600), 1912-1934.

Marcantonio, M., Trost, M., Courcelles, M., Desjardins, M., Thibault, P. (2008). Combined enzymatic and data mining approaches for comprehensive phosphoproteome analyses: application to cell signaling events of interferon-gamma-stimulated macrophages. *Mol Cell Proteomics*, 7(4), 645-660.

McLuckey SA, Goeringer DE. (1997). Slow heating methods in tandem mass spectrometry. *J Mass Spectrom* 32*:* 461–474.

Meyer, H. E., Hoffmann-Posorske, E., Heilmeyer, L. M., Jr. (1991). Determination and location of phosphoserine in proteins and peptides by conversion to S-ethylcysteine. *Methods Enzymol*, 201, 169-185.

Nagahara, H., Latek, R. R., Ezhevsky, S. A., Dowdy, S. F. (1999). 2-D phosphopeptide mapping. *Methods Mol Biol*, 112, 271-279.

Navarro, M. N., Goebel, J., Feijoo-Carnero, C., Morrice, N., Cantrell, D. A. (2011) Phosphoproteomic analysis reveals an intrinsic pathway for the regulation of histone deacetylase 7 that controls the function of cytotoxic T lymphocytes. *Nat Immunol*, 12(4), 352-361.

Nita-Lazar, A., Saito-Benz, H., & White, F. M. (2008). Quantitative phosphoproteomics by mass spectrometry: past, present, and future. Proteomics, 8(21), 4433-4443.

Olsen, J. V., Blagoev, B., Gnad, F., Macek, B., Kumar, C., Mortensen, P. (2006). Global, in vivo, and site-specific phosphorylation dynamics in signaling networks. *Cell*, 127(3), 635-648.

Olsen, J. V., Vermeulen, M., Santamaria, A., Kumar, C., Miller, M. L., Jensen, L. J. (2011) Quantitative phosphoproteomics reveals widespread full phosphorylation site occupancy during mitosis. *Sci Signal*, 3(104), ra3.

Paweletz, C. P., Charboneau, L., Bichsel, V. E., Simone, N. L., Chen, T., Gillespie, J. W. (2001). Reverse phase protein microarrays which capture disease progression show activation of pro-survival pathways at the cancer invasion front. *Oncogene*, 20(16), 1981-1989.

Peng, J., Elias, J. E., Thoreen, C. C., Licklider, L. J., Gygi, S. P. (2003). Evaluation of multidimensional chromatography coupled with tandem mass spectrometry (LC/LC-MS/MS) for large-scale protein analysis: the yeast proteome. *J Proteome Res*, 2(1), 43-50.

Peters, E. C., Brock, A., Ficarro, S. B. (2004). Exploring the phosphoproteome with mass spectrometry. *Mini Rev Med Chem*, 4(3), 313-324.

Pinkse, M. W., Uitto, P. M., Hilhorst, M. J., Ooms, B., Heck, A. J. (2004). Selective isolation at the femtomole level of phosphopeptides from proteolytic digests using 2D-NanoLC-ESI-MS/MS and titanium oxide precolumns. *Anal Chem*, 76(14), 3935-3943.

Ponnikorn, S., Panichakul, T., Sresanga, K., Wongborisuth, C., Roytrakul, S., Hongeng, S. (2011). Phosphoproteomic analysis of apoptotic hematopoietic stem cells from hemoglobin E/beta-thalassemia. *J Transl Med*, 9, 96.

Ptacek, J., Devgan, G., Michaud, G., Zhu, H., Zhu, X., Fasolo, J. (2005). Global analysis of protein phosphorylation in yeast. *Nature*, 438(7068), 679-684.

Reilly JP. (2009). Ultraviolet photofragmentation of biomolecular ions. *Mass Spectrom Rev* 28*:* 425–447.

Reinders, J., & Sickmann, A. (2005). State-of-the-art in phosphoproteomics. *Proteomics*, 5(16), 4052-4061.

Rexer, B. N., Ham, A. J., Rinehart, C., Hill, S., de Matos Granja-Ingram, N., Gonzalez-Angulo, A. M. (2011). Phosphoproteomic mass spectrometry profiling links Src family kinases to escape from HER2 tyrosine kinase inhibition. *Oncogene*. doi: 10.1038/onc.2011.130

Rush, J., Moritz, A., Lee, K. A., Guo, A., Goss, V. L., Spek, E. J. (2005). Immunoaffinity profiling of tyrosine phosphorylation in cancer cells. *Nat Biotechnol*, 23(1), 94-101.

Springer, W. R. (1991). A method for quantifying radioactivity associated with protein in silver-stained polyacrylamide gels. *Anal Biochem*, 195(1), 172-176.

Stahl, S., Branca, R. M., Efazat, G., Ruzzene, M., Zhivotovsky, B., Lewensohn, R. (2011) Phosphoproteomic profiling of NSCLC cells reveals that ephrin B3 regulates pro-survival signaling through Akt1-mediated phosphorylation of the EphA2 receptor. *J Proteome Res*, 10(5), 2566-2578.

Stasyk, T., Dubrovska, A., Lomnytska, M., Yakymovych, I., Wernstedt, C., Heldin, C. H. (2005). Phosphoproteome profiling of transforming growth factor (TGF)-beta signaling: abrogation of TGFbeta1-dependent phosphorylation of transcription factor-II-I (TFII-I) enhances cooperation of TFII-I and Smad3 in transcription. *Mol Biol Cell*, 16(10), 4765-4780.

Stasyk, T., Morandell, S., Bakry, R., Feuerstein, I., Huck, C. W., Stecher, G. (2005). Quantitative detection of phosphoproteins by combination of two-dimensional difference gel electrophoresis and phosphospecific fluorescent staining. *Electrophoresis*, 26(14), 2850-2854.

Stasyk, T., Schiefermeier, N., Skvortsov, S., Zwierzina, H., Peranen, J., Bonn, G. K. (2007). Identification of endosomal epidermal growth factor receptor signaling targets by functional organelle proteomics. *Mol Cell Proteomics*, 6(5), 908-922.

Steen, H., Kuster, B., Mann, M. (2001). Quadrupole time-of-flight versus triple-quadrupole mass spectrometry for the determination of phosphopeptides by precursor ion scanning. *J Mass Spectrom*, 36(7), 782-790.

Steinberg, T. H., Agnew, B. J., Gee, K. R., Leung, W. Y., Goodman, T., Schulenberg, B. (2003). Global quantitative phosphoprotein analysis using Multiplexed Proteomics technology. *Proteomics*, 3(7), 1128-1144.

Syka, J. E., Coon, J. J., Schroeder, M. J., Shabanowitz, J., Hunt, D. F. (2004). Peptide and protein sequence analysis by electron transfer dissociation mass spectrometry. *Proc Natl Acad Sci U S A*, 101(26), 9528-9533.

Tang, W., Deng, Z., Oses-Prieto, J. A., Suzuki, N., Zhu, S., Zhang, X. (2008). Proteomics studies of brassinosteroid signal transduction using prefractionation and two-dimensional DIGE. *Mol Cell Proteomics*, 7(4), 728-738.

Tang, W., Kim, T. W., Oses-Prieto, J. A., Sun, Y., Deng, Z., Zhu, S. (2008). BSKs mediate signal transduction from the receptor kinase BRI1 in Arabidopsis. *Science*, 321(5888), 557-560.

Thompson, A. J., Hart, S. R., Franz, C., Barnouin, K., Ridley, A., Cramer, R. (2003). Characterization of protein phosphorylation by mass spectrometry using immobilized metal ion affinity chromatography with on-resin beta-elimination and Michael addition. *Anal Chem*, 75(13), 3232-3243.

Villen, J., Gygi, S. P. (2008). The SCX/IMAC enrichment approach for global phosphorylation analysis by mass spectrometry. *Nat Protoc*, 3(10), 1630-1638.

Wu, R., Dephoure, N., Haas, W., Huttlin, E. L., Zhai, B., Sowa, M. E. (2011). Correct interpretation of comprehensive phosphorylation dynamics requires normalization by protein expression changes. *Mol Cell Proteomics.* doi: 0.1074/mcp.M111.009654

Wyttenbach, A., Tolkovsky, A. M. (2006). Differential phosphoprotein labeling (DIPPL), a method for comparing live cell phosphoproteomes using simultaneous analysis of (33)P- and (32)P-labeled proteins. *Mol Cell Proteomics,* 5(3), 553-559.

Zetterqvist, O., Ragnarsson, U., Humble, E., Berglund, L., Engstrom, L. (1976). The minimum substrate of cyclic AMP-stimulated protein kinase, as studied by synthetic peptides representing the phosphorylatable site of pyruvate kinase (type L) of rat liver. *Biochem Biophys Res Commun,* 70(3), 696-703.

Zhu, H., Klemic, J. F., Chang, S., Bertone, P., Casamayor, A., Klemic, K. G. (2000). Analysis of yeast protein kinases using protein chips. *Nat Genet,* 26(3), 283-289.

Zubarev, R. A., Kelleher, N.L., McLafferty, F.W. (1998). Electron capture dissociation of multiply charged protein cations. A nonergodic process. *J. Am. Chem. Soc.* 120(13): 3265–66. doi:10.1021/ja973478k

Zubarev, R. A. (2004). Electron-capture dissociation tandem mass spectrometry. *Curr Opin Biotechnol,* 15(1), 12-16.

13

Proteome Kinetics: Coupling the Administration of Stable Isotopes with Mass Spectrometry-Based Analyses

Stephen F. Previs et al.*
Cardiovascular Disease-Atherosclerosis,
Merck Research Laboratories, Rahway, NJ
USA

1. Introduction

Proteins serve many purposes by acting as structural supports, receptors, signaling molecules and enzymes, in addition, they facilitate nutrient transport and maintain immunological responses. Although the concentration of a given protein may not change appreciably over a short interval, proteins are continuously remodeled. In this chapter we consider how to study protein kinetics. Attention is directed towards two critical areas which include (i) the logic behind using different tracers and (ii) how to design and execute experiments that are compatible with proteome-based analyses.

A practical illustration may highlight the importance of using isotope tracers to facilitate research in this area. For example, the concentration of circulating albumin provides a measure of protein nutritional status (and is a predictor of a patient's recovery from disease), however, since the fractional turnover of albumin is relatively slow (~ 3 to 5% of the pool is newly made per day) several weeks of an intervention may be required to affect plasma levels. Recognizing that the concentration of albumin is a delayed-onset marker of nutritional status, investigators have used isotope tracers to determine the acute response of plasma albumin synthesis to a dietary manipulation, accordingly, one can make predictions regarding the efficacy of an intervention. Such studies rely on straightforward experimental designs. Namely, an investigator first decides on what amino acid will be used (e.g. $^2H_3$-leucine) and how will it be administered (e.g. a primed-constant infusion), samples are then collected for a given amount of time and a protein of interest (e.g. albumin) is isolated. Once isolated, the protein of interest is degraded (typically *via* acid hydrolysis) and the labeling of the free amino acid present in the plasma is compared to that of the amino acid that was bound in the protein, i.e. one determines the precursor:product labeling ratio. Although this scenario is relatively straightforward, our review considers the pros and cons surrounding the use of different tracers. In particular, we discuss recent advances in the use of stable

---

* Haihong Zhou[1], Sheng-Ping Wang[1], Kithsiri Herath[1], Douglas G. Johns[1], Thomas P. Roddy[1], Takhar Kasumov[2] and Brian K. Hubbard[1]
[1]*Cardiovascular Disease-Atherosclerosis, Merck Research Laboratories, Rahway, NJ, USA*
[2]*Departments of Gastroenterology and Hepatology and Research Core Services, Cleveland Clinic, Cleveland, OH, USA*

isotope protocols that enable more flexible study designs, including the use of ^{2}H and ^{18}O-labeled water.

The second objective of this chapter aims to consider the utility of modern proteomic methods. For example, in the scenario described above, it is imperative that one purify the protein(s) of interest otherwise a study will reflect the kinetics of a mixture of proteins. Despite the fact that one can extensively purify proteins using immunoprecipitation, gel electrophoresis, etc. those approaches are typically labor intensive. Other methods, e.g. "shotgun" proteomics, can facilitate the resolution of complex mixtures with a minimum of time required for sample preparation, the trade-off is an increase in the amount of time required to process large data sets. It may not be obvious to investigators getting started in this area but the acquisition parameters that are often used in proteome-based studies are not necessarily compatible with the use of stable isotope-based flux protocols. In addition, investigators are often faced with questions such as, is one type of mass spectrometer "better" than another for determining the isotopomer profile? We discuss our experiences in estimating protein flux using proteome-based analyses.

In summary, proteome expression profiles contain information regarding differences between metabolic states yet they are typically of limited value when one aims to explain the nature of those differences. We consider approaches that should allow investigators to perform studies of proteome dynamics and therein move from static expression profiles towards kinetic/mechanistic studies. Where possible, attention is directed towards applications that can be used to advance the study of circulating proteins, especially those that relate to the field of lipoprotein kinetics. We apologize to investigators who do not have their work cited herein, where possible we have tried to identify papers that demonstrate necessary conceptual points and/or represent the initial publications in a given area.

2. Using stable isotope tracers to study protein synthesis and degradation

Rates of protein synthesis can be determined by administering a labeled precursor and then measuring its incorporation into a protein of interest (Figure 1) (Foster et al. 1993;Wolfe and Chinkes 2005). Assuming a simple model, in which there is a well-mixed pool of amino acids and a single product compartment, one can describe the kinetics using equation 1:

$$\text{protein labeling}_{time} = \text{protein labeling}_{max} \times (1 - e^{-FSR \times time}) \quad (1)$$

where protein labeling $_{max}$ represents the asymptotic labeling of a protein and FSR represents its fractional synthetic rate. By measuring the labeling at multiple points in time one can fit the curve and determine FSR. In cases where a steady-state labeling is not reached one typically estimates the kinetics using equation 2:

$$FSR = \text{pseudo-linear change in protein labeling} / (\text{precursor labeling} \times \text{time}) \quad (2)$$

We consider the following example to demonstrate the effect that timing of sample collection can have on estimating the FSR, in this case we have simulated the labeling of proteins with different FSRs (Figure1). Panel A demonstrates that fitting an entire data set to Equation 1 yields the expected FSR. Panel B demonstrates a comparable fit of the data using reduced data sets, fitting the points obtained only at 4 hour intervals to Equation 1 yields the expected FSR. Note that it may not be practical to obtain extensive data sets in all cases (Figure 1A vs B), e.g. one may be limited in regards to blood or tissue sampling, as well,

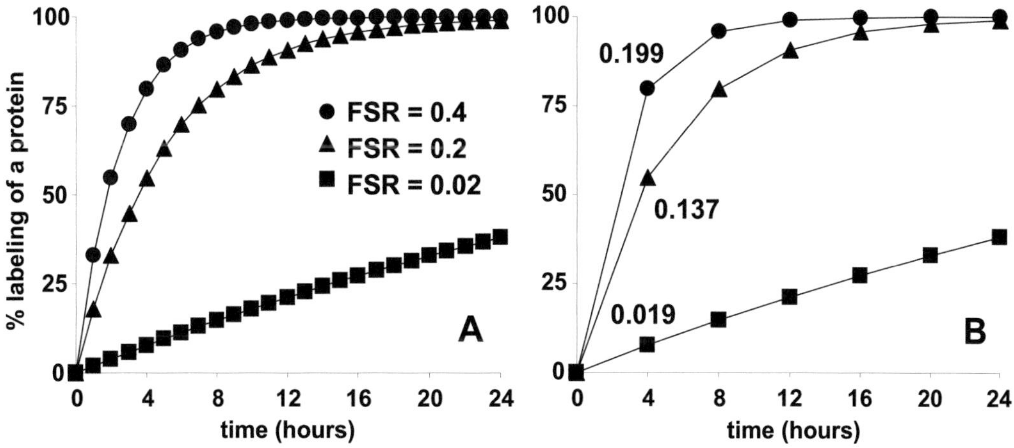

Fig. 1. Effects of modeling on data interpretation. Simulations were run to determine the effect(s) of calculation methods on apparent fractional synthetic rates. Panel A demonstrates a scenario wherein the protein labeling was simulated assuming three rate constants, i.e. FSR = 0.02, 0.2 and 0.4 per hour. Fitting all data points in a given curve using Equation 1 yields the expected rate constants. Panel B demonstrates the effect(s) of using various truncated data sets on apparent FSR. Again, fitting all data in a given curve to Equation 1 yields rate constants that closely agree with the expected values. However, it is possible to substantially underestimate the FSR when using single points, e.g. using Equation 2 and data obtained only at 4 hours leads to estimates of FSR equal to 0.199, 0.137 and 0.019 per hour, as compared to the expected values of 0.4, 0.2 and 0.02 per hour, respectively.

limiting the number of data points increases the throughput since fewer samples need to be analyzed. Panel B also demonstrates the effect of using Equation 2, for example, what would happen if we only obtained data 4 hours after administering a tracer? Clearly there is a reasonable estimate of FSR when the true value is relatively low (~ 0.02) but there is a sizeable underestimate of the FSR in cases where one expects it to equal ~ 0.4 and ~ 0.2. Although the apparent FSR values reported in Panel B are different from the expected values (i.e. when using Equation 2 and the sample obtained at 4 hours), one can still identify differences between the curves, i.e. the expected FSR of 0.4 yields a value of 0.199 whereas the expected FSR of 0.2 yields a value of 0.137. The effect of this error becomes important in cases where one aims to determine the magnitude of an intervention. For example, there is a 2-fold difference between the true FSR values (i.e. 0.4 vs 0.2) yet the apparent FSR values only differ by ~ 1.5-fold (i.e. 0.199 vs 0.137). Therefore, the timing of sample collection has important consequences on the interpretation of the data and the conclusions one may reach regarding physiological homeostasis.

Although the mathematics surrounding tracer kinetics have been described in detail (Foster et al. 1993;Wolfe and Chinkes 2005), there are certain caveats that apply in different fields. For example, investigators working in the area of lipoprotein kinetics have recognized the need to add delays in the modeling (Barrett, Chan, and Watts 2006;Foster et al. 1993;Patterson et al. 2002). Namely, although proteins such as apolipoprotein B are continuously synthesized within liver and/or intestine, they are not immediately secreted into the circulation. Consequently there is a lag time between the administration of a tracer and the appearance of labeled apolipoprotein B in the plasma. Foster *et al.* (Foster et al. 1993) have elegantly outlined the rationale behind different mathematical treatments of a given data set, they demonstrate the impact of various assumptions in regards to the modeling of data on the apparent FSR. It is also important to note the differences when modeling data that are expressed as a tracer-to-tracee ratio vs isotopic enrichment, the former is commonly reported but the latter may be preferred in many instances (Cobelli, Toffolo, and Foster 1992;Ramakrishnan 2006;Toffolo, Foster, and Cobelli 1993).

A second major factor to consider regarding the logic that is applied in kinetic studies centers on heterogeneity in the product pool (note that there are concerns regarding heterogeneity in labeling of the precursor pool, those will be considered in more detail in Section 3) (Foster et al. 1993). To this point we have assumed a simple model in which there is a single pool of product molecules, however, investigators working in the area of lipoprotein kinetics readily recognize the existence of at least two pools of circulating apoB100, one that is associated with VLDL particles and another that is associated with LDL particles (Lichtenstein et al. 1990). While there has been some debate regarding whether or not LDL-apoB100 is made *de novo* or whether it is derived from the delipidation of VLDL it is clear that the labeling curves are dramatically different (Lichtenstein et al. 1990;Shames and Havel 1991). In a classical study, Lichtenstein *et al.*(Lichtenstein et al. 1990) demonstrated that the labeling of VLDL-apoB100 approaches a steady-state in ~ 15 hours whereas the labeling of LDL-apoB100 is still in the pseudo-linear range during the same interval; those studies also demonstrated that there are sizeable differences in the abundance of VLDL-apoB100 vs LDL-apoB100 (Figure 2).

What are the consequences of estimating the FSR of apoB100 from the total labeling, i.e. if one ignores the fact that a small amount of the protein is typically labeled much faster than the bulk pool of apoB? Consider the scenario outlined in Figure 2, the lumped fractional rate constant does not reflect either of the individual fractional rate constants. In addition, although directional changes in the lumped fractional rate constants reflect true changes, the magnitude is underestimated. On the contrary, the ability to measure the absolute flux rate (i.e. the mass of protein made per unit of time) allows one to draw conclusions regarding true changes in the flux, however, one is not able to determine the site of those changes (e.g. Was a single pool affected? If so, which one?). We consider how to estimate protein concentration later.

A final question to consider regarding protein kinetics is centered on quantifying protein breakdown (Figure 3). As noted above, the incorporation of a tracer into a protein of interest can be used to estimate the rate of synthesis, can one estimate the rate of protein breakdown by measuring the elimination of a tracer from a protein of interest? We believe that the answer is "no", or at the very least it is not as straightforward as reports in the literature (Bateman et al. 2006;Bateman et al. 2007). Readers should consider how measurements of isotopic labeling are typically performed and how data are expressed. For example, investigators often use a mass spectrometer to determine isotopic labeling and express data as the ratio of labeled to unlabeled molecules (or the percentage of labeling, i.e. the labeled molecules divided by the sum of labeled and unlabeled molecules) (Dwyer et al. 2002;Lichtenstein et al. 1990;Magkos, Patterson, and Mittendorfer 2007). We agree that in cases where one infuses a labeled amino acid for a given time and then stops the infusion of the tracer that there will be a decrease in the labeling of a given protein over time (Figure 3) (Bateman et al. 2006;Bateman et al. 2007). However, assuming that protein breakdown is a random process, i.e. protein breakdown does not discriminate between labeled and unlabeled molecules, the ratio of labeled to unlabeled protein molecules will not change as

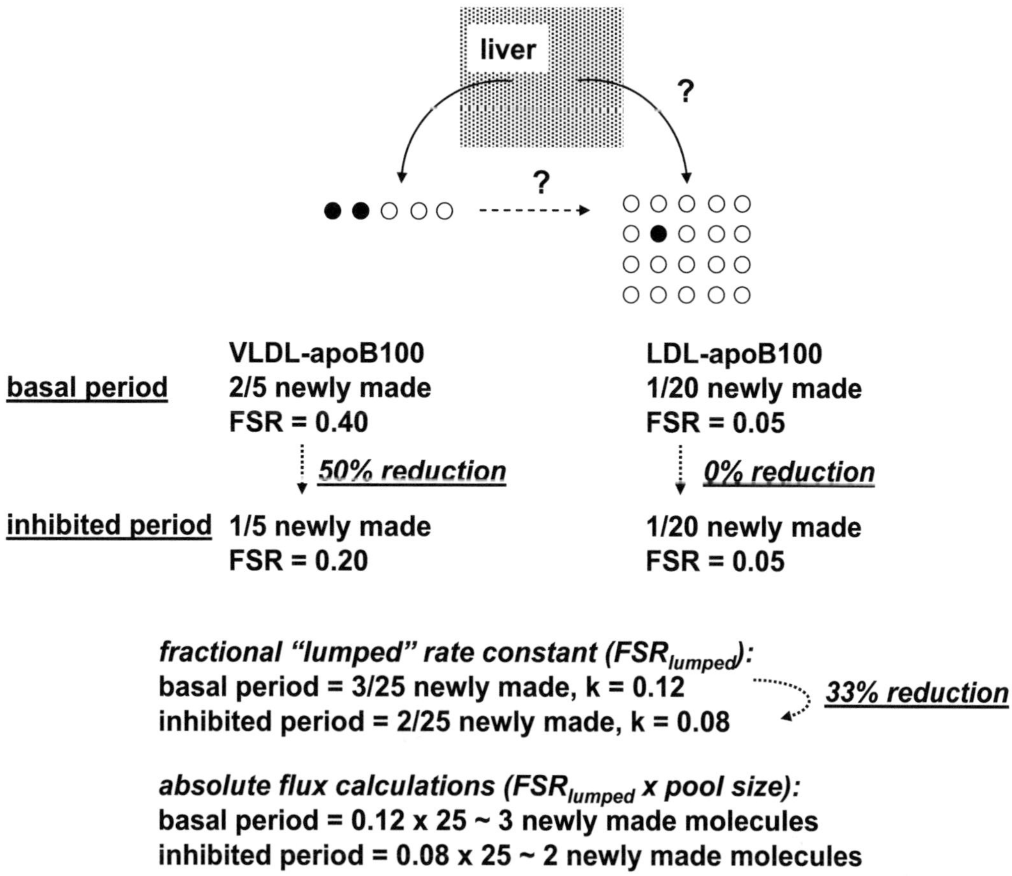

Fig. 2. Effect(s) of lumping pools. Simulations were run to determine the impact of treating a mixed pool as single compartments (open circles represent unlabeled proteins and solid circles represent labeled proteins). For example, apoprotein-B100 is found in VLDL and LDL particles in the plasma. The mass of apoB100 is ~ 5 to 40 times different between these compartments, as well the FSR is considerably different. Assume that VLDL-apoB100 has an FSR of ~ 0.4 and a pool size of ~ 5 molecules whereas LDL-apoB100 has an FSR of ~ 0.05 and a pool size of ~ 20 molecules. If one isolated the individual apolipoprotein pools the aforementioned values would be obtained, however, if one isolated total apoB100 from the plasma the fractional "lumped" synthesis rate would equal ~ 0.12 (3 out of 25 molecules). Now, assume an inhibitor of VLDL-apoB100 is added such that the FSR of VLDL-apoB100 decreases to ~ 0.2 (for simplicity, assume a parallel change occurs in protein degradation so that the pool size remains constant). If one isolates total apoB100 from plasma the fractional "lumped" synthesis rate would equal ~ 0.08 (2 out of 25 molecules). Clearly one would observe a decrease in synthesis but the true effect is substantially underestimated (i.e. the true reduction is 50% in VLDL-apoB100 vs 33% reduction detected in total apoB100). Accounting for the pool size, however, allows one to reliably determine the true change in apoB100 synthesis, i.e. a total 3 molecules are newly made during the basal period vs 2 during the inhibited period.

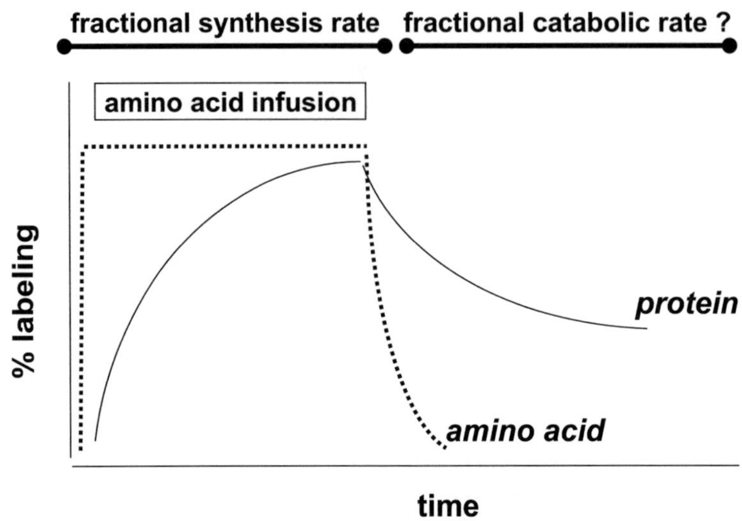

Fig. 3. Tracer-based estimates of protein breakdown. During the infusion of a labeled amino acid (dotted line) one can estimate protein synthesis by determining the change in protein labeling (solid line). Following the infusion of a labeled amino acid one expects a "washout" or a decrease in the labeling. As shown here, however, the rate at which the protein labeling decreases is dependent on the rate of protein synthesis and not protein breakdown. Note that the y-axis is expressed as "% labeling" (consistent with reports in the literature). A major assumption of any tracer method is that the tracer and tracee are indiscriminately metabolized, therefore, after one stops administering a labeled precursor amino acid the labeling in the protein can only decrease if new protein is made in absence of labeled precursor amino acids.

the protein is degraded; the labeling decreases because new proteins are being made from unlabeled precursors (Previs et al. 2004;Waterlow 2006).

We believe that it is possible to estimate protein breakdown using the following logic, changes in the abundance of a protein equal the rate of synthesis minus the rate of breakdown. Protein breakdown can be determined by measuring the abundance of a protein and estimating the rate of synthesis, i.e. one solves the equation for protein breakdown (Bederman et al. 2006). Section 4 considers the merits of different approaches for measuring the abundance of a protein. It should also be emphasized that the ability to measure the abundance of a protein is important in cases where one aims to determine a rate of flux (i.e. the mass of protein that is being renewed per unit of time). For example, to this point we have focused on measuring a fractional rate constant (or a percent of a pool that is turned over per unit of time), one can calculate the absolute amount of newly made protein per unit of time by multiplying the FSR by the pool size (i.e. concentration multiplied by the volume of distribution). In studies of apolipoprotein kinetics, the pool size is typically assumed to equal the plasma volume which is estimated to be 4.5% of body weight (Lichtenstein et al. 1990;Magkos, Patterson, and Mittendorfer 2007). In cases where one aims to study the kinetics of other circulating proteins, for example albumin, it may be necessary to account for distribution between the extravascular and intravascular spaces (Sigurdsson, Shames, and Havel 1981;Wasserman, Joseph, and Mayerson 1955).

3. How can I label the precursor pool?

Our discussion of protein synthesis is entirely focused on the logic of using precursor:product labeling ratios to estimate rates of flux, we are not examining cases in which one injects a pre-labeled protein and then measures its kinetics. Therefore, one should consider how to label the amino acid building blocks used in protein synthesis (Figure 4). Perhaps the most obvious design that comes to mind centers on administering a labeled amino acid (Dudley et al. 1998;Lichtenstein et al. 1990), however, investigators have also administered other labeled precursors (e.g. ^{13}C-glucose, $^{2}H_2O$ and $H_2{}^{18}O$) (Bernlohr 1972;Bernlohr and Webster 1958;Borek, Ponticorvo, and Rittenberg 1958;Busch et al. 2006;De Riva et al. 2010;Rachdaoui et al. 2009;Rittenberg, Ponticorvo, and Borek 1961;Vogt et al. 2005;Wykes, Jahoor, and Reeds 1998). Before discussing the merits of specific approaches we briefly consider the mode of administering the labeled precursor, e.g. a labeled amino acid can be administered as a primed-constant infusion or a single bolus injection (Dwyer et al. 2002;Lichtenstein et al. 1990;Wolfe and Chinkes 2005).

The general logic behind the primed-constant infusion is that one can instantaneously achieve and then maintain a steady-state labeling of the precursor pool (Lichtenstein et al. 1990), whereas a single bolus injection is typically associated with a wave (or pulse) of labeling (Dwyer et al. 2002). A concern with using a primed-constant infusion is that one must have catheterized subjects, while certainly feasible in human studies this is not as practical in many pre-clinical models (especially in drug discovery programs where large numbers of compounds are routinely screened). However, a pro of the primed-constant infusion centers on the degree of product labeling that can be achieved, this can be rather dramatic in studies of apolipoprotein kinetics. For example, when investigators have administered $^{2}H_3$-leucine using a primed-constant infusion the plasma pool can be enriched to nearly 10% for several hours (Lichtenstein et al. 1990). Although some proteins have a rapid turnover others are labeled to a much lesser degree, e.g. the FSR of VLDL-apoB100 and HDL-apoA1 are in the range of ~ 5 and ~ 0.2 pools per day and the labeling typically approaches 7% and 0.75%, respectively.

In contrast to a primed-infusion, when administering a single bolus of $^{2}H_3$-leucine the labeling of VLDL-apoB100 and HDL-apoA1 approaches ~ 2.5% and ~ 0.25%, respectively (Dwyer et al. 2002). These differences in protein flux impact the isotopic labeling and have important implications on the analytical methods that are used to measure the enrichment. One might be able to enhance the use of a bolus injection method by choosing (i) an essential amino acid and/or (ii) an amino acid with a relatively long half-life. For example, one expects less dilution of essential amino acids since they can only be produced by one source (protein breakdown and not *de novo* synthesis), as well, compared to some non-essential amino acids (which participate in rapid inter-organ nitrogen transport) the $t_{1/2}$ of essential amino acids can be relatively slow. It is not surprising that ^{13}C-lysine has been used to make SILAC models (Kruger et al. 2008), since lysine is needed in relatively small amounts complete substitution of unlabeled lysine for ^{13}C-labeled lysine can be managed. The same types of experiments with ^{13}C-alanine would probably be of limited value since alanine is rapidly turned over and it sits at a highly branched point in intermediary metabolism (Wykes, Jahoor, and Reeds 1998). Nevertheless, in limited cases ^{13}C-glucose has been used to quantify protein synthesis (Figure 4). For example, ^{13}C-glucose is converted ^{13}C-pyruvate which readily equilibrates with alanine to yield ^{13}C-alanine, entry of ^{13}C-pyruvate into the citric acid cycle will generate other ^{13}C-labeled amino acids *via* comparable equilibration reactions (Vogt et al. 2005;Wykes, Jahoor, and Reeds 1998).

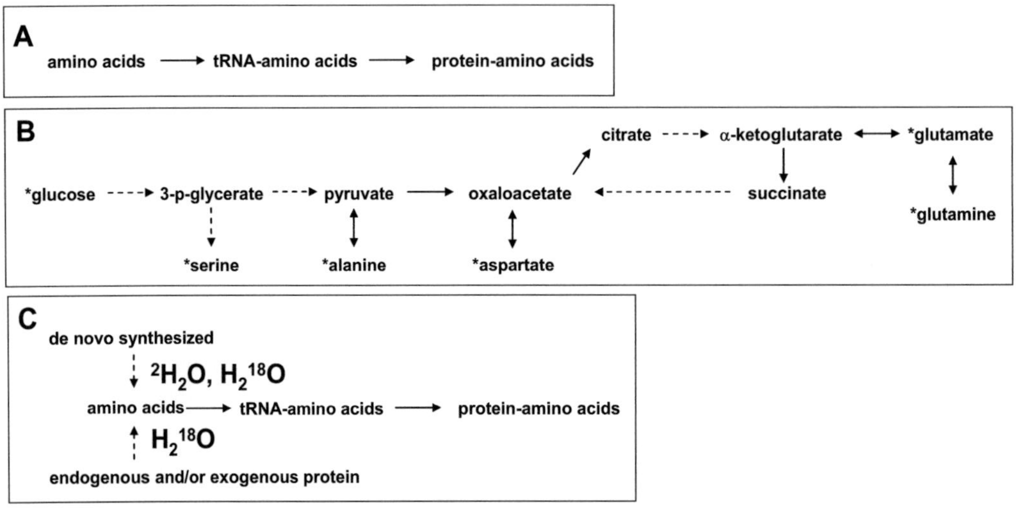

Fig. 4. Approaches to labeling amino acids. Panel A considers a straightforward method in which a labeled amino acid is administered. Panel B considers a scenario in which labeled glucose is administered; glycolytic metabolism will lead to the labeling of several amino acids. Note that an abbreviated metabolic scheme is shown to emphasize certain points of exchange, other amino acids can become labeled as well. Panel C considers the administration of labeled water. In cases where $^{2}H_2O$ is administered it is expected that *de novo* synthesized amino acids will be labeled, as well, amino acids derived from protein breakdown will be labeled provided that amino acid turnover is faster than the rate of amino acid incorporation into newly made protein. In cases where $H_2{}^{18}O$ is administered one expects "instantaneous" labeling of amino acids regardless of their origin.

Another stable isotope that has seen substantial use is ^{15}N-glycine. Historically, this tracer was administered and the excretion of ^{15}N-urea and/or ^{15}N-ammonia was used to estimate the rate of whole-body nitrogen flux (San Pietro and Rittenberg 1953a;San Pietro and Rittenberg 1953b). Note that although investigators administer ^{15}N-glycine, the isotope rapidly mixes (or equilibrates) with other amino acid bound nitrogens which is the rationale for using it to trace "total" nitrogen flux (Matthews et al. 1981;Stein et al. 1980). More recently investigators have fed ^{15}N-labeled diets to animals in an effort to generate heavily labeled proteins that could then be used as internal standards to quantify protein concentrations in other subjects (MacCoss et al. 2005). In clever studies, Price *et al.* (Price et al. 2010) and Zhang *et al.* (Zhang et al. 2011) fed mice ^{15}N-labeled diets and were then able to estimate proteome turnover. The advantage of feeding ^{15}N-labeled diets as compared to a single labeled amino acid (e.g. ^{13}C-lysine) is that numerous protein-bound nitrogens will be labeled, therein increasing the window when measuring shifts in the isotope distribution of a proteolytic-peptide.

A final approach to label the precursor pool centers on the administration of labeled water, either $^{2}H_2O$ or $H_2{}^{18}O$ (Figure 4C) (Cabral et al. 2008;De Riva et al. 2010;Kombu et al. 2009;Rachdaoui et al. 2009;Xiao et al. 2008). The rationale is that cells will generate labeled amino acids in the presence of labeled water, e.g. ^{2}H-labeling can occur *via* transamination and/or *de novo* synthesis. In contrast to the generation of ^{13}C-labeled amino acids from ^{13}C-glucose, which does not label essential amino acids, in the presence of $^{2}H_2O$ one can observe ^{2}H-labeling of essential amino acids (Herath et al. 2011a). Namely, although essential amino acids are not made in a net sense (i.e. ^{13}C-glucose does not yield ^{13}C-leucine), transamination of leucine in $^{2}H_2O$ will label the α-hydrogen. Despite the fact that studies based on the use of labeled water revolutionized our understanding of metabolic biochemistry nearly 80 years ago there appears to have been a dramatic shift away from the use of labeled water in the field of protein dynamics for reasons that remain unclear (Borek, Ponticorvo, and Rittenberg 1958;Schoenheimer and Rittenberg 1938;Ussing 1938;Ussing 1941;Ussing 1980).

We, and others, have recently revisited the use of $^{2}H_2O$ in studies of protein synthesis (Busch et al. 2006;Cabral et al. 2008;De Riva et al. 2010;Kombu et al. 2009;Previs et al. 2004;Rachdaoui et al. 2009;Xiao et al. 2008), we also recognized the potential advantage(s) of using $H_2{}^{18}O$ (Bernlohr 1972;Bernlohr and Webster 1958;Borek, Ponticorvo, and Rittenberg 1958;Rachdaoui et al. 2009;Rittenberg, Ponticorvo, and Borek 1961). Our use of $H_2{}^{18}O$ was based on a classical study in which Rittenberg and colleagues demonstrated that $H_2{}^{18}O$ could be used to study protein synthesis (the outstanding contributions of Bernlohr and others further tested the approach and more clearly outlined the logic) (Bernlohr 1972;Bernlohr and Webster 1958;Borek, Ponticorvo, and Rittenberg 1958;Rittenberg, Ponticorvo, and Borek 1961). Unlike $^{2}H_2O$ which labels amino acids in a less uniform manner, $H_2{}^{18}O$ is expected to label virtually all amino acids to a similar degree. For example, oxygen in the carboxylic group can be labeled during *de novo* production, the degradation of proteins and/or the activation of amino acids (Figure 4C). Indeed, modern quantitative proteomic methods rely on this logic *albeit* for a different purpose, i.e. proteolytic cleavage in the presence of $H_2{}^{18}O$ leads to the generation of ^{18}O-labeled peptides (Miyagi and Rao 2007;Yao et al. 2001).

One point to consider when thinking about using different tracers, e.g. $^{2}H_3$-leucine vs $H_2{}^{18}O$ vs $^{2}H_2O$, is the background labeling over which one measures the incorporation. Since these are all stable isotopes one needs to contend with background labeling, e.g. naturally occurring ^{13}C and ^{15}N account for ~ 1.1% and ~ 0.4% of all carbon and nitrogen, respectively, and make substantial contributions to the isotope profile over which one measures excess labeling from the administered tracer (note that other isotopes also affect the background labeling but to a lesser degree since they are present at lower abundance (e.g. ^{2}H, ^{17}O and/or ^{18}O) and/or are less prevalent (e.g. ^{33}S or ^{34}S) in various proteins. The use of heavily substituted precursors, e.g. $^{2}H_3$-leucine, could be advantageous since the background labeling is lower at the M+3 isotopomer whereas the use of $^{2}H_2O$ and $H_2{}^{18}O$ typically requires that one measure shifts in the M+1/M0 and M+2/M0 ratios, respectively (where the background labeling can be considerably higher). Consequently, the impact of analytical error is expected to be somewhat worse when measuring the M+1/M0 ratio vs the M+3/M0 ratio since the background is higher. One can minimize the effect of analytical error by administering more tracer and/or relying on the fact that multiple copies of a precursor are incorporated into a given protein (e.g. it is possible to incorporate more copies of ^{2}H from body water as compared to $^{2}H_3$-leucine). These points are explained below in more detail. Last, in cases where one administers a pre-labeled amino acid (e.g. $^{2}H_3$-leucine) one is immediately limited when quantifying protein synthesis since it is necessary to identify those peptides that contain the designated amino acid (e.g. $^{2}H_3$-leucine). In contrast, when using a more general tracer, e.g. $^{2}H_2O$ or $H_2{}^{18}O$, it is possible to quantify protein synthesis *via* the labeling of various proteolytic peptides.

4. What should I consider when measuring the labeling of a protein?

In studies of protein synthesis one needs to compare the labeling of the product with that of the precursor. Although this section is primarily centered on the application of proteomic-based analyses for measuring the former, we will first briefly consider measurements of precursor labeling.

Several methods have been developed to measure the labeling of free amino acids; presumably, GC-quadrupole-MS-based methods are so commonplace since the hardware was readily available during the early 1980s when the use of stable isotopes began to dominate the literature (Matthews et al. 1980). In addition, these instruments have reasonable spectral accuracy therein allowing reliable estimates of isotope distributions. Typical protocols require a purification step (often using ion exchange chromatography) followed by derivatization prior to GCMS analyses. Although there are pros and cons to the generation of different derivatives (e.g. tertbutyldimethylsilyl vs N-acetyl-n-propyl, vs oxazolinone derivatives) (Dwyer et al. 2002;Matthews et al. 1980;Patterson, Carraro, and Wolfe 1993) it is clear that excellent precision of the isotope ratios can be achieved using standard equipment, for example, the coefficient of variation in the measured isotope ratios is often $\leq 1.0\%$, ensuring a certain degree of confidence when measuring the labeling of free amino acids. In cases where one decides to administer either $^{2}H_{2}O$ or $H_{2}{}^{18}O$ (and therein allow the subject to generate labeled amino acids) it is necessary to measure the ^{2}H- or ^{18}O-labeling of water (Rachdaoui et al. 2009). Historically, IRMS was used to measure water labeling, however, simple and robust GC-quadrupole-MS-based methods are available for measuring the ^{2}H and ^{18}O-labeling of water (Brunengraber et al. 2002;Shah et al. 2010;Yang et al. 1998).

So then, how can investigators couple isotope tracers with proteomic-based analyses? In our experience we have generally faced two major issues when addressing this question. First, how reproducible are the mass spectrometer-based measurements? Second, what type of instrument is the best? Although the two questions are somewhat related we will consider them separately.

During our earlier work we considered alternative approaches to processing the raw data (Cassano et al. 2007;Wang et al. 2007). For example, our initial studies were conducted with a mostly out-of-date Bruker MALDI-ToF, we devised a strategy in which we would download the raw data and then fit the isotopic distributions to a series of Gaussian peaks (this was done using the commercially available software package "Origin"). One reason for devising this approach centered on the fact that the relatively low resolution achieved on the isotope peaks was not easily integrated using the instrument's software. Please note that the statements made here are not intended to reflect poorly on any vendor, in our previous academic experiences we simply had limited access to state-of-the-art equipment. In developing our earlier work (Cassano et al. 2007;Wang et al. 2007), we performed numerous simulations to ensure the reliability of our approach for integrating the data and therein evaluating how the quality of the primary data would impact the results of the fitting routine, we consider two examples that may be of interest (Figure 5).

Briefly, simulations were run in which 3 Gaussian shaped peaks were generated (e.g. M0, M1 and M2 ions), noise was added using the random number generator in MS Excel; the expected ratios for M1/M0 and M2/M0 were set at 70% and 30%, respectively, and the resolution was set at ~ 30% valley between peaks (this resolution setting was chosen since it compared with what we had observed on the older Bruker MALDI-ToF, which did not

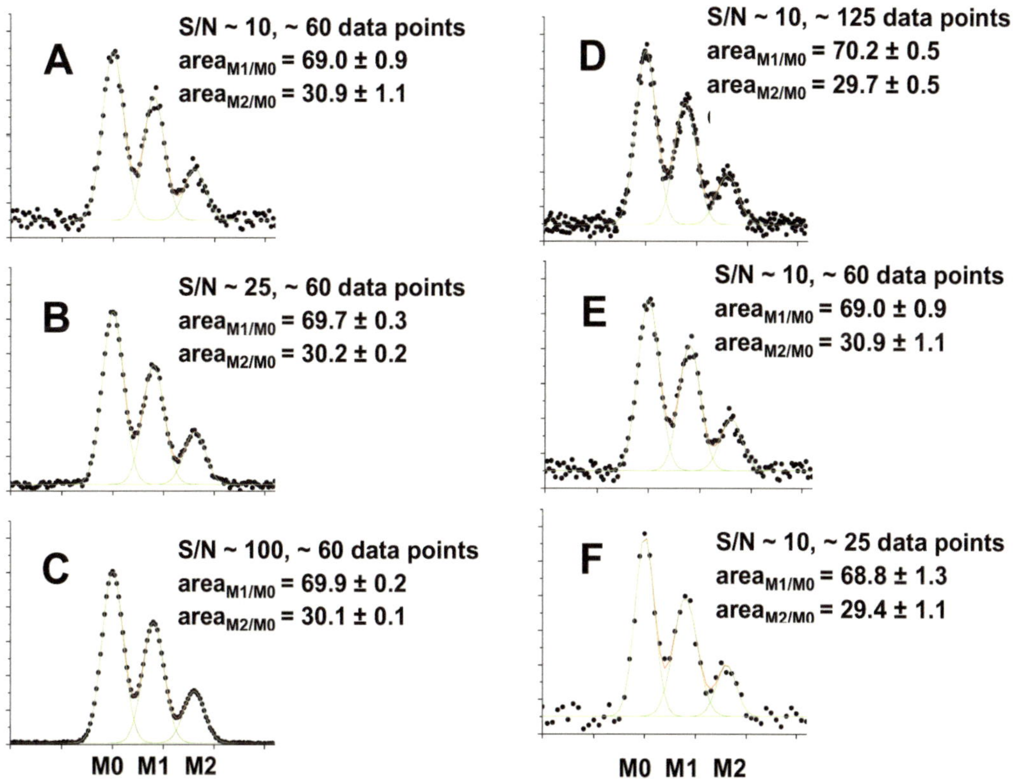

Fig. 5. Determining isotopic distributions. Simulations were run to determine the effect(s) of data quality and fitting on the calculated isotope ratios. In all cases 3 Gaussian shaped peaks were generated (e.g. M0, M1 and M2 ions); noise was added using the random number generator in MS Excel, the expected ratios for M1/M0 and M2/M0 are 70% and 30%, respectively, and the resolution was set at ~ 30% valley between peaks (note that this resolution setting was chosen for our simulations since it corresponded with the data that we were obtaining with an older Bruker MALDI-ToF when run in a linear mode, a somewhat worst-case senario). The simulated data were exported to Origin and fitted assuming a Gaussian model, each simulation was run 5 times, data are expressed as mean ± sem of the measured ratios. In Panels A, B and C we maintained a constant number of data points across the M0, M1 and M2 cluster (~ 60 data points) and we varied the S/N. In panels D, E and F, we maintained a constant and relatively low S/N (~ 10) and varied the number of data points. In all cases, there is reasonably good agreement between the measured and expected ratios.

always cooperate when run in the reflectron, or high-resolution, mode). In each example ~ 60 data points were observed across the 3 peaks, each simulation was run 5 times and data are expressed as mean ± sem of the measured ratios (Figure 5A, B and C). The study demonstrates that our integration method yields a reliable quantification of isotopomer profiles, in all cases there was good agreement between measured:expected ratios. This study is especially useful since protein analyses typically have to contend with peptides at different abundance, e.g. a given digest may contain peptides at S/N ~ 10 whereas others may be present at S/N ~ 100. Thus, we can estimate the level of confidence when determining the isotopic profiles of peptides with low vs high S/N. Although this example implies that a somewhat wide range of abundances can be used to estimate protein labeling, we suggest that it is best to focus quantitations on those peptides that are in greatest abundance since the precision generally improves.

A second scenario to consider in regards to data processing centers on the number of points that one observes across a series of peaks, this can be affected by various factors including the amount of sample that is analyzed, the type of mass analyzer and the analog-to-digital conversion rate. Our previous work mostly relied on the analyses of relatively pure samples, consequently, we primarily used MALDI-ToF (Rachdaoui et al. 2009). In our current work we almost exclusively rely on LC-MS since less purification is required prior to analysis (Kasumov et al. 2011;Zhou et al. 2011). Since one expects that coupling LC to a "discriminating" mass analyzer (e.g. a quadrupole) will reduce the number of data points that are used to describe a peptide's isotopomer profile we ran simulations to determine how the number of data points would affect the fitting/quantitation of the isotopic profile (Figure 5D, E and F). As in the previous example, the expected values of M1/M0 and M2/M0 are 70% and 30%, respectively, and the resolution was set at ~ 30% valley between peaks (the simulation was run 5 times so that data could be expressed as mean ± sem of the measured ratios). Although the simulations were run at a low S/N (~ 10, a somewhat worst-case scenario), it is possible to reasonably fit the peaks even when as few as ~ 25 points are recorded across the 3 isotopes in the profile (Figure 5F).

The examples described above are less about the type of mass spectrometer and more about the processing of raw data. In our current studies, the commercially available software appears to be generally sufficient for obtaining relatively precise measures of isotope clusters. Thus the need for extra effort in regards to data processing may not be justified in all cases. However, an area where data processing may be worth considering centers on using FT-ICR MS (MacCoss et al. 2005). Reports in the literature have discussed a potential bias against isotope peaks present at low abundance (Bresson et al. 1998;Erve et al. 2009), recent efforts by our colleagues have started to address those apparent limitations (Ilchenko et al. 2011). We suspect that LC-FT-ICR MS analyses may offer another unique advantage when quantifying low levels of ^{2}H-labeling. For example, we have demonstrated the ability to quantify low levels of ^{2}H-labeling by resolving the M+1 isotope peak into its ^{13}C and ^{2}H components (Herath et al. 2011b).

To this point we have not considered the acquisition mode under which data would be collected, the examples noted above do not imply MS or MS/MS-based analyses. Indeed, a substantial portion of our previous work was centered around MS-based analyses with less effort towards examining MS/MS-based measurements (Rachdaoui et al. 2009;Wang et al. 2007). Some of the advantages to using MS/MS analyses include (i) enhanced signal:noise, (ii) reduced concerns for overlapping peptides by identifying and characterizing the labeling of numerous fragments and (iii) sequence information on the peptide. The acquisition of MS/MS data to determine isotopic composition on a Q-ToF instrument is demonstrated using an apoA1 derived peptide (Figure 6). The doubly charged parent ion (m/z 520.85) is isolated at low-resolution in the quadrupole and then fragmented, the daughter ions are detected with the ToF analyzer. It is important to note that the relative intensities of the daughter ion profiles are close to the predicted natural abundance and the expected shift in the mass isotopomer distribution to higher isotopic composition with increased mass is

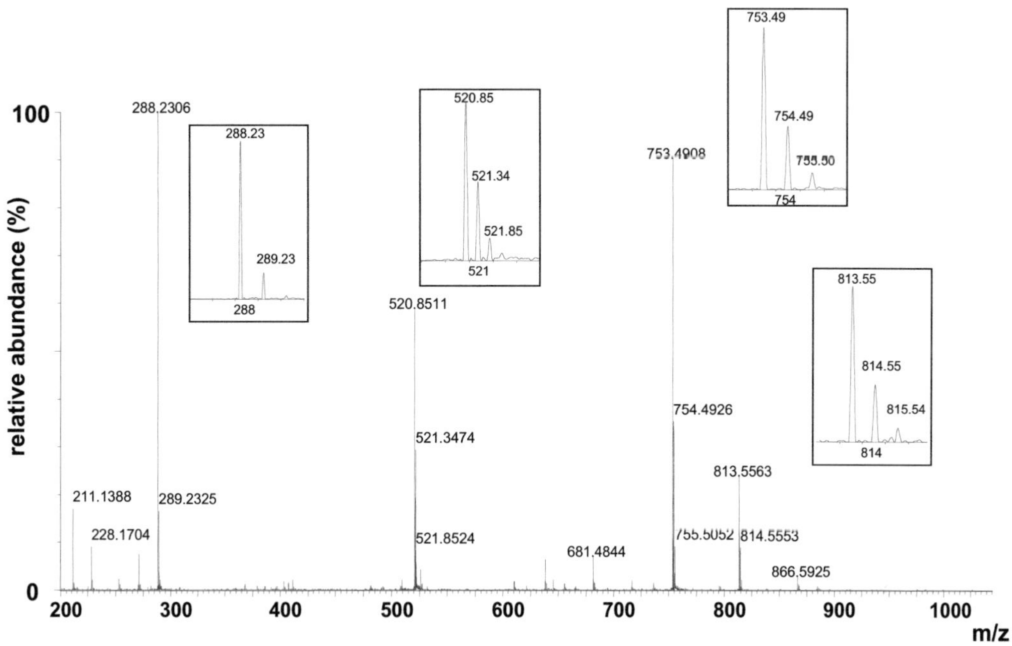

Fig. 6. LC-Q-ToF spectra of ARPALEDLR. The acquisition of MS/MS data to determine isotopic composition on a Q-ToF instrument are demonstrated using the apoA1 derived peptide ARPALEDLR. The doubly charged parent ion (m/z 520.8) is isolated at low-resolution in the quadrupole, fragmented by CID and the daughter ions detected with the ToF analyzer. The relative intensities of the daughter ion profiles are in close agreement with the predicted natural abundance (insets), the expected shift in the mass isotopomer distribution to higher isotopic composition with increased mass is readily apparent by the increase in the M1/M0 ratio of the daughter ions. Note that the insets show changes in 1 amu for isotope clusters at 288.23, 753.49 and 813.55 vs a shift of 0.5 amu for the cluster at 520.85 since these correspond with singly vs doubly charge species, respectively.

readily apparent by the increase in the M1/M0 ratio. These data are in agreement with a recent study in which we demonstrated the ability to measure the labeling of individual amino acids in tryptic peptides (Kasumov et al. 2011). We suspect that MS/MS-based measurements may need to consider the instrument configuration. For example, triple quadrupole measurements are likely to be good but have an inherent bias since one must decide what transitions to monitor. In contrast, Q-ToF measurements have the potential to capture more data and appear to have good reproducibility in regards to quantifying isotope labeling patterns (Castro-Perez et al. 2010).

The next question to address is, can one perform studies of proteome turnover? We consider what this would require for plasma-based analyses. First, although the concentration range of the plasma proteome varies from ~ 35 x 10^9 pg albumin per ml vs ~ 5 pg interleukin-6 per ml, mass spectrometers are flexible enough to identify and quantify analytes across this range (Anderson et al. 2004;Anderson and Anderson 2002). These seemingly positive statements lead into a consideration of the central problem, i.e. assuming that one can detect a protein can one determine its kinetics? Based on our experience, since the signal:noise can play an important role in affecting the apparent labeling the answer is a clear "maybe". We believe that the demands of measuring the mass isotopomer profile of a single peptide conflict with the imperative of identifying the largest number of peptides, making LC-MS protocols employed in proteomic studies less than ideal for some tracer-based protein turnover studies. For example, in preliminary work with an ion trap mass spectrometer, we observed that determination of a peptide's mass isotopomer profile with sufficient precision to quantify ^{2}H-incorporation required that the zoom scan mode be used with multiple scans encompassing an entire peptide chromatographic peak. In principle, this scan sequence (full scan to identify peptides that are present and zoom scan on a desired peptide) conflicts with an emphasis on obtaining data on the largest number of peptides characteristic of proteomic studies. We originally thought that these conflicting demands on the acquisition parameters of the mass spectrometer would limit protein turnover analyses to a smaller number of peptides than are present in the proteome. However, by generating a list of previously identified peptides, from proteins of interest, it should be possible to determine protein turnover rates on 10-100 proteins for a given LC-MS run.

Two recent publications deserve special attention. Namely, Price *et al.* (Price et al. 2010) used a hybrid LTQ/FT instrument to measure turnover of ~2500 proteins in multiple tissues of mice fed with ^{15}N-labeled algae, their MS/MS method consisted of one survey scan followed by several secondary scans of selected ions. Likewise Zhang *et al.* (Zhang et al. 2011) fed mice an *E. coli*-derived ^{15}N-labeled protein mixture. Samples were analyzed using an Orbitrap instrument, full scans at high resolution (~ 60,000 at m/z 400) were used for isotopic distribution analysis; they identified and quantified the kinetics on ~ 700 proteins using a novel software package. It is important to emphasize that in both cases (Price et al. 2010;Zhang et al. 2011), the investigators observed a substantial mass shift because ~ 100% of the diet was labeled, the utility of these analytical approaches needs to be examined when the peptide labeling results in more subtle changes in isotopic distribution. In addition, corrections for inherent spectral error are also needed (Erve et al. 2009). Alternatively, in cases where a complex matrix is obtained, the fractionation of protein classes or the isolation of targeted analytes can be used to enhance the application of this method (Figure 7), e.g. prior to digestion/analyses the samples were subjected to immunodepletion to remove several high abundance proteins.

As discussed earlier, the ability to quantify shifts in the isotopic labeling allow one to estimate the FSR, however, in certain instances it is of interest to determine the absolute rate of synthesis (which requires an estimate of the concentration of a given protein). Numerous techniques can be used to measure the concentration of a protein (or peptides) (Gygi et al. 1999;Gygi et al. 2000;Jaleel et al. 2006;Johnson and Muddiman 2004;van Eijk and Deutz 2003;Yao et al. 2001;Zhang et al. 2001), however, each requires special considerations when applied in combination with a tracer study. First, in regards to labeling methods such as ICAT, one assumes equal generation and recovery of labeled and unlabeled species before mixing and analyzing. We believe that those techniques are of limited value in some studies. For example, if one administers $^{2}H_2O$ to quantify protein synthesis, some reagents (e.g. ICAT or digestion in $H_2{}^{18}O$) may not induce a large enough shift in the peptide mass to allow one to comfortably measure the ^{2}H-labeling profile and determine protein synthesis. For example, suppose that one aims to determine the synthesis and concentration of apoE, which has a $t_{1/2}$ that is estimated to be < 1 hour in rodents (Figure 7). The rate of synthesis can be determined by measuring the ^{2}H-labeling of an apoE-derived peptide. The change in concentration can be determined by digesting a 0 min sample in H_2O and digesting a 60 min

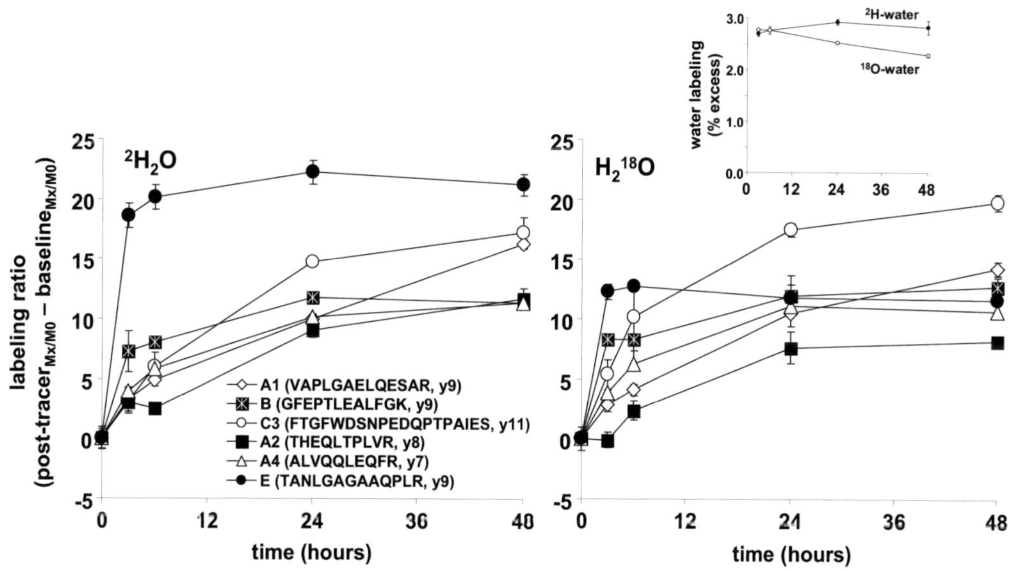

Fig. 7. Labeling of mouse apoproteins. Comparable labeling profiles were observed for several apoproteins in C57BL/6J mice given either ^2H_2O or $H_2{}^{18}O$. Note that animals were given an intraperitoneal bolus of either tracer and then allowed free access to labeled drinking water, as shown in the inset mice exposed to ^2H_2O reached a steady-state labeling whereas mice exposed to $H_2{}^{18}O$ demonstrated a slight decrease in the labeling of body water. As expected, there were sizeable differences in the labeling of the various apoproteins, the relative differences are consistent with the literature, e.g. the FSR of apoE ~ apoB > apoA1. The magnitude of the labeling reflects variation in the amino acid composition of the respective peptides and the $t_{1/2}$. Data are shown as the mean ± standard deviation, n = 3 per time point.

sample in $H_2{}^{18}O$ and then mixing the samples and comparing the relative ^{18}O-labeling in a given peptide. However, the presence of ^{18}O (for quantifying abundance) will likely interfere with measurements of ^{2}H-labeling (protein synthesis). Therefore, each sample would likely need to be analyzed duplicate (first to determine the ^{2}H-labeling to estimate the rate of synthesis and second to determine the ^{18}O- or ICAT-labeling to estimate the concentration).

The use of SILAC methods is more likely to be compatible with the use of tracers in flux studies, i.e. one adds a known amount of a heavily labeled protein mixture and then compares the abundance of the cold peptides with that of heavily labeled SILAC peptides (Ong et al. 2002). While it is clear that SILAC methods are well suited for cell-based and rodent studies (Kruger et al. 2008), a potential drawback centers on the fact that it is not possible to fully label many model systems (e.g. humans). Interestingly, recent studies have demonstrated dynamic SILAC (Andersen et al. 2005;Doherty et al. 2009), i.e. investigators used a SILAC approach for administering a tracer but focused their attention on quantifying the change in labeling of numerous proteins in order to determine their flux. It is important to note that the early reports regarding the SILAC approach (for quantitative proteomics) clearly demonstrated the potential for quantifying proteome kinetics (we refer the readers to Figure 3 of (Ong et al. 2002)). Mann and colleagues monitored the temporal changes in protein labeling to determine when the cells had become fully labeled, from that point they knew that they had generated SILAC cells which could be used to determine the protein expression profiles of other cells (Ong et al. 2002); despite the fact that their major objective was to contrast SILAC and ICAT methods for determining protein expression profiles, they demonstrated the potential for determining proteome turnover.

We believe that a simple and reasonable approach for determining protein abundance, which is compatible with the administration of a tracer for determining proteome kinetics, centers on the use of label-free methods. For example, Wang *et al.*(Wang et al. 2003) reliably quantified numerous peptides by measuring their relative abundance during a given run. Although this approach requires attention to detail during the sample processing and a stable operating system, it is immediately compatible with tracer-based studies since the isotopic labeling patterns are not altered. Clearly, there are factors that may skew the data resulting in estimates of concentrations that are far from the correct value (e.g. ion suppression effects), nevertheless, label-free methods can be used infer relative concentrations and differences between groups (Wang et al. 2003;Wiener et al. 2004). We should note that in cases where one aims to determine the kinetics of a single protein and/or a select group of proteins it is possible to use custom synthesized standards, e.g. this strategy has been used for measuring insulin concentration (Kippen et al. 1997;Stocklin et al. 1997). A related approach would be to use an "isomer dilution" strategy (Thevis et al. 2005), e.g. when studying the kinetics of albumin and/or insulin in rodents one could spike samples with known amounts of human albumin and insulin before processing and analyses.

5. Interpretation of the precursor: Product labeling ratio

Assuming that one has devised a strategy to administer a precursor and one has found a suitable way to measure its incorporation into a protein, there is a final question that must be addressed, how do I interpret the precursor:product labeling ratio? We first consider the scenario in which an investigator has administered a pre-labeled amino acid(s) and later consider the novelty of administering either ^2H_2O or $H_2{}^{18}O$.

As noted earlier, the goal of a primed-infusion is that one will instantaneously achieve and then maintain a steady-state labeling of a given amino acid tracer. Indeed, this was clearly demonstrated by Lichtenstein and colleagues, they simultaneously administered multiple labeled amino acids and observed the incorporation of each into various apoproteins (Lichtenstein et al. 1990). However, although the labeling of VLDL-apoB100 approaches a steady-state by the end of the infusion protocol the enrichment of amino acids in VLDL-apoB100 is substantially lower than the enrichment of those free amino acids in plasma. Although it is not possible to state with certainty the source of this discrepancy, it is clear that the transport of free amino acids into the cell (and/or mixing with the endogenous pool) must be slower than the rate of intracellular protein breakdown (Khairallah and Mortimore 1976) which likely results in marked compartmentation. What are the consequences of this on estimates of protein synthesis? One does not expect problems when the aim is to fit the exponential labeling curve (e.g. collect multiple time points and use Equation 1), in those cases the FSR is estimated from the time it takes to reach steady-state and it does not necessarily matter how labeled the protein is at steady-state (the caveat, however, is that one expects a better fit in cases where the asymptotic value is greatest since there is a large change in labeling over the natural background) (Figure 1A) (Foster et al.

1993). In cases where one aims to determine the synthesis of protein with a small FSR it may be necessary to use Equation 2, therefore, any error in the apparent precursor labeling will have an immediate impact on the estimated FSR. Based on data in the literature, if one assumes that the intracellular labeling equals the plasma labeling one will likely underestimate the FSR of LDL-apoB100 by nearly 2-fold since the labeling of amino acids in plasma is ~ 2 times greater than the estimated intracellular amino acid labeling (Lichtenstein et al. 1990). Note that in many studies, the production of VLDL-apoB100 is not only a parameter of interest but it serves a critical function in estimating LDL-apoB100 production, HDL-apoA1 production, etc. As discussed, the asymptotic labeling of VLDL-apoB100 may be used as a surrogate to estimate the precursor labeling that is needed to calculate LDL-apoB100 and HDL-apoA1 production (Lichtenstein et al. 1990). For example, LDL-apoB100 and HDL-apoA1 have relatively slow rates of synthesis and therefore show pseudo-linear increases in labeling over a short term infusion. As such, it is not practical to model the data and estimate FSR using Eq 1; to estimate the FSR of LDL-apoB100 and/or HDL-apoA1 investigators often use Eq 2 and substitute the asymptotic labeling of VLDL-apoB100 as the precursor labeling (Lichtenstein et al. 1990). The scenario discussed here applies to most cases in which cells are labeled from the outside, e.g. the administration of a pre-labeled amino acid.

One expects more reliable estimates of flux in cases where cells are labeled from the inside provided that one can determine the intracellular precursor labeling. For example, the administration of ^{13}C-glucose leads to the generation of ^{13}C-amino acids (Figure 4) but the labeling of those amino acids is likely to be diluted by carbon exchange (Wykes, Jahoor, and Reeds 1998). In cases where labeled water is used one expects comparable labeling between intracellular and extracullar pools. Dietschy and colleagues clearly demonstrated that water readily distributes in the plasma and that plasma labeling reflects tissue-specific labeling almost instantly (Dietschy and Spady 1984;Jeske and Dietschy 1980). As we have described previously, it is possible to then estimate protein flux by comparing the change in the labeling of proteolytic-peptides with that of body water (Rachdaoui et al. 2009). The caveat is that one must account for the number of copies of the precursor that are incorporated, referred to as *n* (Cabral et al. 2008;Kasumov et al. 2011;Rachdaoui et al. 2009;Xiao et al. 2008). For example, in cases where $H_2{}^{18}O$ is administered, the labeling of the protein will exceed that of the precursor since one expects that each peptide bond will incorporate ^{18}O. Note that in the example shown in the inset for Figure 7 the labeling of water is ~ 2.5 to 3.0% yet the labeling of the various proteins greatly exceeds those levels, therefore one needs to correct the precursor:product labeling ratio by including a constant for *n* (Herath et al. 2011a;Rachdaoui et al. 2009).

6. Summary and conclusions

We believe that it is possible to readily convert static protein expression profiles into dynamic images. Numerous approaches are available for tracing protein synthesis and various strategies have been implemented for measuring the labeling of peptides in complex mixtures. We believe that there is no single best method but certain fundamental points should be recognized. For example, the administration of a labeled precursor can present a challenge for *in vivo* studies. The administration of labeled water may be advantageous in these settings, the tracer can be given orally, it is relatively inexpensive and can be used to study multiple parameters simultaneously (this is especially important in studies of lipoprotein kinetics since questions regarding protein and lipid flux are often of equal importance) (Castro-Perez et al. 2010;Castro-Perez et al. 2011;Dufner and Previs 2003). In contrast, although we have demonstrated the ability to study protein synthesis in cell culture using labeled water (Dufner et al. 2005), we believe that SILAC methods are generally superior for *in vitro* studies since it is trivial to completely substitute fully labeled amino acids for unlabeled amino acids in that setting.

In regards to the analyses of protein mixtures, we believe that there is no single best MS approach. Although our applications have been focused on small groups of proteins, it is clear that the labeling profiles of analytes present in complex mixtures can be sorted out; again, the SILAC literature strongly supports these conclusions. We believe that an area which will likely have an important impact on future studies centers on data processing; in our experience the MS hardware may be limited by the software. As we have demonstrated it is possible to obtain reliable isotopic ratios using commercially available software, however, in some cases alternative methods have been of great value.

7. Acknowledgments

We thank Dr. Vernon E. Anderson for his insight and efforts in developing the early stages of this work, he suggested the possibility of quantifying subtle changes in the labeling profiles of peptides which encouraged us to pursue water-based studies of protein kinetics; our collaborations were great fun.

8. References

Andersen JS, Lam YW, Leung AK, Ong SE, Lyon CE, Lamond AI, and Mann M. 2005. Nucleolar proteome dynamics. *Nature* 433 (7021): 77-83.

Anderson NL, and Anderson NG. 2002. The human plasma proteome: history, character, and diagnostic prospects. *Mol. Cell Proteomics* 1 (11): 845-867.

Anderson NL, Polanski M, Pieper R, Gatlin T, Tirumalai RS, Conrads TP, Veenstra TD, Adkins JN, Pounds JG, Fagan R, and Lobley A. 2004. The human plasma proteome: a nonredundant list developed by combination of four separate sources. *Mol. Cell Proteomics* 3 (4): 311-326.

Barrett PH, Chan DC, and Watts GF. 2006. Thematic review series: patient-oriented research. Design and analysis of lipoprotein tracer kinetics studies in humans. *J Lipid Res.* 47 (8): 1607-1619.

Bateman RJ, Munsell LY, Chen X, Holtzman DM, and Yarasheski KE. 2007. Stable isotope labeling tandem mass spectrometry (SILT) to quantify protein production and clearance rates. *J Am. Soc. Mass Spectrom.* 18 (6): 997-1006.

Bateman RJ, Munsell LY, Morris JC, Swarm R, Yarasheski KE, and Holtzman DM. 2006. Human amyloid-beta synthesis and clearance rates as measured in cerebrospinal fluid in vivo. *Nat. Med.* 12 (7): 856-861.

Bederman IR, Dufner DA, Alexander JC, and Previs SF. 2006. Novel application of the "doubly labeled" water method: measuring CO2 production and the tissue-specific dynamics of lipid and protein in vivo. *American Journal of Physiology-Endocrinology and Metabolism* 290 (5): E1048-E1056.

Bernlohr RW. 1972. 18 Oxygen probes of protein turnover, amino acid transport, and protein synthesis in Bacillus licheniformis. *J Biol. Chem.* 247 (15): 4893-4899.

Bernlohr RW, and WEBSTER GC. 1958. Transfer of oxygen-18 during amino acid activation. *Arch. Biochem. Biophys.* 73 (1): 276-278.

BOREK E, Ponticorvo L, and Rittenberg D. 1958. PROTEIN TURNOVER IN MICRO-ORGANISMS. *Proc. Natl. Acad. Sci. U. S. A* 44 (5): 369-374.

Bresson JA, Anderson GA, Bruce JE, and Smith RD. 1998. Improved isotopic abundance measurements for high resolution Fourier transform ion cyclotron resonance mass spectra via time-domain data extraction. *Journal of the American Society for Mass Spectrometry* 9 (8): 799-804.

Brunengraber DZ, Mccabe BJ, Katanik J, and Previs SF. 2002. Gas chromatography-mass spectrometry assay of the O-18 enrichment of water as trimethyl phosphate. *Analytical Biochemistry* 306 (2): 278-282.

Busch R, Kim YK, Neese RA, Schade-Serin V, Collins M, Awada M, Gardner JL, Beysen C, Marino ME, Misell LM, and Hellerstein MK. 2006. Measurement of protein turnover rates by heavy water labeling of nonessential amino acids. *Biochim. Biophys. Acta* 1760 (5): 730-744.

Cabral CB, Bullock KH, Bischoff DJ, Tompkins RG, Yu YM, and Kelleher JK. 2008. Estimating glutathione synthesis with deuterated water: a model for peptide biosynthesis. *Anal. Biochem.* 379 (1): 40-44.

Cassano AG, Wang B, Anderson DR, Previs S, Harris ME, and Anderson VE. 2007. Inaccuracies in selected ion monitoring determination of isotope ratios obviated by profile acquisition: nucleotide O-18/O-16 measurements. *Analytical Biochemistry* 367 (1): 28-39.

Castro-Perez JM, Previs SF, McLaren DG, Shah V, Herath K, Bhat G, Johns DG, Wang SP, Mitnaul L, Jensen K, Vreeken R, Hankemeier T, Roddy TP, and Hubbard BK. 2010. In-vivo D2O labeling in C57Bl/6 mice to quantify static and dynamic changes in cholesterol and cholesterol esters by high resolution LC mass-spectrometry. *J. Lipid Res.*

Castro-Perez JM, Roddy TP, Shah V, McLaren DG, Wang SP, Jensen K, Vreeken RJ, Hankemeier T, Johns DG, Previs SF, and Hubbard BK. 2011. Identifying static and kinetic lipid phenotypes by high resolution UPLC/MS: Unraveling diet-induced changes in lipid homeostasis by coupling metabolomics and fluxomics. *J Proteome. Res.*

Cobelli C, Toffolo G, and Foster DM. 1992. Tracer-to-tracee ratio for analysis of stable isotope tracer data: link with radioactive kinetic formalism. *Am. J Physiol* 262 (6 Pt 1): E968-E975.

De Riva A, Deery MJ, McDonald S, Lund T, and Busch R. 2010. Measurement of protein synthesis using heavy water labeling and peptide mass spectrometry: Discrimination between major histocompatibility complex allotypes. *Anal. Biochem.* 403 (1-2): 1-12.

Dietschy JM, and Spady DK. 1984. Measurement of rates of cholesterol synthesis using tritiated water. *J. Lipid Res.* 25 (13): 1469-1476.

Doherty MK, Hammond DE, Clague MJ, Gaskell SJ, and Beynon RJ. 2009. Turnover of the human proteome: determination of protein intracellular stability by dynamic SILAC. *J Proteome. Res.* 8 (1): 104-112.

Dudley MA, Burrin DG, Wykes LJ, Toffolo G, Cobelli C, Nichols BL, Rosenberger J, Jahoor F, and Reeds PJ. 1998. Protein kinetics determined in vivo with a multiple-tracer, single-sample protocol: application to lactase synthesis. *Am. J Physiol* 274 (3 Pt 1): G591-G598.

Dufner D, and Previs SF. 2003. Measuring in vivo metabolism using heavy water. *Current Opinion in Clinical Nutrition and Metabolic Care* 6 (5): 511-517.

Dufner DA, Bederman IR, Brunengraber DZ, Rachdaoui N, Ismail-Beigi F, Siegfried BA, Kimball SR, and Previs SF. 2005. Using (H2O)-H-2 to study the influence of feeding on protein synthesis: effect of isotope equilibration in vivo vs. in cell culture. *American Journal of Physiology-Endocrinology and Metabolism* 288 (6): E1277-E1283.

Dwyer KP, Barrett PH, Chan D, Foo JI, Watts GF, and Croft KD. 2002. Oxazolinone derivative of leucine for GC-MS: a sensitive and robust method for stable isotope kinetic studies of lipoproteins. *J Lipid Res.* 43 (2): 344-349.

Erve JCL, Gu M, Wang YD, DeMaio W, and Talaat RE. 2009. Spectral Accuracy of Molecular Ions in an LTQ/Orbitrap Mass Spectrometer and Implications for Elemental Composition Determination. *Journal of the American Society for Mass Spectrometry* 20 (11): 2058-2069.

Foster DM, Barrett PH, Toffolo G, Beltz WF, and Cobelli C. 1993. Estimating the fractional synthetic rate of plasma apolipoproteins and lipids from stable isotope data. *J Lipid Res.* 34 (12): 2193-2205.

Gygi SP, Corthals GL, Zhang Y, Rochon Y, and Aebersold R. 2000. Evaluation of two-dimensional gel electrophoresis-based proteome analysis technology. *Proc. Natl. Acad. Sci. U. S. A* 97 (17): 9390-9395.

Gygi SP, Rist B, Gerber SA, Turecek F, Gelb MH, and Aebersold R. 1999. Quantitative analysis of complex protein mixtures using isotope-coded affinity tags. *Nat. Biotechnol.* 17 (10): 994-999.

Herath K, Bhat G, Miller PL, Wang SP, Kulick A, Andrews-Kelly G, Johnson C, Rohm RJ, Lassman ME, Previs SF, Johns DG, Hubbard BK, and Roddy TP. 2011a. Equilibration of (2)H labeling between body water and free amino acids: enabling studies of proteome synthesis. *Anal. Biochem.* 415 (2): 197-199.

Herath K, Yang J, Zhong W, Kulick A, Rohm RJ, Lassman ME, Castro-Perez JM, Mahsut A, Dunn K, Johns DG, Previs SF, Roddy TP, Attygale A, and Hubbard BK. 2011b. Determination of low levels of 2H-labeling using high-resolution mass spectrometry (HR-MS): Application in studies of lipid flux and beyond. *J Am. Soc. Mass Spectrom.* 22: 154.

Ilchenko S, Previs SF, Rachdaoui N, Chance M, and Kasumov T. 2011. An improved measurement of the isotopic ratio by high resolution mass spectrometry. *J Am. Soc. Mass Spectrom.* 22: 149.

Jaleel A, Nehra V, Persson XM, Boirie Y, Bigelow M, and Nair KS. 2006. In vivo measurement of synthesis rate of multiple plasma proteins in humans. *Am. J Physiol Endocrinol. Metab* 291 (1): E190-E197.

Jeske DJ, and Dietschy JM. 1980. Regulation of Rates of Cholesterol-Synthesis Invivo in the Liver and Carcass of the Rat Measured Using [Water-H-3. *Journal of Lipid Research* 21 (3): 364-376.

Johnson KL, and Muddiman DC. 2004. A method for calculating 16O/18O peptide ion ratios for the relative quantification of proteomes. *J Am. Soc. Mass Spectrom.* 15 (4): 437-445.

Kasumov T, Ilchenko S, Li L, Rachdaoui N, Sadygov RG, Willard B, McCullough AJ, and Previs S. 2011. Measuring protein synthesis using metabolic (2)H labeling, high-resolution mass spectrometry, and an algorithm. *Analytical Biochemistry* 412 (1): 47-55.

Khairallah EA, and Mortimore GE. 1976. Assessment of protein turnover in perfused rat liver. Evidence for amino acid compartmentation from differential labeling of free and tRNA-gound valine. *J Biol. Chem.* 251 (5): 1375-1384.

Kippen AD, Cerini F, Vadas L, Stocklin R, Vu L, Offord RE, and Rose K. 1997. Development of an isotope dilution assay for precise determination of insulin, C-peptide, and proinsulin levels in non-diabetic and type II diabetic individuals with comparison to immunoassay. *J Biol. Chem.* 272 (19): 12513-12522.

Kombu RS, Zhang GF, Abbas R, Mieyal JJ, Anderson VE, Kelleher JK, Sanabria JR, and Brunengraber H. 2009. Dynamics of glutathione and ophthalmate traced with 2H-enriched body water in rats and humans. *Am. J. Physiol Endocrinol. Metab* 297 (1): E260-E269.

Kruger M, Moser M, Ussar S, Thievessen I, Luber CA, Forner F, Schmidt S, Zanivan S, Fassler R, and Mann M. 2008. SILAC mouse for quantitative proteomics uncovers kindlin-3 as an essential factor for red blood cell function. *Cell* 134 (2): 353-364.

Lichtenstein AH, Cohn JS, Hachey DL, Millar JS, Ordovas JM, and Schaefer EJ. 1990. Comparison of deuterated leucine, valine, and lysine in the measurement of human apolipoprotein A-I and B-100 kinetics. *J Lipid Res.* 31 (9): 1693-1701.

MacCoss MJ, Wu CC, Matthews DE, and Yates JR, III. 2005. Measurement of the isotope enrichment of stable isotope-labeled proteins using high-resolution mass spectra of peptides. *Anal. Chem.* 77 (23): 7646-7653.

Magkos F, Patterson BW, and Mittendorfer B. 2007. Reproducibility of stable isotope-labeled tracer measures of VLDL-triglyceride and VLDL-apolipoprotein B-100 kinetics. *J Lipid Res.* 48 (5): 1204-1211.

Matthews DE, Conway JM, Young VR, and Bier DM. 1981. Glycine nitrogen metabolism in man. *Metabolism* 30 (9): 886-893.

Matthews DE, Motil KJ, Rohrbaugh DK, Burke JF, Young VR, and Bier DM. 1980. Measurement of leucine metabolism in man from a primed, continuous infusion of L-[1-3C]leucine. *Am. J. Physiol* 238 (5): E473-E479.

Miyagi M, and Rao KC. 2007. Proteolytic 18O-labeling strategies for quantitative proteomics. *Mass Spectrom. Rev.* 26 (1): 121-136.

Ong SE, Blagoev B, Kratchmarova I, Kristensen DB, Steen H, Pandey A, and Mann M. 2002. Stable isotope labeling by amino acids in cell culture, SILAC, as a simple and accurate approach to expression proteomics. *Mol. Cell Proteomics.* 1 (5): 376-386.

Patterson BW, Carraro F, and Wolfe RR. 1993. Measurement of 15N enrichment in multiple amino acids and urea in a single analysis by gas chromatography/mass spectrometry. *Biol. Mass Spectrom.* 22 (9): 518-523.

Patterson BW, Mittendorfer B, Elias N, Satyanarayana R, and Klein S. 2002. Use of stable isotopically labeled tracers to measure very low density lipoprotein-triglyceride turnover. *J Lipid Res.* 43 (2): 223-233.

Previs SF, Fatica R, Chandramouli V, Alexander JC, Brunengraber H, and Landau BR. 2004. Quantifying rates of protein synthesis in humans by use of (H2O)-H-2: application to patients with end-stage renal disease. *American Journal of Physiology-Endocrinology and Metabolism* 286 (4): E665-E672.

Price JC, Guan S, Burlingame A, Prusiner SB, and Ghaemmaghami S. 2010. Analysis of proteome dynamics in the mouse brain. *Proc. Natl. Acad. Sci. U. S. A* 107 (32): 14508-14513.

Rachdaoui N, Austin L, Kramer E, Previs MJ, Anderson VE, Kasumov T, and Previs SF. 2009. Measuring Proteome Dynamics in Vivo. *Molecular & Cellular Proteomics* 8 (12): 2653-2663.

Ramakrishnan R. 2006. Studying apolipoprotein turnover with stable isotope tracers: correct analysis is by modeling enrichments. *J Lipid Res.* 47 (12): 2738-2753.

Rittenberg D, Ponticorvo L, and BOREK E. 1961. Studies on the sources of the oxygen of proteins. *J Biol. Chem.* 236: 1769-1772.

SAN PIETRO A, and Rittenberg D. 1953a. A study of the rate of protein synthesis in humans. I. Measurement of the urea pool and urea space. *J Biol. Chem.* 201 (1): 445-455.-----. 1953b. A study of the rate of protein synthesis in humans. II. Measurement of the metabolic pool and the rate of protein synthesis. *J Biol. Chem.* 201 (1): 457-473.

Schoenheimer R, and Rittenberg D. 1938. THE APPLICATION OF ISOTOPES TO THE STUDY OF INTERMEDIARY METABOLISM. *Science* 87 (2254): 221-226.

Shah V, Herath K, Previs SF, Hubbard BK, and Roddy TP. 2010. Headspace analyses of acetone: a rapid method for measuring the 2H-labeling of water. *Anal. Biochem.* 404 (2): 235-237.

Shames DM, and Havel RJ. 1991. De novo production of low density lipoproteins: fact or fancy. *J Lipid Res.* 32 (7): 1099-1112.

Sigurdsson G, Shames DM, and Havel RJ. 1981. On the extravascular pool of low-density lipoprotein in rat liver. *Clin. Sci. (Lond)* 61 (5): 611-613.

Stein TP, Leskiw MJ, Buzby GP, Giandomenico AL, Wallace HW, and Mullen JL. 1980. Measurement of protein synthesis rates with [15N]glycine. *Am. J Physiol* 239 (4): E294-E300.

Stocklin R, Vu L, Vadas L, Cerini F, Kippen AD, Offord RE, and Rose K. 1997. A stable isotope dilution assay for the in vivo determination of insulin levels in humans by mass spectrometry. *Diabetes* 46 (1): 44-50.

Thevis M, Thomas A, Delahaut P, Bosseloir A, and Schanzer W. 2005. Qualitative determination of synthetic analogues of insulin in human plasma by immunoaffinity purification and liquid chromatography-tandem mass spectrometry for doping control purposes. *Anal. Chem.* 77 (11): 3579-3585.

Toffolo G, Foster DM, and Cobelli C. 1993. Estimation of protein fractional synthetic rate from tracer data. *Am. J Physiol* 264 (1 Pt 1): E128-E135.

Ussing HH. 1938. Use of amino acids containing deuterium to follow protein production in the organism. *Nature* 142: 399.-----. 1941. The rate of protein renewal in mice and rats studied by means of heavy hydrogen. *Acta Physiol. Scand.* 2: 209-221.-----. 1980. Life with tracers. *Annu. Rev. Physiol* 42: 1-16.

van Eijk HM, and Deutz NE. 2003. Plasma protein synthesis measurements using a proteomics strategy. *J Nutr.* 133 (6 Suppl 1): 2084S-2089S.

Vogt JA, Hunzinger C, Schroer K, Holzer K, Bauer A, Schrattenholz A, Cahill MA, Schillo S, Schwall G, Stegmann W, and Albuszies G. 2005. Determination of fractional synthesis rates of mouse hepatic proteins via metabolic 13C-labeling, MALDI-TOF MS and analysis of relative isotopologue abundances using average masses. *Anal. Chem.* 77 (7): 2034-2042.

Wang B, Sun G, Anderson DR, Jia M, Previs S, and Anderson VE. 2007. Isotopologue distributions of peptide product ions by tandem mass spectrometry: Quantitation of low levels of deuterium incorporation. *Analytical Biochemistry* 367 (1): 40-48.

Wang W, Zhou H, Lin H, Roy S, Shaler TA, Hill LR, Norton S, Kumar P, Anderle M, and Becker CH. 2003. Quantification of proteins and metabolites by mass spectrometry without isotopic labeling or spiked standards. *Anal. Chem.* 75 (18): 4818-4826.

Wasserman K, Joseph JD, and Mayerson HS. 1955. Kinetics of vascular and extravascular protein exchange in unbled and bled dogs. *American Journal of Physiology* 184: 175-182.

Waterlow JC. 2006. *Protein turnover.* Oxfordshire: CABI.

Wiener MC, Sachs JR, Deyanova EG, and Yates NA. 2004. Differential mass spectrometry: a label-free LC-MS method for finding significant differences in complex peptide and protein mixtures. *Anal. Chem.* 76 (20): 6085-6096.

Wolfe RR, and Chinkes DL. 2005. *Isotope tracers in metabolic research: Principles and practice of kinetic analyses.* Hoboken, NJ: Wiley-Liss.

Wykes LJ, Jahoor F, and Reeds PJ. 1998. Gluconeogenesis measured with [U-13C]glucose and mass isotopomer analysis of apoB-100 amino acids in pigs. *Am. J Physiol* 274 (2 Pt 1): E365-E376.

Xiao GG, Garg M, Lim S, Wong D, Go VL, and Lee WN. 2008. Determination of protein synthesis in vivo using labeling from deuterated water and analysis of MALDI-TOF spectrum. *J. Appl. Physiol* 104 (3): 828-836.

Yang D, Diraison F, Beylot M, Brunengraber DZ, Samols MA, Anderson VE, and Brunengraber H. 1998. Assay of low deuterium enrichment of water by isotopic exchange with [U-13C3]acetone and gas chromatography-mass spectrometry. *Anal. Biochem.* 258 (2): 315-321.

Yao X, Freas A, Ramirez J, Demirev PA, and Fenselau C. 2001. Proteolytic 18O labeling for comparative proteomics: model studies with two serotypes of adenovirus. *Anal. Chem.* 73 (13): 2836-2842.

Zhang R, Sioma CS, Wang S, and Regnier FE. 2001. Fractionation of isotopically labeled peptides in quantitative proteomics. *Anal. Chem.* 73 (21): 5142-5149.

Zhang Y, Reckow S, Webhofer C, Boehme M, Gormanns P, Egge-Jacobsen WM, and Turck CW. 2011. Proteome Scale Turnover Analysis in Live Animals Using Stable Isotope Metabolic Labeling. *Anal. Chem.*

Zhou H, McLaughlin T, Herath K, Lassman ME, Rohm RJ, Wang S-P, Dunn K, Kulick A, Johns DG, Previs SF, Hubbard BK, and Roddy TP. 2011. Development of Lipoprotein Synthesis Measurement using LC-MS and Deuterated Water Labeling. *J Am. Soc. Mass Spectrom.* 22: 182.

14

Dynamics of Protein Complexes Tracked by Quantitative Proteomics

Séverine Boulon
Macromolecular Biochemistry Research Centre (CRBM) - CNRS / University of Montpellier France

1. Introduction

Cellular proteins rarely function as individual entities, instead they form multi-molecular complexes that are themselves interconnected in dense functional networks. These networks can perform a diverse range of highly coordinated biological processes. The characterization of protein-protein interaction networks is therefore crucial, not only to elucidate the local function and regulation of single proteins but also, and above all, to capture a comprehensive snapshot of cellular activity as a whole system.

The term "protein complexes" describes structures of varying nature. Protein complexes can be formed both by stable or transient interactions. Stable, long-term interactions can bridge core components of large multi-protein complexes, or molecular machineries, such as the RNA polymerase II complex and the 26S proteasome. On the other hand, interactions that are transient and dynamic in nature are often highly sensitive to regulatory stimuli and signaling events, such as enzyme/substrate complexes.

However, in all cases, protein interactions are prone to strict regulation and vary upon change in cellular environment. In human, the activity and the expression levels of cellular proteins may diverge between various differentiated states and thus lead to specific protein interactome network maps for each cell type. In addition, in a particular cell type, protein interactions are dependent upon physiological and pathological conditions, for example cell proliferation or stress response, and protein interactomes may thus fluctuate, hence reflecting the spatial and temporal complexity of cellular activity. Deciphering the dynamics of these protein interaction networks by assembling sets of various interactomes that echo different cellular conditions, rather than simply draw a comprehensive map of a static protein interactome, remains one of the key challenges in cell biology. Once solved, this will greatly aid understanding of complex mechanisms underlying normal cell behavior and how they are modified by genetic alterations, cancer and other types of diseases. Is this goal completely unrealistic, not to say utopist, given that the mapping of the human protein interactome has not yet been completed? I am eager to believe that the accumulation of outstanding studies and the emergence of new powerful techniques will certainly lead to the achievement of this ambitious project. In fact, various high throughput methodologies have already proven to be very efficient to characterize protein-protein interactions on a proteome scale, yet the scientific community is still far from building a dynamic map of the human protein interactome.

The quest of protein interactome dynamics raises several issues. First, protein interaction analyses must be freed from the non-specific contaminants that are inevitably identified along with genuine, specific protein interactors. This issue represents a major challenge not limited to dynamic studies alone but applicable in the wider field. In addition, it cannot be satisfactory to characterize the existence of interactions only. It is necessary to carry out quantitative studies, which provide scales of interaction intensities and thus help determine the strength of each interaction identified. As a result, protein interaction strengths can be compared between different conditions, which consequently allows for the dynamics of the protein interactome to be, at least partially, dissected. This exploration of the dynamic properties of protein complexes during biological responses is absolutely essential to shift from a descriptive inventory of all possible protein interactions to a more functional pathway analysis.

However, it is necessary to keep in mind that this will not be achieved by a single large-scale experiment but rather by the integration of a large number of individual studies. This will only be possible thanks to an international coordinated effort that will guarantee the collection of high quality protein interaction data that can be efficiently exploited and compiled by the scientific community. This requires that the vast amount of protein interaction data generated is somehow controlled, at the level of experimental procedures, metadata acquisition, data analysis and storage, to ensure reproducibility and reliability. The benefit of compiling such interaction datasets may be immense, both in basic and in clinical research. This will certainly allow for an improved understanding of protein function and regulation in different physiological conditions. This might also provide crucial information to elucidate genotype/phenotype relationships and mechanisms underlying several diseases, including cancer. Finally, this concerted effort will, in a comprehensive manner, help dissect the cell response to different types of drug treatment and chemotherapy, underline all cellular pathways that are altered and therefore grasp the causes of the efficacy and/or the side effects of a treatment.

In this chapter, I will review the different high throughput techniques that are used to study protein-protein interactions, including yeast two-hybrid assays and the combination of affinity purification techniques with mass spectrometry. The latter emerges as a method of choice to analyze the dynamics of protein complexes, especially with the development of MS-based quantitative proteomics strategies. I will particularly describe a workflow that uses affinity purification techniques combined with a triple SILAC labeling, quantitative proteomics approach, which has proven to be efficient, both (i) to reliably distinguish between specific and non-specific interaction partners and (ii) to quantify the changes of protein interactions occurring between different cellular conditions. The workflow illustrated here has been applied to the analysis of RNA polymerase II complex dynamics, which will be explained briefly. I will also present a new methodology, called the Protein Frequency Library (PFL), which can be used as an additional criterion to highlight putative false-positives identified in any pull-down experiment. Finally, the need for the scientific community to compile and standardize protein interaction data of high quality will be discussed, as well as the immense possibilities available to the clinical field through assembling proteome-scale maps of the human protein interactome networks.

2. Towards a comprehensive map of the human protein interactome

2.1 A set of complementary high throughput techniques

Various strategies have been developed to study protein-protein interactions. The goal of this chapter is certainly not to describe all of them. Instead, I will focus on those that have been commonly used in various large-scale studies and have already enabled important insights into the mapping, on a proteome-scale, of the human protein interaction network. Among these strategies, the most common include yeast two-hybrid (Y2-H) and affinity purification coupled to mass spectrometry (MS) techniques, which will be briefly described below, along with a selection of other innovative experimental techniques and *in silico* approaches.

2.1.1 Yeast two-hybrid

Most binary interactions available thus far have been produced by high throughput Y2-H screens that were performed in parallel by many different groups to characterize large interaction networks, or "interactomes", in model organisms and in humans (Parrish et al., 2006). These screens allowed for the identification of several thousand binary protein interactions in *Saccharomyces cerevisiae*, in *Drosophila melanogaster*, in *Helicobacter pilori*, in *Caenorabditis elegans* and in human (reviewed by Ghavidel et al., 2005). Of note, the Center for Cancer Systems Biology (CCSB), hosted at Dana-Farber Cancer Institute, is dedicated to mapping and systematically characterizing protein-protein interactions using Y2-H, an approach referred to as "interactome modeling" (Vidal, 2005). One of the leading projects of the CCSB is based on the comprehensive characterization of the human interactome network, using ORFs contained in the human ORFeome (http://ccsb.dfci.harvard.edu). Y2-H principle is simple and thus it is one of the most standardized techniques used to identify protein interactions (Figeys, 2008). Bait and prey proteins are co-expressed in yeast, the bait being fused to the DNA-binding domain of a transcription factor (usually Gal4 yeast transcription activator) and the prey being fused to the transactivation domain of this transcription factor. The two modules of the transcription factor can only induce the transcription of the reporter gene when brought together by the interaction between the bait and the prey. Therefore, the efficient induction of the reporter gene shows an association bait/prey, which can be considered as a binary interaction, i.e. a direct interaction between the two proteins tested.

2.1.2 Affinity purification combined to mass spectrometry

Despite the early prevalence of the Y2-H strategy, the combination of affinity purification procedures to mass spectrometry (AP-MS) has now emerged as the method of choice for mapping protein interactions. AP-MS provides a highly sensitive technique that enables the comprehensive identification of proteins associated with proteins of interest (baits) in multi-molecular complexes (reviewed by Vasilescu and Figeys, 2006). AP-MS has therefore been widely used for the large-scale characterization of protein complexes in different model organisms, including yeast and *Escherichia coli*. In human, large-scale analyses enabled the identification of thousands of interactions between more than 2,700 proteins, organized into more than 500 distinct complexes (Ewing et al., 2007; Gavin et al., 2006; Krogan et al., 2006).
AP-MS is based on the affinity purification of protein baits and binding partners using antibodies to endogenous or recombinant tagged proteins coupled to affinity matrices, e.g. sepharose and magnetic beads. The co-immunoprecipitation (co-IP) step is followed by the MS identification of proteins present in the eluate. Unlike Y2-H, protein interactions characterized by AP-MS are not necessarily direct interactions, but instead reflect the association between baits and identified interaction partners in multi-protein complexes. Interestingly, quantitative MS-based proteomics strategies hold great promise in the

analysis of protein complex dynamics, as they enable a relative quantification of protein interaction intensities and therefore enable comparison between different cellular conditions (see section 3: SILAC-based Quantitative Proteomics). Another essential point resides in the fact that proteins analyzed by AP-MS techniques are expressed under near-physiological conditions, with correct regulation and post-translational modifications (Kocher and Superti-Furga, 2007).

2.1.3 Other biochemical techniques

Alternative high throughput methods have been developed to analyze protein interactions, however these are not as widely used as Y2-H and AP-MS. Among those, the LUMIER approach (LUminescence-based Mammalian IntERactome mapping) is based on the IP of FLAG-tagged baits that are co-expressed in mammalian cells with putative interaction partners fused to the Renilla Luciferase (RL) enzyme. The intensity of the interaction between the two proteins of interest can then be determined by measuring luciferase enzymatic activity in FLAG immunoprecipitates (Barrios-Rodiles et al., 2005). Applied to the analysis of transforming growth factor-β (TGFβ) pathway, this semi-quantitative methodology was shown to efficiently detect protein interactions dependent on pathway-specific, post-translational modifications (PTMs) and, interestingly, interactions involving membrane proteins, which are usually under-represented in large-scale studies due to their poor recovery in fractionation procedures (Barrios-Rodiles et al., 2005). Both LUMIER and AP-MS approaches identify protein interactions within protein complexes, using quantitative measurements and therefore allow for comparisons between different cellular conditions (for example, in the absence or presence of TGFβ signaling). However, unlike AP-MS, the LUMIER technique requires the overexpression and tagging of both baits and preys, limiting the reliability and the coverage of the results obtained.

Protein-fragment Complementation Assay (PCA) enables the detection of binary protein-protein interactions (PPIs) *in vivo*, in their natural environment. Using the PCA approach described by Tarassov et al (Tarassov et al., 2008), baits and preys are fused to F[1,2] and F[3] complementary N- and C-terminal fragments of a mutant of the mDHFR reporter protein that is insensitive to the DHFR inhibitor methotrexate but retains full catalytic activity. The F[1,2] and F[3] fragment fusions are expressed in *Saccharomyces cerevisiae* MATa and MATα strains, respectively, which are then mated and selected for methotrexate resistance. If the proteins of interest physically interact, the DHFR fragments are brought together and fold into their native structure, thus reconstituting the reporter activity and permitting the survival of the diploid colonies. Binary protein interactions can therefore be directly deduced from the measurement of colony growth. This methodology is an interesting alternative to the Y2-H assay as it enables the identification of direct binary interactions (less than 82 Å between the two proteins of interest) *in vivo* and, unlike Y2-H, is based on proteins in their native subcellular location and post-translationally modified state (Tarassov et al., 2008).

2.1.4 *In silico* approaches

Techniques described above are all based on large-scale experimental methodologies that search for physical interactions *in vivo* and/or *in vitro*. In contrast, *in silico* approaches, which represent an alternative way to obtain protein interaction information, rely upon the curation of all publications in literature that describe either low or high throughput protein interaction studies (where an experiment reporting less than 40 interactions is considered as low throughput). Literature-curated protein interactions are reported in web based interaction databases, most of which are now freely available online. Of note, low-throughput analyses account for a third of the total number of interactions in Biomolecular Interaction Network Database (BIND), i.e. 67,789 low-throughput interactions from a total of 206,859 (Isserlin et al., 2011). Other curated protein interaction resources include the Munich Information center for Protein Sequence (MIPS) protein interaction database, the Molecular INTeraction database (MINT), the Database of Interacting Proteins (DIP), the protein InterAction database (IntAct), the Biological General Repository for Interaction Datasets (BioGRID) and the Human Protein Reference Database (HPRD), which currently report more than 200,000 protein interactions resulting from large-scale curation of several thousand publications.

In silico approaches are precious in compiling, and regularly updating, all protein interaction evidences reported in the literature, and in making them publicly available for the scientific community, however they are not as reliable as generally presumed (Cusick et al., 2009). They indeed rely upon individual studies that are of highly variable quality and often do not provide curators, and readers in general, with essential pieces of information, such as correct gene names, species and precise experimental parameters, making it difficult to decide on interpretations in a reliable manner. In addition, most protein interactions contained in databases (>75%) are only supported by one publication, with only 5% of the total number of interactions being described in three or more publications (Cusick et al., 2009).

However, the main issue raised by these *in silico* approaches resides in the fact that they rely upon small-scale focused studies, which are, by definition, biased towards hypothesis-driven investigations and tend to search for proteins that are already known and have therefore a higher probability of being investigated again. In contrast, high throughput strategies rely upon unbiased discovery-driven explorations, which are absolutely necessary to unveil new unpredictable protein interactions and investigate novel functions (Cusick et al., 2009). Therefore, high throughput methods appear as essential tools to uncover the vast number of protein interactions that remain to be identified for assembling a comprehensive map of the human protein interactome network.

2.2 Comparison of technique limitations

Not a single high throughput method is perfect and, on its own, none will enable the comprehensive characterization of the human protein interactome and its dynamic properties. Drawbacks are inherent to the technique used and are therefore inevitable. For example, in Y2-H, protein interactions are investigated in the yeast system, which does not reflect the native environment of human proteins and is characterized by improper PTMs, processing and regulation of proteins in general. In addition, baits and preys are overexpressed and constrained to the nucleus, which might force interactions that would not occur in natural conditions. These limitations have to be taken into account to interpret results obtained through large-scale Y2-H screens, which might not be ideal to analyze protein interaction dynamics but offer the advantage of identifying binary interactions between protein pairs tested.

In contrast, the AP-MS technique enables the comprehensive identification of all individual protein interaction partners (direct or indirect) for any given bait, thereby leading to the characterization of multi-protein complexes. Interestingly, the combination of AP-MS with quantitative proteomics strategies can efficiently, and quantitatively, reveal changes occurring in protein complexes between different conditions. Limitations of AP-MS techniques reside in the fact that protein complexes have to be "artificially" extracted from their natural environment, e.g. by cell lysis or possibly by cellular fractionation, and immunopurified before MS analysis, leading to putative perturbations of the complex and disruption of weak interactions. To avoid this problem, proteins can be cross-linked before extraction. Alternatively, protein purification may be performed in very poor stringency conditions, to preserve weak interactions, which are often of great biological importance as they can, for example, reveal regulatory mechanisms (discussed in more details below).

From this list of limitations, it is claimed that one assay may be more efficient to capture one type of protein interaction and vice versa. This might explain, at least partially, the small overlap between protein interactions identified by these techniques, due to how fundamentally different they are (Cusick et al., 2005; Figeys, 2008). This does not necessarily reflect the poor reliability of the reported data, but instead, the poor sensitivity of these approaches, which cannot individually cover the whole interactome network due to the technique limitations described above and a still weak sensitivity of detection (Lemmens et al., 2010). Therefore, it may be very powerful to combine these complementary strategies to study protein interactions, each of them providing a partial view of the whole system, with AP-MS identifying protein complexes in their natural environment and Y2-H indicating binary interactions within these protein complexes (Boulon et al., 2010b).

2.3 Reducing the numbers of false positives and false negatives: A major challenge

Before interpreting any high throughput protein interaction data, it is however necessary to discriminate between genuine interactions and non-specific ones, i.e. false positive interactions that are inevitably recovered in all large-scale studies. This represents one of the major challenges in the field, given that non-specific contaminants often represent more than 50% of the identified protein interaction partners. In contrast, false negatives often are transient interactions and/or interactions that only occur in specific conditions and cannot be detected by the experimental setup. These false negatives constitute another important issue in these assays, as low affinity and/or low abundance specific interaction partners are generally of great biological importance to understand protein function, regulation and dynamics. To overcome these issues, it is essential to strive for the highest signal to noise ratio, which encompasses both sensitivity and reliability. High sensitivity, which reduces the number of false negatives, will be achieved through the increase of detection tool performance and the optimization of experimental procedures. High reliability, which reduces the number of false positives or experimental contaminants, also depends on optimal experimental set up but above all, it relies on adequate data analysis.

To date, most data obtained by large-scale studies undergo computational assessments that provide confidence scores or biological significance for each protein interaction identified, based on comparison with other approaches. For example, protein interactions are considered as more reliable if they are supported by data showing either phylogenetic conservations, genetic interactions, subcellular co-localization, similar functional interactions in Gene Ontology (GO) or correlated expression profiles between interacting proteins (Ge et al., 2001; Tong et al., 2004). Indeed, a correlation has been shown between protein interactions and protein localization and expression, with 27% of interacting proteins sharing the same subcellular localization (Ge et al., 2001; Reguly et al., 2006; Tong et al., 2004). In addition, confidence scores also reflect the integration of different properties of the interaction network generated by the analysis, e.g. interaction bi-directionality and network topology (Cloutier et al., 2009; Ewing et al., 2007). However, the calculation of these confidence scores raises two problems. First, these confidence scores are based on the comparison, and the correlation, between new protein interactions and previous observations, thus leading to a bias towards already known data versus unpredictable interactions. Second, confidence scores that rely upon properties of the network can only be calculated when there are a sufficient number of experiments performed, i.e. this analysis pathway is limited to large-scale studies and cannot be applied as efficiently to small-scale studies.

2.4 Analyzing the dynamics of protein complexes

Characterizing protein interactions alone, without mentioning in which specific cellular context the interaction has been found, might be misleading. Indeed, as discussed previously, all protein interactions are dynamic and may be subjected to variations in response to changes in cellular environment, either physiological or pathological. This means that all protein interaction networks are prone to an extreme plasticity, which needs to be taken into account if one wants to draw a faithful map of the human interaction network, or rather faithful maps of the human interaction networks. Does that mean that the scientific community needs to assemble as many protein interaction networks maps as possible conditions? This might seem unrealistic, and it certainly is. But one should definitely aim to notify precise conditions in which each protein interaction was found. This is absolutely essential, when assembling the different pieces of the protein interaction network jigsaw puzzle, to avoid bringing together things that just do not match, e.g. mixing together in a network protein interactions that specifically occur in proliferating cells with interactions that specifically occur in differentiated tissues. In the end, many proteins potentially interact with each other, but the interesting question is "when?". Most researchers are interested in these problems and have already sought for variations in interaction patterns in different cellular contexts, such as cancer and neurodegenerative diseases (Lim et al., 2006).

Interestingly, the combination of affinity purification with quantitative proteomics overcomes most limitations in the sense that it provides high sensitivity, with the development of highly performing MS equipment, and high reliability. Quantitative proteomics indeed enables (i) the identification of protein interactions in their natural environment, with native PTMs and subcellular localization, (ii) the efficient discrimination between specific interaction partners and the non-specific background of contaminants, i.e. proteins that bind non-specifically to the affinity matrices and (iii) the comparison of protein interaction intensities between different conditions.

3. SILAC-based quantitative proteomics: A method of choice to reliably analyze specific protein interactions and protein complex dynamics

MS-based proteomics is not inherently quantitative. Quantitative proteomics strategies that have been developed recently mostly involve isotope labeling via either metabolic incorporation *in vivo/in cellulo* ($^{15}N/^{14}N$ metabolic labeling, Stable Isotope Labeling by

Amino acids in Cell culture (SILAC)) (Ong et al., 2002), chemical modification *in vitro* (ICAT, iTRAQ) (Gygi et al., 1999; Ranish et al., 2003) or enzymatically catalyzed incorporation (^{18}O labeling) (Yao et al., 2001). Alternatively, label-free strategies for protein quantification are based either on the comparison of precursor signal intensity for each peptide across multiple LC-MS data, or on spectral counting (Collier et al., 2010; Wepf et al., 2009). These different methods for quantitative proteomics will not be detailed in this chapter, as they are described in this book by Sap and Demmers (Quantitation in mass spectrometry based proteomics) and Leroy, Matallana and Wattiez (Gel free proteome analysis - isotopic labeling vs. label free approaches for quantitative proteomics). It is noteworthy however that isotope labeling appears to be more sensitive than label free to detect small variations, which means that until label free techniques are more robust and statistically reliable, isotope labeling strategies might be more powerful to analyze subtle changes occurring in protein interaction intensities between different conditions.

3.1 Triple labeling SILAC pull-down workflow

Among isotope labeling strategies, SILAC has emerged as a simple and powerful approach, now widely used to study protein-protein interactions in various organisms and cell types (Boulon et al., 2010a; Mann, 2006; Trinkle-Mulcahy et al., 2008). SILAC methodology relies upon the metabolic labeling of proteins in cell culture, through the incorporation of light, medium and heavy isotope containing amino-acids (arginine and lysine) that can be resolved and quantitated by MS. Light amino-acids refer to naturally occurring environmental isotopes of carbon, nitrogen and hydrogen, i.e. "unlabeled" ^{12}C, ^{14}N and ^{1}H, whereas medium- and heavy-labeled arginine (R) and lysine (K) refer to (i) medium (R6K4): [$^{13}C_6$]arginine (R6) and 4,4,5,5-D4-lysine (K4) and (ii) heavy (R10K8): [$^{13}C_6$, $^{15}N_4$]arginine (R10) and [$^{13}C_6$, $^{15}N_2$]lysine (K8) (Ong et al., 2002). Cells are cultured in SILAC media containing light, medium or heavy amino-acids for at least 5-6 doublings to ensure complete incorporation of isotopic amino-acids.

Various studies have used the SILAC methodology to characterize dynamic changes in protein interactions (reviewed by Dengjel et al., 2010; Gingras et al., 2007; Vermeulen et al., 2008). For example, Blagoev et al identified novel proteins binding to the SH2 domain of the adapter protein Grb2 upon EGF stimulation, by using GST-SH2 fusion protein and GST-based affinity purification (Blagoev et al., 2003). Only two SILAC conditions were used (light and heavy), which allowed for the comparison of SH2 interacting proteins in untreated versus EGF-stimulated cells, with no control IP. Similarly, Foster et al exploited the SILAC method, to identify proteins that interact with GLUT4 in an insulin-dependent manner (Foster et al., 2006). Recently, Kaake et al reported an interesting study based on three SILAC IP experiments performed in parallel in yeast. Their approach was called QTAX (Quantitative analysis of TAP *in vivo* Xlinked protein complexes). TAP-tagged Rpn11 was used as bait, to characterize proteasome interaction partners in three different cell cycle phases (G1, S and M) (Kaake et al., 2010). For each cell cycle phase, a double labeling SILAC strategy was performed, which included an internal negative control and Rpn11 specific pull-down, therefore allowing for the discrimination between putative contaminants and specific interaction partners. However, the comparison of interaction partners between the three cell cycle phases relied upon independent experiments and, therefore, separate MS runs. As a result, subtle variations in protein interaction intensities may not be observed in this study.

The SILAC workflow that is described in this chapter merges advantages from these different strategies. It is based on a triple labeling SILAC pull-down approach, which compares, within a single MS run, an internal negative control, for the identification of putative non-specific contaminants, and two IPs of interest performed in two different conditions, for the direct comparison of protein interaction intensities between the two conditions tested (Figures 1A and 1B). This protocol has proven to be efficient in the characterization of both specific and dynamic interactions and is easily accessible to all laboratories. In brief as summarized in Figure 1A, in the case of a GFP-IP, parental cells are grown in light (R0K0) medium, whereas GFP-protein expressing cells are grown in medium (R6K4) and heavy (R10K8) media. The light condition is used for the control IP, the medium condition for the IP of interest in control conditions (untreated cells) and the heavy condition for the IP of interest in treated cells (chemical inhibitors, stress, etc). Medium and heavy conditions can also be used to compare changes in protein complexes between cell cycle phases, cell types, etc. After cell lysis, extracts from each cell line are precleared before GFP-protein is affinity-purified using GFP_TRAP ® affinity matrix (Chromotek) (Rothbauer et al., 2008). Eluates from each condition are then combined and digested using trypsin. Resulting peptides are analyzed by LC-MS/MS and can be quantified by the MaxQuant software, which has been developed by the Mann group (Cox and Mann, 2008; Cox et al., 2009). As seen in Figure 1C, each peptide identified in the triple labeling SILAC IP experiment shows a typical MS spectrum with three main peaks that correspond to its light (L), medium (M) and heavy (H) isotopic forms, respectively. The relative abundance of each distinct peak area is determined by MaxQuant, which provides M/L, H/L and H/M ratios for each peptide. Protein ratios can then be extrapolated from the median ratio value of all peptides identified for that specific protein.

3.2 Triple labeling SILAC pull-down data analysis

In triple labeling SILAC IP experiments, baits and genuine interaction partners are expected to show high M/L and/or H/L ratios, as opposed to experimental contaminants, e.g. proteins that bind non-specifically to the affinity matrix, which are expected to have M/L and H/L ratios close to 1. Proteins that show M/L and H/L ratios close to zero are likely to be environmental contaminants, such as keratins (Figures 1B and 1C). In contrast, H/M SILAC ratios indicate changes in protein interaction intensities. For example, proteins showing a SILAC H/M ratio<1 are expected to have a decreased interaction with the bait in treated cells versus untreated cells whereas proteins showing a SILAC H/M ratio>1 are expected to have an increased interaction with the bait upon treatment. Of note, it can be very powerful to perform several SILAC IP experiments in parallel to study the dynamics of protein complexes in more than two conditions. In this case, the first experiment can be carried out as described above whereas the other ones can exclude the negative control and directly compare protein interactions in three different conditions (Boulon et al., 2010b). Putative contaminants are thus deduced from the first experiment and many different conditions can be compared in a reliable manner.

Figure 2 shows a method of visualizing triple SILAC IP data, by plotting $\log_2(H/M)$ (y axis) versus $\log_2(M/L)$ (x axis) SILAC ratio values for all proteins identified in the experiment. Interestingly, on this type of graph, most proteins usually cluster around the origin, with M/L and H/M ratios close to 1, and therefore $\log_2$ ratios close to 0 (Figure 2). These proteins

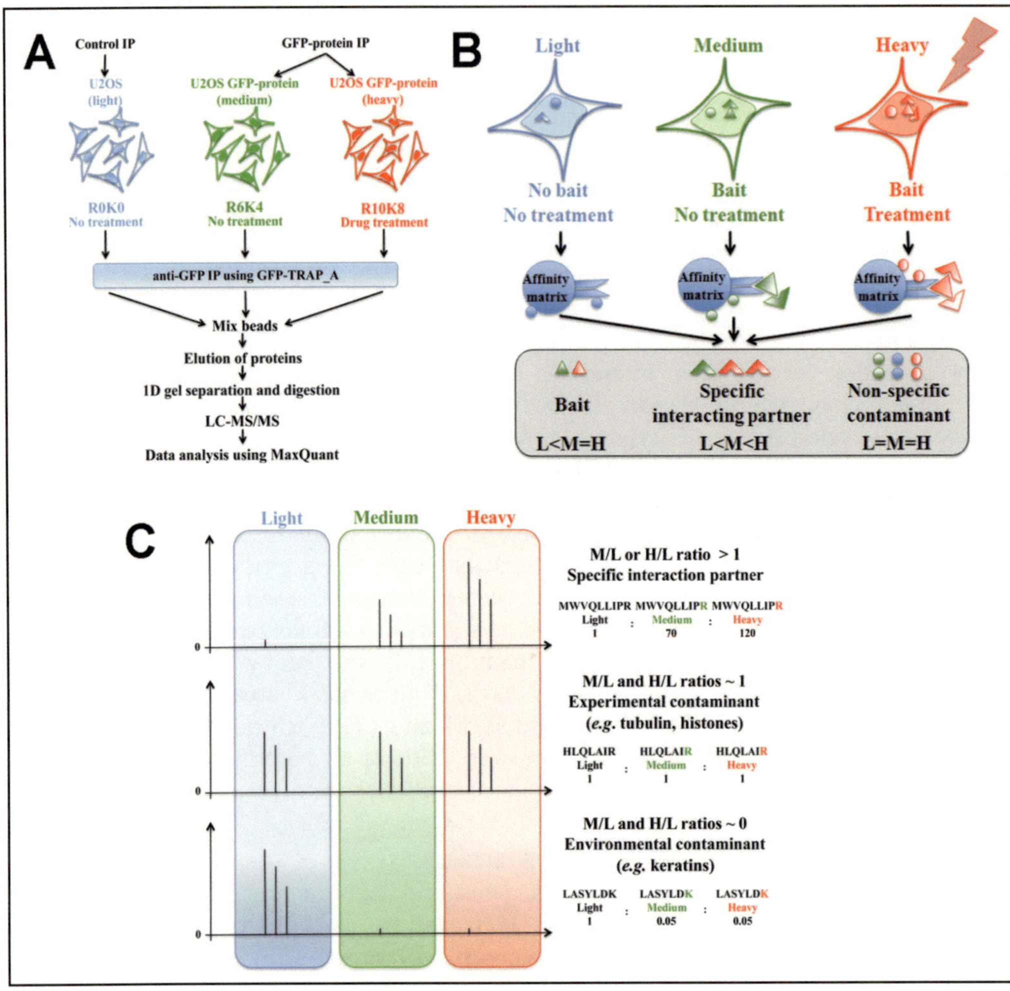

(A) Overview showing the workflow of a representative triple SILAC IP analyzing the changes in specific interaction partners of GFP-tagged bait stably expressed in U2OS cells in response to a drug treatment. References to R0K0, R6K4 and R10K8 culture conditions can be found in the body text. (B) Diagram illustrating the SILAC principle of differential labeling. The bait and its specific interaction partners should only be retrieved in medium and heavy conditions, thereby showing high M/L and/or H/L SILAC ratios, whereas non-specific contaminants are present in all three conditions, thereby showing M/L and H/M close to 1. (C) Typical MS spectra obtained for representative peptides of a specific interaction partner (top), an experimental contaminant binding non-specifically to the affinity matrix (middle) and an external environmental contaminant (bottom). IP: immunoprecipitation; L: light; M: medium; H: heavy; GFP-Trap_A®: GFP binding protein coupled to a monovalent matrix (Chromotek).

Fig. 1. Overview of triple labeling SILAC analysis of protein interaction partners.

are likely to be contaminants, as described above, which often represent more than 50% of all proteins identified in AP-MS experiments. In contrast, putative genuine interaction partners of the bait typically localize to the right side of the graph, with M/L SILAC ratios over a certain threshold, which may vary between experiments. Of note, not a single

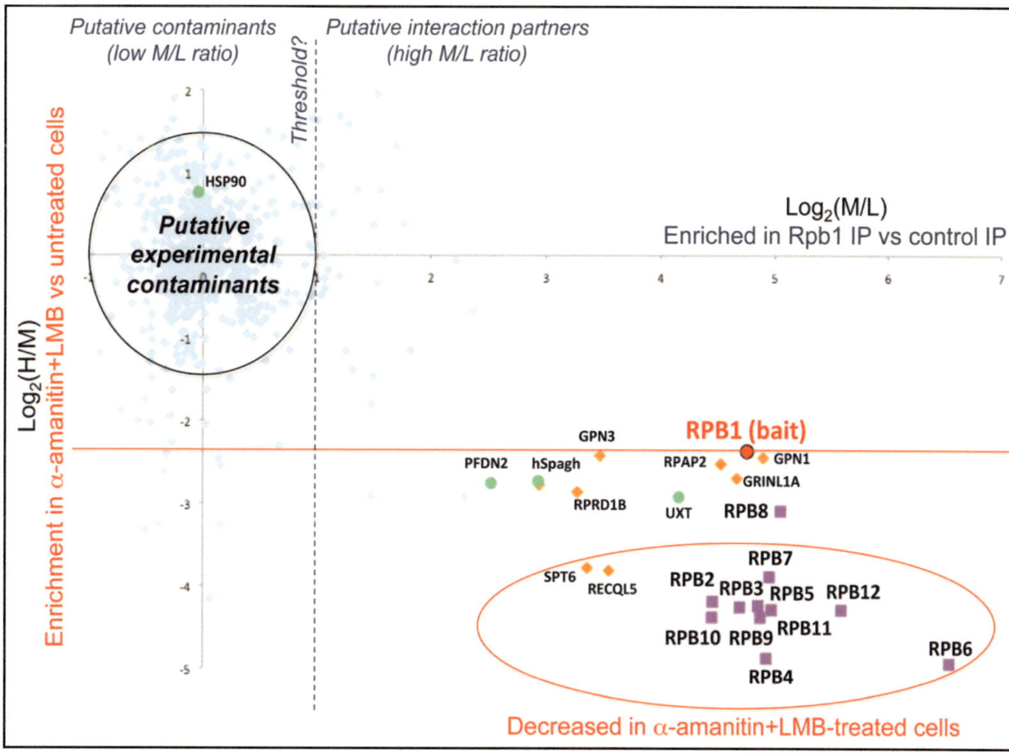

All protein groups identified and quantified by MaxQuant are represented on the graph. The experimental design is similar to Figure 1, with endogenous Rpb1 being used as bait. Cellular extracts from the light condition are incubated with a control antibody (control IP) while cellular extracts from the medium and heavy conditions are incubated with an antibody against endogenous Rpb1. Cells cultured in the heavy condition are treated with α-amanitin and Leptomycin B (LMB) for 15 hours. On the x-axis, $\log_2(M/L)$ ratio correlates with the enrichment of Rpb1 IP versus control IP. Proteins with high $\log_2(M/L)$ ratios are expected to be specific interaction partners. However, not a single threshold can unambiguously separate contaminants from genuine interaction partners. On the y-axis, $\log_2(H/M)$ correlates with the enrichment in α-amanitin+LMB treated cells versus untreated cells. Putative experimental contaminants cluster around the origin. The bait, Rpb1, is spotted in red. The dotted red line shows an alternative x-axis defined by the bait, which separates the proteins whose interaction with the bait is increased after the treatment (above) or decreased (below). Proteins within red oval are proteins whose interaction with the bait is decreased by two-fold or more. Proteins that show a $\log_2(M/L)$ ratio>2 are spotted in orange, RNA polymerase II subunits in purple and R2TP/prefoldin-like complex in green.

Fig. 2. Visualization of the triple SILAC Rpb1 dataset plotted as $\log_2(M/L)$ versus $\log_2(H/M)$ SILAC ratios.

threshold can unambiguously separate contaminants from genuine interaction partners (discussed below).

To analyze protein interaction dynamics, one should focus on those putative specific interaction partners. To start with, the bait itself should display high M/L and H/L SILAC ratios. This indicates that it was efficiently immuno-purified. Otherwise, the IP protocol might need to be optimized. If the efficiency of the IP is the same between the two conditions tested, the $\log_2(H/M)$ ratio of the bait protein should be 0. In practice, this is often not the case, due to changes in expression levels and/or accessibility of the bait induced by the treatment. A way to get around this problem is to draw a second x-axis using the bait protein as a reference. Proteins that locate below this new x-axis reveal a decreased interaction with the bait in treated cells whereas proteins that locate above reveal an increased interaction (Figure 2). This extremely easy to apply visualization method provides in a glimpse an objective conclusion regarding the dynamics of protein complexes.

3.3 Application to the analysis of RNA polymerase II complex dynamics

This triple SILAC IP method has been efficiently applied to the analysis of RNA polymerase II complex dynamics (Boulon et al., 2010b). The RNA polymerase II (RNAPII) complex is an essential multi-protein complex that is involved in the transcription of all mRNAs and capped non-coding RNAs. The structure and subunit composition of this enzyme have been characterized in detail. RNAPII complex is formed by 12 subunits, Rpb1 to Rpb12. Rpb1 and Rpb2, the two largest subunits, form the catalytic core of the enzyme. However, relatively little is known about assembly mechanisms. Recently, a set of RNAPII interacting partners with unknown function was identified by AP-MS (Cloutier et al., 2009; Jeronimo et al., 2007). In collaboration with the Bertrand group, we explored the dynamics of RNAPII complex using the triple SILAC IP strategy described above to capture the function of these different factors. Interestingly, we could show that some of these interaction partners, which are part of the R2TP-prefoldin-like complex, in fact participate to the assembly of the RNAPII holoenzyme in the cytoplasm (Boulon et al., 2010b).

In this work, we took advantage of α-amanitin transcription inhibitor, which is known to induce the degradation of Rpb1, RNAPII largest subunit, and the disassembly of the remaining subunits, which are exported to the cytoplasm (Boulon et al., 2010b; Nguyen et al., 1996). In addition, α-amanitin combined to leptomycin B (LMB) treatment leads to the accumulation of newly synthesized Rpb1 in the cytoplasm, which cannot be imported into the nucleus (Boulon et al., 2010b). Four triple SILAC IP experiments were thus performed in parallel, using endogenous Rpb1, GFP-Rpb3 and GFP-hSpagh as baits. In this chapter, endogenous Rpb1 IPs are described as examples. In brief, in the first experiment, the light condition was used for control IP, whereas medium and heavy conditions were used for endogenous Rpb1 IP in untreated cells ("assembled" Rpb1) versus α-amanitin+LMB treated cells ("unassembled" Rpb1). Eluted Rpb1 and associated partners were digested using trypsin, analyzed by LC-MS/MS and relative SILAC ratios were calculated by MaxQuant. Figure 2 shows Rpb1 IP dataset plotted as $\log_2(H/M)$ against $\log_2(M/L)$ ratios. The identification of specific interaction partners of Rpb1 ($\log_2(M/L) > 2$) revealed the presence of all RNAPII subunits and a set of additional factors, some of which belong to the R2TP/prefoldin-like complex that was previously described by other AP-MS approaches (Cloutier et al., 2009; Jeronimo et al., 2007). RNAPII subunits are marked in purple, whereas R2TP/prefoldin-like complex factors are marked in green. Interestingly, using H/M ratios, we could observe drastic changes in Rpb1 interaction partners between the two conditions tested. We showed (i) that the association between Rpb1 and the other RNAPII subunits is lost upon α-amanitin+LMB treatment (interactions were arbitrarily considered as significantly decreased when a two-fold or greater change was observed upon treatment, as compared to the bait H/M reference ratio) and (ii) that the interaction of Rpb1 with the R2TP/prefoldin-like complex is not affected by the treatment. This indicated both that the holoenzyme is disassembled upon treatment and that R2TP/prefoldin-like factors bind to "unassembled" Rpb1, suggesting that these factors of unknown function might therefore be involved in the stabilization and the assembly of RNAPII subunits, which was later confirmed by other approaches (Boulon et al., 2010b).

A second triple SILAC IP experiment was performed in parallel, using again endogenous Rpb1 as bait, to directly compare three different conditions, i.e. untreated cells versus cells treated with either α-amanitin+LMB or actinomycin D. Actinomycin D is another transcription inhibitor, which induces stalling of the whole RNAPII complex onto DNA within the nucleus. Therefore, actinomycin D is not expected to induce the disassembly of the complex. By comparing untreated versus actinomycin D-treated cells, we could indeed observe that actinomycin D has no effect on Rpb1 association to the other RNA polymerase II subunits. In addition, when directly comparing α-amanitin+LMB versus actinomycin D treatments, it was clear that the association between Rpb1 and the R2TP/prefoldin-like complex is much stronger in cells treated with α-amanitin+LMB. This confirmed that unassembled Rpb1 is specifically associated with the R2TP/prefoldin-like complex and that it is not an indirect consequence of transcription inhibition.

This example shows that the triple SILAC IP strategy can be efficiently applied to the high confidence identification of specific interactions and analysis of protein complex dynamics between several conditions (three different conditions were compared in this study). Here, data mining is enhanced by the integration of several major criteria, including reliable quantitative SILAC ratios. In addition, the quality of MS data highly depends on the number of peptides identified and quantified for each protein and the total sequence coverage. These parameters should therefore also be taken into account to evaluate the reliability of MS results. Interestingly, this SILAC IP strategy can be combined with complementary approaches, including Y2-H, to characterize binary interactions within protein complexes, and fluorescence microscopy, to uncover subcellular localization of protein interactions (Boulon et al., 2010b).

4. Optimization of experimental procedures

As discussed previously, one major challenge of AP-MS experiments is the reliable discrimination between genuine protein interaction partners and non-specific contaminants. This will be facilitated both (i) by the optimization of the experimental procedure, to increase the IP efficiency (high specific signal) and to reduce the background of contaminants (low non-specific noise), therefore tending to a high signal/noise ratio (Boulon et al., 2010a; Trinkle-Mulcahy et al., 2008) and (ii) by an efficient data analysis pathway that allows the reliable "identification" of the putative contaminants. I will first discuss the different important points that need to be taken into account to optimize a triple SILAC IP workflow.

The triple SILAC IP protocol described in Figure 1 is shown for the IP of GFP-tagged baits and the identification of their specific interaction partners in two different conditions, i.e. untreated versus treated cells. However, the triple SILAC co-IP procedure is far from being restricted to the chosen example and can be applied to many different types of investigations, but one has to keep in mind that both the reliability and the sensitivity of the resulting datasets will be extremely dependent on the experimental parameters chosen. It is therefore necessary to think about possible pitfalls and design an "optimal" protocol that will be correlated both to the question asked and to the tools available (ten Have et al., 2011). Important features include the choice of the tag/antibody, the conjugation of antibodies to affinity matrices and the IP protocol.

4.1 Choice of the tag/antibody

Both tag-based and endogenous pull-down experiments have advantages and drawbacks. Whenever possible, the use of antibodies targeted against the endogenous baits should be favored. Indeed, endogenous proteins avoid several problems usually associated with the use of tags, i.e. endogenous proteins are naturally expressed in their native cellular environment, with correct expression regulation, PTMs and above all proper interaction partners. However, this strategy relies on the availability of a specific and high affinity antibody that isolates the endogenous bait protein efficiently, which is often not available. In any case, antibody affinity and specificity should always be checked carefully. Noteworthy, a Swedish project (The Swedish Human Protein Atlas project), funded by the Knut and Alice Wallenberg Foundation, has been initiated to generate, in a high-throughput manner, high quality affinity-purified human antibodies to allow for a systematic exploration of the human proteome using Antibody-Based Proteomics (Uhlen et al., 2010). In May 2011, 11,300 Prestige Antibodies covering more than 50% of the human proteome had been developed (http://www.proteinatlas.org). But not all of them have been tested for IP efficiency, and there might still be a long way before all human proteins can be immunoprecipitated using this antibody library.

In contrast, tagged baits provide a scalable and general method to identify specific protein interaction partners. Different types of tags are commonly used in affinity-purification experiments, such as fluorescent tags (e.g. GFP), His-tag and Flag tag. In addition, a TAP-tag (Tandem Affinity Purification) methodology can be used, rather than a one step procedure (Rigaut et al., 1999). Although this two-step method reduces the amount of contaminants recovered in the IP eluate, it also decreases the general yield of proteins recovered and risks losing biologically relevant low affinity and/or low abundance interaction partners. Alternatively, the GFP tag has proven to be an effective tag for affinity purification procedures, due (i) to its low background of non-specific interactions and (ii) to the efficient recovery possible using recently developed GFP_TRAP ® (Chromotek) affinity matrices (Rothbauer et al., 2008; Trinkle-Mulcahy et al., 2008). In addition, the GFP tag can be used in a dual strategy combining both fluorescence microscopy and affinity-purification (Trinkle-Mulcahy et al., 2008). All tags, however, can potentially affect protein structure, localization and turnover, resulting in alteration of both protein function and association with specific partners. This problem may be countered by trying different locations for the tag, for example C and N terminal positions. The fact that recombinant proteins are usually overexpressed in mammalian cells represents another important perturbation of the system. Interestingly, the BAC TransgeneOmics strategy, developed by the Hyman lab, allows for the expression of GFP-tagged proteins under endogenous promoters and can be used in high throughput approaches for the identification of specific interaction partners, such as QUBIC (QUantitative BAC-green fluorescent protein InteraCtomics) (Hubner et al., 2010; Poser et al., 2008). In all cases, the generation of stable cell lines expressing recombinant proteins, rather than transient transfections, avoids problems linked to the heterogeneity of gene integration and expression levels between cells.

4.2 Conjugation of antibodies to affinity matrices

Antibodies are conjugated, covalently or not, to bead matrices (e.g. sepharose, agarose and magnetic beads). When combined with MS, it is highly recommended to covalently conjugate the antibody to the beads, otherwise a large amount of antibody can be eluted from the beads along with the specific protein complexes and compete with other proteins for further MS identification. The type of beads used for each pull-down experiment is an issue that is worth considering as well, as the efficiency and cleanliness of different types of beads may vary according to the cell type and the type of extract used. In our experience, Dynabeads (Invitrogen) work well for nuclear extracts, whereas Sepharose and Agarose beads (GE-Healthcare) can give lower backgrounds when used with cytoplasmic extracts and whole cell extracts (Trinkle-Mulcahy et al., 2008).

4.3 Cell extraction and immunoprecipitation protocol

Cell lysis and protein extraction may be a challenging part of the procedure, according to the protein complexes of interest. In particular, membrane proteins and proteins attached to macromolecular entities, including chromatin and subnuclear compartments, represent a real challenge to release and are therefore often under-represented in protein interaction studies using "normal" extraction procedures. Specific purification protocols may thus be envisaged, such as the modified chromatin immunopurification (mChIP) method (Lambert et al., 2009).

To reduce the amount of non-specific binding in a co-IP experiment, several options may be considered, including a pre-clearing step (pre-incubation of cellular extracts with bead matrices alone), incubation times kept to their minimum (1h max) and high stringency buffers (for example adequate buffers according to detergent and salt concentrations). Similarly to the TAP-tag strategy, increasing the buffer stringency may reduce the number of false positives identified but also increase the number of false negatives, by losing precious transient protein interaction partners, which are certainly the most difficult, but also the most interesting, proteins to identify.

Therefore, to preserve all genuine protein interaction partners, both stable and transient, medium or low stringency buffers may be favored. As a result, however, many contaminants remain in the analysis, which need to be reliably identified and distinguished from the specific interaction partners.

5. An additional criterion to identify putative contaminants: The Protein Frequency Library

Even though SILAC IP strategies may have proven themselves successful in the identification of stable interaction partners, relying upon isotope labeling ratios alone does not entirely solve the contaminant problem. Indeed, not a single ratio threshold can unambiguously isolate non-specific binders from genuine interaction partners (Figure 2). There is usually no doubt concerning interaction partners identified with high SILAC ratios, which often are genuine stable interaction partners, but in all SILAC IP experiments there are also low abundance and/or low affinity genuine interaction partners (transient interactions) that show low SILAC ratios (between 1 and 1.5 - 2) and are therefore embedded in the background of contaminants. Defaulting to using a high threshold filter eliminates both contaminants and transient interaction partners whereas an overly cautious low threshold will result in keeping both. Hence, it is not possible to rely on SILAC ratios alone to consistently and unambiguously separate contaminants and specific interaction partners.

To address this issue, a new methodology was developed, called the Protein Frequency Library (PFL), which provides an additional objective criterion to the data analysis (Boulon et al., 2010a). The principle of the PFL is based on the knowledge that proteins frequently

found in pull-down experiments are likely to be contaminants binding non-specifically to the affinity matrix. Therefore, the PFL has been generated to annotate all proteins identified at least once in a set of independent MS co-IP experiments with their frequency of detection in these experiments. Hence, the PFL provides a probability estimate for each protein of its likelihood of being a contaminant, which is independent of the information given by the SILAC ratio and therefore can be applied to analyze both SILAC and label-free data. To generate the PFL, a data environment called PepTracker was created by Yasmeen Ahmad (www.peptracker.com). This data environment stores and manages all MS-based proteomics data generated in the Lamond laboratory and quantitated by MaxQuant, currently including more than one hundred SILAC and label-free pull-down experiments. Interestingly, consistent and reliable metadata descriptors are recorded along with datasets. Recorded experimental parameters include organism, cell type, extract type, affinity matrix, protein bait, tag, mass spectrometer, date, user, etc. Due to the high complexity of the analysis and large volumes of data involved, the database was built using a multidimensional data model, which relies upon computational methods drawn from the business intelligence (BI) field designed for rapid interactive responses (Kohn et al., 2005).
In practice, the PFL can be represented as a graph (Figure 3), plotting the frequency of detection (y axis) for each protein identified in any of the pull-down experiments recorded

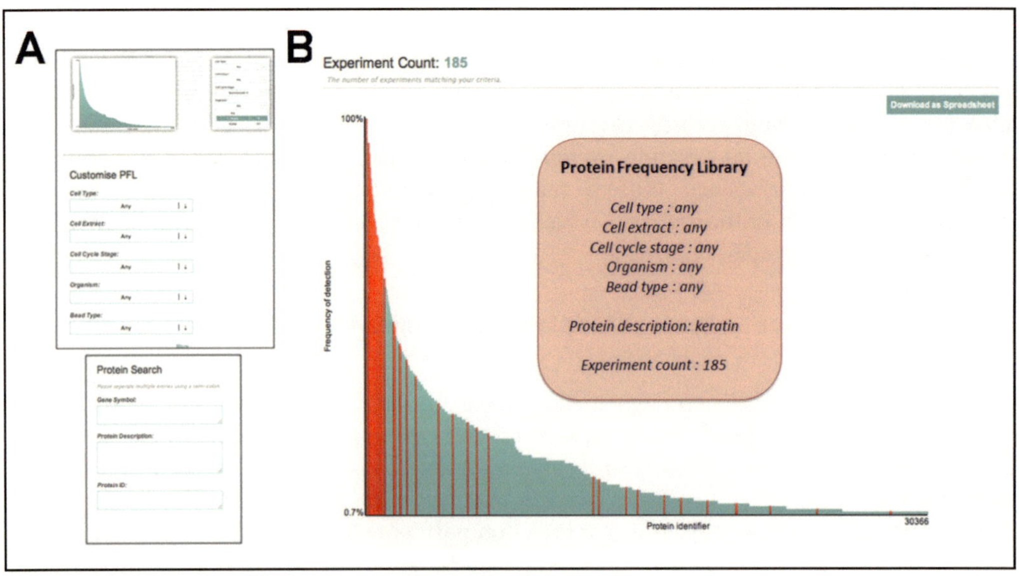

(A) Query interface for the Protein Frequency Library, found at http://www.peptracker.com/datavisual/. The PFL can be filtered on any individual IP experimental parameters recorded in PepTracker, e.g. cell type, cell extract, cell cycle stage, organism, bead type etc. to generate a customized PFL (top). The protein search allows users to specify a gene symbol, protein description or protein ID to be identified in the PFL. (B) Result of "keratin" search in the unfiltered PFL, which currently contains 185 IP experiments. The graph illustrates the frequency of detection (y axis) of all proteins present in the PFL (currently more than 30,000 protein identifiers) (x axis). PFL proteins are ranked from highest to lowest detection frequency (left to right). Most keratins *(red bars)* show high frequencies of detection and can therefore be considered as putative IP non-specific contaminants.

Fig. 3. Web-based search for putative contaminants in IP experiments using the Protein Frequency Library (PFL).

in the data repository (x axis). When proteins are sorted from the highest to the lowest percentage (Figure 3B), the proteins appearing nearest the origin of the graph have the highest probability of being contaminants. The PFL can be applied to analyze data from any MS pull-down experiment, as an additional criterion to evaluate the probability of each protein identified to be a false positive, binding non specifically to the affinity matrix. When applied to SILAC data, it is possible to superimpose the results given by the PFL and the 2D graph, plotting M/L on the x axis and H/M on the y axis, by highlighting on the graph proteins that have a frequency of detection above a threshold value that has to be chosen. The choice of an optimal threshold has to be determined depending on the number of experiments used to generate the PFL and will certainly become lower and lower as new data are added to the repository (Boulon et al., 2010a).

The use of a multidimensional structure, which includes all datasets and associated metadata, allows for possible filtering of the PFL to obtain protein frequencies of detection relevant to each specific set of experimental parameters. Considering all experiments recorded in the database, only those that were performed with the chosen set of experimental parameters are used to generate the PFL, which leads to the generation of a "customized" PFL (Figure 3A). This is of great importance, given that the nature of IP contaminants is highly correlated to the experimental parameters used. For example, contaminants are greatly different according to the bead type chosen, e.g. magnetic or sepharose beads. We have indeed shown that cytoskeleton proteins "stick" to dynamic beads whereas positively charged nuclear proteins are more prone to bind non-specifically to sepharose beads (Trinkle-Mulcahy et al., 2008). Therefore, the PFL can be considered as a dynamic list of "contaminants", which can be filtered for each specific set of experimental parameters. This avoids the need to have a large set of control experiments that exhaustively cover every possible combination of experimental parameters analyzed. The PFL is thus equally applicable to low and high throughput IP experiments.

The use of the PFL is not restricted to the Lamond laboratory. The PFL is now freely accessible online (http://www.peptracker.com/datavisual/) after registration. Figure 3A shows an interface of the PFL that can be used to specify experimental parameters on which the library can be filtered, e.g. organism, cell extract, bead type, etc. All users can therefore select their own experimental parameters and obtain a list of putative contaminants in this specific set of conditions. However, a minimum of 15 independent IP experiments in the experimental count (number of experiments that are taken into account to generate the new customized PFL) might be necessary to provide reliable results. Of note, the PFL is a dynamic tool that is updatable, i.e. the PFL is automatically updated as data from new experiments are added to the data repository, thereby increasing in accuracy. The current PFL is necessarily limited to the experiments performed in the Lamond laboratory. However, it is foreseen that in the future external users will have the ability to upload their own data, and therefore increase the spectrum of experimental parameters available, thereby having a broader impact on the scientific community. From my experience, the PFL is especially helpful in identifying "outsiders", i.e. genuine interaction partners that are of low abundance and/or low affinity, which are otherwise lost among the large, nonspecific background of contaminants and therefore often overlooked in AP-MS studies.

Interestingly, this tool is an example of meta-analysis. Indeed, the PFL is generated through the integration of data from many independent MS IP experiments, performed by independent researchers using various experimental parameters. This process therefore allows for better data mining. One essential point to note is that any analysis relies upon recorded associated metadata, which provide a crucial support to an improved data analysis.

6. Standardization of data analysis and storage

The SILAC IP strategy presented in this chapter can be used in low throughput studies but it can also be scaled up to support large scale surveys of protein interactome dynamics. In any case, the assembly of great interactome maps will require the integration of both low and high throughput protein interaction studies. In fact, small-scale datasets are of great value to protein interaction databases, assuming that they are supported by enough metadata (organism, bait, cell type, treatment etc.), as they often report high resolution analyses that provide details missing in large-scale datasets, such as binding sites and dynamic information, thereby increasing local coverage of large interactome networks (Orchard and Hermjakob, 2011; Sanderson, 2009). The main challenge encountered by the scientific community is not the generation of an increased amount of protein interaction data produced by either low or high throughput studies. In fact, the amount of interaction studies increases in an exponential manner as new technologies emerge, which become increasingly accessible to most international research groups. Instead, major issues reside both in the quality and in the homogeneity of the interaction data generated. Poor quality data that cannot be interpreted or exploited are of low interest for the scientific community. Data that are generated in different groups, using different machines and therefore different file formats and analysis pathways, cannot be accurately compared between each other, thereby leading to the accumulation of independent datasets that cannot be integrated.

As all published data are inherently of variable quality, there is a need to increase the overall reliability of interaction datasets and develop data standards. This will rely on a strict quality control of all data that are uploaded in public repositories. As proposed by Olsen and Mann, the selection of high quality data could result from "social-network like mechanisms", which would calculate confidence scores for each specific result (e.g. each protein-protein interaction) based on the number of times it would be retrieved in various independent studies using different techniques (Olsen and Mann, 2011). This would help eliminate results that are of poor reliability and thereby enhance confidence of protein interaction databases.

Data standardization can probably be considered as one of the main challenges in the field. Currently, interaction data can be found in many different types of format, depending on the vendor and on the analysis pathway. Therefore, creating a common standard data format that could be used by the scientific community would facilitate exchange, comparison and integration of datasets, which is absolutely essential and requires an intense international coordination. Since 2004, a consortium of databases, including BIND, DIP, IntAct, MINT and MIPS, agreed to develop a community standard data model for the representation and exchange of protein interaction data (Hermjakob et al, 2004). These databases were grouped into IMEX (International Molecular interaction EXchange) (Orchard et al., 2007). The standard format called PSI-MI (XML format) was developed by members of the Molecular Interaction (MI) group of the Proteomics Standards Initiative (PSI), which belongs to the Human Proteome Organization (HUPO). Of note, the PSI-MI format cannot handle quantitative MS data yet. In terms of storage of MS data, the PRIDE (PRoteomics IDEntifications) database, hosted at the EBI (European Bioinformatics

Institute), is a centralized, standards compliant, public data repository for MS-based proteomics data that compiles protein and peptide identifications (http://www.ebi.ac.uk/pride).

Along with the standardization of protein interaction data formats, the efficient and reliable recording of metadata is absolutely crucial to better analyze and exploit datasets. Indeed, metadata represent useful information that is required for data mining, comparison and retrospective studies, as it has been shown in the case of the PFL. To address this issue, a HUPO project has led to the development of MIMIx (Minimum Information required for reporting a Molecular Interaction experiment) (Orchard et al., 2007). As described by Orchard et al, MIMIx represents a "compromise" between the vast amount of information that would be necessary to precisely describe and reproduce an interaction experiment, which should be present in any original publication, and the constant load placed on scientists who upload their data into databases. As guidelines, the MIMIx checklist contains several experimental parameters that need to be accurately specified, including the host organism, correct molecule identifiers generated by major databases (Uniprot and RefSeq), detection method etc. In addition, a proper controlled vocabulary should be used (for example, bait/prey), as well as confidence values attributed to the interaction whenever possible (Orchard et al., 2007). These guidelines may (i) help increase the usefulness and the clarity of publications reporting interaction data and (ii) improve systematic recording of protein interaction data in public resources, thereby increasing their access to a wider community.

Finally, efficient protein interaction data analysis relies upon powerful visualization techniques that enable the representation of large and dynamic protein interactome network maps. There is definitely a large demand for this type of tool that will need to be covered by the development of new cutting edge visualization software. Many different tools have been generated already, including the Cytoscape project that has integrated plugins to allow Cytoscape to interact with relational databases, Osprey, which is associated to the BIOGRID database, Genego Metacore and Ingenuity. These tools provide good graphical interfaces to visualize protein interactome networks, although downstream data analysis may also rely on lab oriented tools, such as the PFL.

7. Conclusion

The scientific community has been developing immense efforts to map the human protein interactome network. However the characterization of a static interactome only provides a list of possible interactions, without questioning when these interactions occur, and how they are regulated. It is therefore necessary to focus on a more functional analysis by studying the dynamics of protein interactions. The combination of SILAC-based quantitative proteomics with affinity purification techniques currently provides a reliable strategy to both identify specific protein interaction partners and analyze subtle changes in protein interactions between different conditions. One can envision that the development of new techniques and analysis tools will certainly also favor the use of label-free approaches in the future. However, despite an escalating number of outstanding studies reported in the literature and the increasing performance of technologies, many challenges remain to be faced before a dynamic map of the human interactome can be assembled. In particular, an international coordinated effort to standardize data formats and develop powerful software for data analysis and visualization may allow to efficiently exploit, compare and integrate datasets generated all over the world, thereby resulting in higher reliability and usefulness of protein interaction data. New insights into the human protein interactome dynamics would undoubtedly benefit both basic and clinical sciences, by providing essential information about the function of individual proteins, connections between them and the functional organization of the cell as a whole system. This may rely upon the identification of key protein "hubs", i.e. proteins that are highly connected in interactome networks and may have crucial roles in specific disease pathways. Interestingly, significant correlations have been found between protein interactome maps and disease-associated gene networks, suggesting a potential predictive use of protein interactomes for the identification of non-intuitive disease-related genes and putative drug targets.

8. Acknowledgment

I am very grateful to Yasmeen Ahmad for critical reading of the chapter. I thank Aymeric Bailly for advice and suggestions. I thank the Lamond and Bertrand laboratories for fruitful discussions regarding the development of the strategies described in this chapter. I apologize to those investigators whose studies were not included in this chapter due to space limitations. This work was supported by a Human Frontier Science Program long-term fellowship to the author.

9. References

Barrios-Rodiles, M., K.R. Brown, B. Ozdamar, R. Bose, Z. Liu, R.S. Donovan, F. Shinjo, Y. Liu, J. Dembowy, I.W. Taylor, V. Luga, N. Przulj, M. Robinson, H. Suzuki, Y. Hayashizaki, I. Jurisica, and J.L. Wrana. (2005). High-throughput mapping of a dynamic signaling network in mammalian cells. *Science*. 307:1621-5.

Blagoev, B., I. Kratchmarova, S.E. Ong, M. Nielsen, L.J. Foster, and M. Mann. (2003). A proteomics strategy to elucidate functional protein-protein interactions applied to EGF signaling. *Nat Biotechnol*. 21:315-8.

Boulon, S., Y. Ahmad, L. Trinkle-Mulcahy, C. Verheggen, A. Cobley, P. Gregor, E. Bertrand, M. Whitehorn, and A.I. Lamond. (2010a). Establishment of a protein frequency library and its application in the reliable identification of specific protein interaction partners. *Mol Cell Proteomics*. 9:861-79.

Boulon, S., B. Pradet-Balade, C. Verheggen, D. Molle, S. Boireau, M. Georgieva, K. Azzag, M.C. Robert, Y. Ahmad, H. Neel, A.I. Lamond, and E. Bertrand. (2010b). HSP90 and its R2TP/Prefoldin-like cochaperone are involved in the cytoplasmic assembly of RNA polymerase II. *Mol Cell*. 39:912-24.

Cloutier, P., R. Al-Khoury, M. Lavallee-Adam, D. Faubert, H. Jiang, C. Poitras, A. Bouchard, D. Forget, M. Blanchette, and B. Coulombe. (2009). High-resolution mapping of the protein interaction network for the human transcription machinery and affinity purification of RNA polymerase II-associated complexes. *Methods*. 48:381-6.

Collier, T.S., P. Sarkar, W.L. Franck, B.M. Rao, R.A. Dean, and D.C. Muddiman. (2010). Direct comparison of stable isotope labeling by amino acids in cell culture and spectral counting for quantitative proteomics. *Anal Chem*. 82:8696-702.

Cox, J., and M. Mann. (2008). MaxQuant enables high peptide identification rates, individualized p.p.b.-range mass accuracies and proteome-wide protein quantification. *Nat Biotechnol*. 26:1367-72.

Cox, J., I. Matic, M. Hilger, N. Nagaraj, M. Selbach, J.V. Olsen, and M. Mann. (2009). A practical guide to the MaxQuant computational platform for SILAC-based quantitative proteomics. *Nat Protoc.* 4:698-705.

Cusick, M.E., N. Klitgord, M. Vidal, and D.E. Hill. (2005). Interactome: gateway into systems biology. *Human molecular genetics.* 14 Spec No. 2:R171-81.

Cusick, M.E., H. Yu, A. Smolyar, K. Venkatesan, A.R. Carvunis, N. Simonis, J.F. Rual, H. Borick, P. Braun, M. Dreze, J. Vandenhaute, M. Galli, J. Yazaki, D.E. Hill, J.R. Ecker, F.P. Roth, and M. Vidal. (2009). Literature-curated protein interaction datasets. *Nat Methods.* 6:39-46.

Dengjel, J., L. Jakobsen, and J.S. Andersen. (2010). Organelle proteomics by label-free and SILAC-based protein correlation profiling. *Methods Mol Biol.* 658:255-65.

Ewing, R.M., P. Chu, F. Elisma, H. Li, P. Taylor, S. Climie, L. McBroom-Cerajewski, M.D. Robinson, L. O'Connor, M. Li, R. Taylor, M. Dharsee, Y. Ho, A. Heilbut, L. Moore, S. Zhang, O. Ornatsky, Y.V. Bukhman, M. Ethier, Y. Sheng, J. Vasilescu, M. Abu-Farha, J.P. Lambert, H.S. Duewel, Stewart, II, B. Kuehl, K. Hogue, K. Colwill, K. Gladwish, B. Muskat, R. Kinach, S.L. Adams, M.F. Moran, G.B. Morin, T. Topaloglou, and D. Figeys. (2007). Large-scale mapping of human protein-protein interactions by mass spectrometry. *Mol Syst Biol.* 3:89.

Figeys, D. (2008). Mapping the human protein interactome. *Cell research.* 18:716-24.

Foster, L.J., A. Rudich, I. Talior, N. Patel, X. Huang, L.M. Furtado, P.J. Bilan, M. Mann, and A. Klip. (2006). Insulin-dependent interactions of proteins with GLUT4 revealed through stable isotope labeling by amino acids in cell culture (SILAC). *J Proteome Res.* 5:64-75.

Gavin, A.C., P. Aloy, P. Grandi, R. Krause, M. Boesche, M. Marzioch, C. Rau, L.J. Jensen, S. Bastuck, B. Dumpelfeld, A. Edelmann, M.A. Heurtier, V. Hoffman, C. Hoefert, K. Klein, M. Hudak, A.M. Michon, M. Schelder, M. Schirle, M. Remor, T. Rudi, S. Hooper, A. Bauer, T. Bouwmeester, G. Casari, G. Drewes, G. Neubauer, J.M. Rick, B. Kuster, P. Bork, R.B. Russell, and G. Superti-Furga. (2006). Proteome survey reveals modularity of the yeast cell machinery. *Nature.* 440:631-6.

Ge, Q., V.P. Rao, B.K. Cho, H.N. Eisen, and J. Chen. (2001). Dependence of lymphopenia-induced T cell proliferation on the abundance of peptide/ MHC epitopes and strength of their interaction with T cell receptors. *Proceedings of the National Academy of Sciences of the United States of America.* 98:1728-33.

Ghavidel, A., G. Cagney, and A. Emili. (2005). A skeleton of the human protein interactome. *Cell.* 122:830-2.

Gingras, A.C., M. Gstaiger, B. Raught, and R. Aebersold. (2007). Analysis of protein complexes using mass spectrometry. *Nat Rev Mol Cell Biol.* 8:645-54.

Gygi, S.P., B. Rist, S.A. Gerber, F. Turecek, M.H. Gelb, and R. Aebersold. (1999). Quantitative analysis of complex protein mixtures using isotope-coded affinity tags. *Nat Biotechnol.* 17:994-9.

Hubner, N.C., A.W. Bird, J. Cox, B. Splettstoesser, P. Bandilla, I. Poser, A. Hyman, and M. Mann. (2010). Quantitative proteomics combined with BAC TransgeneOmics reveals in vivo protein interactions. *J Cell Biol.* 189:739-54.

Isserlin, R., R.A. El-Badrawi, and G.D. Bader. (2011). The Biomolecular Interaction Network Database in PSI-MI 2.5. *Database (Oxford).* 2011:baq037.

Jeronimo, C., D. Forget, A. Bouchard, Q. Li, G. Chua, C. Poitras, C. Therien, D. Bergeron, S. Bourassa, J. Greenblatt, B. Chabot, G.G. Poirier, T.R. Hughes, M. Blanchette, D.H. Price, and B. Coulombe. (2007). Systematic analysis of the protein interaction network for the human transcription machinery reveals the identity of the 7SK capping enzyme. *Mol Cell*. 27:262-74.

Kaake, R.M., T. Milenkovic, N. Przulj, P. Kaiser, and L. Huang. (2010). Characterization of cell cycle specific protein interaction networks of the yeast 26S proteasome complex by the QTAX strategy. *J Proteome Res*. 9:2016-29.

Kocher, T., and G. Superti-Furga. (2007). Mass spectrometry-based functional proteomics: from molecular machines to protein networks. *Nat Methods*. 4:807-15.

Kohn, D., G. Murrell, J. Parker, and M. Whitehorn. (2005). What Henslow taught Darwin. *Nature*. 436:643-5.

Krogan, N.J., G. Cagney, H. Yu, G. Zhong, X. Guo, A. Ignatchenko, J. Li, S. Pu, N. Datta, A.P. Tikuisis, T. Punna, J.M. Peregrin-Alvarez, M. Shales, X. Zhang, M. Davey, M.D. Robinson, A. Paccanaro, J.E. Bray, A. Sheung, B. Beattie, D.P. Richards, V. Canadien, A. Lalev, F. Mena, P. Wong, A. Starostine, M.M. Canete, J. Vlasblom, S. Wu, C. Orsi, S.R. Collins, S. Chandran, R. Haw, J.J. Rilstone, K. Gandi, N.J. Thompson, G. Musso, P. St Onge, S. Ghanny, M.H. Lam, G. Butland, A.M. Altaf-Ul, S. Kanaya, A. Shilatifard, E. O'Shea, J.S. Weissman, C.J. Ingles, T.R. Hughes, J. Parkinson, M. Gerstein, S.J. Wodak, A. Emili, and J.F. Greenblatt. (2006). Global landscape of protein complexes in the yeast Saccharomyces cerevisiae. *Nature*. 440:637-43.

Lambert, J.P., L. Mitchell, A. Rudner, K. Baetz, and D. Figeys. (2009). A novel proteomics approach for the discovery of chromatin-associated protein networks. *Mol Cell Proteomics*. 8:870-82.

Lemmens, I., S. Lievens, and J. Tavernier. (2010). Strategies towards high-quality binary protein interactome maps. *J Proteomics*. 73:1415-20.

Lim, J., T. Hao, C. Shaw, A.J. Patel, G. Szabo, J.F. Rual, C.J. Fisk, N. Li, A. Smolyar, D.E. Hill, A.L. Barabasi, M. Vidal, and H.Y. Zoghbi. (2006). A protein-protein interaction network for human inherited ataxias and disorders of Purkinje cell degeneration. *Cell*. 125:801-14.

Mann, M. (2006). Functional and quantitative proteomics using SILAC. *Nat Rev Mol Cell Biol*. 7:952-8.

Nguyen, V.T., F. Giannoni, M.F. Dubois, S.J. Seo, M. Vigneron, C. Kedinger, and O. Bensaude. (1996). In vivo degradation of RNA polymerase II largest subunit triggered by alpha-amanitin. *Nucleic Acids Res*. 24:2924-9.

Olsen, J.V., and M. Mann. (2011). Effective representation and storage of mass spectrometry-based proteomic data sets for the scientific community. *Sci Signal*. 4:pe7.

Ong, S.E., B. Blagoev, I. Kratchmarova, D.B. Kristensen, H. Steen, A. Pandey, and M. Mann. (2002). Stable isotope labeling by amino acids in cell culture, SILAC, as a simple and accurate approach to expression proteomics. *Mol Cell Proteomics*. 1:376-86.

Orchard, S., and H. Hermjakob. (2011). Data standardization by the HUPO-PSI: how has the community benefitted? *Methods Mol Biol*. 696:149-60.

Orchard, S., L. Salwinski, S. Kerrien, L. Montecchi-Palazzi, M. Oesterheld, V. Stumpflen, A. Ceol, A. Chatr-aryamontri, J. Armstrong, P. Woollard, J.J. Salama, S. Moore, J. Wojcik, G.D. Bader, M. Vidal, M.E. Cusick, M. Gerstein, A.C. Gavin, G. Superti-

Furga, J. Greenblatt, J. Bader, P. Uetz, M. Tyers, P. Legrain, S. Fields, N. Mulder, M. Gilson, M. Niepmann, L. Burgoon, J. De Las Rivas, C. Prieto, V.M. Perreau, C. Hogue, H.W. Mewes, R. Apweiler, I. Xenarios, D. Eisenberg, G. Cesareni, and H. Hermjakob. (2007). The minimum information required for reporting a molecular interaction experiment (MIMIx). *Nat Biotechnol.* 25:894-8.

Parrish, J.R., K.D. Gulyas, and R.L. Finley, Jr. (2006). Yeast two-hybrid contributions to interactome mapping. *Curr Opin Biotechnol.* 17:387-93.

Poser, I., M. Sarov, J.R. Hutchins, J.K. Heriche, Y. Toyoda, A. Pozniakovsky, D. Weigl, A. Nitzsche, B. Hegemann, A.W. Bird, L. Pelletier, R. Kittler, S. Hua, R. Naumann, M. Augsburg, M.M. Sykora, H. Hofemeister, Y. Zhang, K. Nasmyth, K.P. White, S. Dietzel, K. Mechtler, R. Durbin, A.F. Stewart, J.M. Peters, F. Buchholz, and A.A. Hyman. (2008). BAC TransgeneOmics: a high-throughput method for exploration of protein function in mammals. *Nat Methods.* 5:409-15.

Ranish, J.A., E.C. Yi, D.M. Leslie, S.O. Purvine, D.R. Goodlett, J. Eng, and R. Aebersold. (2003). The study of macromolecular complexes by quantitative proteomics. *Nat Genet.* 33:349-55.

Reguly, T., A. Breitkreutz, L. Boucher, B.J. Breitkreutz, G.C. Hon, C.L. Myers, A. Parsons, H. Friesen, R. Oughtred, A. Tong, C. Stark, Y. Ho, D. Botstein, B. Andrews, C. Boone, O.G. Troyanskya, T. Ideker, K. Dolinski, N.N. Batada, and M. Tyers. (2006). Comprehensive curation and analysis of global interaction networks in Saccharomyces cerevisiae. *Journal of biology.* 5:11.

Rigaut, G., A. Shevchenko, B. Rutz, M. Wilm, M. Mann, and B. Seraphin. (1999). A generic protein purification method for protein complex characterization and proteome exploration. *Nat Biotechnol.* 17:1030-2.

Rothbauer, U., K. Zolghadr, S. Muyldermans, A. Schepers, M.C. Cardoso, and H. Leonhardt. (2008). A versatile nanotrap for biochemical and functional studies with fluorescent fusion proteins. *Mol Cell Proteomics.* 7:282-9.

Sanderson, C.M. (2009). The Cartographers toolbox: building bigger and better human protein interaction networks. *Brief Funct Genomic Proteomic.* 8:1-11.

Tarassov, K., V. Messier, C.R. Landry, S. Radinovic, M.M. Serna Molina, I. Shames, Y. Malitskaya, J. Vogel, H. Bussey, and S.W. Michnick. (2008). An in vivo map of the yeast protein interactome. *Science.* 320:1465-70.

ten Have, S., S. Boulon, Y. Ahmad, and A.I. Lamond. (2011). Mass spectrometry-based immuno-precipitation proteomics - the user's guide. *Proteomics.* 11:1153-9.

Tong, A.H., G. Lesage, G.D. Bader, H. Ding, H. Xu, X. Xin, J. Young, G.F. Berriz, R.L. Brost, M. Chang, Y. Chen, X. Cheng, G. Chua, H. Friesen, D.S. Goldberg, J. Haynes, C. Humphries, G. He, S. Hussein, L. Ke, N. Krogan, Z. Li, J.N. Levinson, H. Lu, P. Menard, C. Munyana, A.B. Parsons, O. Ryan, R. Tonikian, T. Roberts, A.M. Sdicu, J. Shapiro, B. Sheikh, B. Suter, S.L. Wong, L.V. Zhang, H. Zhu, C.G. Burd, S. Munro, C. Sander, J. Rine, J. Greenblatt, M. Peter, A. Bretscher, G. Bell, F.P. Roth, G.W. Brown, B. Andrews, H. Bussey, and C. Boone. (2004). Global mapping of the yeast genetic interaction network. *Science.* 303:808-13.

Trinkle-Mulcahy, L., S. Boulon, Y.W. Lam, R. Urcia, F.M. Boisvert, F. Vandermoere, N.A. Morrice, S. Swift, U. Rothbauer, H. Leonhardt, and A. Lamond. (2008). Identifying specific protein interaction partners using quantitative mass spectrometry and bead proteomes. *J Cell Biol.* 183:223-39.

Uhlen, M., P. Oksvold, L. Fagerberg, E. Lundberg, K. Jonasson, M. Forsberg, M. Zwahlen, C. Kampf, K. Wester, S. Hober, H. Wernerus, L. Bjorling, and F. Ponten. (2010). Towards a knowledge-based Human Protein Atlas. *Nat Biotechnol.* 28:1248-50.

Vasilescu, J., and D. Figeys. (2006). Mapping protein-protein interactions by mass spectrometry. *Curr Opin Biotechnol.* 17:394-9.

Vermeulen, M., N.C. Hubner, and M. Mann. (2008). High confidence determination of specific protein-protein interactions using quantitative mass spectrometry. *Curr Opin Biotechnol.* 19:331-7.

Vidal, M. (2005). Interactome modeling. *FEBS Lett.* 579:1834-8.

Vidal, M., M.E. Cusick, and A.L. Barabasi. (2011). Interactome networks and human disease. *Cell.* 144:986-98.

Wepf, A., T. Glatter, A. Schmidt, R. Aebersold, and M. Gstaiger. (2009). Quantitative interaction proteomics using mass spectrometry. *Nat Methods.* 6:203-5.

Yao, X., A. Freas, J. Ramirez, P.A. Demirev, and C. Fenselau. (2001). Proteolytic 18O labeling for comparative proteomics: model studies with two serotypes of adenovirus. *Anal Chem.* 73:2456-65.

15

Proteomics Analysis of Kinetically Stable Proteins

Ke Xia, Marta Manning, Songjie Zhang and Wilfredo Colón
Department of Chemistry and Chemical Biology
Center for Biotechnology and Interdisciplinary Studies,
Rensselaer Polytechnic Institute, Troy, New York
USA

1. Introduction

The term "kinetic stability" (KS) is sometimes used to describe proteins that are conformationally trapped by the presence of an unusually high-energy unfolding barrier that considerably decreases their unfolding rate under various conditions. This barrier allows kinetically stable proteins (KSPs) to maintain their fold and activity over longer periods, even in inhospitable environments. KS is likely to play important biological roles, such as the regulation of protein turnover, protection from proteolytic degradation, and blocking access to aggregation-prone conformations. However, the chemical-physical basis and the diversity of biological-pathological roles of protein KS remain poorly understood, in part because for many years the study of KS was limited to individual pure proteins, and involved spectroscopic instrumentation that was not available to most researchers. In this chapter, we will review the discovery of a correlation between a protein's KS and its resistance to the detergent sodium dodecyl sulfate (SDS), and the subsequent development of a diagonal two-dimensional (D2D) SDS-PAGE assay and capillary electrophoresis approaches to identify the proteome of KSPs in any cell or organism.

1.1 Thermodynamics vs kinetic stability

The concept of kinetic stability (KS) as an alternative explanation for protein stability, independent from thermodynamic stability, was introduced in the early 90's (Fig. 1) (Baker & Agard, 1994; Baker, Sohl, & Agard, 1992). KS is conveniently explained by illustrating the unfolding process as an equilibrium reaction between the native folded state (N) and the unfolded state (U), separated by a transition state (TS) (Fig. 1). Since the height of the TS free energy determines the rate of folding and unfolding, the unusually high unfolding free energy barrier of a KSP results in a very slow unfolding rate that practically traps the protein in its native state (Fig 1). It has been suggested that the existence of a high energy barrier separating the folded and unfolded states is an evolutionary feature to preserve protein activity in the severe conditions they might encounter in nature (Cunningham, Jaswal, Sohl, & Agard, 1999). This is consistent with the observation that thermodynamic stability by itself does not fully protect proteins from irreversible denaturation and aggregation arising from denatured conformations that fleetingly form under physiological

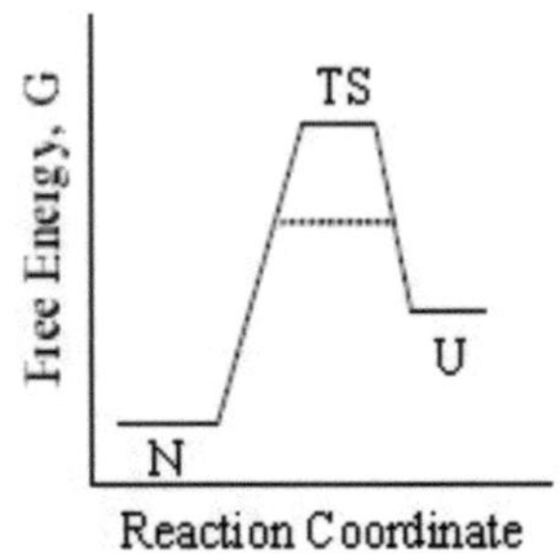

Fig. 1. Free energy diagram illustrating the higher unfolding energy barrier for a kinetically stable protein under native conditions, as compared to that of a normal protein (represented by the dash line). The labels represent the native (N) state, unfolded (U) state, and transition state (TS). Reprinted with permission from (Manning, M., & Colón, W. (2004). Structural basis of protein kinetic stability: Resistance to sodium dodecyl sulfate suggests a central role for rigidity and a bias towards beta sheet structure. *Biochemistry, 43*, 11248-11254). Copyright (2004) American Chemical Society.

conditions (Plaza del Pino, Ibarra-Molero, & Sanchez-Ruiz, 2000). Thus, the presence of an unfolding TS with high energy may protect susceptible proteins against harmful conformations. In summary, KSPs are basically slow-unfolding proteins that are more resistant to aggregation and degradation.

Protein misfolding diseases (PMD) (Johnson, et al., 2005), include some of the most common human ailments, including Alzheimer's, Parkinson's, Type II diabetes, and cancer (Dobson, 2001; Stefani & Dobson, 2003). There is strong evidence that loss or gain of SDS-resistance (correlating to KS), facilitated by mutation, protein damage, or a compromised quality-control system, is linked to some PMD. Aging appears to also play a role, consistent with the late-onset of most PMD. The loss of KS might represent a hazard for the organism, especially for older individuals who have less efficient protein quality control systems (Koga, Kaushik, & Cuervo, 2010; Luce, Weil, & Osiewacz, 2010). In familial amyloid polyneuropathy it is known that missense mutations can compromise the KS of transthyretin (TTR), facilitating tetrameric TTR dissociation and subsequent aggregation into amyloid fibrils (Saraiva, 1995). Remarkably, native state kinetic stabilization of TTR via several strategies (Hammarstrom, Schneider, & Kelly, 2001) can restore the KS of mutated TTR, and is emerging as a therapeutic strategy for TTR amyloidosis (Johnson, et al., 2005).

It is also plausible that some diseases might be associated with protein misfolding into a toxic species with high KS, since such a species would be more difficult to degrade. A striking example is the prion protein, which is linked to various genetic and transmissible diseases (Horwich & Weissman, 1997). The native prion protein lacks KS (Hornemann & Glockshuber, 1998), but the misfolded infectious prion has high KS (Prusiner, Groth, Serban, Stahl, & Gabizon, 1993), thus explaining why it survives the GI track in transmissible prion diseases. Furthermore, the abnormal *in vivo* presence of SDS-resistant (i.e. kinetically stable – see section 3 for SDS-KS correlation) and potentially toxic species is a feature of various PMD, including Alzheimer's disease and Parkinson's disease (Cappai, et al., 2005; Enya, et al., 1999; Funato, Enya, Yoshimura, Morishima-Kawashima, & Ihara, 1999; Haass & Selkoe,

2007; Kawarabayashi, et al., 2004; Lee, et al., 2011; Lesne, et al., 2006; McLean, et al., 1999; Podlisny, et al., 1995; Roher, et al., 1996).

1.2 SDS as a probe for protein kinetic stability

SDS-PAGE was introduced in the 1960s as a method for separating proteins (Shapiro, Vinuela, & Maizel, 1967). Currently SDS-PAGE is perhaps the most fundamental technique in protein biochemistry. The interaction between a protein and SDS is complex and involves nonpolar and electrostatic interactions. In spite of the ubiquitous use of SDS, it is still poorly understood how it denatures proteins when present at above its critical micelle concentration (CMC) (Otzen, 2002). It has been suggested that at concentrations less than 100 mM, (CMC of SDS is ~7 mM in water (Reynolds, Herbert, Polet, & Steinhardt, 1967)) SDS denatures proteins by a mechanism involving ligand-binding-type unfolding kinetics. Furthermore, it was shown that SDS does not alter the transition state energy for protein unfolding (Otzen, 2002), thereby implying that SDS's interaction with a protein's surface has minor effect on the structure and free energy of its native state (Otzcn, 2002).

In 2004, we demonstrated a correlation between KS and the resistance of proteins to denaturation by SDS, resulting in a simple assay that is very effective for probing the KS of proteins (Manning & Colón, 2004). The initial step of our study involved the identification of SDS resistance from a group of 33 proteins. SDS resistance was assayed by comparing the migration on a gel of boiled and unboiled protein samples containing SDS (Fig. 2). Proteins that migrated to the same location on the gel regardless of whether the sample was boiled were classified as not being stable to SDS (Fig. 2B). Those proteins that exhibited a slower migration when the sample was not boiled were classified as being resistant to SDS-induced denaturation (Fig. 2A). The slower migration is a sign of less SDS binding and thereby of a lesser overall negative charge of the SDS-protein complex compared to the fully SDS-bound proteins. Of the proteins tested, eight were found or confirmed to exhibit resistance to SDS, including superoxide dismutase (SOD), streptavidin (SVD), TTR, P22 tailspike protein (TSP), chymopapain (CPAP), papain (PAP), avidin (AVD), and serum amyloid P (SAP) (Fig. 2A).

To probe the KS of our SDS-resistant proteins, we used fluorescence spectroscopy and demonstrated their slow unfolding rates even in 6.6 M guanidine hydrochloride (GuHCl) at 20 °C. To gather further evidence of the KS exhibited by these proteins under native conditions, their unfolding rate constants in the absence of the denaturant were obtained by measuring the unfolding rate at different GuHCl concentrations and extrapolating to 0 M. The native state unfolding rate constants for TTR (Lai, McCulloch, Lashuel, & Kelly, 1997) and SVD (Kurzban, Bayer, Wilchek, & Horowitz, 1991) were obtained from the literature. The unfolding rate in the absence of denaturants for all of the SDS-resistant proteins was found to be very slow (Table 1), with protein half- lives ranging from 79 days to 270 years.

The observation that all of the SDS-resistant proteins were also kinetically stable, suggested that SDS resistance might be caused by KS. To further test the correlation between KS and SDS resistance, we selected a group of six proteins that did not exhibit resistance to SDS and analyzed their unfolding behavior in varying concentrations of GuHCl. The group was chosen to represent a variety of structural characteristics. At 6.6 M, the unfolding of these proteins was too fast to detect with a standard fluorescence spectrophotometer. The lack of

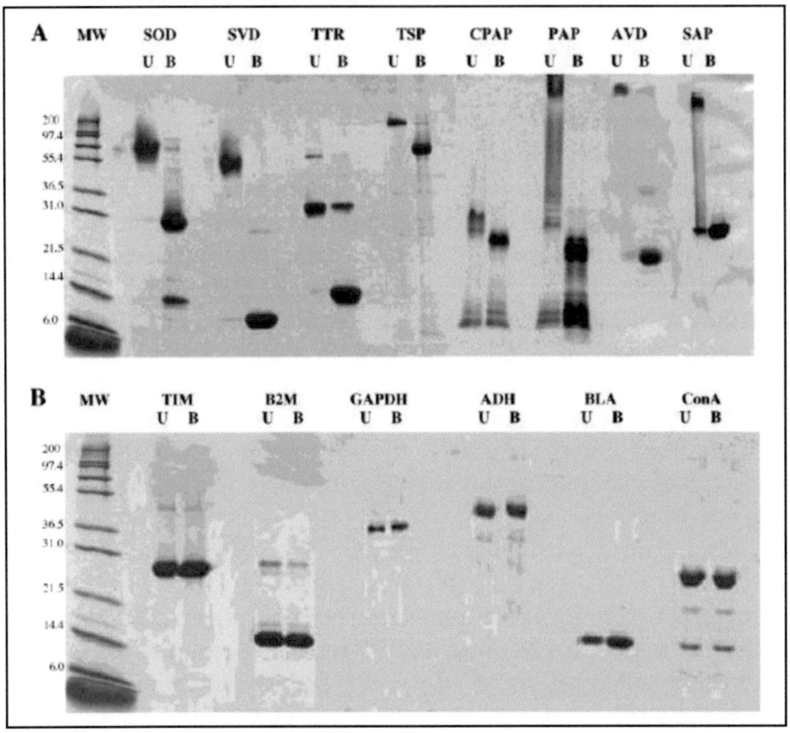

Fig. 2. SDS-PAGE assay to identify SDS-resistant proteins (Manning & Colón, 2004). We tested the proteins (A) papain (PAP), chymopapain (CPAP), avidin (AVD), and superoxide dismutase (SOD), streptavidin (SVD), serum amyloid P (SAP), transthyretin (TTR), Salmonella phage P22 tailspike protein (TSP) and the non-SDS-resistant control group (B) triosephosphate isomerase (TIM), glyceraldehyde 3-phosphate dehydrogenase (GAPDH), concanavalin (ConA), β 2-microglobulin (β2M), bovine alpha-lactalbumin (BLA) and yeast alcohol dehydrogenase(ADH). Identical protein samples were either unheated (U) or boiled (B) for 10 min immediately prior to loading onto the gel. Reprinted with permission from (Manning, M., & Colón, W. (2004). Structural basis of protein kinetic stability: Resistance to sodium dodecyl sulfate suggests a central role for rigidity and a bias towards beta sheet structure. *Biochemistry, 43*, 11248-11254). Copyright (2004) American Chemical Society.

KS exhibited by these proteins was confirmed by their native unfolding half-lives, which ranged from 14 min to 19 h (Table 1). The above results support the existence of a correlation between KS and resistance to SDS-induced denaturation. Therefore, SDS-PAGE could serve as a simple method for identifying and selecting KSPs. This method has the advantage that proteins can be easily tested for kinetic stability without having to carry out unfolding experiments. Also, only microgram amounts of sample are needed, and the method is potentially suitable for identifying KSPs present in cell extracts without need for purification. From an application perspective, this assay has the potential of being adaptable for high-throughput applications to enhance the KS of proteins of interest. This could lead to proteins with greater shelf life and/or decreased tendency to aggregate, consistent with the suggestion that the deterioration of an energy barrier between native and pathogenic states as a result of mutation might be a key factor in the misfolding and aggregation of some proteins linked to amyloid diseases (4, 18).

SDS-Resistant			Not SDS-Resistant		
proteins	k_{unf} (s^{-1}) in 0 M GdnHCl	unfolding half-life	proteins	k_{unf} (s^{-1}) in 0 M GdnHCl	unfolding half-life
AVD	$8.1E^{-11}$	270 years	ADH	$8.1E^{-5}$	19 hours
TTR	$9.0E^{-11}$	244 years	TIM	$9.0E^{-5}$	15 hours
PAP	$1.3E^{-10}$	165 years	BLA	$1.3E^{-5}$	12 hours
TSP	$1.6E^{-9}$	13 years	β2M	$1.6E^{-4}$	24 min
SOD	$6.0E^{-9}$	3.7 years	ConA	$6.0E^{-4}$	22 min
CPAP	$8.8E^{-9}$	2.5 years	GAPDH	$8.8E^{-4}$	14 min
SVD	$2.5E^{-8}$	318 days			
SAP	$1.0E^{-7}$	79 days			

Table 1. Unfolding rate constant and half-lives of proteins resistant and not resistant to SDS. Adapted with permission from (Manning, M., & Colón, W. (2004). Structural basis of protein kinetic stability: Resistance to sodium dodecyl sulfate suggests a central role for rigidity and a bias towards beta sheet structure. *Biochemistry, 43,* 11248-11254). Copyright (2004) American Chemical Society.

2. Diagonal two-dimensional (D2D) SDS-PAGE: A proteomics tool for identifying kinetically stable proteins

In the last section, we discussed the correlation between KS and the resistance of proteins to denaturation by SDS, resulting in a simple SDS-PAGE-based assay that is very effective for identifying proteins that have high KS as demonstrated by their resistance to SDS. It is a simple and fast method that could be applied in any lab to test whether a protein is kinetically stable or not. However, the resolution of 1D SDS-PAGE is not sufficient for proteomic research and the KSP bands in a protein mixture are hard to differentiate from non-KSP bands. Therefore, we combined the non-heating and heating SDS-PAGE steps within a single experiment, resulting in a method that we named diagonal two-dimensional (D2D) SDS-PAGE, which combined with mass spectrometry allows the identification of potential KSPs present in complex mixtures. This D2D SDS-PAGE method is similar to previous ones used for the detection of protease susceptibility (Nestler & Doseff, 1997) and to identify stable oligomeric protein complexes in the inner membrane of *E. coli* (Spelbrink, Kolkman, Slijper, Killian, & de Kruijff, 2005). We applied D2D SDS-PAGE to the cell lysate of *E. coli*, and upon proteomics analysis we identified many putative KSPs, thereby giving some insight about potential structural and functional biases in favor and against KS (Xia, et al., 2007).

2.1 Diagonal two-dimensional (D2D) SDS-PAGE method for identifying kinetically stable proteins

In the first step of our D2D SDS-PAGE assay, the unheated sample containing a mixture of proteins is analyzed in the first dimension by SDS-PAGE (Fig. 3A). The gel lane containing the proteins is then cut out and the gel strip is incubated in SDS-PAGE sample buffer and boiled for 10 min (Fig. 3B) before placing above a larger gel for the second dimension run (Fig. 3C). Most proteins will be denatured by SDS even without heating, and thus will migrate the same distance in both gel dimensions, resulting in a diagonal line of spots with a

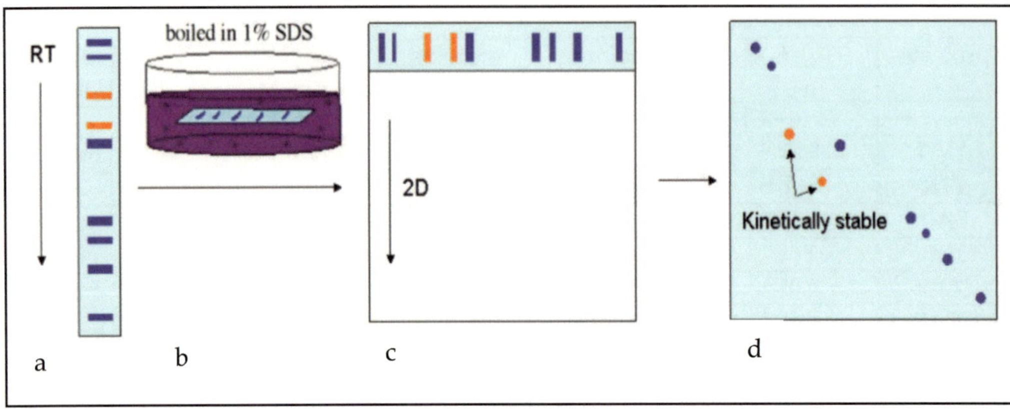

Fig. 3. D2D SDS-PAGE assay for detecting KSPs ((Xia, et al., 2007; Xia, Zhang, Solina, Barquera, & Colon, 2010). After 1D SDS-PAGE, the gel strip is excised and incubated in boiling SDS. The strip is then placed on top of a new gel, followed by a 2nd dimension separation. A diagonal pattern results from the equal migration of non-KSPs in both dimensions. KSPs migrate less in the 1st dimension due to their resistance to SDS, and therefore show up left of the gel diagonal. Reprinted with permission from (Xia, K., Zhang, S., Solina, B. A., Barquera, B., & Colon, W. (2010). Do prokaryotes have more kinetically stable proteins than eukaryotic organisms? *Biochemistry, 49*(34), 7239-7241). Copyright (2010) American Chemical Society.

negative slope across the gel (Fig. 3D). However, SDS-resistant proteins will travel a shorter distance in the first dimension gel and therefore, after the second dimension SDS-PAGE they will end up migrating to a region below the gel diagonal, separated from the bulk proteins. It should be noted that the distance of the spots from the diagonal should not correlate with KS, but rather will depend on several factors, including the oligomeric state, the MW, and the overall charge of the protein.

2.2 D2D SDS-PAGE validation: Identifying the proteome of kinetically stable proteins in *E. coli*

To confirm whether the D2D SDS-PAGE method could detect KSPs from complex mixtures, we applied it to analyze the cell lysate of *E. coli* (Xia, et al., 2007). The D2D SDS-PAGE gel showed the anticipated diagonal pattern arising from the same migration in both dimensions of non-SDS-resistant (i.e. non-kinetically stable) proteins (Fig. 4). Nevertheless, many spots were present below the gel diagonal, and these represent the most abundant KSPs present in the cell lysate of *E. coli*. To identify these proteins, each spot was cut out and subjected to trypsin digestion and proteomics analysis using LC-MS/MS. The resulting MS/MS data were searched against the *E. coli* protein database using the algorithm Mascot 2.1 (Perkins, Pappin, Creasy, & Cottrell, 1999). As reasonable criteria for the correct identification of proteins, we solely included proteins that had at least two peptide hits with a p-value of <0.05, leading to the identification of 50 non-redundant proteins (Table 2). *E. coli* expresses ~884 water-soluble proteins that are observable on a typical 2D gel (Sigdel, Cilliers, Gursahaney, & Crowder, 2004), and therefore, our results indicate that most *E. coli* proteins lack KS. Interestingly, Fig. 4 shows a few unexpected bands and some smearing above the gel diagonal. By adding DTT just

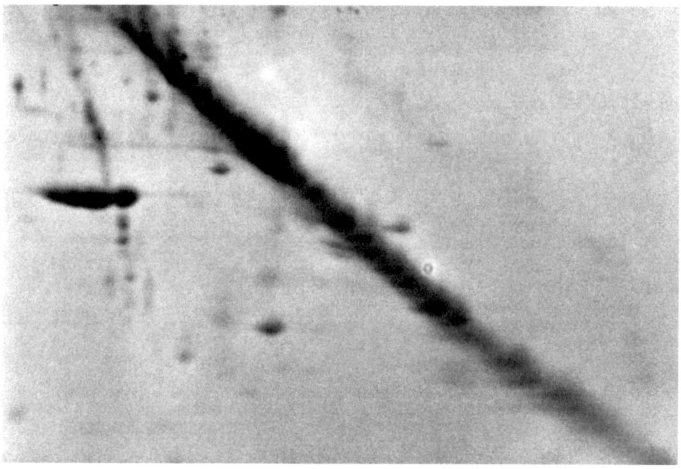

Fig. 4. Analysis of the cellular lysate of *E. coli* by D2D SDS-PAGE (Xia, et al., 2007). The *E. coli* cell lysate was diluted 5-fold and incubated for 5 min in SDS sample buffer (pH 6.8) to a final concentration of 45 mM Tris-HCl, 1% SDS, 10% glycerol, and 0.01% bromophenol blue). A 250 μL aliquot of the lysate solution was loaded without prior heating onto a well of a 12% acrylamide gel (16cm x 14cm x 3mm). The visible spots to the left of the gel diagonal represent the soluble putative KSPs in *E. coli.*

GenBank identifier	name	# . of res	2°	4°	function
15804817	inorganic pyrophosphatase	176	α/β	6	To catalyze the reaction Diphosphate + H_2O = 2 phosphate
15802070	superoxide dismutase, iron(ii)	193	α/β	2	To destroy toxic radicals which are normally produced within the cells
9507572	chloramphenycol acetyltransferase	219	α/β	3	To covalentely attach an acetyl group from acetyl coA to the chloramphenicol molecule.
443293	triosephosphate isomerase Tim	255	α/β	2	To catalyze the reversible interconversion of triose phosphates isomers dihydroxyacetone phosphate and D-glyceraldehyde 3-phosphate.
16131680	uridine phosphorylase	253	α/β	6	To catalyze the reversible phosphorylytic cleavage of uridine and deoxyuridine to uracil and ribose- or deoxyribose-1-phosphate.
14488510	ompf Porin	340	β	3	To form passive diffusion pores to allow low molecular weight hydrophilic materials across the outer membrane
51247607	glycerophosphoryl diester phosphodiesterase	336	α/β	2	To hydrolyze deacylated phospholipids to form glycerol-3-phosphate and the corresponding alcohols
75196280	periplasmic glycerophosphoryl diester phosphodiesterase	371	α/β	2	To hydrolyze deacylated phospholipids to form glycerol-3-phosphate and the corresponding alcohols
11514297	elongation factor, Tu	393	β + α/β	1	To mediate the entry of the aminoacyl tRNA into a free site of the ribosome.

GenBank identifier	name	# . of res	2°	4°	function
15804731	aspartate ammonia-lyase (aspartase)	493	α	4	To catalyze the reversible deamination of the amino acid L-aspartic acid to produce fumaric acid and ammonium ion.
6435772	outer membrane protein Ompx	148	β	3	To neutralize host defense mechanisms
15803823	30S ribosomal protein S4	206	α/β	21	one of the six primary binding proteins to 16s rRNA
16131215	bacterioferritin	158	α	24	iron storage and detoxification
112489962	modulator of drug activity B (MdaB)	204	α/β	2	The MdaB-QuMo operon might protect the cell primarily from more complex quinone compounds.
13786833	pyridoxine 5'-phosphate synthase	242	α/β	8	To catalyze the terminal step in *E. coli* de novo vitamin B6 biosynthesis
2914323	enoyl reductase with bound NAD and benzo-diazaborine	261	α/β	4	To reduce unsaturated acyl carrier protein by reduced pyridine nucleotide
6730179	reduced thioredoxin reductase	320	α/β	2	To reduce thioredoxin
26248038	glyceraldehyde-3-phosphate dehydrogenase	334	α/β	4	Involved in the glycolysis pathway
1421289	S-adenosylmethionine synthetase	383	α/β	2	To catalyze the formation of S-adenosylmethionine
4557950	beta-ketoacyl-acp synthase II	412	α/β	2	To catalyze the condensation reaction of fatty acid synthesis by the addition to an acyl acceptor of two carbons from malonyl-ACP
14278152	2-amino-3-ketobutyrate CoA ligase	401	α/β	2	To catalyze the reaction of Acetyl-CoA + glycine = CoA + 2-amino-3-oxobutanoate.
15804373	transcription termination factor Rho	419	α+β	12	To facilitate transcription termination.
1310928	maltoporin lamb	421	β	3	Involved in the transportation of maltose and maltodextrin
9256952	outer membrane protein Tolc	428	α+β	3	Outer membrane channel and required for proper expression of outer membrane protein genes.
15804852	leucyl aminopeptidase	503	α/β	6	To catalyze the removal of the N-terminal amino acid from most L-peptides
42146	IF2, IF1, and tRNA of E. coli 70S initiation complex	733	α/β	2	Function by binding to the small subunit of the ribosome during the initiation of protein synthesis
91213467	glycerol kinase	537	β	8	To catalyze the formation of glycerol 3-phosphate from ATP and glycerol.
146264	xanthine guanine phosphoribosyltransf erase	152	α/β	4	Involved in purine salvage pathway
30065622	purine nucleoside phosphorylase	239	α/β	6	To cleave guanosine or inosine to respective bases and sugar-1- phosphate molecules
223571	protein L12	272	α/β	31	Might be the binding site for factors involved in protein synthesis and be important for accurate translation
75175990	transaldolase	317	α/β	1	To play a role in the balance of metabolites in the pentosephosphate pathway

GenBank identifier	name	# . of res	2°	4°	function
42810	the *E. coli* RNA polymerase alpha subunit amino-terminal domain	329	α+β	2	To catalyze the transcription of DNA into RNA with the four ribonucleoside triphosphates as substrates.
16132149	isoaspartyl dipeptidase	390	α/β	8	To break down β linkages, which are the peptide bonds between the side chain of an aspartate residue and another amino acid
38491472	GroEL	548	α/β	7	A chaperone required for the proper folding of many proteins in prokaryotes, chloroplasts, and mitochondria.
75233972	L-fucose isomerase and related proteins	591	α/β	3	To convert the aldose L-fucose into the corresponding ketose L-fuculose using Mn2+ as a cofactor.
110643069	fructose-bisphosphate aldolase class II	421	α/β	2	To brake down fructose 1,6-bisphosphate into glyceraldehyde 3-phosphate and dihydroxyacetone phosphate (DHAP)
15800320	alkyl hydroperoxide reductase, C22 subunit;	187	α/β	10	To reduce organic hydroperoxides in its reduced dithiol form
15804455	glutamine synthetase	469	α/β	12	To catalyze the condensation of glutamate and ammonia to form glutamine.
15799927	phosphoheptose isomerase	192	α/β	4	To catalyze the isomerization of sedoheptulose 7-phosphate in D-glycero-D-manno-heptose 7-phosphate
15804539	catalase; hydroperoxidase HPI(I)	726	α	2	To exhibit both catalase and broad-spectrum peroxidase activities
15799800	dihydrolipoamide dehydrogenase	474	α/β	2	To degrade lipoamide and produce dihydrolipoamide
42377	unnamed protein	549	α+β	2	Unknown
15803853	elongation factor EF-2	704	β + α/β	1	To promote the GTP-dependent translocation of the nascent protein chain from the A site to the P site of the ribosome
15799862	(3R)-hydroxymyristoyl ACP dehydratase	151	α+β	6	Involved in saturated fatty acid biosynthesis
75237743	predicted GTPases	490			Unknown
148247	proline dipeptidase	443			To hydrolyze Xaa-Pro dipeptides and also acts on aminoacylhydroxyproline analogs
15802566	galactitol-1-phosphate dehydrogenase	346			To react with NAD^+ to produce L-Tagatose 6-phosphate And NADH and H^+
38704050	fructose-bisphosphate aldolase	350			Involved in the glycolysis pathway
15801691	putative receptor	353			Unknown
91211384	hypothetical protein UTI89_C2371	374			Unknown

The various columns describe the GenBank identifier number, the name of the protein, the number of residues per subunit, the secondary (2°) structure content, the quaternary (4°) structure content, and the main known function of the protein. This table is adapted from Xia, et al., 2007.

Table 2. Nonredundant subset of SDS-resistant/KSPs proteins in *E. Coli.*

before the heating step, we have confirmed that these bands result from disulfide bond formation during heating. (Xia, et al., 2007)

2.3 Mass spectrometry and identification of KSPs

Protein off-diagonal spots were excised from the gel, washed, reduced, alkylated, and digested in-gel with trypsin overnight. The peptide mixture was extracted, dried and dissolved in 10 μl of 5% formic acid. A Q-TOF 2 mass spectrometer (Waters, Milford, MA) equipped with the CapLC system was used for the LC-MS/MS experiments. We used a trap column of 180 μm ID × 50 mm packed with 10 μm R2 resin (Applied Biosystems, Foster City, CA) connected in series with a 100 μm ID × 160 mm capillary column packed with 5 μm C18 particles. 10μl of the peptide mixture was injected into the trap column at speed of 12 μl/min and desalted for 6 min before being eluted to the capillary column. The peptides were then eluted with final flow rate 250 nl/min by a series of mobile phase B gradients (5 to 10% B in 4 min, 10 to 30% B in 61 min, 30 to 85% B in 5 min, 85 to 85% B in 5 min). Mobile phase A consisted of 0.1% formic acid, 3% acetonitrile and 0.01% TFA, whereas mobile phase B consisted of 0.075% formic acid, 0.0075% TFA in 98/2 acetonitrile/water solution. The mass spectrometer setup was in a data dependent acquisition mode. Ions were selected for MS/MS analysis based on their intensity and charge state +2 to +4. The MS survey scan range is m/z 400-1600 with an acquisition time of 1 sec, whereas the MS/MS fragmentation scan range is m/z 100-2000 with an acquisition time of 2.4 sec. Mascot 2.1 (Matrix Science, London, UK) was used to search all of the MS/MS spectra against the *E. coli* protein database from NCBINR. MS and MS/MS mass tolerance was setup as 1.2 Da and 0.6 Da respectively. PKL files were created by the software Masslynx 3.5 from Waters. The searching parameters setup was as follows: trypsin-specificity restriction with 1 missing cleavage site and variable modifications including oxidation (M), deamidation (NQ), and alkylation (C).

Unlike chemical denaturation, SDS appears to denature proteins by irreversibly trapping them during the transient times in which proteins are unfolded (Manning & Colón, 2004), and since KSPs rarely escape their native state, they are virtually immune to SDS-induced denaturation. Since our initial study (Manning & Colón, 2004) we have analyzed dozens of other proteins and have not observed an exception to this observation. However, there may be other reasons independent of KS that may result in SDS-resistance. For example, proteins that are highly negatively charged may repel SDS. The 50 SDS-resistant *E. coli* proteins we identified in this study have isoelectric points that range from 4-10, and therefore none is expected to electrostatically repel SDS. Also, proteins that are not KS in themselves, but may be part of kinetically stable complexes could lead to false-positives in our assay. A literature search of the proteins listed in Table 1 revealed several proteins that form complexes with GroEL, including S-adenosylmethionine synthase, elongation factor Tu, RNA polymerase α-chain and 50S ribosomal protein L7/L12 (Houry, Frishman, Eckerskorn, Lottspeich, & Hartl, 1999). Interestingly, the GroEL complexes have been shown to be SDS-resistant, whereas GroEL itself and some of its binding partners are known to lack SDS-resistance (Houry, et al., 1999). Thus, D2D SDS-PAGE might also be useful for identifying kinetically stable complexes resulting from the interaction of non-KSP proteins.

2.4 Kinetically stable proteins in *E. coli* have a bias towards enzymatic function

The functions of the KSPs identified by D2D SDS-PAGE were compared with a non-redundant subset of the *E. coli* proteome to determine whether KS is more or less common in proteins with particular functions (Table 2). Interestingly, whereas ~ 32% of all the proteins in *E. coli* are enzymes, this percentage is ~ 70% in KSPs (Fig. 5A), although there was no preference or aversion for a particular type of enzyme family (Fig. 5B). A larger database will be needed to determine whether this is a general observation. However, it seems plausible that some functions might be more compatible with KS. For example, oxidoreductases might have a predisposition towards KS because they often contain co-factors and metals, and are frequently exposed to potentially harmful free radicals. In contrast, ligase function might require high regulation and flexibility that might be incompatible with KS. (Verdecia, et al., 2003). The absence of kinetically stable transporters and regulators (Fig. 5A) is in agreement with the efficient regulation requirement for these proteins. In particular, KS seems incompatible with transcription factors, which must be quickly turned on and off. Future proteomic analyses of other organisms will increase the number of known KSPs and might provide new insight about the link between protein function and KS.

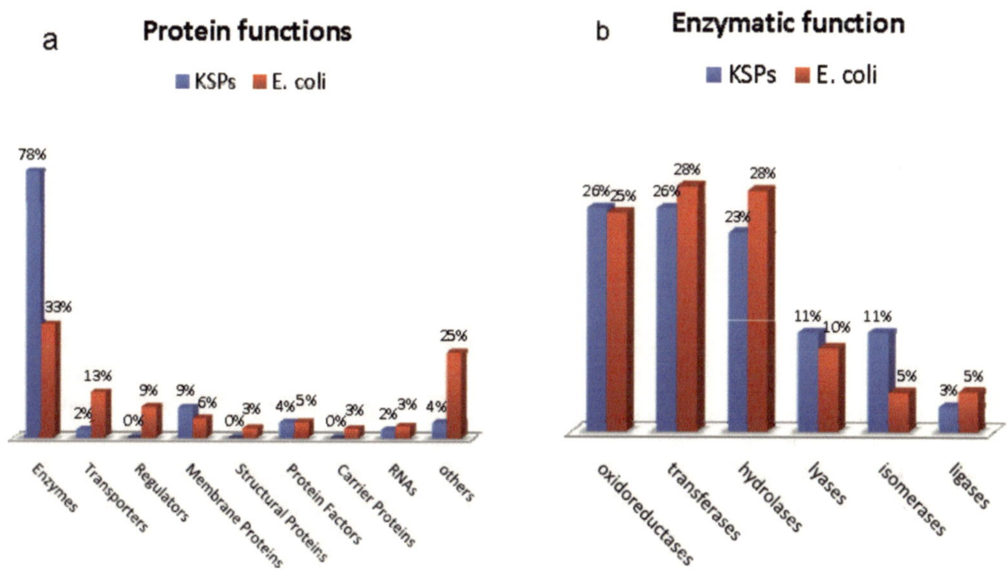

Fig. 5. Different protein (a) and enzymatic (b) functions for a non-redundant subset of the *E. coli* proteome compared to the KSPs identified by D2D SDS-PAGE (Xia, et al., 2007) (a) The kinetically stable subproteome has significantly more enzymes ($p < 0.0001$), but fewer regulators ($p = 0.0082$) and transporters ($p = 0.0076$). Other changes were not statistically significant at the 95% confidence level. Functional assignments were made using the *E. coli* genome and proteome database (GenProtEC). "Other" refers to other functions, including: leader peptides, external origin, cell processes, lipoproteins, pseudogenes, phenotypes, unknown functions, unclassified proteins and sites. (b) Comparison of the six most common enzyme functions does not show statistically significant differences at the 95% confidence level. Enzyme functions were obtained using the BRENDA web site. This figure is adapted from Xia, et al., 2007.

2.5 Monomeric and alpha helical proteins have lower probability of possessing high kinetic stability

Structural analysis of the 50 KSPs identified in *E. coli* (Table 2), yielded 44 that have known 3D structures or are linked to homologs of known structures. To identify potentially structural features among these KSPs, their secondary (2°) structures were compared to the *E. coli* proteome using the classification obtained by the CATH database (CATH). As shown in Fig. 6A, there was a modest difference in the percentage of β structures compared to the proteins in *E. coli*, but a clear difference in the percentages of α and α/β proteins. Remarkably, very few proteins with all alpha-helical structure were kinetically stable, and none of these were monomeric. Thus it appears that monomeric alpha helical structures might be incompatible with the topological complexity that might be required for KS. Perhaps, α/β proteins might be more likely to possess KS because mixtures of 2° structure lead to more complex topologies.

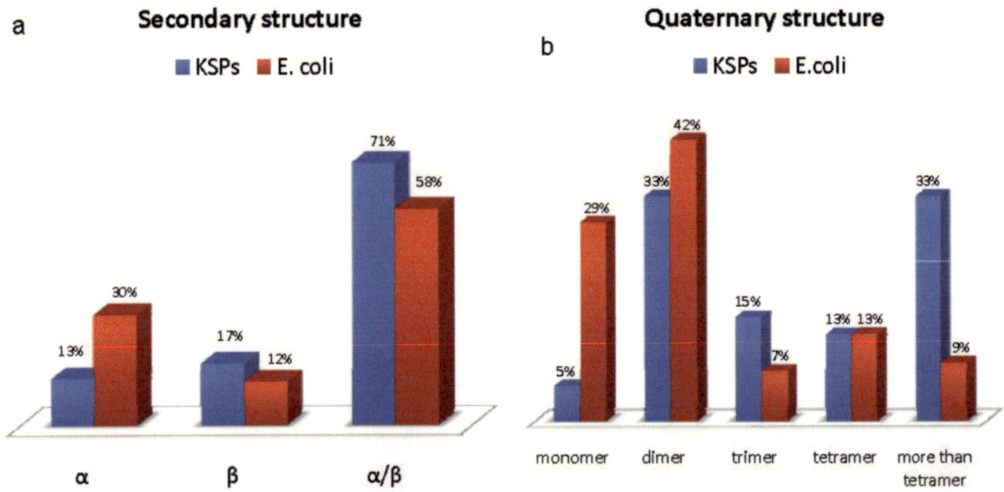

Fig. 6. Secondary (a) and quaternary (b) structure distribution of the non-redundant subset of the *E. coli* proteome compared to the KSPs identified by D2D SDS-PAGE. (a) The KSPs have fewer (p=0.0034) all alpha-helical proteins compared to the rest of the *E. coli* proteome. Structure classifications were made using the CATH database (CATH). (b) The KSPs include only a few monomers (p=0.0002), and significantly more large oligomeric structures with at least five subunits (p<0.001). Dimers and tetramers occur at approximately the same frequencies. Quaternary structure information was obtained from the PQS Protein Quaternary Structure database ("EMBL-EBI PQS Protein Quaternary Structure database "). This figure is adapted from Xia, et al., 2007.

Analysis of the oligomeric/quaternary (4°) structures show that the percentage of monomeric KSPs in *E.coli* is much lower compared to the whole proteome, whereas the percentage of oligomeric proteins with 5 or more subunits is significantly higher (Fig. 6B). Although it is not clear why higher oligomeric structures might favor KS, they might confer greater rigidity and protection of surface residues from water.

2.6 Thermophilic and mesophilic prokaryotes have a greater number of kinetically stable proteins than eukaryotic organisms

D2D SDS-PAGE provides a unique opportunity to investigate the proteome of KSPs in diverse organisms. Therefore, we studied the cell lysates of the thermophilic bacteria *Thermus thermophilus* and *Thermus aquaticus*, and the archaea *Sulfolobus acidocaldarius*, which grow at optimal temperatures of 65, 70, and 80°C, respectively (Xia, Zhang, Solina, Barquera, & Colon, 2010) These thermophiles exhibited high number of SDS-resistant (i.e.

KSPs) proteins (Fig. 7A). In contrast, the mesophilic bacteria *Escherichia coli* (Xia, et al., 2007), *Vibrio cholerae*, and *Bacillus subtilis*, showed significant variation and fewer KSPs than the thermophiles, especially in the upper left area of the gel where the higher molecular weight proteins migrate (Fig. 7B). We also studied three very different eukaryotic organisms from separate kingdoms, including *Saccharomyces cerevisiae*, maize, and *Tetrahymena thermophila*. Remarkably, these eukaryotic organisms exhibited very few, if any, KSPs (Fig. 7C). Therefore, our results clearly showed that thermophiles and prokaryotes have more KSPs than mesophiles and eukaryotes, respectively(Xia, et al., 2010)

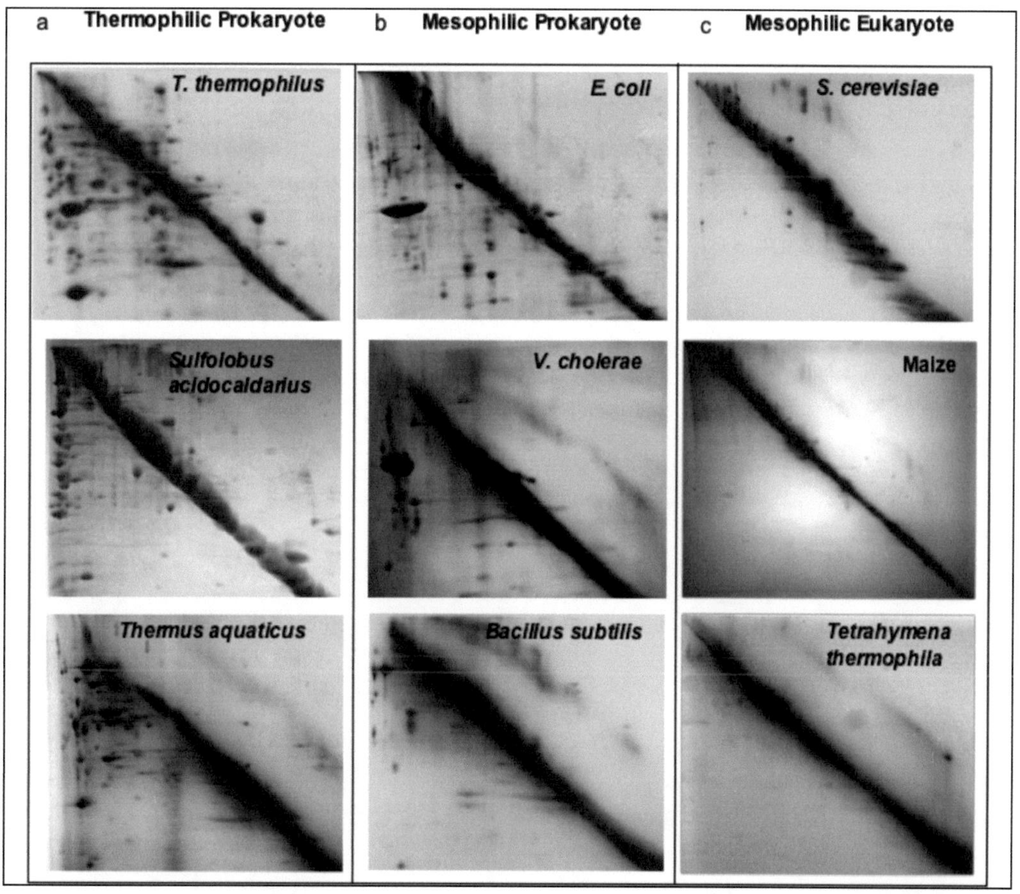

Fig. 7. D2D SDS-PAGE of the lysate of various organisms to probe the extent of kinetically stable proteins present (Xia, Zhang, Solina, Barquera, & Colon, 2010). (a) Thermophilic prokaryotes (*T. thermophilus, S. acidoldarius,* and *T. aquaticus*) exhibited significantly more spots migrating to the left of the diagonal than (b) mesophilic prokaryotes (*E. coli, V. cholerae,* and *B. subtilis*). (c) Mesophilic eukaryotes (*Sa. cerevisiae*, maize, and *Te. thermophila*) showed the fewest number of KSPs spots. Because of differences in background staining, the pictures were slightly enhanced by Microsoft office picture manager through linear adjustment of contrast, brightness, and color applied to the entire image in each case. Reprinted with permission from (Xia, K., Zhang, S., Solina, B. A., Barquera, B., & Colon, W. (2010). Do prokaryotes have more kinetically stable proteins than eukaryotic organisms? *Biochemistry, 49*(34), 7239-7241). Copyright (2010) American Chemical Society.

The results of this study suggest that KS might be a very significant feature of certain proteins required for the adaptation and survival of microbial organisms, which lack cellular sub-compartments, and possess a primordial defense system. In contrast, eukaryotes might be less dependent on KS for survival, and this property might not be generally compatible with the regulatory demands of these more sophisticated organisms. Thus, the presence of many KSPs in prokaryotes, especially thermophiles, suggests that this property is essential for the survival of these simpler organisms. Proteomics analysis of KSPs in thermophilic organisms might reveal a subset of critical proteins that play a major role in determining the ability of these organisms to live and thrive at higher temperatures.

3. Proteomics analysis of KSPs by capillary electrophoresis

3.1 Capillary electrophoresis as an effective method to detect KSPs

Based on the previous correlation between KS and a protein's SDS-resistance, we set out to explored whether SDS-capillary electrophoresis (CE) would be suitable for identifying KSPs (Zhang, Xia, Chung, Cramer, & Colon, 2010). We used eight control proteins, including four KSPs and four non-KSPs. The unheated samples of the non-KSPs α-chymotrypsin (CHT), glucose dehydrogenase (GD), concanavalin A (ConA) and myoglobin (MYO) were denatured by SDS, resulting in identical migration on the gel as the respective samples that were boiled. In contrast, the unheated samples of the KSPs glucose oxidase (GO), streptavidin (SVD), superoxide dismutase (SOD), and subtilisin carlsberg (SCA) were resistant to SDS and exhibited a slower migration on the gel. Analysis of these proteins by CE, which is based on the same electrophoretic principles as SDS-PAGE, showed results consistent with SDS-PAGE (Fig. 8). The CE data for the 4 non-KSPs showed that all boiled

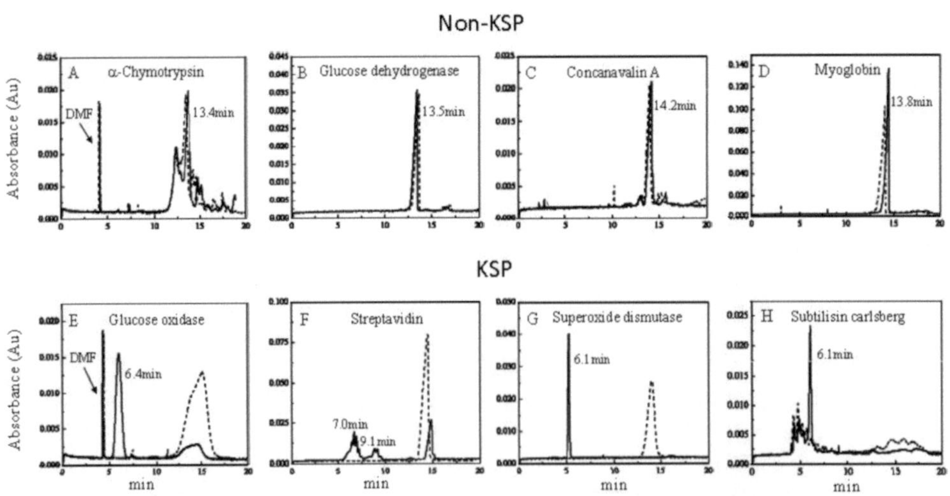

Fig. 8. Electropherograms showing the migration of unboiled and boiled samples of non-KSPs (A–D) and KSPs (E–H). Black solid lines and dash lines represent the data of samples incubated in SDS that were not boiled or boiled, respectively. Samples were incubated in 20 mM sodium phosphate buffer (pH 7.4) containing 1% (w/v) SDS for 10 min. The electropherograms of unboiled and boiled non-KSPs showed little difference, but unboiled KSPs had significantly faster migration. This figure is adapted from Zhang, Xia, Chung, Cramer, & Colon, 2010.

and unboiled non-KSPs had similar migration times of 13.4 - 14.2 min (Fig. 8A-D). Since KSPs bind few SDS molecules, the KSPs had faster CE migration time of 6-7 min. (Figure 8E-H). In most cases there were also one or more smaller peaks in the 14-15 min region, similar to that of non- KSP, suggesting that the protein was partially denatured by SDS. Interestingly, the broader and/or multiple peaks observed in Figs. 8F and H suggest conformational heterogeneity and might arise by the presence of different population of species.

We used a fused silica capillary to separate the proteins, and therefore, the positively charged cations of the buffer solution interact with the negatively charged silanoate groups and form a mobile cation layer. Under normal polarity with the anode (+) at the sample inlet and the cathode (-) at the sample outlet, the mobile cation layer is pulled in the direction of the negatively charged cathode. The solvation of these cations cause the bulk buffer solution to migrate with the mobile layer, producing the electro-osmotic flow (EOF). Thus, the protein:SDS complexes of denatured proteins are highly negatively charged and experience more repulsion from the cathode (outlet), resulting in slower migration than the relatively SDS-free KSPs.

3.2 Using capillary electrophoresis to identify the proteome of KSPs

In CE, proteins are typically detected using UV absorption, laser induced fluorescence (LIF), or by coupling to a mass spectrometer (MS) (Fonslow & Yates, 2009; Garcia-Campana, Taverna, & Fabre, 2007; Herrero, Ibanez, & Cifuentes, 2008; Kasicka, 2008; Stutz, 2005). A limitation when using UV detection is that the limit of detection (LOD) is in the micromolar range. However, this disadvantage can be overcome by using other detectors. The sensitivity of LIF is subnanomolar (Gutman & Kessler, 2006) and a MS detector could provide amol-range sensitivity (Gaspar, Englmann, Fekete, Harir, & Schmitt-Kopplin, 2008; Haselberg, de Jong, & Somsen, 2007; Hernandez-Borges, Borges-Miquel, Rodriguez-Delgado, & Cifuentes, 2007; Tempels, Underberg, Somsen, & de Jong, 2007). Thus, it is possible to interface a CE instrument with a LTQ orbitrap MS using an Agilent sheath-flow adapter kit that can be used with any ESI-MS instrument. The method of choice for coupling CE to ESI/MS is the coaxial sheath-flow interfacing, which is stable, and provides the best sensitivity (in the amol range). Stability and sensitivity of this interfacing has been confirmed by a number of studies (Gaspar, et al., 2008; Haselberg, et al., 2007; Hernandez-Borges, et al., 2007; Tempels, et al., 2007).

We have shown that under certain conditions (Zhang, et al., 2010), KSPs will move faster in CE than non-KSPs. Since most proteins in any organism are not kinetically stable, they will migrate slower and together due to their similar z/m value. In contrast, the KSPs will have a lower and variable z/m value that will allow CE to separate with high resolution the low abundant KSPs from the bulk of the non-KSPs. A description of the general approach of the proposed CE experiment is shown in Fig. 9. In terms of the CE instrument, the SDS concentration, the voltage, the capillary length, and the loading amount should be optimized to achieve the best separation.

3.3 Limitations of D2D SDS-PAGE and advantages of CE

Although 2D electrophoresis (2DE) is often used to separate proteins for proteomics analysis, it has several disadvantages. Since MS analysis is limited to proteins that can be visualized, only abundant proteins can be seen. 2DE is also time-consuming and does not lend itself

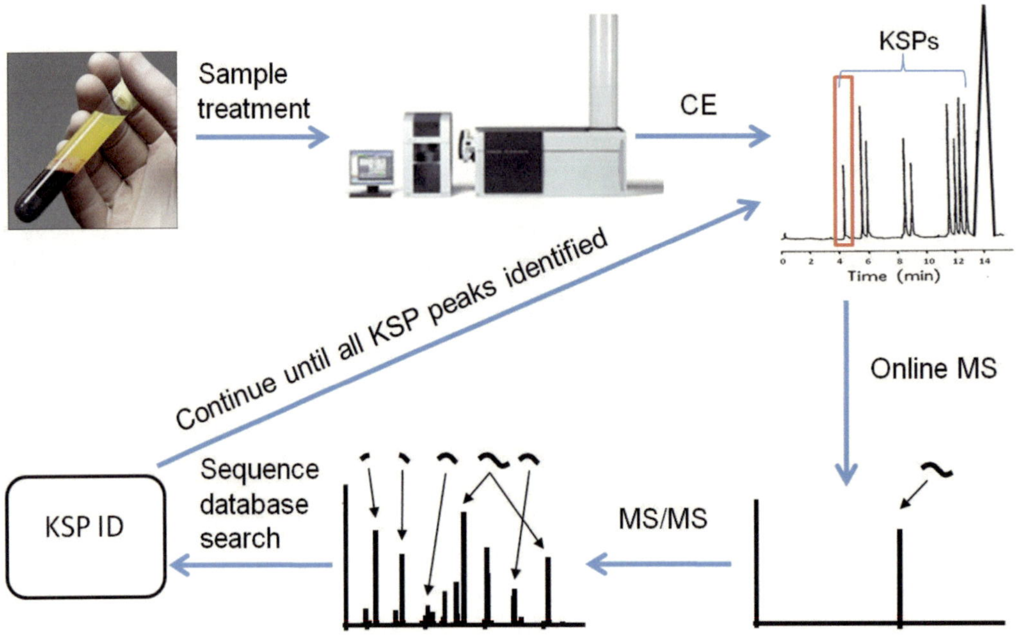

Fig. 9. Identifying the KSPs of an organism by CE-MS. KSPs will be separated from non-KSPs by CE and eluted directly into an online orbitrap mass spectrometer. Following measurement of its mass, intact KSP are directly fragmented in the machine. The mass of its daughter fragments is measured and then analyzed using a database to determine the identity of the KSPs.

to automation. In addition, each step requires lengthy optimization and user intervention. Furthermore, sample handling can easily introduce protein loss and artifacts, such as oxidation and other side-chain modifications. Since CE is based on the same general electrophoretic principle as PAGE, we were able to show that CE can also be used to identify KSPs (Fig. 8) (Zhang, et al.,2010). CE is an efficient and highly sensitive separation method that is widely used in biochemical and pharmaceutical research (Kostal, Katzenmeyer, & Arriaga, 2008; Little, Paquette, & Roos, 2006; McEvoy, Marsh, Altria, Donegan, & Power, 2008). CE is fast and cost-effective, and allows high sample throughput, easy automation, separation efficiency, precision, and only requires nanoliter volumes of sample (Dolnik, 2006, 2008). The small diameter of the capillaries allows better heat dissipation than gel electrophoresis, thereby minimizing band broadening. CE, especially when using an MS detector, could analyze very small amounts of sample with high sensitivity. For example, MS can identify a 1 pmol spot on a gel, whereas in the case of CE, it is possible to identify a 1 fmol amount of protein, and perhaps up to 1 amol with an orbitrap MS instrument (Dolnik, 2006, 2008). Table 2 provides a side-by-side comparison between D2D SDS-PAGE and CE. Thus, although D2D SDS-PAGE is accessible and affordable for identifying KSPs in complex organisms, CE-MS is more promising for faster analysis and for identifying KSPs that have low abundance. Furthermore, the sharpness of the peaks observed by CE might reveal valuable and unique information about the conformational heterogeneity of KSPs and the extent of protein KS.

Comparison	D2D-SDS-PAGE	Capillary Electrophoresis
Instrument cost	$2k	≥ $70k
Time cost (one sample)	more than 10h plus digestion and MS time	less than 1h
Technical demands	laborious, need experience and good hands	automated procedure
Reproducibility of result	could be variable depending on technical expertise	reproducible
Sample amount	0.3~1mg protein	nl~ml (1ng-1mg) protein
Resolution	10,000 for 2DE, up to one hundred KSP spots could be separated in a single gel.	>1000 peptides
Sensitivity*	coomassie stain - pmol florescence stain - fmol	MS detector - fmol
MS coupling	N/A	online connection
Extra information	spot pattern	retention time

*A 1 pmol coomassie stain can be identified by MS, but a fmol fluorescence spot cannot due to protein loss (e.g. crosslinking to the gel, getting trapped by the surface of pipette tip or tube) during the in-gel digestion process.

Table 3. Comparison of D2D-SDS-PAGE and CE

4. Conclusion

The D2D SDS-PAGE method described here is simple and accessible for the proteomics-level identification of KSPs. In contrast the SDS-CE methods is fast, sensitive, and has the potential to be applied in high throughput fashion. The key feature of both methods is their ability to separate SDS-resistant (i.e. KSPs) from non-SDS-resistant (i.e. non-KSPs) proteins. Therefore, mild conditions must be employed during the separation step to preserve the conformational integrity of the proteins. Afterwards, conventional proteomics analysis may be carried out to identify KSPs.

Living organisms have a diversity of sub-proteomes that are involved in different pathways or functions. The sub-proteome of KSPs is likely to include proteins that must have longer half-lives and possess resistance to degradation for the benefit of the organism. The SDS-based methods we have developed will make it possible to study a variety of systems, including the cellular lysates of microorganisms, human plasma and other biological fluids, normal and diseased cells, and all types of plants and food materials. Such studies will increase the database of KSPs and will facilitate investigation of the structural basis and the diverse functional roles of KS. Furthermore, they will stimulate research to understand the biological and pathological roles of the abnormal gain or loss of KS in proteins.

5. Acknowledgment

The work described in this chapter was supported by grants (MCB 0519507 and 0848120) from the US National Science Foundation

6. References

Baker, D., & Agard, D. A. (1994). Kinetics versus thermodynamics in protein folding. *Biochemistry, 33*(June 21), 7505-7509.

Baker, D., Sohl, J. L., & Agard, D. A. (1992). A protein folding reaction under kinetic control. *Nature, 356*(Mar 19,6366), 263-265.

Cappai, R., Leck, S. L., Tew, D. J., Williamson, N. A., Smith, D. P., Galatis, D., et al. (2005). Dopamine promotes alpha-synuclein aggregation into SDS-resistant soluble oligomers via a distinct folding pathway. *FASEB J, 19*(8), 1377-1379.

CATH. *http://www.cathdb.info/latest/index.html.*

Cunningham, E. L., Jaswal, S. S., Sohl, J. L., & Agard, D. A. (1999). Kinetic stability as a mechanism for protease longevity. *Proc. Natl. Acad. Sci. U S A, 96*(20), 11008-11014.

Dobson, C. M. (2001). Protein folding and its links with human disease. *Biochem Soc Symp*(68), 1-26.

Dolnik, V. (2006). Capillary electrophoresis of proteins 2003-2005. *Electrophoresis, 27*(1), 126-141.

Dolnik, V. (2008). Capillary electrophoresis of proteins 2005-2007. *Electrophoresis, 29*(1), 143-156.

EMBL-EBI PQS Protein Quaternary Structure database *http://pqs.ebi.ac.uk/.*

Enya, M., Morishima-Kawashima, M., Yoshimura, M., Shinkai, Y., Kusui, K., Khan, K., et al. (1999). Appearance of sodium dodecyl sulfate-stable amyloid beta-protein (Abeta) dimer in the cortex during aging. *Am J Pathol, 154*(1), 271-279.

Fonslow, B. R., & Yates, J. R., III. (2009). Capillary electrophoresis applied to proteomic analysis. *Journal of Separation Science, 32*(8), 1175-1188.

Funato, H., Enya, M., Yoshimura, M., Morishima-Kawashima, M., & Ihara, Y. (1999). Presence of sodium dodecyl sulfate-stable amyloid beta-protein dimers in the hippocampus CA1 not exhibiting neurofibrillary tangle formation. *Am J Pathol, 155*(1), 23-28.

Garcia-Campana, A. M., Taverna, M., & Fabre, H. (2007). LIF detection of peptides and proteins in CE. *Electrophoresis, 28*(1-2), 208-232.

Gaspar, A., Englmann, M., Fekete, A., Harir, M., & Schmitt-Kopplin, P. (2008). Trends in CE-MS 2005-2006. *Electrophoresis, 29*(1), 66-79.

GenProtEC. *http://genprotec.mbl.edu.*

Gutman, S., & Kessler, L. G. (2006). The US Food and Drug Administration perspective on cancer biomarker development. *Nat. Rev. Cancer, 6*(7), 565-571.

Haass, C., & Selkoe, D. J. (2007). Soluble protein oligomers in neurodegeneration: lessons from the Alzheimer's amyloid beta-peptide. *Nature Reviews Molecular Cell Biology, 8*(2), 101-112.

Hammarstrom, P., Schneider, F., & Kelly, J. W. (2001). Trans-suppression of misfolding in an amyloid disease. *Science, 293*(5539), 2459-2462.

Haselberg, R., de Jong, G. J., & Somsen, G. W. (2007). Capillary electrophoresis-mass spectrometry for the analysis of intact proteins. *J. Chromatogr., A, 1159*(1-2), 81-109.

Hernandez-Borges, J., Borges-Miquel, T. M., Rodriguez-Delgado, M. A., & Cifuentes, A. (2007). Sample treatments prior to capillary electrophoresis-mass spectrometry. *J. Chromatogr., A, 1153*(1-2), 214-226.

Herrero, M., Ibanez, E., & Cifuentes, A. (2008). Capillary electrophoresis-electrospray-mass spectrometry in peptide analysis and peptidomics. *Electrophoresis, 29*(10), 2148-2160.

Hornemann, S., & Glockshuber, R. (1998). A scrapie-like unfolding intermediate of the prion protein domain PrP(121-231) induced by acidic pH. *Proc Natl Acad Sci U S A, 95*(11), 6010-6014.

Horwich, A. L., & Weissman, J. S. (1997). Deadly conformations-Protein misfolding in prion diseases. *Cell, 89*, 499-510.

Houry, W. A., Frishman, D., Eckerskorn, C., Lottspeich, F., & Hartl, F. U. (1999). Identification of in vivo substrates of the chaperonin GroEL. *Nature, 402*(6758), 147-154.

Johnson, S. M., Wiseman, R. L., Sekijima, Y., Green, N. S., Adamski-Werner, S. L., & Kelly, J. W. (2005). Native state kinetic stabilization as a strategy to ameliorate protein misfolding diseases: a focus on the transthyretin amyloidoses. *Acc Chem Res, 38*(12), 911-921.

Kasicka, V. (2008). Recent developments in CE and CEC of peptides. *Electrophoresis, 29*(1), 179-206.

Kawarabayashi, T., Shoji, M., Younkin, L. H., Wen-Lang, L., Dickson, D. W., Murakami, T., et al. (2004). Dimeric amyloid beta protein rapidly accumulates in lipid rafts followed by apolipoprotein E and phosphorylated tau accumulation in the Tg2576 mouse model of Alzheimer's disease. *J Neurosci, 24*(15), 3801-3809.

Koga, H., Kaushik, S., & Cuervo, A. M. (2010). Protein homeostasis and aging: The importance of exquisite quality control. *Ageing Res Rev, 10*(2), 205-215.

Kostal, V., Katzenmeyer, J., & Arriaga, E. A. (2008). Capillary Electrophoresis in Bioanalysis. *Anal. Chem. (Washington, DC, U. S.), 80*(12), 4533-4550.

Kurzban, G. P., Bayer, E. A., Wilchek, M., & Horowitz, P. M. (1991). The quaternary structure of streptavidin in urea. *J Biol Chem, 266*(22), 14470-14477.

Lai, Z., McCulloch, J., Lashuel, H. A., & Kelly, J. W. (1997). Guanidine hydrochloride-induced denaturation and refolding of transthyretin exhibits a marked hysteresis: equilibria with high kinetic barriers. *Biochemistry, 36*(33), 10230-10239.

Lee, H. J., Baek, S. M., Ho, D. H., Suk, J. E., Cho, E. D., & Lee, S. J. (2011). Dopamine promotes formation and secretion of non-fibrillar alpha-synuclein oligomers. *Exp Mol Med, 43*(4), 216-222.

Lesne, S., Koh, M. T., Kotilinek, L., Kayed, R., Glabe, C. G., Yang, A., et al. (2006). A specific amyloid-beta protein assembly in the brain impairs memory. *Nature, 440*(7082), 352-357.

Little, M. J., Paquette, D. M., & Roos, P. K. (2006). Electrophoresis of pharmaceutical proteins: status quo. *Electrophoresis, 27*(12), 2477-2485.

Luce, K., Weil, A. C., & Osiewacz, H. D. (2010). Mitochondrial protein quality control systems in aging and disease. *Adv Exp Med Biol, 694*, 108-125.

Manning, M., & Colón, W. (2004). Structural basis of protein kinetic stability: Resistance to sodium dodecyl sulfate suggests a central role for rigidity and a bias towards beta sheet structure. *Biochemistry, 43*, 11248-11254.

McEvoy, E., Marsh, A., Altria, K., Donegan, S., & Power, J. (2008). Capillary electrophoresis for pharmaceutical analysis. *Handb. Capillary Microchip Electrophor. Assoc. Microtech. (3rd Ed.)*, 135-182.

McLean, C. A., Cherny, R. A., Fraser, F. W., Fuller, S. J., Smith, M. J., Beyreuther, K., et al. (1999). Soluble pool of Abeta amyloid as a determinant of severity of neurodegeneration in Alzheimer's disease. *Ann Neurol, 46*(6), 860-866.

Nestler, H. P., & Doseff, A. (1997). A two-dimensional, diagonal sodium dodecyl sulfate-polyacrylamide gel electrophoresis technique to screen for protease substrates in protein mixtures. *Anal. Biochem., 251*(1), 122-125.

Otzen, D. E. (2002). Protein unfolding in detergents: effect of micelle structure, ionic strength, pH, and temperature. *Biophys J, 83*(4), 2219-2230.

Perkins, D. N., Pappin, D. J., Creasy, D. M., & Cottrell, J. S. (1999). Probability-based protein identification by searching sequence databases using mass spectrometry data. *Electrophoresis, 20*(18), 3551-3567.

Plaza del Pino, I. M., Ibarra-Molero, B., & Sanchez-Ruiz, J. M. (2000). Lower kinetic limit to protein thermal stability: a proposal regarding protein stability in vivo and its relation with misfolding diseases. *Proteins, 40*(1), 58-70.

Podlisny, M. B., Ostaszewski, B. L., Squazzo, S. L., Koo, E. H., Rydell, R. E., Teplow, D. B., et al. (1995). Aggregation of secreted amyloid beta-protein into sodium dodecyl sulfate-stable oligomers in cell culture. *J Biol Chem, 270*(16), 9564-9570.

Prusiner, S. B., Groth, D., Serban, A., Stahl, N., & Gabizon, R. (1993). Attempts to restore scrapie prion infectivity after exposure to protein denaturants. *Proc Natl Acad Sci U S A, 90*(7), 2793-2797.

Reynolds, J. A., Herbert, S., Polet, H., & Steinhardt, J. (1967). The binding of divers detergent anions to bovine serum albumin. *Biochemistry, 6*(3), 937-947.

Roher, A. E., Chaney, M. O., Kuo, Y. M., Webster, S. D., Stine, W. B., Haverkamp, L. J., et al. (1996). Morphology and toxicity of Abeta-(1-42) dimer derived from neuritic and vascular amyloid deposits of Alzheimer's disease. *J Biol Chem, 271*(34), 20631-20635.

Saraiva, M. J. M. (1995). Transthyretin mutations in health and disease. *Human Mutations, 5*, 191-196.

Shapiro, A. L., Vinuela, E., & Maizel, J. V., Jr. (1967). Molecular weight estimation of polypeptide chains by electrophoresis in SDS-polyacrylamide gels. *Biochem Biophys Res Commun, 28*(5), 815-820.

Sigdel, T. K., Cilliers, R., Gursahaney, P. R., & Crowder, M. W. (2004). Fractionation of soluble proteins in Escherichia coli using DEAE-, SP-, and phenyl sepharose chromatographies. *J. Biomol. Tech., 15*(3), 199-207.

Spelbrink, R. E., Kolkman, A., Slijper, M., Killian, J. A., & de Kruijff, B. (2005). Detection and identification of stable oligomeric protein complexes in Escherichi coli inner membranes: a proteomics approach. *J. Biol. Chem., 280*(31), 28742-28748.

Stefani, M., & Dobson, C. M. (2003). Protein aggregation and aggregate toxicity: new insights into protein folding, misfolding diseases and biological evolution. *J Mol Med, 81*(11), 678-699.

Stutz, H. (2005). Advances in the analysis of proteins and peptides by capillary electrophoresis with matrix-assisted laser desorption/ionization and electrospray-mass spectrometry detection. *Electrophoresis, 26*(7-8), 1254-1290.

Tempels, F. W. A., Underberg, W. J. M., Somsen, G. W., & de Jong, G. J. (2007). On-line coupling of SPE and CE-MS for peptide analysis. *Electrophoresis, 28*(9), 1319-1326.

Verdecia, M. A., Joazeiro, C. A., Wells, N. J., Ferrer, J. L., Bowman, M. E., Hunter, T., et al. (2003). Conformational flexibility underlies ubiquitin ligation mediated by the WWP1 HECT domain E3 ligase. *Mol. Cell., 11*(1), 249-259.

Xia, K., Manning, M., Hesham, H., Lin, Q., Bystroff, C., & Colon, W. (2007). Identifying the subproteome of kinetically stable proteins via diagonal 2D SDS/PAGE. *Proc Natl Acad Sci U S A, 104*(44), 17329-17334.

Xia, K., Zhang, S., Solina, B. A., Barquera, B., & Colon, W. (2010). Do prokaryotes have more kinetically stable proteins than eukaryotic organisms? *Biochemistry, 49*(34), 7239-7241.

Zhang, S., Xia, K., Chung, W. K., Cramer, S. M., & Colon, W. (2010). Identifying kinetically stable proteins with capillary electrophoresis. *Protein Sci, 19*(4), 888-892.

16

Vinyl Sulfone: A Multi-Purpose Function in Proteomics

F. Javier Lopez-Jaramillo, Fernando Hernandez-Mateo
and Francisco Santoyo-Gonzalez
*Instituto de Biotecnologia,
Universidad de Granada, Granada
Spain*

1. Introduction

The outstanding development attained in the actual state-of-the-art on Proteomics has been reached not only by the integration of a panel of sophisticated analytical and bioinformatics techniques and instrumentations but also by the intelligent application of classical and advanced synthetic methodologies used in protein chemistry (Lundblad, 2005; Tilley *et al.*, 2007). Covalent modification of proteins is a powerful way to modulate their macromolecular function. Nature accomplishes such alterations through a range of post-translational modifications that in turn mediate protein activity. Artificial covalent modification of proteins is an arduous but fruitful task of major interest for the biophysics and biochemistry communities that normally pursue as goals the detection or purification of the protein itself in order to have a more thorough understanding of molecular mechanisms and the expansion of the applicability of such biomolecules. Despite the intrinsic difficulties associated to perform those chemical modifications of proteins, the attachment of analytical or engineered probes for protein tracking (labelling) (Giepmans *et al.*, 2006; Waggoner, 2006; Wu & Goody, 2010) or protein profiling (chemical proteomics) (Evans & Cravatt, 2006; Cravatt *et al.*, 2008), the introduction of affinity tags for separation-isolation of proteins (affinity chromatography) (Azarkan *et al.*, 2007; Fang & Zhang, 2008) or for mass spectroscopy-based protein identification and characterization (chemical tagging) (Leitner & Lindner, 2006), the immobilization onto solid supports (microarray technologies) (Wong *et al.*, 2009) and the conjugation with other biomolecules (post-translational modifications) (Gamblin *et al.*, 2008b; Walsh, 2009; Heal & Tate, 2010) are among some of the most useful and frontier techniques and methodologies used in Proteomics.
For the chemical modification of proteins, a large number of strategies are nowadays available (Hermanson, 2008). The straightforward and probably most used of those strategies takes advantage of the chemical reactivity of the endogenous amino acid side chains commonly by using the nucleophilic character of some of them in a nucleophile-to-electrophile reaction pattern that leads to specific functional outcomes (Baslé *et al.*, 2010). This classical residue-specific modification chemistry, however, is rarely sufficiently selective to distinguish one residue within a sea of chemical functionality and for this reason more intricate approaches have been developed in recent times to introduce a unique chemical handle in the target protein that is orthogonal to the remainder of the proteome

(Hackenberger & Schwarzer, 2008). Direct incorporation of non-canonical amino acids into proteins via the subversion of the biosynthetic machinery is an attractive means of introducing selectively new functionality by either a site-specific or residue-specific manner (Beatty & Tirrell, 2009; de Graaf *et al.*, 2009; Johnson *et al.*, 2010; Liu & Schultz, 2010; Voloshchuk & Montclare, 2010; Young & Schultz, 2010) that in combination with recent and notorious advances in bioorthogonal reactions (nucleophilic addition to carbonyl, 1,3-dipolar cycloaddition reactions, Diels-Alder reactions, olefin cross-metathesis reactions and palladium-catalyzed cross-coupling reactions) has allowed an exquisite level of selectivity in the covalent modification of proteins (Wiltschi & Budisa, 2008; Sletten & Bertozzi, 2009; Lim & Lin, 2010; Tiefenbrunn & Dawson, 2010). In spite that major technical challenges have been overcome, a prodigious amount of lab work and the concurrently optimization of a larger set of parameters is normally required for those advanced and selective methodologies in comparison with conventional organic reaction development.

In this general frame, the purpose of the present chapter is to provide a general outlook on the applications on Proteomics of a particular methodology, the vinyl sulfone chemistry (Simpkins, 1990; Forristal, 2005; Meadows & Gervay-Hague, 2006), with a particular emphasis in some recent advances that illustrate the multi-purpose character of this chemical function in this field. Vinyl sulfones readily forms covalent adducts with many nucleophiles ("hard" and "soft") via a Michael-type 1,4-addition. Two prominent characteristics of this reactive behaviour have allowed its implementation on Proteomics: the possibility to perform those reactions in physiological conditions (aqueous media, slightly alkaline pH and room temperature) that preserves the biological function of the proteins and the absence of catalysts and by-products. In addition, the introduction of the vinyl sulfone is not a difficult task and the resulting functionalized reagents or intermediates are stable.

The chapter is organized in three sections. In a first instance, a general overview of the vinyl sulfone chemistry in terms of the most relevant methods of synthesis and aspects of their reactivity will be followed by a discussion of the application of this chemical behaviour with proteins. Their advantages and disadvantages with other currently available methodologies to modify amine and thiol groups naturally present in proteins will be compared. In a second section the applications of vinyl sulfones to Proteomics will be enumerated. Finally, the wide scope of the vinyl sulfone chemistry in other omic sciences will be discussed.

2. Vinyl sulfone chemistry

Vinyl sulfones (α,β-unsaturated sulfones) are productive and widely used intermediates in organic synthesis that also have a remarkable biomedical significance owed to their capability to act as irreversible inhibitors of many types of cysteine proteases through conjugate addition of the thiol group of the active site cysteine residue. This feature is the basis of some modern applications of this chemical function to Proteomics as it will be discussed below (section 3.2). Currently, there exists a solid body of knowledge on the chemical reactivity of the vinyl sulfone that allows the functionalization of any organic substrate.

2.1 Synthesis of vinyl sulfones

Vinyl sulfone is a functional group accessible by a broad variety of traditional synthetic methods and other contemporary reactions that have been comprehensively reviewed (Simpkins, 1990; Forristal, 2005; Meadows & Gervay-Hague, 2006).

For Proteomics, the most relevant of these procedures are those that used 2-halo or 2-hydroxyethylthioethers as starting materials (Fig. 1). From these compounds formation of a vinyl sulfone is feasible by three alternative strategies: sequential elimination and oxidation in either order (routes a and b) or simultaneous oxidation-elimination (route c). When the elimination step is firstly performed (route a), the vinyl thioether intermediate obtained can be easily oxidized to the corresponding vinyl sulfones by common oxidizing agents (H_2O_2-acetic acid, m-chloroperbenzoic acid -mCPBA- or periodic acid -HIO_4-) or the commercial Oxone® reagent. The slow kinetic showed by the method based in H_2O_2 (Bordwell & Pitt, 1955) has been overcome by the concomitant use of some catalysts ($MnSO_4$ or tetrakis(pentafluorophenyl)porphyrin) in order to exploit the goodness of this methodology: low cost and toxicity, and high yields (Alonso *et al.*, 2002; Baciocchi *et al.*, 2004).

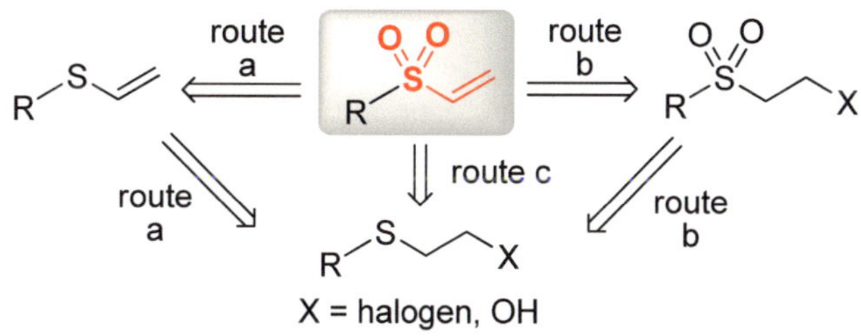

Fig. 1. General retrosynthetic pathway for the synthesis of vinyl sulfones from 2-halo or 2-hydroxyethylthioethers

In the alternative sequence (route b), the sulfone is obtained previously by using the reagents just mentioned followed by the elimination step that is favoured by the strong electron-withdrawing effect of the sulphur function, being only necessary a weak base (triethylamine) in case of the dehydrohalogenation (Brace, 1993) or the conversion on a good leaving group, usually a sulfonic ester, in the dehydration option (Lee *et al.*, 2000; Galli *et al.*, 2005).

On the other hand, ammonium molybdate in the presence of H_2O_2 or ozone allows the formation of vinyl sulfones in one-step from derivatized ethylthioethers with satisfactory results (route c) (Krishna *et al.*, 2003).

In addition to the methodologies commented, the ionic and radical addition to unsaturated compounds (alkenes, alkynes and allenes), the addition of sulfonyl-stabilized carbanions to carbonyl compounds, the manipulation of acetylenic sulfones and the use of organometallic reagents are other routes for the synthesis of vinyl sulfones (Simpkins, 1990; Forristal, 2005; Meadows & Gervay-Hague, 2006) that in practice have found limited applications in Proteomics up to the present.

2.2 Reactivity of vinyl sulfones

Vinyl sulfones as sulfonyl-containing compounds readily undergo a variety of cycloaddition reactions and conjugate additions as excellent Michael acceptors because of the electron poor nature of their double bond owed to the sulfone's electron withdrawing capability that make them good electrophiles. The cycloadditions reactions have been reviewed in detail (De Lucchi & Pasquato, 1988; Simpkins, 1990; Forristal, 2005) but their applications in Proteomics have been null. For this reason these relevant reactions are considered out of the scope of the present chapter and an interested reader is referred to those articles. However, conjugate additions to vinylsulfones involving both "hard" and "soft" nucleophiles are of paramount importance in Proteomics, and for this reason a general outlook of this sort of reactions is given.

A significant body of work has been devoted to the conjugate additions of vinyl sulfones with carbon nucleophiles with both non-stabilised organometallics and stabilised anions (including enolates). In addition, vinyl sulfones have been widely exploited as acceptors in radical conjugated additions with a variety of nucleophilic radicals (Srikanth & Castle, 2005) and have been used in organocatalytic methodologies where they have demonstrated their versatility and power in asymmetric reactions for the construction of carbon-carbon bonds with exceptional levels of enantioselectivity (Alba *et al.*, 2010). Aside from these reactions with carbon nucleophiles, heteroatomic nucleophiles involving nitrogen, sulphur and oxygen can participate efficiently in conjugate addition reactions with vinyl sulfones in a protic environment where the incipient carbanion is quickly quenched. In these reactions, base catalysts are often unnecessary for amines because of the strong nucleophility of the nitrogen atom. However, although thiols are generally more nucleophilic than amines, weak bases are often used to deprotonate them due to their comparatively higher acidity (Bednar, 1990).

All the conjugate additions with vinyl sulfones share a similar reaction pattern by addition at the β-position of the sulfone and, on this basis, these reactions are a well-established method of creating β-heterosubstituted sulfones (Fig. 2). In all cases, the resulting 1,4-addition products contain either the sulfonyl moiety which can undergo subsequent functional group transformations or can be easily removed (by means of Mg or Hg/Na) making these compounds a perfect choice to afford easily naked alkyls (Nájera & Yus, 1999).

Fig. 2. General conjugated Michael-type addition of vinyl sulfones and nucleophiles

Heteroatomic nucleophiles differ in the kinetic of their conjugate addition to Michael acceptors including vinyl sulfones, fact that is relied to their nucleophilicity. Studies on model compounds, including amino-acids, were performed to evaluate the influence on the reaction rates of these α,β–unsaturated compounds of different factors either inherent to the nucleophiles (charge, electronic structure and size) or depending on the environment (interactions with neighbouring ionisable groups, steric factors and pH) (Friedman *et al.*, 1965; Morpurgo *et al.*, 1996; Lutolf *et al.*, 2001). As a general rule, it is observed a direct correlation between the reaction rates and the anion concentration which is determined by the pK_a values and the pH of the medium in such a way that rates increase with pH due to the increased concentration of the anion. However, comparative studies performed in these pioneering contributions concerning the relative nucleophilic reactivities of amino groups and mercaptide ions showed that at comparable pK_a values and steric environments vinyl sulfones react with thiols significantly quicker than with amines or other nucleophiles. From these results, it has been assumed that vinyl sulfones are selective in the reactions with thiol groups relative to reaction with amino groups providing that the reaction is not carried out at alkaline pH. The implementation of these observations in protein chemistry is on the rationale behind numerous chemoselective modifications of cysteine-containing peptides

and proteins by the Michael-type addition reaction of vinyl sulfone derivatives. However, given the multifunctional character and complexity of proteins, the preference of vinyl sulfones for thiol groups should be considered with precaution as recent findings have demonstrated (*vide infra* section 2.3.).
Considering that selectivity is a key point in bioconjugation and particularly in Proteomics, the next section is devoted to give a general overview of the different strategies currently available for the modification of the side groups of amino and thiol-containing amino acids to put in context vinyl sulfone-based strategies in relation with those methodologies.

2.3 Vinyl sulfones and other methodologies for chemical modification of proteins at amino and thiol-containing residues

The most popular but one of the least site-specific and residue-specific strategies for modification of proteins targets the lysine residue because of its predominant presence (up to 6% of the overall amino acid sequence, the 11th most frequent residue) (Villar & Kauvar, 1994; Villar & Koehler, 2000; UniProtKB/TrEMBL database, 2011-06), the reactivity of the ε-amine group of its side chain, its minor relevance from a biological point of view and its accessibility at the surface of those biomolecules.
Although the primary amine group of lysine is protonated under physiological pH, it can still react as a nucleophile (Fig. 3). Amine reactive electrophilic reagents used with proteins are usually acylating agents, such as succinimidyl esters, sulfonyl chlorides and isothiocyanates (**1**, **2** and **3**, respectively). However they are not exempt of drawbacks. Succinimidyl esters (**1**) are the best suited amine reactive compounds as they react with lysines without exogenous reagents such as bases. More soluble but less reactive sulfosuccinimidyl esters have been used to overcome their poor water solubility (Staros *et al.*, 1986). Sulfonyl chlorides (**2**) are highly reactive but are also unstable in water (Lefevre *et al.*, 1996), specially at the high pH required for the reaction with aliphatic amines, and they can also react with phenols (tyrosine), aliphatic alcohols (serine, threonine), thiols (cysteine) and imidazole (histidine). Isothiocyanates (**3**) are stable in water although their reactivity is only moderate and the degradation of the resulting thiourea has been reported (Banks & Paquette, 1995). In addition, the optimal pH needed for the reaction with lysine of these reagents (pH 9-9.5) is higher than for the formation of succinimidyl esters (pH 8-9) and may be unsuitable for modifying alkaline-sensitive proteins.
Other approaches are: a) the reductive amination of an aldehyde (**4**) using water compatible hydrides, a two-step procedure that make this route more challenging (Jentoft & Dearborn, 1979); b) the amidination with imidoesters (**5**) at elevated pH (~9) or with iminothiolane (**6**, Traut's reagent) near pH 8, reagents that conserve the overall charge of the side group (Means & Feeney, 1990), and c) the use of thioesters or dithioesters (**7**, X=O or S, respectively), being these last mild reagents for lysine residues in the absence of competing cysteine residues (Wieland *et al.*, 1953) that reacts very fast, specifically and irreversibly although they have a limited solubility in water.
In contrast with lysine, cysteine residues are perhaps the most convenient target of the proteogenic amino acids for selective modification of proteins owing to their low natural abundance (the second less abundant amino acid in proteins with a frequency of 1.36%) (Villar & Kauvar, 1994; Villar & Koehler, 2000; UniProtKB/TrEMBL database, 2011-06) and the strong nucleophilic character of the sulfhydryl side chain higher than a primary amine, especially at pH below 9, that results in a general kinetic selective modification of cysteine over lysine residues. Despite thiols often form disulphide

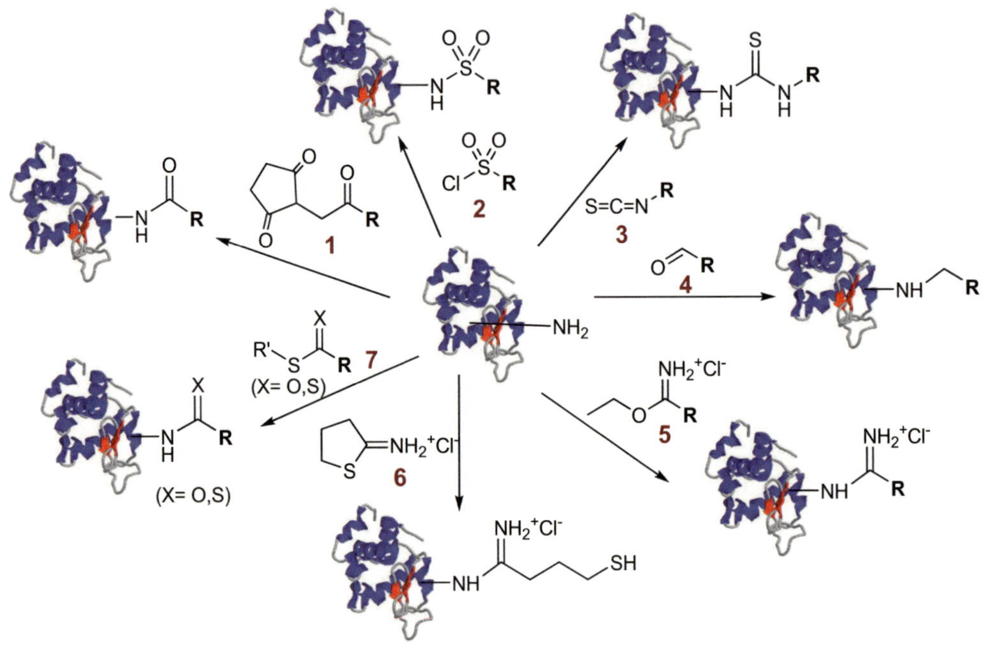

Fig. 3. Reagents for the chemical modification of Lysine residues in proteins

oxidized dimers, the enduring utility of this amino acid in protein modification is evidenced for the wide panel of methodologies that allow the mild, selective, rapid and quantitative reaction at cysteine and their derivatives under appropriate conditions in either a reversible or irreversible way (Fig. 4) (Chalker *et al.*, 2009). Direct alkylation methods with a variety of electrophilic reagents such as α–halocarbonyls (**8**, iodoacetamide), Michael acceptors (including maleimides **9**, vinyl sulfones **10** and related α,β-unsaturated systems) and β–haloethylamine (**11**) (Lindley, 1956) are common techniques for cysteine modification. More specific reactions of the sulfhydryl groups that do not interfere with other amino acids are oxidation and desulfuration of cysteine. Protein modifications via oxidative disulfide bond formation is one of the simplest methods that can be accomplished by simple air oxidation, disulfide exchange with Ellman's reagent (5,5'-dithiobis-(2-nitrobenzoic acid) -DTNB-) or some others activated reagents (iodine or sulfenyl halides) (**12-14**) (Anson, 1940; Fontana *et al.*, 1968). Desulfurization at cysteine may involve its transformation into a thioether, the reductive removal of the thiol group to yield alanine (Yan & Dawson, 2001) or the oxidative elimination of cysteine to yield dehydroalanine (Bernardes *et al.*, 2008), which behaves as a Michael acceptor with thiol nucleophilic reagents. Finally, some metal-mediated reactions (cross-metathesis and Kirmse-Doyle reactions) performed in ally sulfide derivatives have recently extended the panoply of chemical modifications at cysteine (Lin *et al.*, 2008).
However, cysteine modification is not exempt of some drawbacks because besides the low frequency of cysteine in proteins and its relevance for the function, the difference of nucleophility between amine and thiol groups in proteins is dependent on surrounding residues (Bednar, 1990), the selectivity being compromised, and the use of specific reaction on the sulfhydryl group may be limited by the compatibility of the reaction conditions with the functionality of the protein.

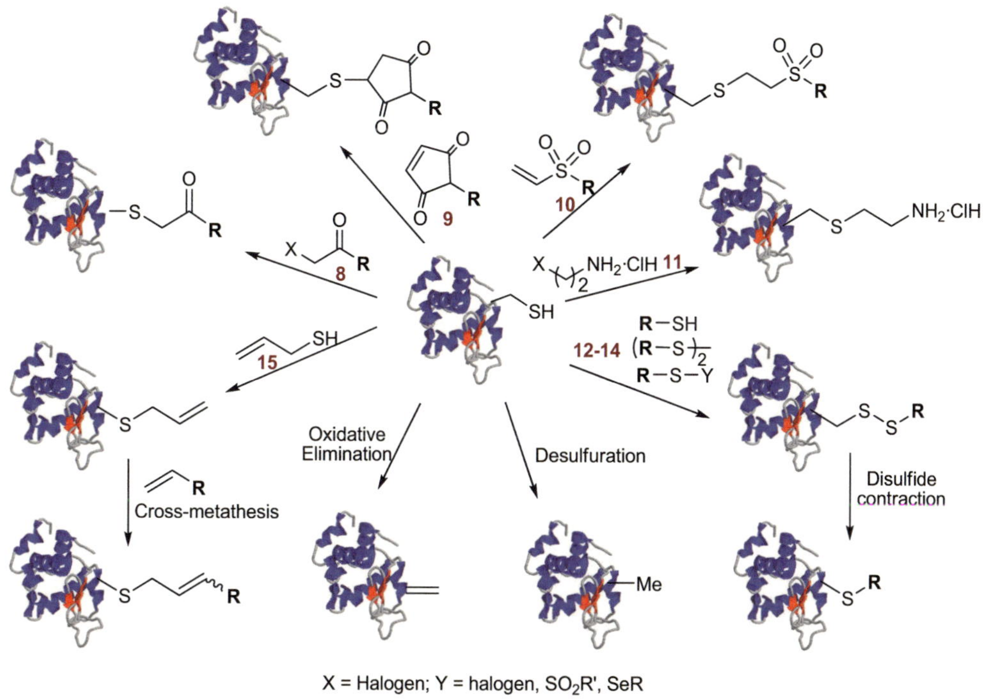

Fig. 4. Reagents for the chemical modification of Cysteine residues in proteins

In this context, the excellent capability of vinyl sulfones to act as Michael acceptors has been used but not fully exploited up to the present for protein modification, despiste attractive characteristics offered by this methodology such as water stability of the sulfur function for extended periods, particulary at neutral pH where they are resistant to hydrolysis, the lack of by-products in conjugated reactions, the needless use of organometallic catalysts, and the stability of the linkages formed.

It is generally accepted that i) the larger nucleophilic character of thiol makes cysteine residues the preferential target of vinyl sulfone derivatized reagents, ii) the ε-amino groups of lysine and to a lesser extent the imidazole ring of histidine side chain are secondary targets and iii) the pH of the reaction medium may be use to control the relative reactivity of these funtional group (Friedman & Finley, 1975; Masri & Friedman, 1988). Studies on the reactivity of poly(ethylene glycol) vinyl sulfone toward reduced ribonuclease (Morpurgo *et al.*, 1996) found that the reaction with cysteine groups is rapid and selective at pH 7-9 and with lysines proceeds slowly at pH 9.3. Other residues were described as not reactive. These results have been the dogma of the reactivity of vinyl sulfone with proteins. However, as early as 1965, it was reported that at comparable pKa values and steric enviroment thiols are 280 times more reactive than amine groups but also that the reactivity of the thiol group in an aminothiol acid is influenced by the presence of charge on neighboring amino groups and caution in the use of specific sulfhydryl specific reagents in proteins was recommended (Friedman *et al.*, 1965). A systematic study of the thio Michael additions confirmed the importance of the charges close to the cysteine and the existence of a linear correlation between thiolate concentration and kinetic constants (Lutolf *et al.*, 2001). More recently, the authors' group also found unexpected reactivity of His at pH 7.7 in the reaction of lysozyme with sugar vinyl sulfone derivatives and a double addition to a single Lys while other Lys residues remained unreacted (Lopez-Jaramillo *et al.*, 2005). In fact, the reaction of lysozyme proceeds very fast even at pH 5. At this point, it is important to recall that the non equivalence of identical residues present in proteins is an important concept frequently overlooked. The different nucleophilic character of identical residues is a well illustrated concept. Thus, it has been reported pKa values for internal lysines as low as 5.3 (Isom *et al.*, 2011), the standard pKa being ~10.4, and also pKa values for histidine ranging from 9.2 (His72 in tyrosine phosphatase) to 4.6 (His40 of bovine chymiotrypsinogen), the standard pKa value being 6.6 (Edgcomb & Murphy, 2002). Thus, the presence of a plethora of potential reactive groups in proteins and the dependence of their reactivity on the neighboring residues make group-specific modification chemistry unsuited as a general strategy for the selective modification of a particular residue but still valid for many omics applications.

Finally, it should be mentioned that in comparison with maleimides, one of the most widely-used conjugated reagent for chemical modifications of thiol-containing proteins, vinyl sulfones offers as advantages the aforementioned enhanced stability in aqueous alkaline conditions and the fact that the reaction product is a single stereoisomer, unlike conjugation with maleimides, which produces two potential stereoisomers.

3. Application of vinyl sulfones in proteomics

Vinyl sulfones have found application in most of the subdomains of modern Proteomics. Overall, these applications can be group in two main areas: labeling in their different variants (attachment of analytical or engineered probes for protein tracking, protein identification or protein profiling) and immobilization with different purposes (affinity chromatography and microarray technologies), two of the cornerstones of any omic science. In addition, vinyl sulfones have found applications in the conjugation of proteins with other biomolecules to yield post-translational modifications.

3.1 Vinyl sulfone-based labeling and chemical tagging

Proteomic often requires labeling of compounds for detection/isolation. Mass spectrometry offers a label-less method currently used in quantitative Proteomics. *Stricto sensu* it involves an isotopic labeling, usually referred as isotope tagging (Nakamura & Oda, 2007; Iliuk *et al.*, 2009), that can be carried out *in vivo* in the cell culture by metabolic incorporation or alternatively after protein extraction by chemical labeling, the former highlighting the importance of chemical tagging reactions (Leitner & Lindner, 2006). The reactivity of the vinyl sulfone group toward amino acids naturally occurring in proteins is conceptually an attractive derivatization strategy to promote the covalent attachment of labels to proteins. Despite that bibliographic references are scarce, vinyl sulfone derivatized dyes, fluorophores and other tags (biotin) have been already described and implemented in Proteomics.

The use of vinyl sulfone derivatized reagent for detection in Proteomic dates back to 1972 when Remazol dyes were reported as prestaining reagents during denaturation prior to sodium dodecyl sulfate polyacrylamide gel electrophoresis (SDS PAGE) that also allow the tracking by eye of the migration of the protein during the electrophoretic separation (Griffith, 1972). Remazol dyes are easily converted to vinyl sulfone derivatives at alkaline pH and upon reaction the electrophoretic mobility of the sample is not seriously affected since the dodecyl sulfate bound to the protein renders irrelevant any small charge difference in the proteins. Ulterior studies found that pre-stained proteins eluted from the gels retained immunological reactivity and were suitable to raise monospecific antibodies (Saoji *et al.*, 1983). The original idea is still valid and Remazol dyes are currently used in prestained color-coded molecular weight markers for gel electrophoresis (Compton *et al.*, 2002).

In a mass spectrometry-driven proteomic scenario gel electrophoresis is still part of the workflow, the in gel digestion of proteins being a cornerstone (Shevchenko *et al.*, 2006). However, the staining of the gel, selection and extraction, in-gel reduction, alkylation and destain for the subsequent tryptic digestion is a time and labor demanding process that represents a bottleneck. Pre-electrophoresis staining is an attractive approach that has not been extensively used because of the slight mobility differences that have been reported, despite the availability of fluorescent dyes that are charge-matched to preserve the pI of the proteins upon labeling (Miller *et al.*, 2006). In this context, the use of Uniblue A (**16**, Fig. 5), the vinyl sulfone derivative of Remazol Brilliant Blue R colorant, has been proposed as a straightforward strategy that i) yields the covalently stain of both simple and complex protein samples within 1 minute and ii) does not compromised protein profiles on the gels (Mata-Gomez, 2010). Another application in this area proposed the use of the reactivity of divinyl sulfone with the α-amino groups of N-terminal residues in Proteomics since it enhances the abundance of the a1 fragments, defining the N-terminal residue and providing a "one step Edman like information"(Boja *et al.*, 2004).

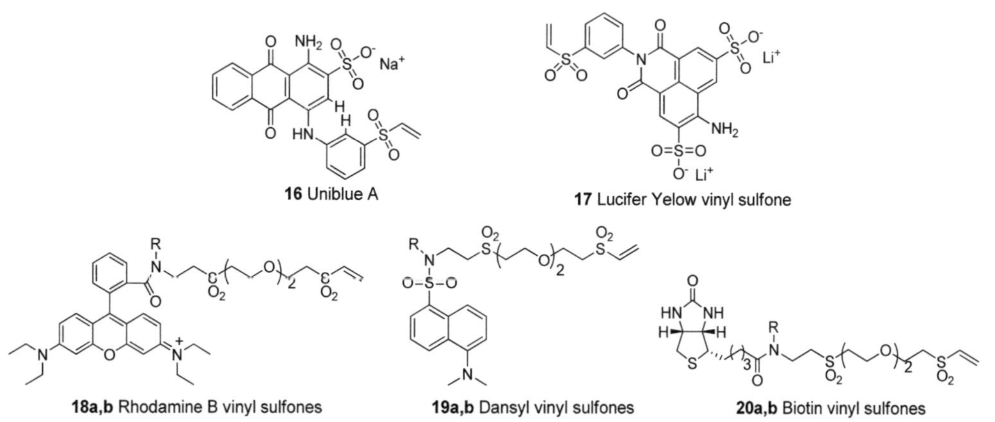

16 Uniblue A

17 Lucifer Yelow vinyl sulfone

18a,b Rhodamine B vinyl sulfones

19a,b Dansyl vinyl sulfones

20a,b Biotin vinyl sulfones

a R = $-CH(CH_3)CH_2CH_3$; **b** R = $-CHC{\equiv}CH$ **(AVST)**

Fig. 5. Some vinyl sulfone derivatized dyes, fluorescent probes and tags (biotin) used in Proteomics

On that concerning fluorophores, to our knowledge Lucifer Yellow vinyl sulfone (**17**, Fig. 5) was the first vinyl sulfone derivatized fluorophore applied to protein studies. This compound was the fluorescent probe used for fluorescence resonance energy transfer experiments on the chloroplast coupling factor 1 that showed that ATP induces changes on the nucleotide binding site and switches properties (Shapiro & McCarty, 1988; 1990) and allowed to gain insight into the asymmetry of the α subunit of CF1 (Lowe & McCarty, 1998). Lucifer Yellow vinyl sulfone was also used to study the interaction between Rod G-protein α subunit and cGMP-phosphodiesterase γ-subunit (Artemyev *et al.*, 1992). More recently the synthesis of vinyl sulfone derivatized rhodamine B and dansyl (**18a** and **19a**,

Fig. 5) and their reactivity with a pool of commercial proteins has been reported (Morales-Sanfrutos *et al.*, 2010b). The results showed that the protein itself influences the extend of coupling and that the labeling is feasible regardless of isoelectric point, number of potential nucleophiles or presence of glycosylation in the protein. The study also showed the potential of the rhodamine B vinyl sulfone as a prestaining reagent since the labelling does not affect the electrophoretic mobility or post-electrophoresis silver stain. The analysis of the influence of the reaction conditions on the reactivity between vinyl sulfone and Henn Egg-white (HEW) lysozyme revealed that the reaction takes place even in acidic media and that slight variations of pH or temperature exert a clear direct effect on the number of labels coupled to lysozyme. In a later work the same authors developed a series of alkyne vinyl sulfone derivatized tags (AVST reagents) bearing rhodamine B or dansyl (**18b** and **19b**, Fig. 5) and demonstrated their applicability as self-reporter reagents for monitoring the introduction of the alkyne function and their potential to carry out further functionalization in any scenario based on click-chemistry (Morales-Sanfrutos *et al.*, 2010a).

Vinyl sulfone derivatization has been used to attach other tags to proteins such as biotin. The authors' group has described (Morales-Sanfrutos *et al.*, 2010b) the conjugation of biotin vinyl sulfone (**20a**, Fig. 5) to promote the coupling of the biotinylated protein to avidin in the context of what is known as *avidin-biotin technology* (Savage et al., 1992). In the more advanced contribution mentioned above (Morales-Sanfrutos *et al.*, 2010a), the synthesis of vinyl sulfone bifunctional tags bearing simultaneously biotin and a fluorophore as a single-attachment-point reagents (BTSAP, **22** and **23** Fig. 6) was easily performed by click copper-catalyzed azide-alkyne cycloaddition (CuAAC) attachment of the AVST fluorophores reagents (**18-20b**, Fig. 5 and Fig. 6). The combination of vinyl sulfone as reactive group, biotin as an anchor point and a fluorophore as a reporter group in the BTSAP reagents made of versatile compounds with a clear potential in Proteomics as illustrated in the labeling of the low reactive protein horseradish peroxidase (HRP) (Fig. 6, route a). Alternatively, the dual labeling of this protein was also attained by a CuAAC-based sequential approach consisting in the labeling with an AVST reagent and ulterior click conjugation with an azide-containing biotin derivative (**21**) (Fig. 6, route b).

3.2 Vinyl sulfone-based chemical proteomics

Vinyl sulfones have been used in activity-based protein profiling (ABPP) (Evans & Cravatt, 2006; Hagenstein & Sewald, 2006), a methodology of interest in the so-called chemical Proteomics subdomain devoted to measure the activity of proteins to gain insight into the functional role of proteins in cell physiology and pathology. ABPP is a chemical strategy based on the use of activity-based probes (ABPs), small molecules that form activity dependent covalent bonds to a target enzyme (Fig. 7). These probes contain three main elements: (1) a warhead or reactive functional group that forms the covalent bond with the active site catalytic residue of a target (2) a linker that can be used to control the specificity of binding interactions between the probe and target enzyme and (3) a tagging group that allows probe labeled targets to be isolated, biochemically characterized or imaged. The majority of ABPs contain electrophilic warheads derived from well-known irreversible enzyme inhibitors. Many of the most versatile ABPs represent the simple conjugation of well-characterized covalent inhibitors to reporter tags such as fluorophores and biotin. The research efforts performed in this field have engendered ABPP probes for numerous enzyme classes.

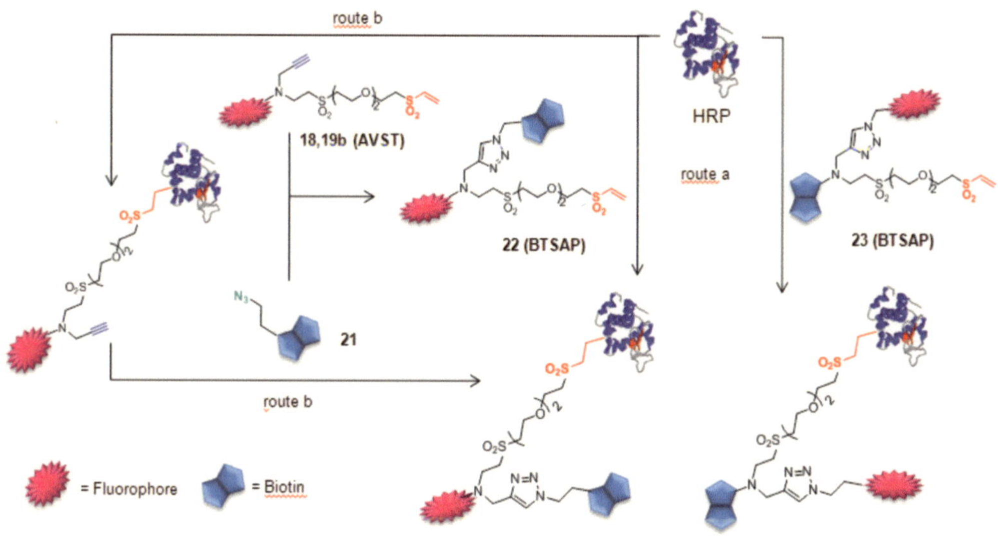

Fig. 6. Vinyl sulfone derivatized bifunctional tag single-attachment-point reagents (BTSAP)

Traditionally vinyl sulfones have been recognized as Cys protease inhibitors (Palmer *et al.*, 1995; Wang & Yao, 2003) and hence a large number of vinyl sulfone-containing peptides have been synthesized and exploited to inhibit them. On this basis, the reactivity of the vinyl sulfone function toward thiols has been used in ABPs to address cysteine proteases. As representative examples of these vinyl sulfone-based ABPs, it can be mentioned the studies performed on deubiquitinating enzymes with ABPs consisting of a truncated ubiquitinn or ubiquitin-like probe and biotin as reporter (**24**, Fig. 7A) (Borodovsky *et al.*, 2005). In addition, aryl vinyl sulfone and sulfonate probes have been developed to investigate the activity of protein tyrosine phosphatases (PTP) (Fig 7B.). In this case an azide group has been incorporated to the tag (**25**) to attach by click chemistry alkyne labels (**26**) such as biotin in order to facilitate the analysis (Liu *et al.*, 2008). Finally, in other studies dipeptidyl peptidase I (i.e. cathepsin C) has been selectively labeled by a vinyl sulfone norvaline-homophenylalanine dipeptide ABP (**27**, Fig. 7C) (Yuan *et al.*, 2006).

Vinyl sulfone-based ABPs have been also used for other class of hydrolases (Fig. 8). Thus, a series of tripeptide and tetrapeptide vinyl sulfone has been used as proteasome-directed ABPs to selectively engage the catalytic threonine nucleophile within proteasome active sites (Bogyo *et al.*, 1998; Nazif & Bogyo, 2001). By varying the peptide portion of the probes in a positional scanning library (**29**), the researchers gained insights into the substrate recognition properties of specific proteasomal subunits, culminating in the development of Z-subunit specific inhibitors that were used to identify this subunit as the principal trypsin-like activity of the proteasome. More recently, azide versions of vinyl sulfone probes (**30**) were used as tag-free ABPs to profile proteasomal activities in living cells, detection accomplished by tandem labeling strategies using highly specific bioorthogonal Staudinger ligation with a phosphine reporter tag (Ovaa *et al.*, 2003).

3.3 Vinyl sulfone-based affinity chromatography applications

Two dimensional gels resolve no more than several thousand proteins, only the most abundant ones being visualized, and Proteomics also includes the analysis of post-

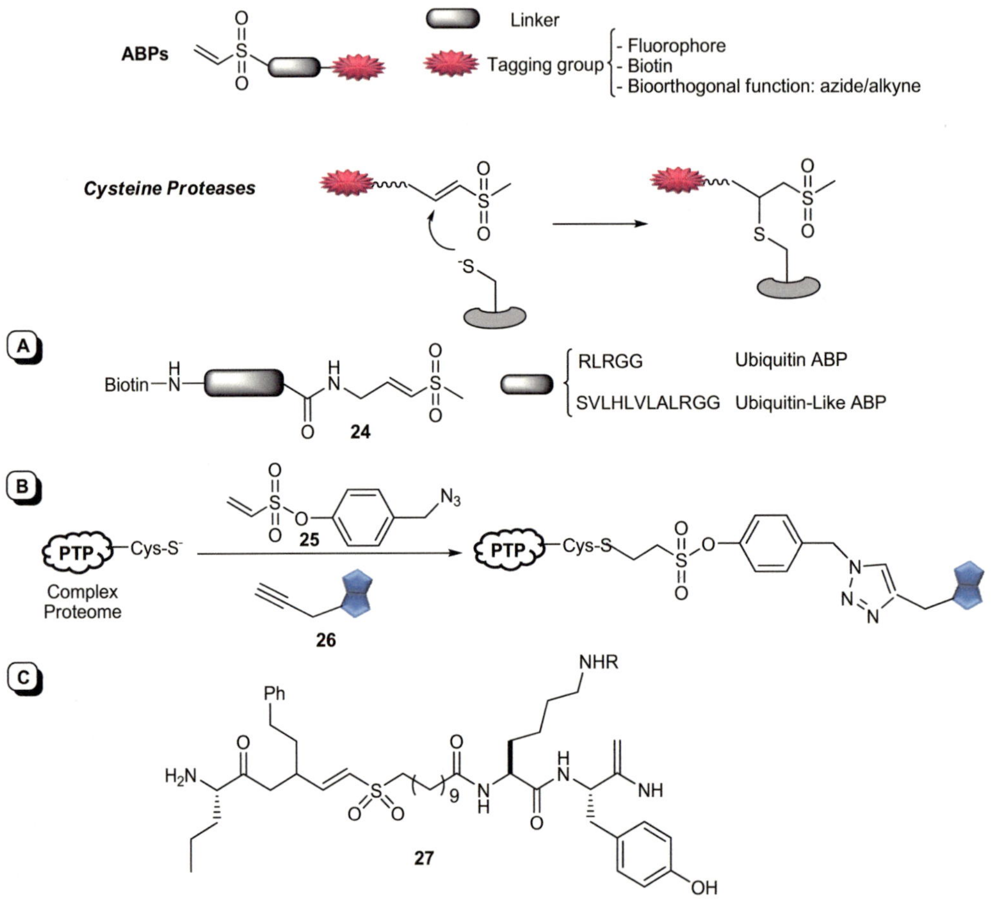

Fig. 7. General structure of vinyl sulfone-based ABPs and representative examples of those targeting cysteine proteases.

translational modifications (Mann & Jensen, 2003) and protein-protein interactions (Blagoev *et al.*, 2003). In this context, the immobilization of ligands plays a central role either for bioseparation and concentration of biomolecules (Lee & Lee, 2004; Azarkan *et al.*, 2007), for pull-down assays and mass spectrometry analysis (Bécamel *et al.*, 2002) or for high-throughput screening in array format (Cahill, 2000). However, examples of vinyl sulfone functionalized supports either for affinity chromatography or arrays are scarce.

Still in use, divinyl sulfone-activated agarose was the first support bearing the vinyl sulfone function to turn it out into an affinity support upon reaction with a wide variety of ligands. Described in 1975 (Porath *et al.*, 1975), Lihme *et al.* (Lihme *et al.*, 1986) were who reported its application in affinity chromatography as an alternative to CNBr-activated gels. They coupled i) rabbit immunoglobulin for preparation of goat anti-rabbit immunoglobulin, ii) goat anti-rabbit immunoglobulin for preparation of rabbit immunoglobulin, iii) lectins and iv) L-fucose. Remarkably beads coupled to lectins or saccharides are currently used in glycomics (Kaji *et al.*, 2003; Bunkenborg *et al.*, 2004; Yang & Hancock, 2004). Vinyl sulfone activated agarose has been the bead of choice to study the interaction of pepsin with aromatic amino acids (Frydlova *et al.*, 2004; Frydlova *et al.*, 2008) or for the isolation of

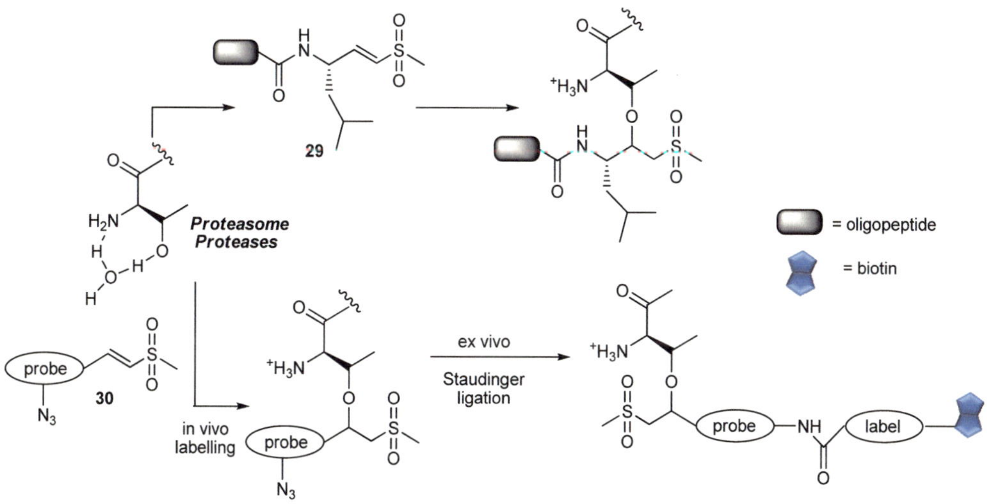

Fig. 8. Representative examples of vinyl sulfone-based ABPs targeting proteasomal proteases.

phophorylcholine-binding proteins (Liberda *et al.*, 2002a) by affinity chromatography. To our knowledge the only work using vinyl sulfone activated sepharose that may resemble Proteomics is that published by Liberda et al, (Liberda *et al.*, 2002b) who immobilized mannan to isolate mannan-binding bull seminal proteins that were identified by N-terminal amino acid sequencing.

In principle, the use of silica in Proteomics is discouraged since as predicted by Arai and Norde (Arai & Norde, 1990) macromolecules are adsorbed onto silica via strong electrostatic interactions and the secondary structure of the proteins can be distorted. However, the authors' group has reported the functionalization of silica with vinyl sulfone (**31**, Fig. 9) to yield a novel "ready to use" pre-activated material that reacts with biomolecules in mild conditions, preserves the activity of enzymes and can be used as an open support in Proteomics (Morales-Sanfrutos *et al.*, 2010b; Ortega-Munoz *et al.*, 2010). In a recent work (Traverso *et al.*, 2010), the application of this hybrid organic-inorganic material to Proteomics was further validated in a pull down experiment that demonstrated the different affinity of two pea h-type thioredoxins for proteins from a crude extract: thioredoxin h2 interacted with classical antioxidant proteins whereas thioredoxin h1 was able to capture a transcription factor, suggesting a regulatory role. These results support the use of vinyl sulfone silica in Proteomics for the study of protein-protein interactions.

3.4 Vinyl sulfone-based microarray technologies

Arrays are another important tool in Proteomics. They rely on the interaction between an immobilized probe and the molecules in the sample being analyzed. Immobilization is an important variable and different methods of both covalent and non-covalent immobilization are used with their pros and cons. Up to the present, a limited number of reports have described the preparation and use of different vinyl sulfone-modified surfaces in the construction of microarrays, the majority of them focused on potential applications in other omics (*vide infra* section 4). Only one of these contributions describes a gelatin-based substrate functionalized with vinyl sulfone groups for fabricating protein arrays (Fig. 10)

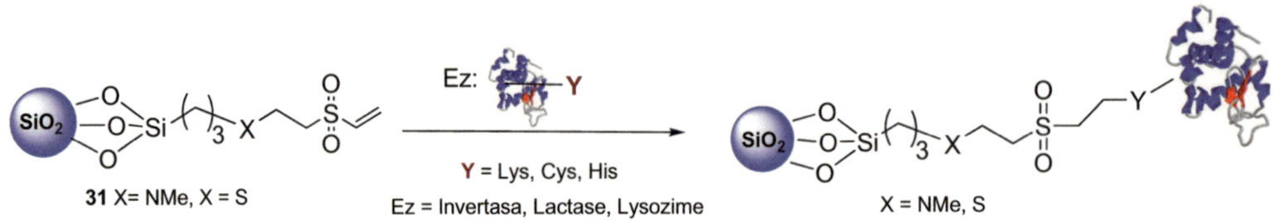

Fig. 9. Immobilization of enzymes onto vinyl sulfone silica

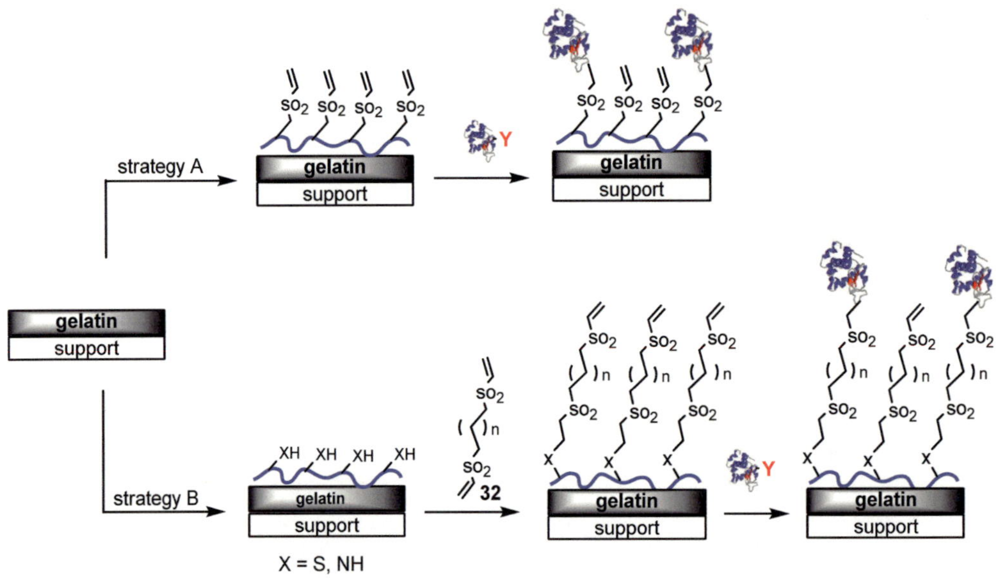

Fig. 10. Vinyl sulfone-gelatine protein microarrays

(Qiao *et al.*, 2003). The rationale behind the design of these materials are the use of gelatin coating to eliminate non-specific protein binding and the affixing to this gelatin surface of a vinyl sulfone derivatized polymer scaffold to enable them for the direct immobilization of proteins (strategy A). In an alternatively strategy, the gelatin surface is first affixed with a polymer scaffold rich in thiols or amine groups, then reacted with a bis(vinylsulfonyl) compound (**32**) and finally bonded to a protein capture agent such as an antibody (strategy B).

3.5 Vinyl sulfone-based post-translational modifications

Protein post-translational modifications increase the functional diversity of the proteome and the access to pure protein derivatives is essential in order to gain insight into structure-activity relationships and their biological role. Among the different post-translational modifications of proteins, glycosylation is the most prevalent one, occurring in at least 50% of all proteins (Apweiler *et al.*, 1999). However, the fact that glycosylation is not template driven makes the large scale production of glycoproteins a challenging task that has been approached by biological, enzymatic and chemical strategies (Davis, 2002; Bennett & Wong, 2007; Gamblin *et al.*, 2008a; Bernardes *et al.*, 2009). The authors' group has already demonstrated the feasibility of the vinyl sulfone functionalization of the anomeric carbon on different carbohydrates (**33**) as a procedure for the chemical glycosylation of proteins (Fig. 11) (Lopez-Jaramillo *et al.*, 2005) and current work is focused on its application in the context of glycoscience to explore protein-carbohydrate interactions. In this context, a model system comprising four monosaccharides (L-fucose, D-glucose, D-mannose and N-acetyl-D-glucosamine) and three disaccharides (lactose, maltose and melibiose) with a vinyl sulfone group at the anomeric carbon were reacted with four model proteins (lysozyme, BSA, concanavalin A and lumazine) (unpublished results). Enzyme-linked lectine assays (ELLA) of the resulting neoglycoconjugates with lectins revealed that the extent of binding of the lectins was consistent with their carbohydrate-binding specificity: Concanavalin A (ConA) showed binding with proteins derivatized with vinyl sulfone D-mannose while peanut agglutinin (PNA), ulex europeaus aggluttinin (UEA) and wheat germ agglutinin (WGA) interacted with those proteins reacted with vinyl sulfone derivatized lactose, L-fucose and N-acetyl-D-glucosamine, respectively.

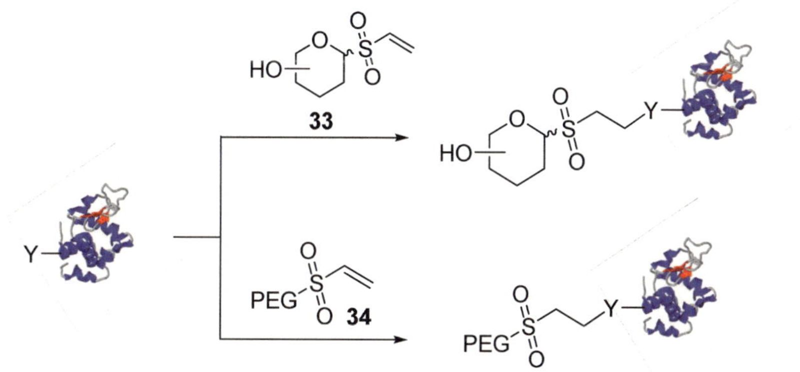

Fig. 11. Vinyl sulfone based glycosylation and PEGylation of proteins

Although *stricto sensu* PEGylation (covalent attachment of polyethylene glycol -PEG-chains) is not a post-translational modification, it is important for pharmaceutical and biological applications (Brannon-Peppas, 2000). Covalent attachment of PEG to proteins shields their antigenic and immunogenic epitopes, interferes with the receptor mediated uptake and prevents recognition and degradation by proteolytic enzymes. Vinyl sulfone chemistry has been exploited in this field. Poly(ethylene glycol) vinyl sulfone (**34**, Fig. 11) was synthesized and its highly selective reaction with thiol groups relative to amino groups at pH lower than 9 was described (Morpurgo *et al.*, 1996). The idea of using vinyl sulfone derivatives for PEGylation at cysteine residues is still accepted and later contributions reported the use of these methodology although being aware of the side reaction with lysine at elevated pH (Roberts *et al.*, 2002).

4. Vinyl sulfones in other omics sciences

The complete sequencing of the human genome has led to a new era referred to as omic sciences that comprise a wide range of disciplines aims at analyzing the relationships among the different elements of various omes. A common characteristic is the use of innovative technology platforms that allow the high-throughput detection and identification of the large amount and variety of molecules expressed in living organisms. Both immobilization on a solid surface either for affinity chromatography applications or as arrays and coupling to other biomolecules are important elements shared by all omic sciences.

In the context of immobilization, vinyl sulfone activated sepharose and vinyl sulfone silica are two open affinity chromatographic supports valid not only in Proteomics (*vide supra* section 3.4) but also in glycomics to isolate glycoproteins if lectins are immobilized or in genomics if amine or thiol functionalized oligonucleotides are used. In the particular case of glycomics, divinyl sulfone (DVS) has been used for the surface functionalization of either the wells of microtiter plates containing primary amino groups (Hatakeyama *et al.*, 1996; Hatakeyama *et al.*, 1997) or hydroxyl-terminated self-assembled monolayers (SAMs) on Au (Cheng *et al.*, 2011). Both materials have demonstrated their capability for the direct chemical immobilization of natural and chemically derived carbohydrates as well as glycoproteins and their applicability for the development of a simple assay to determine lectin activity, in case of the vinyl sulfone functionalized microplates, and for the fabrication of a glycan microarray, in case of the vinyl sulfone derivatized SAMs. On the other hand, the activation of molecules via vinyl sulfone functionalization is a wide scope strategy for labeling (colorants and fluorophores) and tagging (biotin) not limited to Proteomics. Finally, in the particular case of glycomics, vinyl sulfone derivatization of sugars is especially appealing since as described above (section 3.5) it is suitable for the synthesis of neoglycoconjugates that are recognized by lectins.

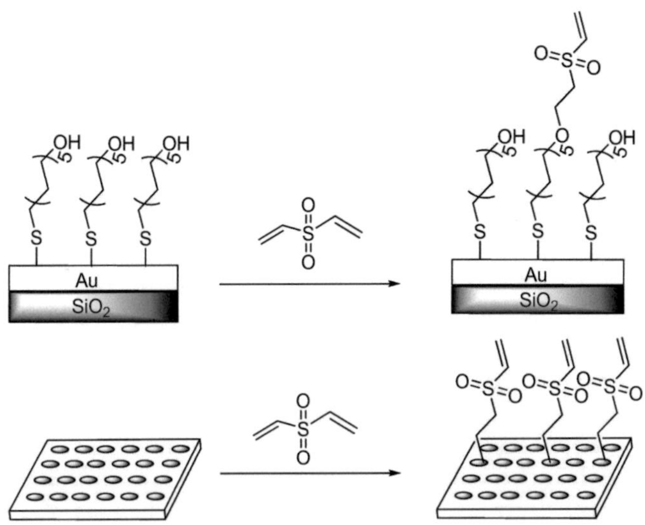

Fig. 12. Divinyl sulfone (DVS) functionalization of surfaces (SAM and microtiter plates) for applications in glycomics

In lipidomics, immobilized lipids are a valuable tool for the characterization and study of the lipid-protein interaction. This issue is not new in pharmaceutical industry where some of the most famous drugs target lipid-metabolizing enzymes. For example, atorvastatin (Lipitor from Pfizer) is a competitive inhibitor of 3-hydroxy-3-methylglutaryl-coenzyme A reductase (HMG-CoA), the rate controlling enzyme involve in the metabolic pathway of cholesterol, or Celecoxib (Celebrex from Pfizer) is a selective inhibitor of cyclooxygenase-2, enzyme responsible for the conversion of arachidonic acid into prostaglandin that is the molecule involved in inflammation and pain. Thus, lipid profiling for the identification of metabolic pathways and enzymes involved is an area of interest in lipidomics (Wenk, 2005). Both covalent and non-covalent immobilization strategies do not seem to compromise the activities of the group of phosphoinostides (Feng, 2005). In order to promote the covalent immobilization with surfaces, reactive groups including amine among others are introduced in the lipid molecule and in this context the above mentioned microarrays based in vinyl sulfone derivatized monolayers (SAMs) (Cheng *et al.*, 2011) can be applied to the synthesis of lipid microarrays. Another important issue is protein lipidization where vinyl sulfone chemistry can play a role. In general it is assumed that the hydrophobic acyl groups are involved in protein-membrane interaction and protein-protein interactions (McCabe & Berthiaume, 1999; Taniguchi, 1999). Historically, fatty acylation has been divided into two classes: cotranslational addition of myristate to N-terminal glycine through amide linkage (myristoylation) and post-translational addition of palmitate through a thioester linkage to cystein. Both N-terminal and thiol groups can be targeted by vinyl sulfone chemistry. Finally, it should be mentioned that, although for a different purpose, the authors' group has reported the synthesis of alkyl vinyl sulfones and vinyl sulfone functionalization of cholesterol and their reaction with poly(amidoamine) (PAMAM) dendrimers for the preparation of dendrimers-based nonviral gene delivery vectors with improved transfection efficiencies (Fig. 13) (Morales-Sanfrutos *et al.*, 2011).

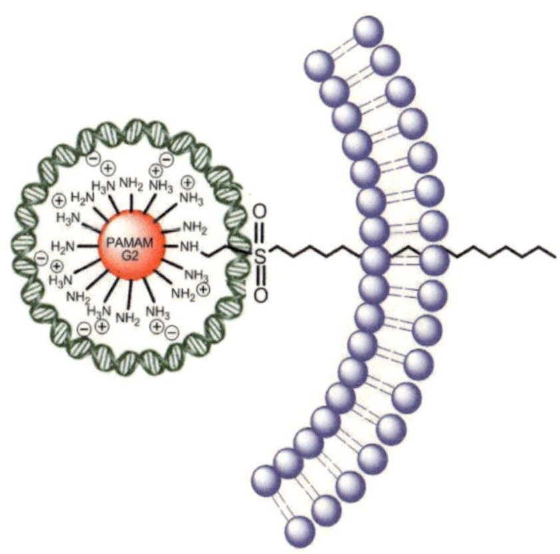

Fig. 13. Alkyl sulfonyl derivatized PAMAM-G2 dendrimers engineered by vinyl sulfone chemistry as nonviral gene delivery vectors with improved transfection efficiencies.

In the field of genomics, a method for gene analysis by simultaneously performing the polymerase chain reaction (PCR) reaction and the hybridization reaction of an oligonucleotide, a polynucleotide or a peptide nucleic acid fixed on a vinylsulfonyl functionalized silicate glass micro-array obtained by a tandem treatment with an amino silane coupling agent and a bis(vinylsulfonyl) compound has been reported (Iwaki *et al.*, 2004). This method avoids traditional operations where PCR and hybridization reactions are separately performed for gene analysis.

5. Conclusion

The reactivity of the vinyl sulfone function toward thiol and amine groups that are naturally present or routinely introduced in most of biomolecules makes it a wide scope strategy for functionalization with a clear potential in omic sciences. The examples in the previous sections are indicative to the usefulness of vinyl sulfone reactivity in Proteomics owed to their excellent capability to act as Michael acceptors in physiological conditions (aqueous media, slightly alkaline pH and room temperature) that preserves the biological function of the proteins with no formation of by-products. However, despite the existence of a body of knowledge in bibliography, the applications of vinyl sulfones are only partially exploited and the vast potential of these compounds for targeting biological macromolecules is yet to be unearthed. For the particular case of Proteomics it is important to recall the presence of a panoply of potential reactive groups in proteins and the dependence of their reactivity on the neighboring residues. Nevertheless, vinyl sulfone group is appealing despite the modification of a particular residue is far from trivial since this is not a critical issue for many applications in Proteomics. Its impact in other sciences is promising but still unexplored.

6. Acknowledgment

The authors acknowledge Direccion General de Investigacion Cientıfica y Tecnica (DGICYT) (CTQ2008-01754) and Junta de Andalucia (P07-FQM-02899) for financial support.

7. References

Alba, A.-N. R.; Companyo, X. & Rios, R. (2010). Sulfones: new reagents in organocatalysis. *Chem. Soc. Rev.*, 39, 6, 2018-2033, ISNN 0306-0012

Alonso, D. A.; Nájera, C. & Varea, M. (2002). Simple, economical and environmentally friendly sulfone synthesis. *Tetrahedron Lett.*, 43, 19, 3459-3461, ISNN 0040-4039

Anson, M. L. (1940). The reactions of iodine and iodoacetamide with native egg albumin. *J. Gen. Physiol.*, 23, 3, 321-331, ISNN 0022-1295

Apweiler, R.; Hermjakob, H. & Sharon, N. (1999). On the frequency of protein glycosylation, as deduced from analysis of the SWISS-PROT database. *Biochim. Biophys. Acta*, 1473, 1, 4-8, ISNN 0304-4165

Arai, T. & Norde, W. (1990). The behavior of some model proteins at solid-liquid interfaces 1. Adsorption from single protein solutions. *Colloid Surface*, 51, 1-15, ISNN 0166-6622

Artemyev, N. O.; Rarick, H. M.; Mills, J. S.; Skiba, N. P. & Hamm, H. E. (1992). Sites of interaction between rod G-protein α-subunit and cGMP-phosphodiesterase γ-subunit. Implications for the phosphodiesterase activation mechanism. *J. Biol. Chem.*, 267, 35, 25067-25072, ISNN 0021-9258

Azarkan, M.; Huet, J.; Baeyens-Volant, D.; Looze, Y. & Vandenbussche, G. (2007). Affinity chromatography: A useful tool in proteomics studies. *J. Chromatogr. B*, 849, 1-2, 81-90, ISNN 1570-0232

Baciocchi, E.; Gerini, M. F. & Lapi, A. (2004). Synthesis of Sulfoxides by the Hydrogen Peroxide Induced Oxidation of Sulfides Catalyzed by Iron Tetrakis (pentafluorophenyl) porphyrin: Scope and Chemoselectivity. *J. Org. Chem.*, 69, 10, 3586-3589, ISNN 0022-3263

Banks, P. R. & Paquette, D. M. (1995). Comparison of 3 common amine reactive fluorescent probes used for conjugation to biomolecules by capillary zone electrophoresis. *Bioconjugate Chem.*, 6, 4, 447-458, ISNN 1043-1802

Baslé, E.; Joubert, N. & Pucheault, M. (2010). Protein Chemical Modification on Endogenous Amino Acids. *Chem. Biol.*, 17, 3, 213-227, ISNN 1074-5521

Beatty, K. E. & Tirrell, D. A. (2009). Noncanonical Amino Acids in Protein Science and Engineering. In: *Protein Engineering* Köhrer, C., RajBhandary, U. L. (Eds.). pp. 127-153, Springer, ISBN 978-3-540-70941-1, Berlin

Bécamel, C.; Galéotti, N.; Poncet, J.; Jouin, P.; Dumuis, A.; Bockaert, J. & Marin, P. (2002). A proteomic approach based on peptide affinity chromatography, 2-dimensional electrophoresis and mass spectrometry to identify multiprotein complexes interacting with membrane-bound receptors. *Biol. Proced. Online*, 4, 1, 94-104, ISNN 1480-9222 (Electronic)

Bednar, R. A. (1990). Reactivity and pH dependence of thiol conjugation to N-ethylmaleimide: detection of a conformational change in chalcone isomerase. *Biochemistry*, 29, 15, 3684-3690, ISNN 0006-2960

Bennett, C. S. & Wong, C.-H. (2007). Chemoenzymatic approaches to glycoprotein synthesis. *Chem. Soc. Rev.*, 36, 8, 1227-1238, ISNN 0306-0012

Bernardes, G. J. L.; Castagner, B. & Seeberger, P. H. (2009). Combined Approaches to the Synthesis and Study of Glycoproteins. *Chem. Biol.*, 4, 9, 703-713, ISNN 1554-8929

Bernardes, G. J. L.; Grayson, E. J.; Thompson, S.; Chalker, J. M.; Errey, J. C.; ElOualid, F.; Claridge, T. D. W. & Davis, B. G. (2008). From disulfide- to thioether-linked glycoproteins. *Angew. Chem., Int. Ed.*, 47, 12, 2244-2247, ISNN 1433-7851

Blagoev, B.; Kratchmarova, I.; Ong, S.-E.; Nielsen, M.; Foster, L. J. & Mann, M. (2003). A proteomics strategy to elucidate functional protein-protein interactions applied to EGF signaling. *Nat. Biotechnol.*, 21, 3, 315-318, ISNN 1087-0156

Bogyo, M.; Shin, S.; McMaster, J. S. & Ploegh, H. L. (1998). Substrate binding and sequence preference of the proteasome revealed by active-site-directed affinity probes. *Chem. Biol.*, 5, 6, 307-320, ISNN 1074-5521

Boja, E. S.; Sokoloski, E. A. & Fales, H. M. (2004). Divinyl Sulfone as a Postdigestion Modifier for Enhancing the a1 Ion in MS/MS and Postsource Decay: Potential Applications in Proteomics. *Anal. Chem.*, 76, 14, 3958-3970, ISNN 0003-2700

Bordwell, F. G. & Pitt, B. M. (1955). The Formation of α-Chloro Sulfides from Sulfides and from Sulfoxides. *J. Am. Chem. Soc.*, 77, 3, 572-577, ISNN 0002-7863

Borodovsky, A.; Ovaa, H.; Meester, W. J. N.; Venanzi, E. S.; Bogyo, M. S.; Hekking, B. G.; Ploegh, H. L.; Kessler, B. M. & Overkleeft, H. S. (2005). Small-molecule inhibitors and probes for ubiquitin- and ubiquitin-like-specific proteases. *ChemBioChem*, 6, 2, 287-291, ISNN 1439-4227

Brace, N. O. (1993). An economical and convenient synthesis of phenyl vinyl sulfone from benzenethiol and 1,2-dichloroethane. *J. Org. Chem.*, 58, 16, 4506-4508, ISNN 0022-3263

Harris, J. M. & Zalipsky, S (Eds.) (1997) *Poly(ethylene glycol): Chemistry and biological applications* ACS Symposium Series 680, ISBN 0841235376, Washington.

Bunkenborg, J.; Pilch, B. J.; Podtelejnikov, A. V. & Wisniewski, J. R. (2004). Screening for N-glycosylated proteins by liquid chromatography mass spectrometry. *Proteomics*, 4, 2, 454-465, ISNN 1615-9853

Cahill, D. J. (2000). Protein arrays: a high-throughput solution for proteomics research? *Trends Biotecnol.*, 18, 47-51, ISNN 0167-7799

Compton, M. M.; Lapp, S. A. & Pedemonte, R. (2002). Generation of multicolored, prestained molecular weight markers for gel electrophoresis. *Electrophoresis*, 23, 19, 3262-3265, ISNN 0173-0835

Cravatt, B. F.; Wright, A. T. & Kozarich, J. W. (2008). Activity-based protein profiling: From enzyme chemistry to proteomic chemistry. *Annu. Rev. Biochem.*, 77, 383-414, ISNN 0066-4154

Chalker, J. M.; Bernardes, G. J. L.; Lin, Y. A. & Davis, B. G. (2009). Chemical Modification of Proteins at Cysteine: Opportunities in Chemistry and Biology. *Chem-Asian J.*, 4, 5, 630-640, ISNN 1861-4728

Cheng, F.; Shang, J. & Ratner, D. M. (2011). A Versatile Method for Functionalizing Surfaces with Bioactive Glycans. *Bioconjugate Chem.*, 22, 1, 50-57, ISNN 1043-1802

Davis, B. G. (2002). Synthesis of Glycoproteins. *Chem. Rev.*, 102, 2, 579-602, ISNN 0009-2665

de Graaf, A. J.; Kooijman, M.; Hennink, W. E. & Mastrobattista, E. (2009). Nonnatural Amino Acids for Site-Specific Protein Conjugation. *Bioconjugate Chem.*, 20, 7, 1281-1295, ISNN 1043-1802

De Lucchi, O. & Pasquato, L. (1988). The role of sulfur functionalities in activating and directing olefins in cycloaddition reactions. *Tetrahedron*, 44, 22, 6755-6794, ISNN 0040-4020

Edgcomb, S. P. & Murphy, K. P. (2002). Variability in the pKa of histidine side-chains correlates with burial within proteins. *Proteins*, 49, 1, 1-6, ISNN 0887-3585

Evans, M. J. & Cravatt, B. F. (2006). Mechanism-based profiling of enzyme families. *Chem. Rev.*, 106, 106, 3279-3301, ISNN 0009-2665

Fang, X. & Zhang, W.-W. (2008). Affinity separation and enrichment methods in proteomic analysis. *J. Proteomics*, 71, 3, 284-303, ISNN 1874-3919

Feng, L. (2005). Probing lipid-protein interactions using lipid microarrays. *Prostag. Oth. Lipid M.*, 77, 1-4, 158-167, ISNN 1098-8823

Fontana, A.; Scoffone, E. & Benassi, C. A. (1968). Sulfenyl halides as modifying reagents for polypeptides and proteins. II. Modification of cysteinyl residues. *Biochemistry*, 7, 3, 980-986, ISNN 0006-2960

Forristal, I. (2005). The chemistry of α,β-unsaturated sulfoxides and sulfones: an update. *J. Sulfur Chem.*, 26, 2, 163-195, ISNN 1741-5993

Friedman, M.; Cavins, J. F. & Wall, J. S. (1965). Relative Nucleophilic Reactivities of Amino Groups and Mercaptide Ions in Addition Reactions with α,β-Unsaturated Compounds. *J. Am. Chem. Soc.*, 87, 16, 3672-3682, ISNN 0002-7863

Friedman, M. & Finley, J. W. (1975). Reactions of proteins with ethyl vinyl sulfone. *Int. J. Pept. Prot. Res.*, 7, 6, 481-486, ISNN 0367-8377

Frydlova, J.; Kucerova, Z. & Ticha, M. (2004). Affinity chromatography of porcine pepsin and pepsinogen using immobilized ligands derived from the specific substrate for this enzyme. *J. Chromatogr. B*, 800, 1-2, 109-114, ISNN 1570-0232

Frydlova, J.; Kucerova, Z. & Ticha, M. (2008). Interaction of pepsin with aromatic amino acids and their derivatives immobilized to Sepharose. *J. Chromatogr. B*, 863, 1, 135-140, ISNN 1570-0232

Galli, U.; Lazzarato, L.; Bertinaria, M.; Sorba, G.; Gasco, A.; Parapini, S. & Taramelli, D. (2005). Synthesis and antimalarial activities of some furoxan sulfones and related furazans. *Eur. J. Med. Chem.*, 40, 12, 1335-1340, ISNN 0223-5234

Gamblin, D. P.; Scanlan, E. M. & Davis, B. G. (2008a). Glycoprotein Synthesis: An Update. *Chem. Rev.*, 109, 1, 131-163, ISNN 0009-2665

Gamblin, D. P.; van Kasteren, S. I.; Chalker, J. M. & Davis, B. G. (2008b). Chemical approaches to mapping the function of post-translational modifications. *FEBS J.*, 275, 9, 1949-1959, ISNN 1742-464X

Giepmans, B. N. G.; Adams, S. R.; Ellisman, M. H. & Tsien, R. Y. (2006). The Fluorescent Toolbox for Assessing Protein Location and Function. *Science*, 312, 5771, 217-224, ISNN 0036-8075

Griffith, I. P. (1972). Immediate visualization of proteins in dodecyl sulfate-polyacrylamide gels by prestaining with remazol dyes. *Anal. Biochem.*, 46, 2, 402-412, ISNN 0003-2697

Hackenberger, C. P. R. & Schwarzer, D. (2008). Chemoselective Ligation and Modification Strategies for Peptides and Proteins. *Angew. Chem., Int. Ed.*, 47, 52, 10030-10074, ISNN 1521-3773

Hagenstein, M. C. & Sewald, N. (2006). Chemical tools for activity-based proteomics. *J. Biotechnol.*, 124, 1, 56-73, ISNN 0168-1656

Heal, W. P. & Tate, E. W. (2010). Getting a chemical handle on protein post-translational modification. *Org. Biomol. Chem.*, 8, 4, 731-738, ISNN 1477-0520

Hermanson, G. T. (Ed.) (2008). *Bioconjugate Techniques* (2nd). Academic Press, ISBN 9780123705013, San Diego

Iliuk, A.; Galan, J. & Tao, W. A. (2009). Playing tag with quantitative proteomics. *Anal. Bioanal. Chem.*, 393, 2, 503-513, ISNN 1618-2642

Isom, D. G.; Castaneda, C. A.; Cannon, B. R. & Garcia-Moreno, B. E. (2011). Large shifts in pKa values of lysine residues buried inside a protein. *P. Natl. Acad. Sci. USA*, 108, 13, 5260-5265, ISNN 0027-8424

Iwaki, Y.; Shinoki, H. & Seshimoto, O. (2004). Detection of genes by simultaneous PCR/reverse transcription and hybridization on a covalently immobilized DNA probe microarray, U.S. Patent 7,169,583

Jentoft, N. & Dearborn, D. G. (1979). Labeling of proteins by reductive methylation using sodium cyanoborohydride. *J. Biol. Chem.*, 254, 11, 4359-4365, ISNN 0021-9258

Johnson, J. A.; Lu, Y. Y.; Van Deventer, J. A. & Tirrell, D. A. (2010). Residue-specific incorporation of non-canonical amino acids into proteins: recent developments and applications. *Curr. Opin. Chem. Biol.*, 14, 6, 774-780, ISNN 1367-5931

Kaji, H.; Saito, H.; Yamauchi, Y.; Shinkawa, T.; Taoka, M.; Hirabayashi, J.; Kasai, K.; Takahashi, N. & Isobe, T. (2003). Lectin affinity capture, isotope-coded tagging and mass spectrometry to identify N-linked glycoproteins. *Nat. Biotechnol.*, 21, 6, 667-672, ISNN 1087-0156

Krishna, P. R.; Lavanya, B.; Jyothi, Y. & Sharma, G. (2003). Radical Mediated Diastereoselective Synthesis of Benzothiazole Sulfonyl Ethyl C-Glycosides. *J. Carbohydr. Chem.*, 22, 6, 423-431, ISNN 0732-8303

Lee, J. W.; Lee, C.-W.; Jung, J. H. & Oh, D. Y. (2000). Facile Synthesis of Vinyl Sulfones from β-Bromo Alcohols. *Synth. Commun.*, 30, 16, 2897 - 2902, ISNN 0039-7911

Lee, W. C. & Lee, K. H. (2004). Applications of affinity chromatography in proteomics. *Annal. Biochem.*, 324, 1, 1-10, ISNN 0003-2697

Lefevre, C.; Kang, H. C.; Haugland, R. P.; Malekzadeh, N. & Arttamangkul, S. (1996). Texas Red-X and rhodamine Red-X, new derivatives of sulforhodamine 101 and lissamine rhodamine B with improved labeling and fluorescence properties. *Bioconjugate Chem.*, 7, 4, 482-489, ISNN 1043-1802

Leitner, A. & Lindner, W. (2006). Chemistry meets proteomics: The use of chemical tagging reactions for MS-based proteomics. *Proteomics*, 6, 20, 5418-5434, ISNN 1615-9861

Liberda, J.; Manaskova, P.; Svestak, M.; Jonakova, V. & Ticha, M. (2002a). Immobilization of L-glyceryl phosphorylcholine: isolation of phosphorylcholine-binding proteins from seminal plasma. *J. Chromatogr. B*, 770, 1-2, 101-110, ISNN 1570-0232

Liberda, J.; Ryslava, H.; Jelinkova, P.; Jonakova, V. & Ticha, M. (2002b). Affinity chromatography of bull seminal proteins on mannan-Sepharose. *J. Chromatogr. B*, 780, 2, 231-239, ISNN 1570-0232

Lihme, A.; Schafer-Nielsen, C.; Larsen, K. P.; Muller, K. G. & Bog-Hansen, T. C. (1986). Divinylsulphone-activated agarose. Formation of stable and non-leaking affinity matrices by immobilization of immunoglobulins and other proteins. *J. Chromatogr. B*, 376, 299-305, ISNN 0021-9673

Lim, R. K. V. & Lin, Q. (2010). Bioorthogonal chemistry: recent progress and future directions. *Chem. Commun.*, 46, 10, 1589-1600, ISNN 1359-7345

Lin, Y. A.; Chalker, J. M.; Floyd, N.; Bernardes, G. J. L. & Davis, B. G. (2008). Allyl sulfides are privileged substrates in aqueous cross-metathesis: Application to site-selective protein modification. *J. Am. Chem. Soc.*, 130, 30, 9642-9643, ISNN 0002-7863

Lindley, H. (1956). A New Synthetic Substrate for Trypsin and its Application to the Determination of the Amino-acid Sequence of Proteins. *Nature*, 178, 4534, 647-648, ISNN 0028-0836

Liu, C. C. & Schultz, P. G. (2010). Adding New Chemistries to the Genetic Code. *Annu. Rev. Biochem.*, 79, 1, 413-444, ISNN 0066-4154

Liu, S.; Zhou, B.; Yang, H.; He, Y.; Jiang, Z.-X.; Kumar, S.; Wu, L. & Zhang, Z.-Y. (2008). Aryl Vinyl Sulfonates and Sulfones as Active Site-Directed and Mechanism-Based Probes for Protein Tyrosine Phosphatases. *J. Am. Chem. Soc.*, 130, 26, 8251-8260, ISNN 0002-7863

Lopez-Jaramillo, F. J.; Perez-Banderas, F.; Hernandez-Mateo, F. & Santoyo-Gonzalez, F. (2005). Production, crystallization and X-ray characterization of chemically glycosylated hen egg-white lysozyme. *Acta Crystallogr. F*, F61, 4, 435-438, ISNN 1744-3091

Lowe, K. M. & McCarty, R. E. (1998). Asymmetry of the α Subunit of the Chloroplast ATP Synthase as Probed by the Binding of Lucifer Yellow Vinyl Sulfone. *Biochemistry*, 37, 8, 2507-2514, ISNN 0006-2960

Lundblad, R. I. (2005). *The Evolution from Protein Chemistry to Proteomics: Basic Science to Clinical Application* (1st). CRC Press, ISBN 9780849396786, Boca Raton

Lutolf, M. P.; Tirelli, N.; Cerritelli, S.; Cavalli, L. & Hubbell, J. A. (2001). Systematic Modulation of Michael-Type Reactivity of Thiols through the Use of Charged Amino Acids. *Bioconjugate Chem.*, 12, 6, 1051-1056, ISNN 1043-1802

Mann, M. & Jensen, O. N. (2003). Proteomic analysis of post-translational modifications. *Nat. Biotechnol*, 21, 3, 255-261, ISNN 1087-0156

Masri, M. S. & Friedman, M. (1988). Protein reactions with methyl and ethyl vinyl sulfones. *J. Protein Chem.*, 7, 1, 49-54, ISNN 0277-8033

Mata-Gomez, M. Y., M.; Winkler, R. (2010). Rapid pre-gel visualization of proteins with mass spectrometry compatibility, In: *Nature Preceeding*, Availabe from: http://hdl.handle.net/10101/npre.2010.5163.1

McCabe, J. B. & Berthiaume, L. G. (1999). Functional Roles for Fatty Acylated Amino-terminal Domains in Subcellular Localization. *Mol. Biol. Cell*, 10, 11, 3771-3786, ISNN 1059-1524

Meadows, D. C. & Gervay-Hague, J. (2006). Vinyl sulfones: Synthetic preparations and medicinal chemistry applications. *Med. Res. Rev.*, 26, 6, 793-814, ISNN 1098-1128

Means, G. E. & Feeney, R. E. (1990). Chemical Modifications of Proteins: History and Applications. *Bioconjugate Chem.*, 1, 1, 2-12, ISNN 1043-1802

Miller, I.; Crawford, J. & Gianazza, E. (2006). Protein stains for proteomic applications: Which, when, why? *Proteomics*, 6, 20, 5385-5408, ISNN 1615-9853

Morales-Sanfrutos, J.; Lopez-Jaramillo, F. J.; Hernandez-Mateo, F. & Santoyo-Gonzalez, F. (2010a). Vinyl Sulfone Bifunctional Tag Reagents for Single-Point Modification of Proteins. *J. Org. Chem.*, 75, 12, 4039-4047, ISNN 0022-3263

Morales-Sanfrutos, J.; Lopez-Jaramillo, J.; Ortega-Munoz, M.; Megia-Fernandez, A.; Perez-Balderas, F.; Hernandez-Mateo, F. & Santoyo-Gonzalez, F. (2010b). Vinyl sulfone: a versatile function for simple bioconjugation and immobilization. *Org. Biomol. Chem.*, 8, 3, 667-675, ISNN 1477-0520

Morales-Sanfrutos, J.; Megia-Fernandez, A.; Hernandez-Mateo, F.; Giron-Gonzalez, M. D.; Salto-Gonzalez, R. & Santoyo-Gonzalez, F. (2011). Alkyl sulfonyl derivatized PAMAM-G2 dendrimers as nonviral gene delivery vectors with improved transfection efficiencies. *Org. Biomol. Chem.*, 9, 3, 851-864, ISNN 1477-0520

Morpurgo, M.; Veronese, F. M.; Kachensky, D. & Harris, J. M. (1996). Preparation and Characterization of Poly(ethylene glycol) Vinyl Sulfone. *Bioconjugate Chem.*, 7, 3, 363-368, ISNN 1043-1802

Nájera, C. & Yus, M. (1999). Desulfonylation reactions: Recent developments. *Tetrahedron*, 55, 35, 10547-10658, ISNN 0040-4020

Nakamura, T. & Oda, Y. (2007). Mass spectrometry-based quantitative proteomics. *Biotechnol. Genet. Eng. Rev.*, 24, 147-163, ISNN 0264-8725

Nazif, T. & Bogyo, M. (2001). Global analysis of proteasomal substrate specificity using positional-scanning libraries of covalent inhibitors. *P. Natl. Acad. Sci. USA*, 98, 6, 2967-2972, ISNN 0027-8424

Ortega-Munoz, M.; Morales-Sanfrutos, J.; Megia-Fernandez, A.; Lopez-Jaramillo, F. J.; Hernandez-Mateo, F. & Santoyo-Gonzalez, F. (2010). Vinyl sulfone functionalized silica: a "ready to use" pre-activated material for immobilization of biomolecules. *J. Mater. Chem.*, 20, 34, 7189-7196, ISNN 0959-9428

Ovaa, H.; van Swieten, P. F.; Kessler, B. M.; Leeuwenburgh, M. A.; Fiebiger, E.; van den Nieuwendijk, A. M. C. H.; Galardy, P. J.; van der Marel, G. A.; Ploegh, H. L. & Overkleeft, H. S. (2003). Chemistry in Living Cells: Detection of Active Proteasomes by a Two-Step Labeling Strategy. *Angew. Chem., Int. Ed.*, 42, 31, 3626-3629, ISNN 1521-3773

Palmer, J. T.; Rasnick, D.; Klaus, J. L. & Bromme, D. (1995). Vinyl Sulfones as Mechanism-Based Cysteine Protease Inhibitors. *J. Med. Chem.*, 38, 17, 3193-3196, ISNN 0022-2623

Porath, J.; Laas, T. & Janson, J. C. (1975). Agar derivatives for chromatography, electrophoresis and gel-bound enzymes : III. Rigid agarose gels cross-linked with divinyl sulphone (DVS). *J. Chromatogr.*, 103, 1, 49-62, ISNN 0021-9673

Qiao, T. A.; Leon, J. W.; Penner, T. L. & Yang, Z. (2003). Substrate for protein microarray containing gelatin-based functionalized polymer, U.S. Patent 6,815.078

Roberts, M. J.; Bentley, M. D. & Harris, J. M. (2002). Chemistry for peptide and protein PEGylation. *Adv. Drug. Deliv. Rev.* , 54, 4, 459-476, ISNN 0169-409X

Saoji, A. M.; Jad, C. Y. & Kelkar, S. S. (1983). Remazol brilliant blue as a pre-stain for the immedite visualization of human serum proteins on polyacrylamide gel disc electrophoresis. *Clin. Chem.*, 29, 1, 42-44, ISNN 0009-9147

Savage, M. D.; Mattson, G.; Desai, S.; Nielander, G. W.; Morgensen, S. & Conklin, E. J. (1992). *Avidin-Biotin Chemistry: A Handbook*, Pierce Chemical Company, ISBN 9780935940114, Rockford.

Shapiro, A. B. & McCarty, R. E. (1988). Alteration of the nucleotide-binding site asymmetry of chloroplast coupling factor 1 by catalysis. *J. Biol. Chem.*, 263, 28, 14160-14165, ISNN 0021-9258

Shapiro, A. B. & McCarty, R. E. (1990). Substrate binding-induced alteration of nucleotide binding site properties of chloroplast coupling factor 1. *J. Biol. Chem.*, 265, 8, 4340-4347, ISNN 0021-9258

Shevchenko, A.; Tomas, H.; Havlis, J.; Olsen, J. V. & Mann, M. (2006). In-gel digestion for mass spectrometric characterization of proteins and proteomes. *Nat. Protoc.*, 1, 6, 2856-2860, ISNN 1750-2799

Simpkins, N. S. (1990). The chemistry of vinyl sulphones. *Tetrahedron*, 46, 20, 6951-6984, ISNN 0040-4020

Sletten, E. M. & Bertozzi, C. R. (2009). Bioorthogonal Chemistry: Fishing for Selectivity in a Sea of Functionality. *Angew. Chem., Int. Ed.*, 48, 38, 6974-6998, ISNN 1521-3773

Srikanth, G. S. C. & Castle, S. L. (2005). Advances in radical conjugate additions. *Tetrahedron*, 61, 44, 10377-10441, ISNN 0040-4020

Staros, J. V.; Wright, R. W. & Swingle, D. M. (1986). Enhancement by N-hydroxysulfosuccinimide of water-soluble carbodiimide-mediated coupling reactions. *Anal. Biochem.*, 156, 1, 220-222, ISNN 0003-2697

Taniguchi, H. (1999). Protein myristoylation in protein-lipid and protein-protein interactions. *Biophys. Chem.*, 82, 2-3, 129-137, ISNN 0301-4622

Tiefenbrunn, T. K. & Dawson, P. E. (2010). Chemoselective ligation techniques: Modern applications of time-honored chemistry. *Peptide Sci.*, 94, 1, 95-106, ISNN 1097-0282

Tilley, S. D.; Joshi, N. S.; Francis, M. B. & Begley, T. P. (2007). *Proteins: Chemistry and Chemical Reactivity*, On line, John Wiley & Sons, Inc., ISBN 9780470048672.

Traverso, J. A.; Lopez-Jaramillo, F. J.; Serrato, A. J.; Ortega-Munoz, M.; Aguado-Llera, D.; Sahrawy, M.; Santoyo-Gonzalez, F.; Neira, J. L. & Chueca, A. (2010). Evidence of non-functional redundancy between two pea h-type thioredoxins by specificity and stability studies. *J. Plant Physiol.* , 167, 6, 423-429, ISNN 0176-1617

UniProtKB/TrEMBL database (2011-06) (15400876 sequence entries comprising 4982458690 amino acids)

Villar, H. O. & Kauvar, L. M. (1994). Amino acid preferences at protein binding sites. *FEBS Lett.*, 349, 1, 125-130, ISNN 0014-5793

Villar, H. O. & Koehler, R. T. (2000). Amino acid preferences of small, naturally occurring polypeptides. *Biopolymers*, 53, 3, 226-232, ISNN 0006-3525

Voloshchuk, N. & Montclare, J. K. (2010). Incorporation of unnatural amino acids for synthetic biology. *Mol. Biosyst.*, 6, 1, 65-80, ISNN 1742-206X

Waggoner, A. (2006). Fluorescent labels for proteomics and genomics. *Curr. Opin. Chem. Biol.*, 10, 1, 62-66, ISNN 1367-5931

Walsh, G. (Ed.) (2009). *Post-Translational Modification of Protein Biopharmaceuticals*, Wiley-VCH Verlag GmbH & Co. KGaA, ISBN: 9783527320745, Weinheim

Wang, G. & Yao, S. Q. (2003). Combinatorial synthesis of a small-molecule library based on the vinyl sulfone scaffold. *Org. Lett.*, 5, 23, 4437-4440, ISNN 1523-7060

Wenk, M. R. (2005). The emerging field of lipidomics. *Nat. Rev. Drug Discov.*, 4, 7, 594-610, ISNN 1474-1776

Wieland, T.; Bokelmann, E.; Bauer, L.; Lang, H. U.; Lau, H. & Schafer, W. (1953). Polypeptide syntheses. VIII. Formation of sulfur containing peptides by the intramolecular migration of aminoacyl groups. *Justus Liebigs Ann. Chem.*, 583, 129-149, ISNN 0075-4617

Wiltschi, B. & Budisa, N. (2008). Bioorthogonal chemical transformations in proteins by an expanded genetic code. In *Probes and Tags to Study Biomolecular Function, Miller L. W., pp.139-162,* Wiley-VCH Verlag GmbH & Co. KGaA, ISBN 9783527315666, Weinheim

Wong, L. S.; Khan, F. & Micklefield, J. (2009). Selective Covalent Protein Immobilization: Strategies and Applications. *Chem. Rev.*, 109, 9, 4025-4053, ISNN 0009-2665

Wu, Y.-W. & Goody, R. S. (2010). Probing protein function by chemical modification. *J. Pept. Sci.*, 16, 10, 514-523, ISNN 1099-1387

Yan, L. Z. & Dawson, P. E. (2001). Synthesis of peptides and proteins without cysteine residues by native chemical ligation combined with desulfurization. *J. Am. Chem. Soc.*, 123, 4, 526-533, ISNN 0002-7863

Yang, Z. P. & Hancock, W. S. (2004). Approach to the comprehensive analysis of glycoproteins isolated from human serum using a multi-lectin affinity column. *J. Chromatogr. A*, 1053, 1-2, 79-88, ISNN 0021-9673

Young, T. S. & Schultz, P. G. (2010). Beyond the Canonical 20 Amino Acids: Expanding the Genetic Lexicon. *J. Biol. Chem.*, 285, 15, 11039-11044, ISNN 0021-9258

Yuan, F.; Verhelst, S. H. L.; Blum, G.; Coussens, L. M. & Bogyo, M. (2006). A Selective Activity-Based Probe for the Papain Family Cysteine Protease Dipeptidyl Peptidase I/Cathepsin C. *J. Am. Chem. Soc.*, 128, 17, 5616-5617, ISNN 0002-7863

17

Gel-Free Proteome Analysis Isotopic Labelling Vs. Label-Free Approaches for Quantitative Proteomics

Baptiste Leroy[1], Nicolas Houyoux[1], Sabine Matallana-Surget[2,3] and Ruddy Wattiez[1]

[1]*Dept. of Proteomics and Microbiology, University of Mons (UMONS)*

[2]*UPMC Univ Paris 06, UMR7621, Laboratoire d'Océanographie Microbienne, Observatoire Océanologique, Banyuls/mer*

[3]*CNRS, UMR7621, Laboratoire d'Océanographie Microbienne, Observatoire Océanologique, Banyuls/mer*

[1]*Belgium*

[2,3]*France*

1. Introduction

For more than three decades, proteomics have been a crucial tool for deciphering the intricate molecular systems governing biology. O'Farrel was the first to utilise 2 dimensional gel electrophoresis (2DE) to perform actual complex proteomic analyses (O'Farrell, 1975). 2DE has very quickly emerged at the forefront of this rapidly growing field of research and has allowed for thousands of studies in widely varied domains. The development of 2D Fluorescence Gel Electrophoresis (2D DIGE) has provided more accurate and reliable proteins quantification due to the simultaneous migration on a same gel of samples to be compared, avoiding gel-to-gel variation. More recently, technological improvements in liquid chromatography and mass spectrometry have made it possible to develop so called "gel-free proteomics" in which, after total proteome enzymatic digestion, the produced peptides are separated with a high resolution chromatographic system and identified using tandem mass spectrometry. A gel-free approach presents a number of advantages over 2DE, such as a higher sensitivity, an easier automation of procedures to provide a better reproducibility and a reduced influence of intrinsic protein characteristics (pI, molecular weight, etc.). Nevertheless, a high complementarity between 2DE and gel-free approaches has been extensively reported (Finamore et al., 2010; Charro et al., 2011; Matallana-Surget et al., submitted), which suggests that both methods will continue to be considered together for a long time. Furthermore, 2DE also presents some advantages over a gel-free workflow approach in particular contexts. Indeed, 2DE presents the important benefit of allowing for the detection of protein isoforms, which is still complicated using gel-free approaches.

Another example is an immunoproteomic workflow in which 2DE is followed by immunodetection, which allows for the targeted analysis and detection of antigen candidates.
An additional important difference between gel-free approaches and gel-based workflow is found in the quality of the quantitative data obtained. Indeed, 2DE quantification relies on spot volume measurement, which implies each protein is quantified based on a single data point. In contrast, multiple peptides from the same protein can be used for quantification using gel-free approaches. This major difference clearly indicates that gel-free workflow-derived quantitative data are more statistically robust. However, before gel-free workflow approaches can be used in differential proteomics, the intrinsic limitation of mass spectrometry-based peptide analyses, the ion suppression effect, must first be addressed.

2. Ion suppression effect

The ion suppression effect can be defined as a negative influence of the chemical environment of a compound upon its ionisation. In other words, in addition to the chemical characteristics of a compound, the molecules present around it during the ionisation process will influence its ionisation. Even if this phenomenon is also observed under MALDI conditions, this chapter will focus on electrospray ionisation (ESI)-based LC MS/MS workflow, and thus, only ESI ion suppression effects will be discussed here.
Electrospray ionisation results from a complex process that is not fully understood today but most likely relies on ion ejection from a droplet due to electric field strength and on solvent evaporation leading to charge acquisition and gas phase transfer. More details on the ionisation principle are available in a recent review by Wilm (Wilm, 2011). The importance of ion suppression effects in ESI have been mainly investigated in toxicology analyses in the context of LC MS/MS detections and the quantification of target compounds in biological matrices. A mechanistic investigation in 2000 (King et al., 2000) concluded that the gas phase reaction of charge transfer was likely less important than the solution phase processes into ion suppression effect under electrospray conditions. The results of this study point out that the modification of small droplet formation due to non-volatile compounds is the main cause of ionisation suppression, given that other mechanisms can also play minor roles in analyte ionisation. Using LC MS/MS-based toxicological analyses of target compounds in biological matrices, Muller *et al.* (Muller et al., 2002) also confirmed that the majority of the observable ion suppression effect was limited to the early period of reverse phase (RP) chromatography when unretained polar compounds were present in the electrospray solution. In RP-LC MS/MS-based proteomic analyses, the ion suppression effect due to these unretained polar compounds should not be a major concern because peptides will generally be significantly slowed on a C18 reversed phase column and thus, not co-elute with such compounds. Nevertheless, the ion suppression effect also originates from in-solution competition between the co-eluting analytes for charge acquisition and gas phase ejection during ESI. This decreased ionisation efficiency is sometimes referred to as the matrix effect and can be understood using the equilibrium partitioning model (Enke, 1997). The equilibrium partitioning model describes that ESI nano-droplets consist of two phases: an electrically neutral phase containing solvent molecules at its centre and an excess charge containing surface layer. An analyte is distributed between these two phases based on factors such as its hydrophobicity, its charge density or its basicity. Only analytes present at the surface layer will be amenable to ionisation and consequently, a competition can exist

between analytes for distribution in the charged surface. In other words, co-eluting compounds enter into a competition for this distribution and therefore, for ionisation. This model helps explain why the increased non-polar character of peptides, which leads to an increased affinity for the surface phase, results in a more successful competition for surface localisation and thus, ionisation (Cech and Enke, 2000). Ion suppression effects and in particular, the observed competition of co-eluting peptides for ionisation (the matrix effect) have made difficult the quantitative use of gel-free proteomic techniques. The first strategy to address this issue has been to analyse samples that have to be compared simultaneously after mixing them and thus, ensuring that the ionisation of the quantified peptides has been performed with exactly the same matrix effect. In order to achieve this goal, proteins or peptides from the samples that are to be compared are labelled using an isotopic coded tag that will not influence their behaviour during LC MS/MS but rather introduce a mass shift between samples that will allow the discrimination of the origin of the peptide.
The second strategy for quantitative gel-free proteomics relies on an early demonstration (Voyksner and Lee, 1999) that peptide peak intensity correlates with its concentration in a sample and could thus be used to compare one run to another. Nevertheless, to analyse a complex peptide mixture, one must take into account the matrix effect. Indeed, in such a mixture, to be able to compare the run-to-run peak intensities of a peptide, one needs to be able to exactly reproduce the same chromatographic separation for all the samples that are being compared so that all the peptides are always ionised with the same co-eluting peptides. If this prerequisite is not satisfied, the competition between the peptides for ionisation will not be conserved, and a difference in ionisation efficiency will introduce biases into the quantitative data. This manner of interpreting data has led to the development of gel-free quantitative proteomics, which rely on a highly reproducible chromatographic separation and have developed very quickly since the recent apparition of ultra pressure chromatography.

3. Quantification strategies

Quantitative proteomic can be classified in two major approaches: the stable isotope labelling and the label-free techniques (Figure 1).

3.1 Isotope-coded labelling

As suggested above, isotope-coded labelling allows for mass shift introduction between the proteins/peptides of the samples to be compared, which makes it possible to mix them before an LC MS/MS analysis. As the peptides contained within the samples to be compared have been ionised under exactly the same conditions, their intensities can be compared in order to achieve a relative quantification. The first developed method based on this principle was ICAT (isotope-coded affinity tag;(Gygi et al., 1999)), which relies on cysteine tagging followed by the affinity-based enrichment of tagged peptides. Initially, the ICAT tag consisted of a biotin moiety used for affinity enrichment and a thiol-specific reactive group for cysteine labelling. These two groups were separated from each other by a linker group, which contained 8 hydrogens in the light tag and 8 deuteriums in the heavy tag (Gygi et al., 1999). Thus, this tag introduces a mass shift of 8 Da, which will allow the peptides in the samples to be distinguished from one another and compared based on their mass spectrum; this tag also makes it possible to measure the relative abundances of the corresponding proteins in the two samples. This strategy has been referred to as non-

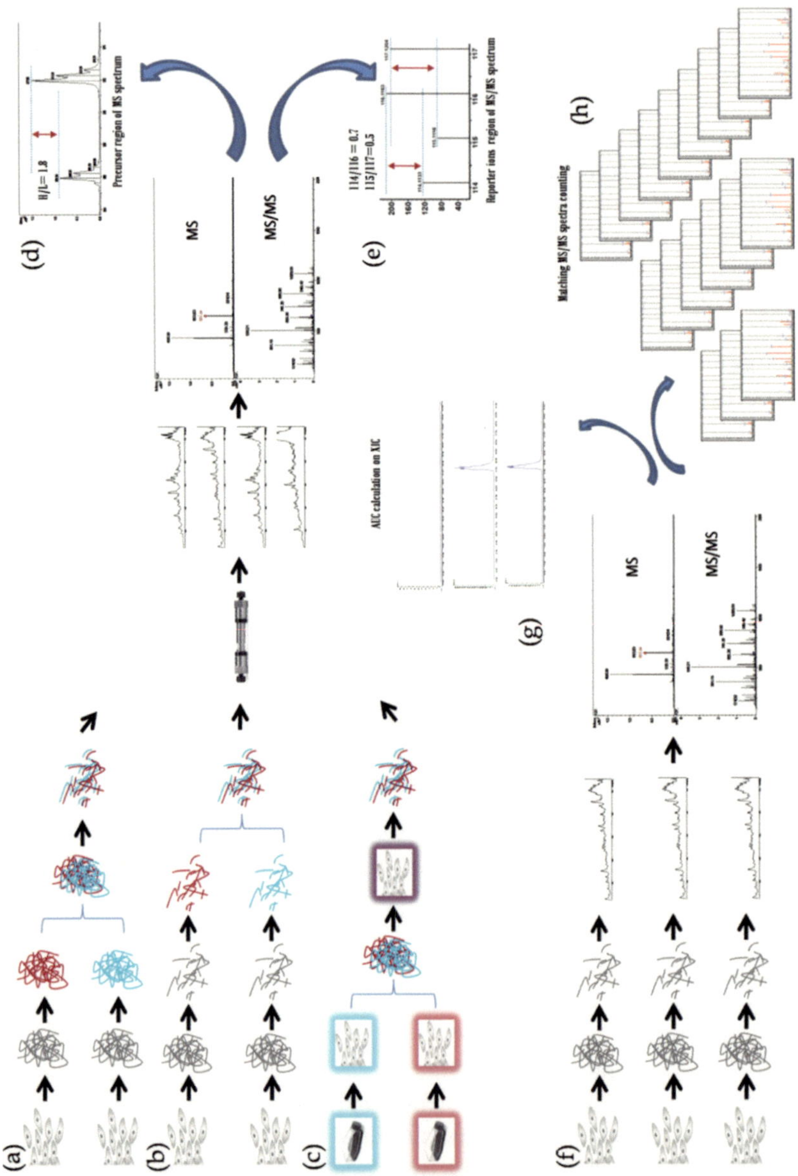

Fig. 1. Schematic representation of the main quantification workflow. (a-e) Isotopic labelling relies on the introduction of a discriminative mass shift, which allows sample mixing before analyses with 2D-LC MS/MS. Quantitative data are obtained in the MS spectrum in the case of non-isobaric labelling (d) or in MS/MS mode due to the release of a reporter group upon fragmentation during isobaric labelling (e). Isotopic labelling can be performed at the protein (a), peptide (b) or cell culture level (c). (f-h) In a label-free workflow, samples are prepared and analysed separately by LC MS/MS (f). Quantitative data can be obtained either from the Area under the curve (AUC) calculated from an extracted ion chromatogram for the representative peptides of a protein (g) or from a number of matching MS/MS associated with a protein (h).

isobaric labelling. ICAT has been continuously improved by first replacing deuterium coding with the C^{13} isotope in order to minimise the chromatographic resolution of the isotope-coded peptides (Zhang and Regnier, 2002) and then introducing a disulphide bond in the linker so affinity-trapped labelled peptides can be more efficiently eluted from the avidin affinity matrix by reductive cleavage of the linker (Hansen et al., 2003). Several non-isobaric tags have also been developed that similarly rely on the quantification of the peptides in the MS due to the isotopically introduced mass shift.
Following the development of non-isobaric labelling, isobaric tags such as iTRAQ (isobaric tag for relative and absolute quantification) were introduced (Ross et al., 2004). Isobaric tags are composed of an amine-specific reactive group, which allows for the tagging of proteins on lysines and peptides on their N-termini and a reporter group that has a different isotopic composition when comparing the different versions of the tag and thus, different masses. A balance group, which has an isotopic composition complementary to the reporter so that the global mass of reporter + balance group is constant between the different versions of the tag, is placed between the reactive and the reporter group. Thus, the tagged peptides are not discriminated in the MS. Upon peptide fragmentation by MS/MS, the reporter group is released and appears in the low mass range of the MS/MS spectrum. Separated from their balance group, the reporter ions are distinguishable from each other because their different isotopic composition introduces a 1- Da mass shift between them. The relative abundance of the peptides/proteins in the samples to be compared is deduced from the relative abundance of the corresponding reporter ions.

3.1.1 Non-isobaric labelling

ICAT was the first commercially available isotopic-labelling reagent, with a thiol-specific tag to target low abundance amino acids and due to enrichment, enabling a significant decrease of sample complexity (Figure 1a). However, a significant proportion of proteins could not be quantified with ICAT because of a lack of cysteines and hence, a high number of proteins were only quantified based on single peptides. This main limitation, observed with ICAT technique, encouraged researchers to develop alternative tags. ICPL (isotope-coded protein labelling) is one of these tags and was mainly developed to solve the low sequence coverage drawback of ICAT (Schmidt et al., 2005). ICPL, an amine reactive tag targeting lysines on intact proteins, was supposed to address this issue and reduce the proportion of unquantifiable proteins by increasing the number of quantified peptides per protein (Schmidt et al., 2005). Using ICPL, we and other groups have demonstrated that an important proportion of identified peptides in trypsin digested samples still lack a lysine and were not tagged and not quantifiable (Mastroleo et al., 2009b; Paradela et al., 2009). In the protein-labelling conformation, ICPL only allows for the quantification of approximately 70% of the identified proteins. Using the amine reactivity of ICPL, we have developed and optimised a peptide level labelling strategy called post-digest ICPL (Figure 1b), which allows for the tagging of the N-termini of all peptides, making them amenable to quantification (Leroy et al., 2010). This strategy is still currently used in our lab and has allowed for a significant number of successful analyses, some of which are presented below. While increasing the global amount of obtained quantitative data, labelling at the peptide level also implies that a highly cautious sample preparation technique must be employed to avoid bias introduction because samples will be mixed very late in the workflow process. It also impairs the possibility of protein-based sample fractionation, and a high resolution peptide chromatographic separation (using 2D-LC) or high throughput data acquisition system will be required. It can be assumed that the earlier the sample labelling and mixing the lower the chance of bias introduction. Therefore, the best solution is to mix the samples even before protein extraction so that the chance of bias introduction is extremely decreased, and all protein fractionation methods can be easily applied to the sample. Such a procedure exists and is based on the introduction of a mass shift between the samples to be compared through the metabolic incorporation of isotope-coded amino acids during cell culture (Figure 1c; (Ong et al., 2002)). The two most common metabolic labelling are the ^{15}N labelling, usually used for microorganisms (Li et al., 2007; Ting et al., 2009) whereas the stable isotope labelling in cell culture (SILAC) is mostly used for mammalian cells (Ong et al., 2002; Mann et al., 2006). This method SILAC allows all tryptic peptides to be labelled and quantified if lysine and arginine are used as the isotope-coded amino acids. During such a workflow, harvested cells from treated samples versus control can be mixed immediately and extracted together, which ensures a perfectly unbiased sample treatment. Obviously, all methods suffer from some limitations. This method is only practical for auxotrophically cultivable organisms. Therefore, most bacteria as well as tissue samples are excluded from this workflow (Bantscheff et al., 2007). Recently, metabolically-labelled mice have been introduced to the market, which makes it possible to perform some tissue analyses using this type of workflow (Wu et al., 2004).

As an alternative to chemical (ICAT, ICPL, etc.) and metabolic (SILAC, ^{15}N, etc.) non-isobaric labelling, the enzymatic introduction of isotopic differences between samples has also been developed. In this case, the hydrolysis of the peptide bond during enzymatic digestion is realised in presence of regular water for one sample but with ^{18}O-containing water for the second sample, which results in the exchange of 2 ^{16}O for 2 ^{18}O at the C-terminus of the produced peptide in the latter case (Ye et al., 2009). A 4-Da mass shift will be introduced and used to discriminate between peptides originating from samples that are to be compared. This method is very straightforward, but differences in the rate of oxygen exchange between different peptides are sometimes problematic.

Most non-isobaric labelling strategies were developed as duplex strategies in which two samples can be compared. In order to increase analytical throughput, multiplexing is being introduced, notably with an ICPL tag (SERVA) for which a triplex and a quadruplex version were recently released. In this new version, the introduced mass shift is only 2 Da, and triply charged peptides will only be separated by 0.66 m/z. Under low resolution, such as using an ion trap mass spectrometer, the isotopic pair will become difficult to discriminate, and multiplexing sample analyses using non-isobaric labelling definitely require a high resolution mass spectrometer. On the other hand, a multiplexing capability is clearly an advantage of isobaric labelling strategies in which it can be more easily implemented.

3.1.2 Isobaric labelling

ITRAQ has been described by Ross and co-workers (Ross et al., 2004). Here, low mass reporter ions produced after peptide fragmentation in a mass spectrometer are used for quantifications (Figure 1). In this low mass range, the 1-Da mass shift between the singly charged reporter ions is easily discriminated and can be used for quantification, even with a low resolution instrument. This facilitates multiplexing analysis and, hence, ITRAQ exists in a 8-plex version and TMT (tandem mass tag; (Dayon et al., 2008) exists in a 6-plex version. This multiplexing capability undoubtedly represents a major advantage of isobaric labelling because as many as 4 pairs of control/case samples can be simultaneously analysed. Alternatively, multiplexing can be used to perform technical replicates at the same time to increase the statistical power of the dataset. To date, ITRAQ represents the most commonly used quantification strategy with more than 500 entries found in a Pubmed bibliographic search engine using the keyword "ITRAQ" versus less than 350 for "SILAC". As in the post-digest ICPL, ITRAQ and TMT rely on peptides labelled at their N-termini due to an amine reactive group and also allows for the quantification of all identified peptides. Nevertheless, ITRAQ also has its own limitations, mainly due to the necessity of analysing a low mass range of the MS/MS spectra, which is generally not performed using a quadrupole /ion trap mass spectrometer.

Non-isobaric labelling also presents another advantage over isobaric tagging, *i.e.*, the ability to include differences in relative abundances for an isotopic pair in the precursor selection criteria to determine which ion will be selected for fragmentation. In other words, the mass spectrometer could preferentially select peptides for which a differential abundance has been detected and can virtually decrease the sample complexity and focus on the differentially abundant proteins. This is obviously not possible with isobaric tags because quantitative data are only available after precursor selection and fragmentation have occurred.

3.1.3 Examples of isotopic labelling applications

The major benefit of using isotopic labelling workflow in differential proteomics is the high accuracy of the obtained quantitative data. As discussed below, isotopic labelling definitely surpasses label-free approaches in this aspect. In this section, we will emphasise case studies in which high precision data were obtained and validated using alternative methods.

In our lab, we are currently involved in the analysis of naive T cell activation through anti-CD3/CD28 in the presence or absence of co-activating interleukins, notably IL-6. This project aims to better understand the mechanisms that underlie T cell differentiation, particularly T follicular helpers (Eddahri et al., 2009). In this study using post-digest ICPL and 2D-LC MS/MS, some obvious markers of Th2 polarisation of IL6-activated T cells were detected, as was expected (unpublished data). In addition, slight differences were also observed for proteins related to cellular trafficking. As T cell cellular trafficking is already known to be important for T cell differentiation (Tanaka et al., 2007), the validation of these observations is essential. We are focusing our efforts on a microtubule (Mi) polymerisation factor for which only a slight increase of abundance could be observed in IL6-activated T cells. The fold change observed in two biological replicates were only 1.33 and 1.48 with 2 and 5 peptides being used for quantification, respectively (Figure 2). This protein was selected because its means fold change (calculated on the 2 biological replicates) was statistically different from 1 based on t-student analysis ($t < 0.05$). Western blot analysis was used to quantify the relative abundance of this protein on a third biological replicate using image based quantification (Figure 2). A fold change of 1.4, obtained by western blotting, confirmed the accuracy as well as the reproducibility of the observation made using post-digest ICPL.

A second example comes from the analysis of the fear-conditioning influence on neuronal plasticity. In this context, fear-conditioned rat cerebral tissue was compared to unconditioned controls using post-digest ICPL and 2D-LC MS/MS. In this analysis, the abundance of very few proteins was altered between the samples and control and only very

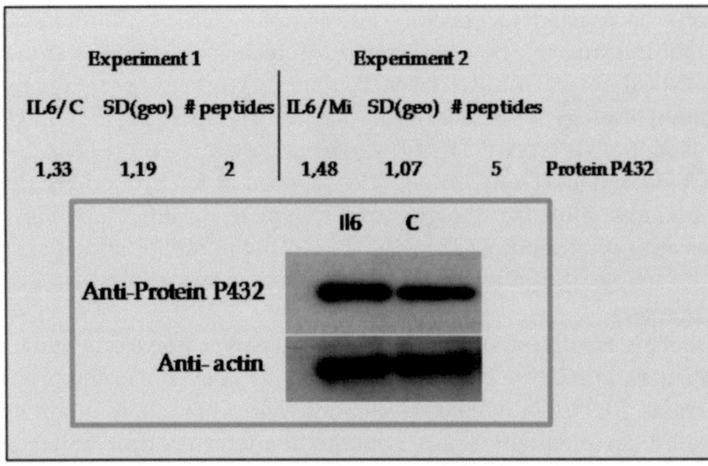

Fig. 2. Western blot analysis was used to validate the data obtained by post-digest ICPL. C : control; Mi, Microtubules; SD(geo), geometric standard deviation.

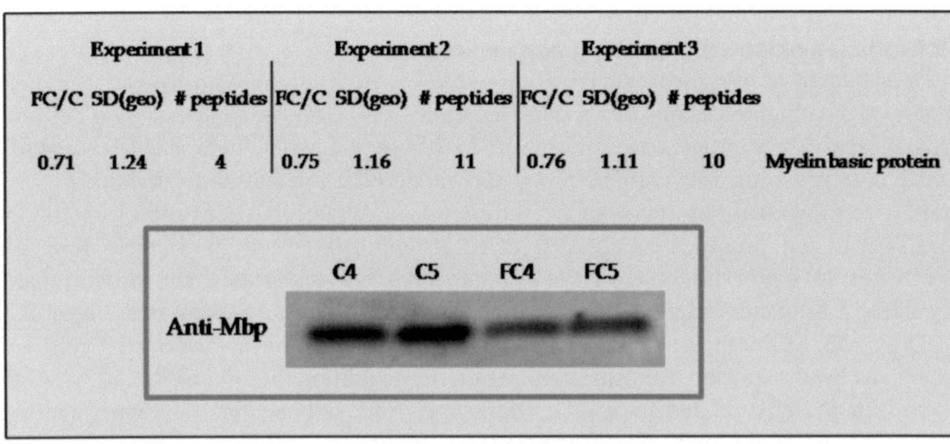

Fig. 3. Western blot analysis was used to validate data obtained by post-digest ICPL. Mbp: Myelin basic protein

slight changes were observed (unpublished data). Among the three biological replicates analysed, a protein was always modified with the same fold change of around 0.75 (Figure 3, t <0.05), which meant a slightly lowered abundance in the fear-conditioned animals. As this protein is known to be related to neuronal plasticity, it was mandatory to be able to confirm the 2D-LC MS/MS-obtained data. Here, again using a western blot (Figure 3), this protein has been selected for validation, which made it possible to confirm, after image-based quantification, a 30% decrease in the abundance of this protein.

The third example has been recently published by a Finnish group and is also related to the T cell differentiation mechanism but in presence of an alternative interleukin, namely IL4 (Moulder et al., 2010). In this study, the nuclear fraction was analysed 6 and 24 hrs after IL4 supplementation or control anti-CD3/CD28 activation of naive T cells. As observed in our study, the differences between the IL4-activated cells and control activation were very scarce and of a low amplitude. Moulder and co-workers considered 3 biological replicates, and all were analysed three times using an ITRAQ 4-plex kit. In this study, a random effect meta-analysis model was used to estimate the representative expression ratios for each protein. Thanks to this elaborate study design they were able to apply a fold change cut-off of 1.2 (a 20% abundance variation) to their dataset and highlight abundance modifications for important proteins. Moreover, their observations can also be confirmed by fluorescence-based western blot analysis.

Finally, there is another example of a very well-designed study in which ITRAQ was proven to be highly reproducible. Uwin and co-workers (Unwin et al., 2006) analysed the differences in proteomes of two lineages of stem cells LSK+ (Lin+, Sca+, Kit+) and LSK- (Lin-, Sca+, Kit-) and also applied a cut-off of 1.2 to their obtained dataset, even though only 2 biological replicates were analysed. The use of such a low fold change threshold was justified by filtering their dataset based on intra-condition variability limits. Indeed, a 4-plex ITRAQ kit was used to label and analyse the two LSK+ biological replicates together and the two LSK- biological replicates. For a protein to be considered of a different abundance, the LSK+1 vs. LSK+2 as well as the LSK-1 vs. LSK-2 ratios of that protein had to first be between 1.10 and 0.92 (minimal intra-condition variability), and in addition, both the LSK+1 vs. LSK-1 and LSK-2 and LSK+2 vs. LSK-1 and LSK-2 ratios had to be higher than 1.2 with a $p<0.05$ in a pairwise Student's *t*-test analysis. This analysis clearly indicates the very high value of the multiplexing capability of isobaric labelling workflow and the extreme accuracy of the quantitative data that can be obtained using isotopic labelling strategies.

3.2 Label-free approaches

Label-free approaches fundamentally demonstrate that a MS signal observed for a peptide correlates very well with its abundance in the sample (Chelius and Bondarenko, 2002). A difficulty arises from the effect of the matrix, which may differ between the two separate LC MS/MS runs, and thus impair a fair comparison of the data sets acquired consecutively. Bondarenko and co-workers (Bondarenko et al., 2002) were the first to demonstrate that such a comparison of MS signals between individually acquired datasets was possible even with complex protein mixtures like serums. Thus, it appears that if a chromatographic separation is sufficiently controlled, the matrix effect is not that different between successive runs, and MS data can be used to quantify MS/MS- identified peptides (Figure 1g).

In addition, relying on the assumption that the matrix effect can be controlled, other approaches for label-free protein abundance comparisons have been described that rely on MS/MS data. Indeed, Liu *et al.* (Liu et al., 2004) demonstrated that the number of MS/MS spectra acquired by LC MS/MS for a defined protein correlates over 2 orders of magnitude with its abundance in the sample (Figure 1h). This very simple measurement relies on the principle that the more we see a protein the more abundant it should be in the sample. This type of strategy is termed spectral counting and can be opposed to "MS spectral intensity measurements" as "MS/MS features analyses".

3.2.1 MS-based label-free analysis

Since the first demonstration of the linearity of the MS signal and protein abundance relationship by Chelius and Bondarenko (Chelius and Bondarenko, 2002) as well as Wang and co-workers (Wang et al., 2003), who performed a large scale demonstration of the applicability of this finding on large numbers of samples, MS-based label-free analyses have continuously been optimised and used more frequently in biological studies and especially in clinical research. Although sample preparation and MS data acquisition must be performed very cautiously as they represent a mandatory step, processing the data from a MS-based label-free approach is far for being trivial. Indeed, for all ions to be quantified, an area under the curve (AUC), based on the m/z and chromatographic retention time of the ion, has to be determined for all samples. Even with the best chromatographic system, this step will first necessitate a realignment of multiple chromatograms to compensate for the long analysis-induced retention time drift. This aspect is critically dependent on the quality of the chromatographic system and in particular, on its stability. The implementation of an ultra-HPLC system presenting excellent stability in terms of retention time now alleviates this step and will probably become mandatory to achieve a high quality MS-based label-free quantitative analysis. Once the chromatograms are suitably aligned, the AUC can be calculated for a particular ion based on its measured m/z. Of course, the accuracy of the data will depend on the ability to calculate the AUC for particular peptides and to avoid contamination by co-eluting peptides with similar m/z values. In that aspect, mass spectrometric resolution is critical and can help narrow the AUC calculation windows and, thus, eliminate most of the contaminating signal (figure 4).

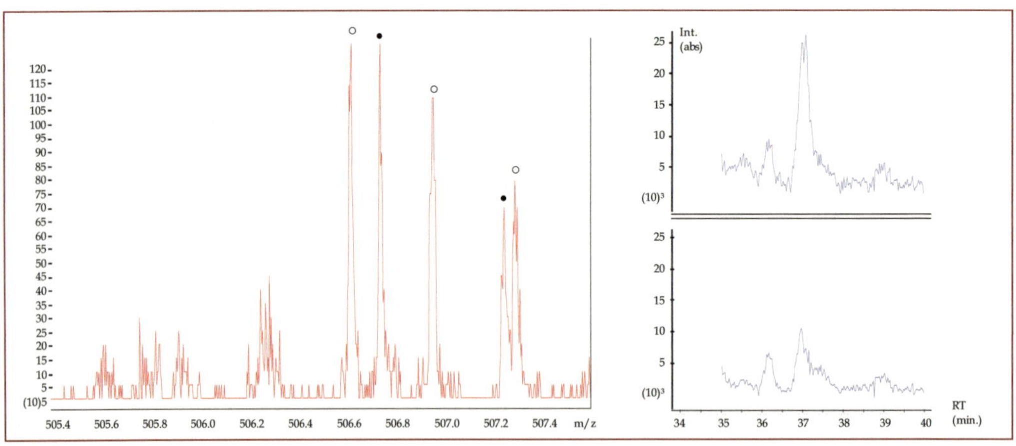

Fig. 4. High resolution mass spectrometry allows AUC calculations based on narrow m/z windows. In the case of co-eluting peptides of similar m/z values (left panel), the calculated AUC can be very different if narrow m/z windows (0.05; right lower panel) are used or if larger m/z windows (0.15; right higher panel) are taken into account due to a lower mass spectrometer resolution. Personal data obtained using the Triple TOF5600 (ABSciex).

A large amount of software has been developed for MS-based label-free quantification, and new tools are frequently released, which indicates a keen interest in these methods. A description of these softwares is beyond the scope of this chapter and has recently been performed by Neilson *et al.* (Neilson et al., 2011). However, it is interesting to note a

dichotomy in MS-based label-free data processing strategies. Indeed, if most of the early implemented data processing tools relied on an identified peptide list for AUC calculation, software now exists that allows for an unbiased total ion quantification independent of positive identification during the database search process. Quantitative data are calculated for all detected m/z notwithstanding an identified ion or even selected for fragmentation. Such an approach allows for the circumvention of a low sampling drawback of data-dependent acquisition in MS, which generally results in missing low level peptides. Here, all detectable ions are quantified and ions for which differential abundances have been observed can be identified by a subsequent targeted analysis. Obviously, as in this case, a quantification step only relies on accurate masses and RT measurements (AMRT or AMT workflow) without prior confirmation by MS/MS, such a workflow is only applicable to high resolution mass spectrometry-based platforms.
Another way to address the sampling bias of data-dependent acquisition has been proposed by Plumb and co-workers (Plumb et al., 2006). These authors developped the first real data-independent acquisition (DIA) workflow called MS^E (E states for elevated energy) and is available as an acquisition mode with WATERS instruments. This strategy is aimed to obtain the fragmentation data for all detectable ions by avoiding precursor selection (as in data-dependent acquisition, DDA) and isolation and rather acquiring alternatively low and high collision energy mass spectra for a full mass range. Using multiple criteria, a tremendous algorithm is then charged to associate a precursor mass deduced from low energy spectra and its fragment ions obtained in the high collision energy spectra. The grouping of fragment ions with their parent ions mainly relies on intensity and the elution profile. This theoretically comprehensive quantification and identification of all detectable ions has triggered significant interest and already been used in numerous publications (Blackburn et al., 2010; Herberth et al., 2011; Mbeunkui and Goshe, 2011). A variant of this workflow has recently been implemented by ABSciex (MS^{All}) on its triple TOF5600, and it can be assumed that all MS vendors will implement a DIA-like workflow on their instruments.

3.2.2 MS/MS-based label-free analysis

MS/MS-based label-free derived data represent the simplest process to quantify information. Indeed, there is no need to align a chromatogram to calculate AUC or to detect isotopic pairs, and everything required is contained in the database search results. MS/MS-based label-free quantification relies on the assumption that in data-dependent acquisition (DDA) analysis the sampling probability of a protein (*i.e.*, the number of MS/MS spectra related to a protein) is a function of the protein's abundance in the sample (Liu et al., 2004), which can be estimated by the so-called "spectral count" of a protein. MS/MS-based label-free quantification has been diversified using different parameters such as peptide counts (the number of unique peptides; (Gao et al., 2003)), sequence coverage and several tentatively normalised indices (NSAF, Normalised spectral abundance factor, etc.;(Ishihama et al., 2005; Florens et al., 2006)).
The accuracy of MS/MS-based label-free quantification has also been extensively investigated and proven unexpectedly high given the extreme simplicity of the measurement. In 2006, Zhang *et al.* (Zhang et al., 2006) compared spectral count, peptide count and sequence coverage in terms of reliability and also investigated the statistical relevance of such measurements. Interestingly, they linked the fold change, which can be

perceived statistically, with the actual number of spectral counts. This analysis showed that below 15 spectral counts only fold change higher than 2 could be detected no matter what statistical test was used. However, if more than 50 spectral counts were obtained, a fold change of 1.5 was detected. More recently, Colaert and co-workers (Colaert et al., 2011) estimated the global standard deviation of three different MS/MS-based label-free techniques and concluded that all of them had global SD of around 0.5. If a simple threshold in the format of mean +/- 2 SD is applied to such a dataset, fold changes higher than 2 were generally measurable.

Spectral counting has also been modified to allow for comparisons of the abundances of different proteins and absolute quantification. emPAI (**e**xponentially **m**odified **p**rotein **a**bundance **i**ndex) normalises the number of identified peptides of a protein by the number of theoretically observable peptides to account for differences in sequence characteristics between different proteins and allows for their quantitative comparison. More recently, APEX (**a**bsolute **p**rotein **ex**pression; (Lu et al., 2007)) profiling was developed to measure the absolute protein concentration per cell from the proportionality between the protein abundance and the number of peptides observed, by using a correction factor that correlates the likelihood of peptides observed to their intrinsic characteristics (length, amino acid composition, etc.).

3.2.3 Examples of label-free analysis

Recently, we had the opportunity to challenge the label-free analytical platform from WATERS using the ion mobility-implemented synapt G2 mass spectrometer (unpublished data) and its MS^E data-independent acquisition features. We analysed three biological replicates of crude protein extracts from *Variovorax* sp. SRS16 cultured in the presence or absence of the phenylurea herbicide linuron. This strain has already been shown to catabolise linuron (Breugelmans et al., 2007), and we have already performed gel-based (Breugelmans et al., 2010) as well as isotopic labelling gel-free proteomic analyses on these samples (Bers et al., *In Press*), indicating us what changes should be expected. All three biological replicates were injected three times, and only proteins identified in at least 2 out of the 3 technical replicates as well as in each biological replicate were considered for quantification. A statistical analysis was performed on the mean linuron/control ratio and, thus, only proteins with a rejected null hypothesis (ratio = 1) and a p-value <0.05 were accepted as modified in abundance. An arbitrary cut-off of 1.5 and 0.66 was additionally applied. Using these stringent criteria (identification in all three biological replicates and in at least 2 out of 3 technical replicates with a p-value <0.05), 83 proteins (33 up, 50 down) out of the 1500 identified were considered to differ in amount between the linuron and control condition. This label-free analysis gives us a tremendous increase in proteome coverage, multiplying the number of detected and quantified proteins by a factor of 3.

A particular feature of label-free analysis is its higher throughput, which facilitates large sample size analysis in clinical research and biomarker discovery. Moon and co-workers (Moon et al., 2011) recently used a MS^E-based label-free strategy efficiently in order to discover biomarker candidates. From the urinary exosome proteome of IgA nephropathy (IgAN), **t**hin **b**asement **m**embrane **n**ephropathy (TBMN) and healthy patients, they were able to identify and quantify more than 1800 proteins, among which 83 differed in amount between IgAN and TBMN. Four IgAN/TBMN-discriminating biomarkers were selected among this dataset (aminopeptidase N, vasorin precursor ceruloplasmin and alpha-1-antitrypsin). These candidate biomarkers were submitted to western blot analysis on an independent set of samples, which failed to confirm differential expression for one of them but was validated in the other three. ROCs for differentiation between IgAN and TBMN indicated a high potential use for ceruloplasmin in this context because it provided a specificity of 91% for a sensitivity of 100%.

Using spectral counting, Saydam and co-workers (Saydam et al., 2010) recently analysed differences between human meningioma cells and primary arachnoidal cells. Proteins were separated by SDS-PAGE, the gels were cut into ten bands, submitted to in-gel digestion and the peptides were analysed by LC MS/MS. For all identified proteins, the spectral count was determined, and the differences between the samples were evaluated for statistical relevance using beta-binomial test. In this very simple workflow, 2800 proteins were identified (protein prophet probability >99% and at least 2 peptides), and 10% of them were statistically different in amount between the archnoidal cells and meningioma. Proteins belonging to the **mi**ni**c**hromosome **m**aintenance (MCM) family were observed in a higher abundance in meningioma cells and were submitted for further validation by qRT-PCR and western blotting.

4. Isotopic labelling or label-free approaches?

4.1 Relative quantification

Most proteomic studies aim to compare different states of a proteome rather than obtaining absolute quantitative data. When designing such a differential proteomic analysis, one has to face, with Cornelian dilemma, the question of which method would be most suitable for obtaining valuable data useful for better characterising a biological system. This is a very difficult and important topic for which many parameters must be considered. In the above section, we have tried to describe and exemplify the main existing methods for relative comparisons of protein abundances. Here, we will try to summarise, based on their pros and cons, which methods best suit which needs.

A first principle could be to use the most straightforward and simple method possible. In regards to this aspect, MS/MS-based label-free analysis clearly comes first. This method only requires that sample preparation and data acquisition are reproducible, which is usually expected. Here, there is no requirement for time-consuming sample labelling or for an analytical platform using ultra-HPLC and high resolution MS. This type of analysis can be applied to a variety of samples, such as very large sample cohort, often required for clinical research, and no limitation exists concerning the number of conditions that can be compared at a time. Finally, assuming a convenient correction factor is used, absolute quantitative data can be obtained, which allows not only for a comparison of the abundance of a protein in different samples, but also ranks the proteins in a defined proteome based on their abundance (Mastroleo et al., 2009a). Of course, as no ideal method exists, a MS/MS-based label-free approach also has a major drawback, data accuracy. As already described above, numerous analyses have been conducted to estimate this accuracy and concluded this technique does not easily detect fold changes lower than 2 (Zhang et al., 2006; Colaert et al., 2011). Nevertheless, it is important to replace this accuracy in the context of a biological question. Indeed, if only major changes are of interest or if the samples to be compared are expected to be highly different, MS/MS-based quantitative data could be sufficient. Equivalently, if one goal is to discover biomarker candidates able to discriminate diseased from healthy patients, it is not mandatory to be able to detect very slight fold changes. On the contrary, only proteins presenting major differences between controls and clinical cases will ultimately be useful for physicians to help them in their diagnostics or prognostics. It appears that pure biomarker discovery studies can be typically performed using MS/MS-based label-free approaches, and a more elaborate workflow would only be helpful if functional data could also be gained or are needed.

If not only discriminative but also functional data are to be obtained, acquiring accurate quantitative data is absolutely required. In these cases, both MS-based label-free approaches and isotopic labelling could be suitable. Nevertheless, the pros and cons of both strategies can help in the decision.

First, MS-based label-free techniques are only able to reach the isotopic labelling accuracy of quantitative data (CV>20%) if the analyses are performed on the latest generation mass spectrometers. Although MS vendors are continuing their efforts to allow access to such pieces of equipment to an increasing number of labs, to date, they are not considered as a benchtop device that is easily handled and accessible. On the other hand, isotopic labelling is easily amenable to high accuracy studies using a first generation Q-TOF device, or quadrupole ion trap.

The number of conditions to be analysed needs to be carefully considered. Indeed, isotopic labelling is limited in its multiplexing capacity, since so far only TMT and iTRAQ allow the comparison of multiple (up to 6 and 8 respectively) samples at the same time. For non-isobaric labelling, multiplexing capacities are, to date, limited at 4 samples, and in this case again, high resolution MS would be required.

Analysis throughput is generally considered to be lower when isotopic labelling is used because 2D-LC peptide separation is usually necessary to avoid the co-elution of peptides with similar m/z values, which might introduce errors in quantification. Label-free approaches, which rely on high resolution MS systems and ultra-HPLC, generally more efficiently deal with co-eluting peptides with similar m/z values and can be performed using 1D-LC. Nevertheless, it must be kept in mind that if the analysis of two mixed samples using 2D-LC requires around 12 hours, no gain in machine time will be obtained if the same samples are analysed using a 2-hour gradient in 1D-LC because, to obtain statistical relevance using data from a label-free analysis, a triplicate injection of all samples is generally required.

Another advantage of isotopic labelling is that when tagging occurs at the protein or even at the cell culture level, such as in SILAC, samples can be mixed very early in the workflow, and, thus, potential biases are avoided. During label-free approaches, full sample processing is performed separately, and the risk of a biased treatment is obviously increased.

In some cases, a high accuracy will not be sufficient and ultimate precision will be required. This is the case if very slight modifications are expected, such as in the example of T cell differentiation we have highlighted above. Post-translational modifications of proteins can also dramatically change a protein's function even if the fold changes are extremely small. In regards to this aspect, isotopic labelling still surpass label-free approaches and is the method of choice if fold changes lower than 1.4 must be efficiently characterised, as described above.

4.2 Absolute quantification

It has already been described that MS/MS-based label-free approaches can be used to reach an absolute quantification, but here again the precision is generally low and only orders of magnitude can be determined. Isotopic labelling is more easily amenable to the accurate absolute quantification of targeted proteins in a MS workflow. Absolute quantitative methods aim to measure the absolute protein level using a standard peptide to the corresponding protein. This is achieved by mixing a known amount of the synthesized isotope-coded form of a peptide from the protein to be quantified and using it as an internal standard to calculate the endogenous amount of the protein (method AQUA, Absolute quantification) (Gerber et al., 2003). This principle has been diversified in order to multiplex the proteins being quantified as well as to decrease biases introduced during sample treatment. In a new workflow termed concat, a chimeric protein composed of concatenated isotope-coded peptides to be quantified is introduced in the sample before enzymatic digestion. The common enzymatic digestion of reference peptides and endogenous peptides ensures a higher accuracy and allows for easy multiplexing.

Until recently, MS-based label-free approaches did not support absolute quantification. Nevertheless, Silva and co-workers from WATERS Corporation (Silva et al., 2006) reported that MS^E was the most accurate label-free technique for estimating absolute abundance by using average of the three most abundant tryptic peptides, which was reported to be proportional to protein molarity. This discovery used a unique internal standard to obtain absolute quantitative data for 6 exogenous standard proteins spiked into serum with a relative error below 15%. Moreover, 11 proteins of the serum matrix could also be quantified, and the obtained data correlates very well with the values available in the literature. To date, this absolute quantification feature has, to our knowledge, only been implemented in WATERS software packages.

5. Conclusion

In this chapter, we have described the most widely used strategies for quantitative proteomics studies. All have their pros and cons, which makes the choice of one of them difficult for non-proteomic researchers. Different criteria can be used in order to distinguish which method is best-suited to a given biological question. Among these, the data accuracy level required is probably the most interesting. With numerous proteomic analyses focusing on biomarker discovery, MS/MS-based label-free workflows are, to date, underutilised. When accurate data must be obtained, isotopic labelling methods and label-free approaches work equally well. Isotopic labelling will nevertheless still be of interest when high precision is required. It is expected that, in the future, easier access and development of highly reproducible nano-HPLC separation, high resolution mass spectrometer, and efficient computational tools will greatly improve the reliability and the use of label-free workflows.

6. Acknowledgements

The authors thank F. Andris, O. Leo, N. Mari and S. Denanglaire for their collaboration in T cell differentiation. This work was co-funded by the Walloon Region and the European

Regional Development Fund. Data related to neuronal plasticity were acquired in the lab by N. Houyoux under collaboration with L. Ris and E. Godaux.

7. References

Bantscheff, M., M. Schirle, G. Sweetman, J. Rick & B. Kuster (2007). "Quantitative mass spectrometry in proteomics: a critical review." *Anal Bioanal Chem* 389(4): 1017-31.

Bers, K., B. Leroy, P. Breugelmans, P. Albers, R. Lavigne, S. Sorensen, J. Amaand, W. Dejonghe, R. De Mot, R. Wattiez & D. Springael (*In Press*). "Identification of gene functions involved in mineralization of the phenylurea herbicide linuron in Variovorax sp. strain SRS16." *AEM*.

Blackburn, K., F. Mbeunkui, S. K. Mitra, T. Mentzel & M. B. Goshe (2010). "Improving protein and proteome coverage through data-independent multiplexed peptide fragmentation." *J Proteome Res* 9(7): 3621-37.

Bondarenko, P. V., D. Chelius & T. A. Shaler (2002). "Identification and relative quantitation of protein mixtures by enzymatic digestion followed by capillary reversed-phase liquid chromatography-tandem mass spectrometry." *Anal Chem* 74(18): 4741-9.

Breugelmans, P., P. J. D'Huys, R. De Mot & D. Springael (2007). "Characterization of novel linuron-mineralizing bacterial consortia enriched from long-term linuron-treated agricultural soils." *FEMS Microbiol Ecol* 62(3): 374-85.

Breugelmans, P., B. Leroy, K. Bers, W. Dejonghe, R. Wattiez, R. De Mot & D. Springael (2010). "Proteomic study of linuron and 3,4-dichloroaniline degradation by Variovorax sp. WDL1: evidence for the involvement of an aniline dioxygenase-related multicomponent protein." *Res Microbiol* 161(3): 208-18.

Cech, N. B. & C. G. Enke (2000). "Relating electrospray ionization response to nonpolar character of small peptides." *Anal Chem* 72(13): 2717-23.

Charro, N., B. L. Hood, D. Faria, P. Pacheco, P. Azevedo, C. Lopes, A. B. de Almeida, F. M. Couto, T. P. Conrads & D. Penque (2011). "Serum proteomics signature of cystic fibrosis patients: a complementary 2-DE and LC-MS/MS approach." *J Proteomics* 74(1): 110-26.

Chelius, D. & P. V. Bondarenko (2002). "Quantitative profiling of proteins in complex mixtures using liquid chromatography and mass spectrometry." *J Proteome Res* 1(4): 317-23.

Colaert, N., J. Vandekerckhove, K. Gevaert & L. Martens (2011). "A comparison of MS2-based label-free quantitative proteomic techniques with regards to accuracy and precision." *Proteomics* 11(6): 1110-3.

Dayon, L., A. Hainard, V. Licker, N. Turck, K. Kuhn, D. F. Hochstrasser, P. R. Burkhard & J. C. Sanchez (2008). "Relative quantification of proteins in human cerebrospinal fluids by MS/MS using 6-plex isobaric tags." *Anal Chem* 80(8): 2921-31.

Eddahri, F., S. Denanglaire, F. Bureau, R. Spolski, W. J. Leonard, O. Leo & F. Andris (2009). "Interleukin-6/STAT3 signaling regulates the ability of naive T cells to acquire B-cell help capacities." *Blood* 113(11): 2426-33.

Enke, C. G. (1997). "A predictive model for matrix and analyte effects in electrospray ionization of singly-charged ionic analytes." *Anal Chem* 69(23): 4885-93.

Finamore, F., L. Pieroni, M. Ronci, V. Marzano, S. L. Mortera, M. Romano, C. Cortese, G. Federici & A. Urbani (2010). "Proteomics investigation of human platelets by shotgun nUPLC-MSE and 2DE experimental strategies: a comparative study." *Blood Transfus* 8 Suppl 3: s140-8.

Florens, L., M. J. Carozza, S. K. Swanson, M. Fournier, M. K. Coleman, J. L. Workman & M. P. Washburn (2006). "Analyzing chromatin remodeling complexes using shotgun proteomics and normalized spectral abundance factors." *Methods* 40(4): 303-11.

Gao, J., G. J. Opiteck, M. S. Friedrichs, A. R. Dongre & S. A. Hefta (2003). "Changes in the protein expression of yeast as a function of carbon source." *J Proteome Res* 2(6): 643-9.

Gerber, S. A., J. Rush, O. Stemman, M. W. Kirschner & S. P. Gygi (2003). "Absolute quantification of proteins and phosphoproteins from cell lysates by tandem MS." *Proc Natl Acad Sci U S A* 100(12): 6940-5.

Gygi, S. P., B. Rist, S. A. Gerber, F. Turecek, M. H. Gelb & R. Aebersold (1999). "Quantitative analysis of complex protein mixtures using isotope-coded affinity tags." *Nat Biotechnol* 17(10): 994-9.

Hansen, K. C., G. Schmitt-Ulms, R. J. Chalkley, J. Hirsch, M. A. Baldwin & A. L. Burlingame (2003). "Mass spectrometric analysis of protein mixtures at low levels using cleavable 13C-isotope-coded affinity tag and multidimensional chromatography." *Mol Cell Proteomics* 2(5): 299-314.

Herberth, M., D. Koethe, Y. Levin, E. Schwarz, N. D. Krzyszton, S. Schoeffmann, H. Ruh, H. Rahmoune, L. Kranaster, T. Schoenborn, M. F. Leweke, P. C. Guest & S. Bahn (2011). "Peripheral profiling analysis for bipolar disorder reveals markers associated with reduced cell survival." *Proteomics* 11(1): 94-105.

Ishihama, Y., Y. Oda, T. Tabata, T. Sato, T. Nagasu, J. Rappsilber & M. Mann (2005). "Exponentially modified protein abundance index (emPAI) for estimation of absolute protein amount in proteomics by the number of sequenced peptides per protein." *Mol Cell Proteomics* 4(9): 1265-72.

King, R., R. Bonfiglio, C. Fernandez-Metzler, C. Miller-Stein & T. Olah (2000). "Mechanistic investigation of ionization suppression in electrospray ionization." *J Am Soc Mass Spectrom* 11(11): 942-50.

Leroy, B., C. Rosier, V. Erculisse, N. Leys, M. Mergeay & R. Wattiez (2010). "Differential proteomic analysis using isotope-coded protein-labeling strategies: comparison, improvements and application to simulated microgravity effect on Cupriavidus metallidurans CH34." *Proteomics* 10(12): 2281-91.

Liu, H., R. G. Sadygov & J. R. Yates, 3rd (2004). "A model for random sampling and estimation of relative protein abundance in shotgun proteomics." *Anal Chem* 76(14): 4193-201.

Lu, P., C. Vogel, R. Wang, X. Yao & E. M. Marcotte (2007). "Absolute protein expression profiling estimates the relative contributions of transcriptional and translational regulation." *Nat Biotechnol* 25(1): 117-24.

Mastroleo, F., B. Leroy, R. Van Houdt, C. s' Heeren, M. Mergeay, L. Hendrickx & R. Wattiez (2009a). "Shotgun proteome analysis of Rhodospirillum rubrum S1H: integrating data from gel-free and gel-based peptides fractionation methods." *J Proteome Res* 8(5): 2530-41.

Mastroleo, F., R. Van Houdt, B. Leroy, M. A. Benotmane, A. Janssen, M. Mergeay, F. Vanhavere, L. Hendrickx, R. Wattiez & N. Leys (2009b). "Experimental design and environmental parameters affect Rhodospirillum rubrum S1H response to space flight." *Isme J.*

Matallana-Surget, S., B. Leroy, J. Derock & R. Wattiez (submitted). "Proteome-wide analysis and diel proteomic profiling in the cyanobacterium Arthrospira platensis PCC 8005." *ISME J.*

Mbeunkui, F. & M. B. Goshe (2011). "Investigation of solubilization and digestion methods for microsomal membrane proteome analysis using data-independent LC-MSE." *Proteomics* 11(5): 898-911.

Michel, B., B. Leroy, V. Stalin Raj, F. Lieffrig, J. Mast, R. Wattiez, A. F. Vanderplasschen & B. Costes (2010). "The genome of cyprinid herpesvirus 3 encodes 40 proteins incorporated in mature virions." *J Gen Virol* 91(Pt 2): 452-62.

Moon, P. G., J. E. Lee, S. You, T. K. Kim, J. H. Cho, I. S. Kim, T. H. Kwon, C. D. Kim, S. H. Park, D. Hwang, Y. L. Kim & M. C. Baek (2011). "Proteomic analysis of urinary exosomes from patients of early IgA nephropathy and thin basement membrane nephropathy." *Proteomics* 11(12): 2459-75.

Moulder, R., T. Lonnberg, L. L. Elo, J. J. Filen, E. Rainio, G. Corthals, M. Oresic, T. A. Nyman, T. Aittokallio & R. Lahesmaa (2010). "Quantitative proteomics analysis of the nuclear fraction of human CD4+ cells in the early phases of IL-4-induced Th2 differentiation." *Mol Cell Proteomics* 9(9): 1937-53.

Muller, C., P. Schafer, M. Stortzel, S. Vogt & W. Weinmann (2002). "Ion suppression effects in liquid chromatography-electrospray-ionisation transport-region collision induced dissociation mass spectrometry with different serum extraction methods for systematic toxicological analysis with mass spectra libraries." *J Chromatogr B Analyt Technol Biomed Life Sci* 773(1): 47-52.

Neilson, K. A., N. A. Ali, S. Muralidharan, M. Mirzaei, M. Mariani, G. Assadourian, A. Lee, S. C. van Sluyter & P. A. Haynes (2011). "Less label, more free: approaches in label-free quantitative mass spectrometry." *Proteomics* 11(4): 535-53.

O'Farrell, P. H. (1975). "High resolution two-dimensional electrophoresis of proteins." *J Biol Chem* 250(10): 4007-21.

Ong, S. E., B. Blagoev, I. Kratchmarova, D. B. Kristensen, H. Steen, A. Pandey & M. Mann (2002). "Stable isotope labeling by amino acids in cell culture, SILAC, as a simple and accurate approach to expression proteomics." *Mol Cell Proteomics* 1(5): 376-86.

Paradela, A., A. Marcilla, R. Navajas, L. Ferreira, A. Ramos-Fernandez, M. Fernandez, J. F. Mariscotti, F. Garcia-del Portillo & J. P. Albar (2010). "Evaluation of isotope-coded protein labeling (ICPL) in the quantitative analysis of complex proteomes." Talanta 280 (4):1496-502.

Plumb, R. S., K. A. Johnson, P. Rainville, B. W. Smith, I. D. Wilson, J. M. Castro-Perez & J. K. Nicholson (2006). "UPLC/MS(E); a new approach for generating molecular fragment information for biomarker structure elucidation." *Rapid Commun Mass Spectrom* 20(13): 1989-94.

Ross, P. L., Y. N. Huang, J. N. Marchese, B. Williamson, K. Parker, S. Hattan, N. Khainovski, S. Pillai, S. Dey, S. Daniels, S. Purkayastha, P. Juhasz, S. Martin, M. Bartlet-Jones, F. He, A. Jacobson & D. J. Pappin (2004). "Multiplexed protein quantitation in Saccharomyces cerevisiae using amine-reactive isobaric tagging reagents." *Mol Cell Proteomics* 3(12): 1154-69.

Saydam, O., O. Senol, T. B. Schaaij-Visser, T. V. Pham, S. R. Piersma, A. O. Stemmer-Rachamimov, T. Wurdinger, S. M. Peerdeman & C. R. Jimenez (2010). "Comparative protein profiling reveals minichromosome maintenance (MCM) proteins as novel potential tumor markers for meningiomas." *J Proteome Res* 9(1): 485-94.

Schmidt, A., J. Kellermann & F. Lottspeich (2005). "A novel strategy for quantitative proteomics using isotope-coded protein labels." *Proteomics* 5(1): 4-15.

Silva, J. C., M. V. Gorenstein, G. Z. Li, J. P. Vissers & S. J. Geromanos (2006). "Absolute quantification of proteins by LCMSE: a virtue of parallel MS acquisition." *Mol Cell Proteomics* 5(1): 144-56.

Tanaka, Y., S. Hamano, K. Gotoh, Y. Murata, Y. Kunisaki, A. Nishikimi, R. Takii, M. Kawaguchi, A. Inayoshi, S. Masuko, K. Himeno, T. Sasazuki & Y. Fukui (2007). "T helper type 2 differentiation and intracellular trafficking of the interleukin 4 receptor-alpha subunit controlled by the Rac activator Dock2." *Nat Immunol* 8(10): 1067-75.

Unwin, R. D., D. L. Smith, D. Blinco, C. L. Wilson, C. J. Miller, C. A. Evans, E. Jaworska, S. A. Baldwin, K. Barnes, A. Pierce, E. Spooncer & A. D. Whetton (2006). "Quantitative proteomics reveals posttranslational control as a regulatory factor in primary hematopoietic stem cells." *Blood* 107(12): 4687-94.

Voyksner, R. D. & H. Lee (1999). "Investigating the use of an octupole ion guide for ion storage and high-pass mass filtering to improve the quantitative performance of electrospray ion trap mass spectrometry." *Rapid Commun Mass Spectrom* 13(14): 1427-37.

Wang, W., H. Zhou, H. Lin, S. Roy, T. A. Shaler, L. R. Hill, S. Norton, P. Kumar, M. Anderle & C. H. Becker (2003). "Quantification of proteins and metabolites by mass spectrometry without isotopic labeling or spiked standards." *Anal Chem* 75(18): 4818-26.

Wilm, M. (2011). "Principles of electrospray ionization." *Mol Cell Proteomics* 10(7): M111 009407.

Wu, C. C., M. J. MacCoss, K. E. Howell, D. E. Matthews & J. R. Yates, 3rd (2004). "Metabolic labeling of mammalian organisms with stable isotopes for quantitative proteomic analysis." *Anal Chem* 76(17): 4951-9.

Ye, X., B. Luke, T. Andresson & J. Blonder (2009). "18O stable isotope labeling in MS-based proteomics." *Brief Funct Genomic Proteomic* 8(2): 136-44.

Zhang, B., N. C. VerBerkmoes, M. A. Langston, E. Uberbacher, R. L. Hettich & N. F. Samatova (2006). "Detecting differential and correlated protein expression in label-free shotgun proteomics." *J Proteome Res* 5(11): 2909-18.

Zhang, R. & F. E. Regnier (2002). "Minimizing resolution of isotopically coded peptides in comparative proteomics." *J Proteome Res* 1(2): 139-47.

18

Quantitative Proteomics Using iTRAQ Labeling and Mass Spectrometry

H. R. Fuller[1,2] and G. E. Morris[1,2]
[1]*Wolfson Centre for Inherited Neuromuscular Disease, RJAH Orthopaedic Hospital, Oswestry*
[2]*Institute for Science and Technology in Medicine, Keele University*
UK

1. Introduction

Proteomics research involves the identification and characterisation of proteins in order to elucidate their function and interactions with other proteins. Since the composition of protein mixtures can vary between cell types and can change under certain physiological conditions, one aim is often to quantify up- or down-regulation of individual proteins. Characterisation of proteomic changes associated with disease often helps to shed light on disease mechanisms and identify useful biomarkers and therapeutic targets. It is rarely the case that such proteins are either "present" or "absent", but more likely that they vary in abundance to different degrees. It is therefore important to have a sensitive and accurate method to measure these changes using an unbiased approach.

Shotgun proteomics approaches enable identification of proteins that are up-regulated or down-regulated under specific conditions and this can be studied in different cell and tissue lysates. Isobaric tags for relative and absolute quantification (iTRAQ™) make it possible to both identify and quantify proteins simultaneously. iTRAQ™ can easily be multiplexed, enabling analysis of up to 8 different samples within the same experiment. Our objectives in this chapter are to place iTRAQ™ (isobaric tags for relative and absolute quantification) in context in the history of attempts to bring quantitative studies to proteomics, to explain what it can do, to describe in some detail the protocol that we use in this laboratory and to illustrate the application of iTRAQ™ to medical and clinically-relevant problems, including our own work on the proteomic effects of common drug treatments.

2. A brief history of quantitative proteomics

Over the last two decades, the emergence of vast genomic databases has completely revolutionized the way in which mass spectrometry is used to analyze proteins. Many proteins are now well represented in databases, and their annotations are increasingly becoming more detailed to include information such as sites of post-translational modification. However, this information is only qualitative, which means that differential comparisons of protein expression in a perturbed system, with reference to "control" proteins in a database, are not yet possible. It is possible, however, to perform parallel comparisons of protein expression in different systems using approaches that require staining or labeling of proteins.

The traditional 2-dimensional gel approach, where differentially expressed stained spots are excised and identified by mass spectrometry has many limitations. The wide range of protein abundance often obscures low abundance proteins and not all types of proteins are amenable to gel electrophoresis. Reproducibility is often an issue due to gel-dependent variation and this means that quantitation is often difficult and unreliable (reviewed by Issaq and Veenstra, 2008).

Shotgun proteomics methods involving isotope labeling of proteins have been developed during the last decade and overcome some of the difficulties associated with quantification using gel-based approaches (Wu *et al.*, 2005). One strategy, called SILAC (stable isotope labeling by amino acids in cell culture), involves metabolic incorporation of specific amino acids into proteins (Ong *et al.*, 2002). Two cell populations are grown in culture media that are identical except that one of them contains a 'light' and the other a 'heavy' form of a particular amino acid (e.g. ^{12}C and ^{13}C labeled L-lysine, respectively). Both samples are combined after the cells are harvested and the proteins are identified by mass spectrometry. Metabolic incorporation of the amino acids into the proteins results in a mass shift of the corresponding peptides and the ratio of peak intensities in the mass spectrum reflects the relative protein abundance. Whilst SILAC is a highly efficient technique, a major drawback is that it relies on endogenous labeling of cell lines, so it is not suitable for use with primary tissue such as patient samples (e.g. muscle and serum).

Another strategy, Isotope Coded Affinity Tags (ICAT®), is a cysteine specific, protein-based labeling strategy designed to compare two different sample states (Gygi *et al.*, 1999). One sample is labeled with a light isotope and the other with a heavy isotope, and then the samples are combined and analyzed by mass spectrometry. The ratios of signal intensities of the ICAT-tagged peptide pairs are quantified to determine the relative levels of proteins in the two samples. The specificity of ICAT reagents for cysteine residues means that the approach is sometimes preferred because it reduces sample complexity. However, this also creates a drawback in that peptides lacking cysteine residues will not be labeled, so many important peptides, including those with post-translational modifications (PTMs) will be discarded.

Isobaric tagging strategies overcome some of the major limitations of isotope tagging. One such method was developed by Applied Biosystems (now AB Sciex) and is called iTRAQ™: isobaric tags for relative and absolute quantification (Ross *et al.*, 2004). The reagents were originally designed for the simultaneous multiplexed analysis of up to 4 samples, but are now available as an 8-plex kit (Choe *et al.*, 2007). The iTRAQ™ tags react with all primary amines of peptides, which means that all peptides are labeled and information about their post-translational modifications are retained. The isobaric nature of the tags also means that the same peptide from each of the samples being compared appears as a single peak in the mass spectrum. This reduces the complexity of the data when compared to isotopic labeling strategies where "heavy" and "light" versions of each peptide are detected in each mass spectrum.

3. iTRAQ™: Isobaric tags for relative and absolute quantification

3.1 iTRAQ™ reagent chemistry

The iTRAQ™ tags are isobaric labels that react with primary amines of peptides including the N-terminus and ε-amino group of the lysine side-chain. Each label has a unique charged reporter group, a peptide reactive group, and a neutral balance group to maintain an overall mass of 145Da (Figure 1). When a peptide is fragmented by MS/MS fragmentation, the iTRAQ™ reporter groups break off and produce distinct ions at *m/z* 114, 115, 116, 117, 118, 119, 121 and 122. The relative intensities of the reporter ions are directly proportional to the relative abundances of each peptide in the samples that being compared. In addition to producing strong reporter ion signals for quantification, MS/MS fragmentation of iTRAQ™-tagged peptides also produces strong y- and b-ion signals for more confident identification. During the design of the iTRAQ™ tags, the reporter ion masses were carefully selected in order to minimize interference from noise in the low mass region such as matrix ions, immonium and fragment ions. This is the reason that the 8-plex reagents skip from 119 to 121, since the phenylalanine immonium ion appears at *m/z* 120.

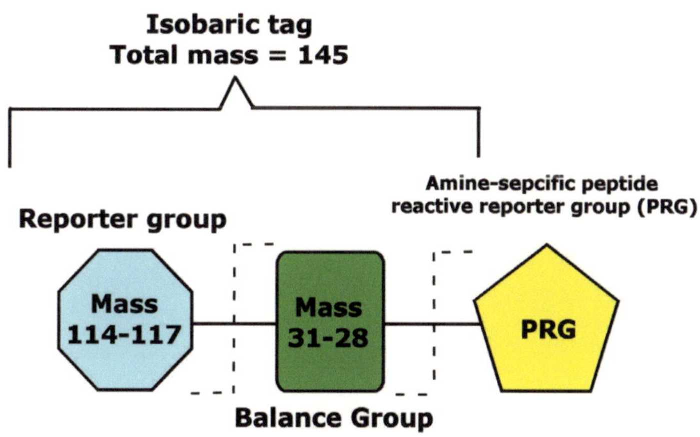

Fig. 1. Structure of the iTRAQ™ reagents.

Each isobaric tag has a unique charged reporter group, a peptide reactive group, and a neutral balance group to maintain an overall mass of 145Da.

3.2 iTRAQ™ work-flow

The general workflow for an iTRAQ™ experiment with 4 tags is shown in Figure 2. Each sample is reduced, alkylated, and digested with trypsin. Each set of peptides is then labeled with a different one of the 4 (or 8) iTRAQ™ tags, pooled, separated by liquid chromatography (LC), and the resulting fractions are analysed using mass spectrometry.

3.3 Digging deeper

It is not always essential to separate proteins before digestion, but some form of fractionation will be needed in order to detect relatively-low abundance components. A simple one-dimensional LC separation of peptides from a whole proteome will overwhelm the mass spectrometer, and highly abundant peptides will mask detection of others. By separating proteins and/or peptides in more than one dimension, it starts to become possible to "see the wood for the trees". Multidimensional protein identification technology (MudPIT) is a common technique for whole proteomic analysis such as iTRAQ™ comparisons, and can be performed off-line or coupled directly to the mass spectrometer (Washburn *et al.*, 2001). There are many choices of chromatography techniques, including affinity chromatography, ion exchange chromatography, reversed-phase chromatography and size-exclusion chromatography.

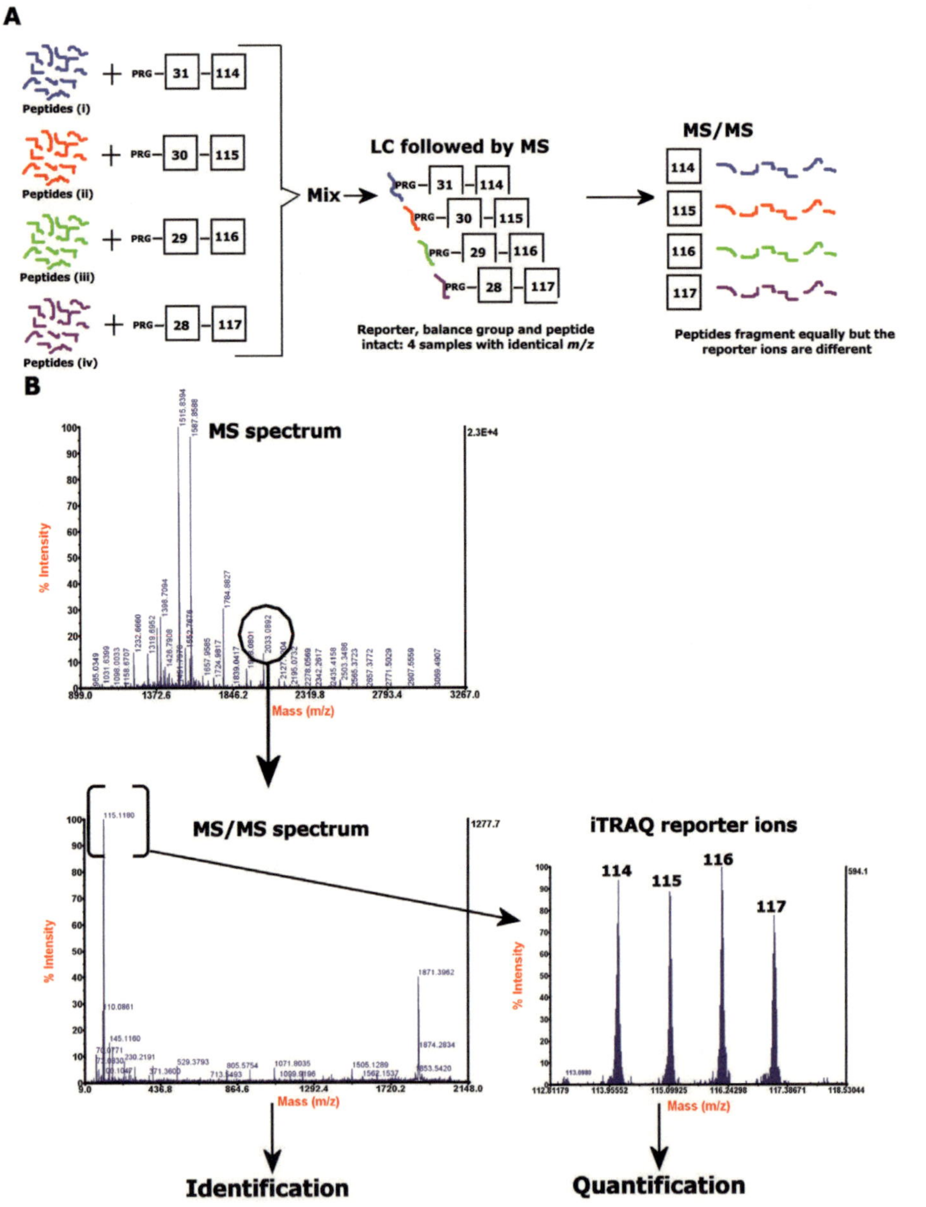

Fig. 2. A general scheme and example data for a 4-plex iTRAQ experiment.

A. Based on a figure by Zieske (2006), this illustration shows the general scheme of a 4-plex iTRAQ experiment. Each of the four sets of peptides are labeled with one of the iTRAQ reagents, mixed together and separated by liquid chromatography. In MS analysis, each identical peptide from the four sets appears as a single precursor (the iTRAQ balance group ensures that all tags have the same overall *m/z*). Following fragmentation in MS/MS, the iTRAQ reporter ions break off and their relative intensities are used for quantification. Each of the four peptides fragments in the same way and give rise to b- and y-ions for identification.

B. shows some example MS and MS/MS data, with an expanded view of the low-mass region of the MS/MS spectrum to show the resolved iTRAQ reporter ions.

Even after LC separation in two dimensions, we are still only able to scratch the top 10-20% of the surface of complex mammalian proteomes using standard instrumentation (Fuller *et al.*, 2010). For a global, unbiased view of the proteome, this is a good starting point and can often yield clues to follow up further. If particular types of low abundant proteins are of interest then enrichment such as subcellular fractionation or immunodepletion of abundant proteins will be necessary.

3.4 Instrumentation

Matrix-assisted laser desorption ionisation (MALDI) MS/MS and electrospray ionisation (ESI) MS/MS are the most common types of mass spectrometer used for iTRAQ™ analysis, and there have been several comparisons of the two types of instrument for accuracy and performance of iTRAQ™ quantification. Shirran and Botting (2010) analysed a fixed concentration of a six-protein mix and concluded that MALDI MS/MS gave the most accurate results. In contrast, two other studies where more complex biological samples were analysed concluded that analyses by MALDI and ESI are comparable in terms of accuracy and performance (Kuzyk *et al.*, 2009 and Scheri *et al.*, 2008). Whilst it is possible to re-analyse archived LC-separated samples by MALDI MS/MS (and so has the potential to yield more data), the trade-off is that MALDI analysis usually takes longer than ESI analysis.

Under standard MS/MS fragmentation (collision-induced dissociation (CID)), an ion trap is unable to analyze small product ions because of their low mass cut-off limitation. This meant that traditionally, iTRAQ™-based quantification was not possible using an ion trap or hybrid instrument containing an ion trap such as the LTQ-Orbitrap. Recently developed fragmentation methods now make it possible to perform iTRAQ™-based quantification on an LTQ-Orbitrap and include Pulsed Q Dissociation (PQD) (Bantscheff *et al.*, 2008) and higher energy C-trap dissociation (HCD) (Zhang et al., 2009). Both fragmentation methods are less suited for protein identification at a proteomic scale than CID fragmentation, but when combined with CID, HCD allows sensitive and accurate iTRAQ™ quantification of whole proteomes (Köcher *et al.*, 2009).

3.5 Accuracy of iTRAQ™-based quantification

There are many reports in the literature that demonstrate the reliability of iTRAQ™ to measure changes spanning up to two orders of magnitude accurately on MALDI and ESI platforms using low- and high-complexity protein mixtures (Fuller *et al.*, 2010, Scheri *et al.*, 2008 and Yang *et al.*, 2007). Even in whole proteome protein mixtures, it is possible to achieve good correlation between iTRAQ™ ratios and those measured biochemically by methods such as quantitative western blotting and immunofluorescence microscopy, providing appropriate statistical analysis of the iTRAQ™ data is carried out (Fuller *et al.*, 2010).

There are, however, instances where this is not the case. Low-signal data have higher relative variability, irrespective of the instrumentation used (Karp *et al.*, 2010). Since low abundance proteins are usually detected with fewer peptides, they are often disregarded from datasets when statistics-based filtering approaches are used. Several bioinformatics-based models have been suggested to help resolve this problem about heterogeneity of variance, and include an additive-multiplication error model for peak intensities (Karp *et al.*, 2010) and IsobariQ software that employs variance stabilizing normalization (VSN) algorithms (Arntzen *et al.*, 2011).

There are also an increasing number of reports that there is a degree of underestimation of iTRAQ™ ratios, seen especially with larger changes (Ow *et al.*, 2009, Karp *et al.*, 2010 and Ow *et al.*, 2011). "Ratio compression", as it has been termed, is thought to arise from several factors including isotopic contamination and background interference. Providing accurate isotope factors are available, it is possible to correct for impurities from chemical enrichment and natural isotope abundance in the iTRAQ™ reagents using data processing software (e.g. this is a standard function in GPS Explorer software, AB Sciex). The bigger problem arises from background interference: if two peptides have a very similar *m/z* and cannot be resolved by the mass spectrometer during precursor ion selection, the resulting MS/MS spectrum will contain fragment ions and iTRAQ™ reporter ions from both peptides. One of the two peptides may be identified using this data, but its iTRAQ™ ratios may have been "diluted" by those arising from the other peptide. This issue is currently very difficult to minimise but it has been suggested that it can be partly alleviated using high-resolution sample fractionation (Ow *et al.*, 2011).

4. Example protocol for iTRAQ™ analysis using a MALDI TOF/TOF

The following protocol is one we routinely use for analysis of iTRAQ™ samples on an AB Sciex 4800 MALDI-TOF/TOF instrument, but the method could be used with other mass spectrometers since we have omitted any instrument-specific information.

4.1 Cell / tissue extraction

- Extract cell pellets in 10 volumes of extraction buffer (w/v) containing 6M Urea, 2M thiourea, 2% CHAPS and 0.5% SDS in HPLC-grade water.
- Sonicate extracts briefly to disrupt DNA and leave on ice for 10 minutes, followed by centrifugation at 13,000 x g for 10 minutes at 4°C to pellet any insoluble material.
- To remove detergents that may interfere with iTRAQ™ labeling, precipitate the proteins by the addition of 6 volumes of ice cold acetone overnight at -20°C.
- Pellet the acetone precipitates by centrifugation at 13,000 x g for 10 minutes at 4°C and then carefully remove and discard the supernatant.
- Allow the acetone to evaporate and resuspend the pellets in 6M Urea in 50mM triethylammonium bicarbonate (TEAB).*
- Determine the protein concentration in each sample and balance them carefully so that all samples contain the same amount of total protein. Each tag is capable of labeling 100µg of protein but it is best to aim for slightly less than this (i.e. no more than 85µg) to allow for protein estimation errors. It is better to have slightly less protein than to have unlabeled peptides appearing in the mass spectrometer.

*It is important to avoid using buffers containing primary amines such as Tris buffers.

4.2 iTRAQ™ labeling

- Perform reduction and alkylation steps using the reagents and instructions provided in the iTRAQ™ labeling kit (AB Sciex). Detailed instructions can be found on the iTRAQ™ chemistry reference guide, available on the AB Sciex website.
- Dilute the extracts in 50mM TEAB so that the urea concentration is less than 1M before the addition of trypsin.
- Digest with sequencing grade trypsin (1:20 w/w in 50 mM TEAB) and incubate overnight at 37°C.

- Dry down digests in a vacuum centrifuge (in order to maximise iTRAQ™ labeling efficiency the volume of each sample should be less than 50µl).
- For the iTRAQ™ labeling step follow instructions provided with the iTRAQ™ reagents kit.

4.3 Dimension I: Strong cation-exchange (SCX) chromatography

- Pool the iTRAQ™-labeled peptides and make up to a total volume of 2.5mls in SCX buffer A (10mM phosphate, pH3 in 20% acetonitrile (MeCN) (Romil, UK)). The volume can be adjusted depending on the sample loop size. In order to ensure efficient binding to the SCX column, the final pH should as close to pH3 as possible. This can be achieved by the addition of orthophosphoric acid, whilst being careful to ensure that the overall phosphate concentration is not increased significantly (as this will also affect the binding efficiency).
- The following flow rates and conditions are optimized for use with a polysulfoethyl A, SCX column (300A, 5uM (PolyLC))
    - Load the pooled peptides (2.5mls) onto a SCX column at a flow rate of 400ul/minute.
    - Following sample injection wash the column with SCX buffer A until the baseline returns (this usually takes about 10-15 minutes).
    - Run the gradient as follows: 0-50% SCX buffer B (10mM phosphate, 1M NaCl, pH3 in 20% acetonitrile) over 25 minutes followed by a ramp up from 50% to 100% SCX buffer B over 5 minutes. Finally, wash the column in 100% SCX buffer B for 5 minutes before equilibrating for 10 minutes with SCX buffer A.
    - Collect 400ul fractions during the elution period (this usually yields about 20 fractions) and dry down completely in a vacuum centrifuge.
- Once dry, fractions can be stored at -20°C until the next step.

*Polysulfoethyl A columns work best at ambient temperature so if you have a column oven you should remember to turn it off. Use of 0.1% TFA or high concentrations of formic acid in the mobile phase is not recommended so it is best to equilibrate the system with SCX buffer A before connecting the column.

4.4 Dimension II: Reversed-phase chromatography

Prior to mass spectrometry analysis, separate fractions by reversed-phase liquid chromatography. The following flow rates and conditions are optimised for use with a Pepmap C18 column, 200µm x 15cm (LC Packings).

- Resuspend fractions in 30µl of RP buffer A (2% acetonitrile, 0.05% TFA in water (Sigma Chromasolv plus)). The order of fractions should be randomized in order to minimise effects from sample carry-over on the column.
- Perform reversed-phase chromatography separation of each fraction using the following gradient:
    - Load fractions at a flow rate of 3µl/minute
    - 10 minutes isocratic pre-run at 100% RP buffer A (0.05% TFA in 2% acetonitrile in water),
    - followed by a linear gradient from 0-30% RP buffer B (0.05% TFA in 90% acetonitrile in water) over 100 minutes,
    - followed by another linear gradient from 30%-60% RP buffer B over 35minutes.

- Wash the column in 100% RP buffer B for a further 10 minutes, before a final equilibration step in 100% RP buffer A for 10 minutes.
- During the elution gradient, spot the eluate at 10 second intervals using a Probot (LC Packings) with α-cyano-4-hydroxycinnamic acid (CHCA) at 3mg/ml (70% MeCN, 0.1% TFA) at a flow rate of 1.2µl/min.

4.5 MALDI-TOF/TOF analysis

Instrument settings will of course vary, depending on the type of MALDI TOF/TOF instrument used. Even two identical machines from the same vendor may need to be tuned and optimised slightly differently for optimal performance. For this reason, we have just highlighted some important general issues to consider, rather than suggesting exact instrument settings:

- **Internal standards:** MALDI TOF/TOF analysis of fractions from one reversed-phase LC run can take many hours and so it is important to be sure that the instrument is calibrated for the duration. By spiking a known standard into the MALDI matrix, with an internal calibration processing method specified, you can ensure that every fraction contains an internal reference. Glu-1-Fibrinopeptide is a common standard of choice, but it is important to ensure that you optimise the amount you spike in so that it is detectable in MS but not so abundant that it masks detection of your iTRAQ™ peptides.
- **Ion statistics:** in order to get good ion statistics for iTRAQ™ quantification it is important to acquire enough data from each spectrum. A total of 1000 shots per MS spectrum (with no stop conditions) and at least 2500 shots per MS/MS spectrum (no stop conditions) should be acquired.
- **CID:** The MS/MS acquisition operating mode should specify that collision-induced dissociated (CID).
- **Precursor ion selection:** it is important to set a limit on the number of precursors selected per spot for MS/MS analysis to enable you to get maximum of data out the spot, without burning the spot out. The order in which precursor ions are selected and analysed is also an important consideration. For an unbiased view of a proteome, a common approach is to acquire weaker precursors first (i.e. those with a lower signal/noise) as these are harder to get good fragmentation data from when spots start to burn out. Another approach is to skip a selected number of the strongest precursors in each spot, in an attempt to negate masking of low abundant peptides by those that are in high abundance.

4.6 Bioinformatics

There are several different software packages for performing database searches with iTRAQ™ data and many utilize MASCOT as the search engine. Software that supports iTRAQ™ quantification will have several particular features: the ability to exclude the iTRAQ™ reporter ion masses from the search, identify spectra with fixed iTRAQ™ modifications (N-term (iTRAQ™), lysine (iTRAQ™) and methyl methanethiosulfonate (MMTS) modification of cysteine residues) and to apply correction factors to the peak areas of the iTRAQ™ reporter peaks in peptide spectra identified. Although it is possible to manually calculate relative quantification, many software packages will also be able to perform this function automatically. GPS Explorer (AB Sciex), for example, is able to calculate iTRAQ™ protein and peptide ratios for all identified peptides in the database search. The ratio is calculated by selecting one tag as the reference mass and applying the following calculation: ratio = fragment corrected area / reference corrected area. A normalization factor is usually also applied, and can be useful to normalize any deviances in iTRAQ™ ratios due to unequal total protein in each sample set and impurities in the iTRAQ™ tags themselves (normalized iTRAQ™ Ratio = Ratio / median iTRAQ™ Ratio of all found pairs).

4.7 Validation

Quantitative proteomic experiments such as iTRAQ™ are performed in an unbiased fashion and are intended to provide us with clues for further study, rather than to provide definitive answers. In order to extract useful information from the masses of data that are produced in iTRAQ™ experiments, it is important to have a system in place to interrogate and validate the data carefully. The approach used will depend on the aim of the experiment and the type of comparison that is being done (e.g. pair-wise, 4-plex but with samples in duplicate, 4-plex but with 4 different samples), but some points for consideration are listed:

- **Cut-off values:** data can be simplified dramatically by first applying a cut-off to remove proteins that are detected with less than a certain number of peptides, and also with less than a certain total ion score confidence interval. A good starting point is to apply a very stringent filter and reduce if necessary (i.e. discard proteins detected with less than 95% total ion score confidence interval and less than 2 peptides). The data may then be filtered further to leave proteins that up- or down-regulation by a defined minimum amount.
- **Statistics:** the iTRAQ™ protein ratio is an average ratio, calculated using the individual peptide ratios for each peptide used to identify the protein. One inaccurate peptide ratio may dramatically skew the average ratio for the whole protein, so for validation, it is important to also look closely at the individual peptide ratios for each protein. One way to do this is to perform statistical tests that compare normalized peptide ratios from each sample in a pair-wise fashion.
- **Biochemistry:** providing antibodies are available, iTRAQ™-predicted changes in protein levels can be confirmed by biochemical methods such as western blotting, ELISA or immunohistochemistry.
- **Mass spectrometry approaches:** although they often take time to optimise, high through-put mass spectrometry-based assays such as multiple-reaction monitoring (MRM) can be especially useful for validating biomarkers, especially when antibodies are not available (Anderson and Hunter, 2006).

5. Applications of iTRAQ™ in medical research

Since 2005, several hundred papers have been published that describe applications of iTRAQ™ to many areas of medical research, including, but not limited to: various cancers, neurodegenerative disorders, liver and kidney problems, pre-eclampsia, diabetes, host-pathogen interactions, pancreatitis and autoimmune disorders. The majority of these studies were designed to discover biomarkers in order to understand disease mechanisms, to improve methods for early and sensitive diagnosis, to identify potential therapeutic targets, or to understand the mechanism of action of drugs. A smaller number of studies also attempted to identify biomarkers that could be useful for predicting the prognosis of patients with various types of cancer (Rehman *et al.*, 2008; Matta *et al.*, 2009; Tripathi *et al.*, 2010).

5.1 Examples of clinically-relevant iTRAQ™ applications

An early, clinically-relevant application of iTRAQ™ was in 2005 when DeSouza *et al.* identified nine potential biomarkers for endometrial cancer. In 2007, they performed a much larger 40-sample iTRAQ™ study in an attempt to verify these earlier findings, and found that none of the nine previously identified potential biomarkers had the sensitivity and specificity to be used individually to discriminate between normal and cancer samples. They did however, find that a panel of three of these proteins: pyruvate kinase, chaperonin 10 and α_1-antitrypsin, gave good results with sensitivity, specificity, predictive value and positive predictive value of 0.95 in a logistic regression analysis (DeSouza *et al.*, 2007). Glen *et al.* (2008) used iTRAQ™ to identify tumor regression antigen, gp96, as a highly-significant marker to distinguish benign from malignant prostate cancers. Rudrabhatla *et al.* (2010) used iTRAQ™ to identify amino-acid residues on neurofilament proteins that were more highly-phosphorylated in Alzheimer Disease patients, while Abdi *et al.* (2006) reported potential biomarkers in cerebrospinal fluid to distinguish Alzheimer's disease, Parkinson's disease and dementia with Lewy body (DLB). The greatest improvement of iTRAQ™ over 2D-gels is observed with membrane proteins and Han *et al.* (2008) were able to use it to identify potential therapeutic targets for autosomal dominant polycystic kidney disease by comparing kidney plasma membranes from wild-type and diseased mouse models. Grant *et al.* (2009) used iTRAQ™ to study the effects of aging on the proteome of cardiac left ventricles and obtain clues to the mechanism of loss of diastolic function with age. Pendyala *et al.* (2010) used iTRAQ™ to show that the vitamin E binding protein, afamin, is down-regulated after viral infection in a study of HIV-1-associated neurocognitive disorder (HAND). Although serum samples present a problem for iTRAQ™ because of high concentrations of a few major proteins, the work of Dwivedi *et al.* (2009) illustrates how this was overcome in a study of the proteomic effects of anti-TNF-alpha treatment of rheumatoid arthritis patients. These examples illustrate the wide variety of applications of iTRAQ™ to the most common of all human health problems.

5.2 Considerations when comparing patients

With adequate care, meaningful iTRAQ™ comparisons of diseased versus control tissues are possible, as illustrated in the previous section. However, iTRAQ™ comparisons of patient-derived material such as skin fibroblasts, serum, CSF, saliva and other tissue types present additional problems: in particular, they may display differences due to the age, sex or genetic background of the original donors, rather than specifically due to a genetic mutation or disease state. For example, Miike *et al.* (2010) used iTRAQ™ to show that there are gender differences in serum protein composition and Truscott *et al.* (2010) used iTRAQ™ analysis of human lenses to show that protein-membrane interactions change significantly with age. Our own work on the inherited neuromuscular disease, spinal muscular atrophy (SMA), can also be used to illustrate some of the issues associated with comparing patient-derived material. The widely-used GM03813 primary skin fibroblasts from a spinal muscular atrophy (SMA) patient (Coriell Cell Repositories) have a genetic mutation that causes a large reduction in the levels of SMN protein. Using iTRAQ™ labeling technology, followed by two-dimensional liquid chromatography and MALDI TOF/TOF analysis, we quantitatively compared the proteomes of a variety of SMA and control skin fibroblast lines. Comparison of SMA patient fibroblasts with an unrelated control of similar age showed that the largest differences reflected their different genotypes (i.e. HLA and MHC antigens). This was largely overcome by comparison with fibroblasts from the child's mother, an unaffected SMA carrier (GM03814). However, myogenic cells present in one primary cell line (GM03813) but not the other resulted in an apparent increase in the myoblast-specific protein, desmin in the SMA cells (Figure 3). This observation enabled us to obtain a myoblast-free fibroblast population for further studies by immortalizing and cloning this primary cell line (Fuller et al, 2010).

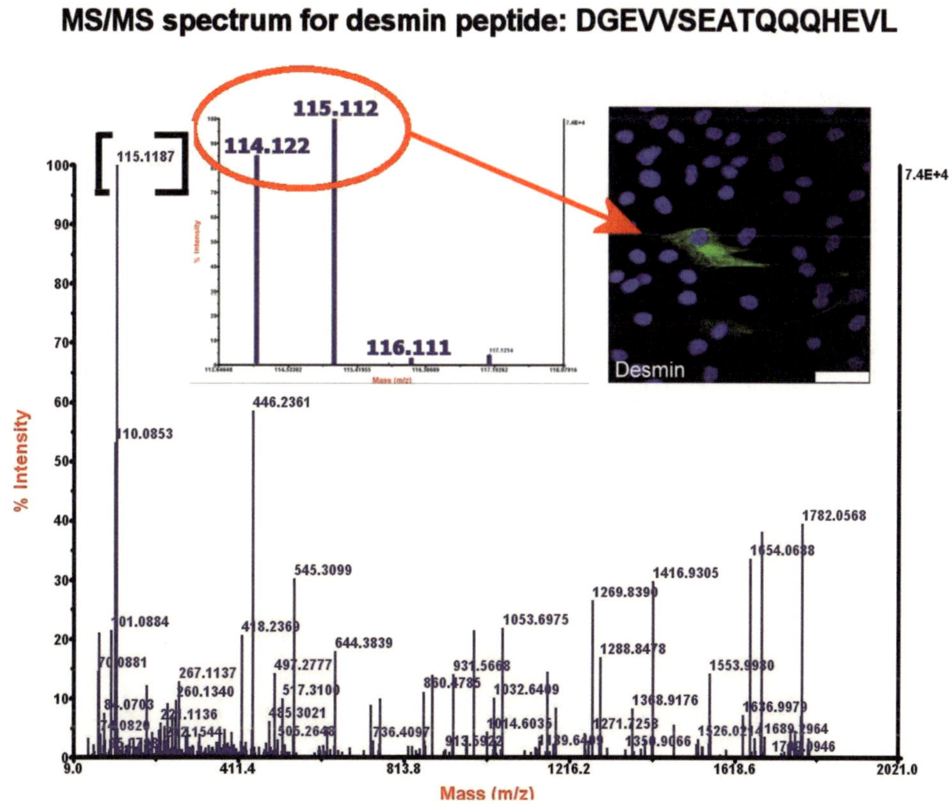

Fig. 3. A non-homogeneous patient cell line gave false positive iTRAQ™ results.

Peptides from an SMA patient cell line were analysed in duplicate (labeled with 114 and 115 iTRAQ tags) and compared to a control cell line, also analysed in duplicate (labeled with 116 and 117 iTRAQ tags). An example MS/MS spectrum is shown for a peptide identified as the muscle-specific protein, desmin. The image inset on the top left is an expanded MS/MS spectrum showing that only the 114 and 115 reporter ions were detected. The suspicion that the SMA patient cell line contained myogenic cells, absent from control cells, was confirmed by immunofluorescence microscopy with an anti-desmin antibody (green in the inset image).

5.3 Proteomic effects of drug treatments

In-vivo studies that monitor the therapeutic effect of drugs on patients over time are very complicated to design and involve considering many factors such as: the time of day tissue sample is taken, change in diet, infection and secondary effects caused by the disease or aging. However, an iTRAQ™ comparison of a single cell line, with and without a drug, is a much more straightforward general approach to understanding the mechanisms of action of drugs and their side-effects. Wang *et al.* (2010) used this approach to examine the effect of the beta blocker Carvedilol in vascular smooth muscle cells and found 13 proteins that were altered in expression. Another example is the work of Bai *et al.* in 2010, when they used iTRAQ™ to look at the effects of the anti-coagulation drug warfarin on HepG2 cells and identified two proteins, DJ-1 and 14-3-3 Protein, that were altered in expression.

We recently used this approach to identify possible side-effects of drugs for spinal muscular atrophy (SMA) (Fuller *et al.*, 2010). Valproate is commonly used as an anticonvulsant in epilepsy and as a mood stabilizer, but its long-term side-effects can include bone loss. As a histone deacetylase (HDAC) inhibitor, valproate has also been considered for treatment of SMA. Using iTRAQ labeling, we performed a quantitative comparison of the proteome of an SMA skin fibroblast cell line, with and without valproate treatment. The most striking change was a reduction in collagens I and VI, while over 1000 other proteins remained unchanged. The collagen-binding glycoprotein, osteonectin (SPARC, BM-40) was one of the few other proteins that were significantly reduced by valproate treatment. Collagen I is the main protein component of bone matrix and osteonectin has a major role in bone development, so the results suggest a possible molecular mechanism for bone loss following long-term exposure to valproate. An example MS/MS spectrum showing reduction of a collagen I peptide after treatment valproate is shown in Figure 4.

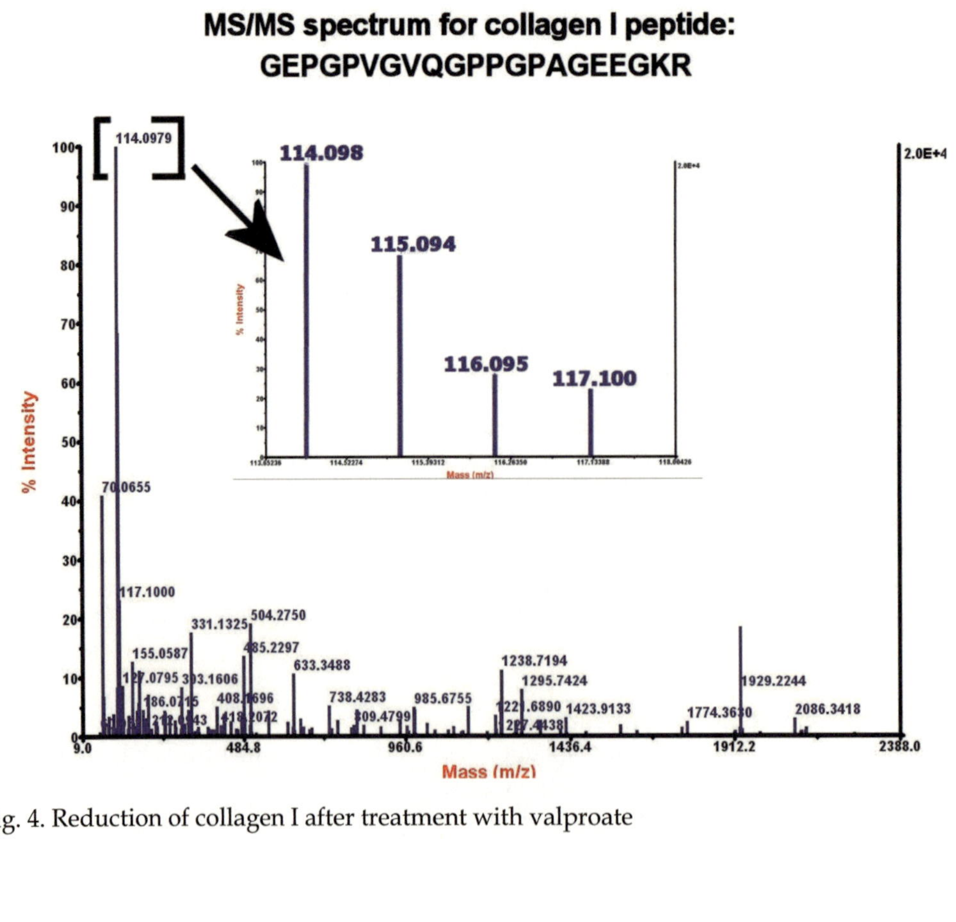

Fig. 4. Reduction of collagen I after treatment with valproate

Peptides from an SMA patient cell line treated with valproate were analysed in duplicate (labeled with 116 and 117 iTRAQ tags) and compared to the same cell line without valproate treatment, also analysed in duplicate (labeled with 114 and 115 iTRAQ tags). An example MS/MS spectrum is shown for a peptide identified as collagen. The image inset on the top left is an expanded MS/MS spectrum showing that the 116 and 117 iTRAQ reporter ions were much lower in intensity than the 114 and 115 iTRAQ reporter ions. Biochemical studies confirmed that collagen I is reduced after treatment with valproate (Fuller *et al.*, 2010).

5.4 Bringing biomarkers to the bedside

Although there are many reports in the literature using iTRAQ™ to identify potential biomarkers of disease, very few biomarkers ever get fully validated to the stage where they can be used in a clinical setting to benefit patients. The low rate of transition from the laboratory to the clinic is something that is seen with biomarkers in general, and not just those identified by iTRAQ™ or other quantitative proteomic approaches. In order for a new biomarker to be introduced into routine clinical practice, a slow and detailed process is required to obtain evidence that it is robust, precise and reproducible, in addition to demonstrating that it will improve patient management and outcome, and have audit and cost benefits (reviewed in detail by Sturgeon *et al.*, 2010).

6. Summary and future prospects

Without a doubt, iTRAQ™ labeling of peptides has had a significant impact on the development of quantitative proteomics over the last 8 years. The ability to multiplex and analyze up to 8 samples within the same experiment adds flexibility to the experimental design without complicating MS data analysis. In 2008, Thermo Fisher in-licensed an isobaric mass tagging technology called TMT, which can be multiplexed to allow analysis of up to 6 samples, further confirmation of the wide acceptance of this technique.

The discovery of new biomarkers will help us to understand disease mechanisms and prognosis better, to improve methods for early and sensitive diagnosis, to identify therapeutic targets, or to understand the mechanism of action of drugs. Although iTRAQ™ has been very useful for potential biomarker discovery, issues regarding analytical and experimental variability need to be addressed before the benefit of iTRAQ™ reaches routine analysis in the clinical laboratory. With further developments to address issues affecting accuracy of iTRAQ quantification and improving data analysis tools, medical research may benefit greatly from iTRAQ-based quantitative proteomics over the coming years.

7. Acknowledgements

Our own research was supported by grants from the Jennifer Trust for SMA, the Muscular Dystrophy Association (USA) and the RJAH Institute of Orthopaedics, UK.

8. References

Abdi, F., Quinn, J.F., Jankovic, J., McIntosh, M., Leverenz, J.B., Peskind, E., Nixon, R., Nutt, J., Chung, K., Zabetian, C., Samii, A., Lin, M., Hattan, S., Pan, C., Wang, Y., Jin, J., Zhu, D., Li. G.J., Liu, Y., Waichunas, D., Montine, T.J. & Zhang, J. (2006). Detection of biomarkers with a multiplex quantitative proteomic platform in cerebrospinal fluid of patients with neurodegenerative disorders. *J Alzheimers Dis,* Vol.9, No.3, (December 2006), pp.293-348.

Anderson, L. & Hunter, C.L. (2006). Quantitative mass spectrometric multiple reaction monitoring assays for major plasma proteins. *Mol Cell Proteomics,* Vol.5, No.4, (April 2006), pp.573-88.

Arntzen, M.Ø., Koehler, C.J., Barsnes, H., Berven, F.S., Treumann, A. & Thiede, B. (2011). IsobariQ: software for isobaric quantitative proteomics using IPTL, iTRAQ, and TMT. *J Proteome Res,* Vol.10, No.2, (February 2011), pp.913-20.

Bai, J., Sadrolodabaee, L., Ching, C.B., Chowbay, B. & Ning Chen, W. (2010). A comparative proteomic analysis of HepG2 cells incubated by S(-) and R(+) enantiomers of anti-coagulating drug warfarin. *Proteomics.* Vol.10, No.7, (April 2010), pp.1463-73.

Bantscheff, M., Boesche, M., Eberhard, D., Matthieson, T., Sweetman, G. & Kuster, B. (2008). Robust and sensitive iTRAQ quantification on an LTQ Orbitrap mass spectrometer. *Mol Cell Proteomics,*Vol.7, No.9, (September 2008), pp.1702-13.

Choe, L., D'Ascenzo, M., Relkin, N.R., Pappin, D., Ross, P., Williamson, B., Guertin, S., Pribil, P. & Lee, K.H. (2007). 8-plex quantitation of changes in cerebrospinal fluid protein expression in subjects undergoing intravenous immunoglobulin treatment for Alzheimer's disease. *Proteomics,* Vol.7, No.20, (October 2007), pp.3651-60.

DeSouza, L., Deihl, G., Rodrigues, M.J., Guo, J., Romaschin, A.D., Colgan, T.J. & Siu, K.W. (2005). Search for cancer markers from endometrial tissues using differentially labeled tags iTRAQ and cICAT with multidimensional liquid chromatography and tandem mass spectrometry. *Journal of Proteome Research,* Vol.4, No.2, (March-April 2005), pp.377-86.

DeSouza, L.V., Grigull, J., Ghanny, S., Dube, V., Romaschin, A.D., Colgan T.J. & Siu, K.W. (2007). Endometrial carcinoma biomarker discovery and verification using differentially tagged clinical samples with multidimensional liquid chromatography and tandem mass spectrometry. *Molecular and Cellular Proteomics,* Vol.6, No.7, (July 2007), pp.1170-82.

Dwivedi, R.C., Dhindsa, N., Krokhin, O.V., Cortens, J., Wilkins, J.A. & El-Gabalawy, H.S. (2009). The effects of infliximab therapy on the serum proteome of rheumatoid arthritis patients. *Arthritis Res Ther,* Vol.11, No.2, R32.

Fuller, H.R., Man, N.T., Lam, le. T., Shamanin, V.A., Androphy, E.J. & Morris, G.E. (2010). Valproate and bone loss: iTRAQ proteomics show that valproate reduces collagens and osteonectin in SMA cells. *Journal of Proteome Research,* Vol.9, No.8, (August 2010),pp.4228-33.

Glen, A., Gan, C.S., Hamdy, F.C., Eaton, C.L., Cross, S.S., Catto, J.W., Wright, P.C. & Rehman, I. (2008). iTRAQ-facilitated proteomic analysis of human prostate cancer cells identifies proteins associated with progression. *J Proteome Res,* Vol.7, No.3, (March 2008), pp.897-907.

Grant, J.E., Bradshaw, A.D., Schwacke, J.H., Baicu, C.F., Zile, M.R. & Schey, K.L. (2009). Quantification of protein expression changes in the aging left ventricle of Rattus norvegicus. *J Proteome Res.* Vol.8, No.9, (September 2009), pp.4252-63.

Gygi, S.P., Rist, B., Gerber, S.A., Turecek, F., Gelb, M.H. & Aebersold, R. (1999). Quantitative analysis of complex protein mixtures using isotope-coded affinity tags. *Nature Biotechnology,* Vol. 17, No.10, (October 1999), pp. 994–9.

Han, C.L., Chien, C.W., Chen, W.C., Chen, Y.R., Wu, C.P., Li, H. & Chen, Y.J. (2008). A multiplexed quantitative strategy for membrane proteomics: opportunities for mining therapeutic targets for autosomal dominant polycystic kidney disease. *Mol Cell Proteomics*. Vol.7, No.10, (October 2008), pp.1983-97.

Issaq, H.J., Veenstra, T.D. (2008). Two-dimensional polyacrylamide gel electrophoresis (2D-PAGE): advances and perspectives. *Biotechniques,* Vol.44, No.5, pp.697-700.

Karp, N.A., Huber, W., Sadowski, P.G., Charles, P.D., Hester, S.V. & Lilley, K.S. (2010). Addressing accuracy and precision issues in iTRAQ quantitation. *Mol Cell Proteomics*. Vol.9, No.9, (September 2010), pp.1885-97.

Köcher, T., Pichler, P., Schutzbier, M., Stingl, C., Kaul, A., Teucher, N., Hasenfuss, G., Penninger, J.M. & Mechtler K. (2009). High precision quantitative proteomics using iTRAQ on an LTQ Orbitrap: a new mass spectrometric method combining the benefits of all. *J Proteome Research,* Vol.8, No.10, (October 2009), pp. 4743-52.

Kuzyk, M.A., Ohlund, L.B., Elliott, M.H., Smith, D., Qian, H., Delaney, A., Hunter, C.L. & Borchers, C.H. (2009) A comparison of MS/MS-based, stable-isotope-labeled, quantitation performance on ESI-quadrupole TOF and MALDI-TOF/TOF mass spectrometers. *Proteomics,* Vol.9, No.12, (June 2009), pp.3328-3340.

Matta, A., Tripathi, S.C., DeSouza, L.V., Grigull, J., Kaur, J., Chauchan, S.S., Thakar, A., Shukla, N.K., Duggal, R., DattaGupta, S., Ralhan, R., Michael Siu, K.W. (2009). Heterogeneous ribonucleprotein K is a marker of oral leukoplakia and correlates with poor prognosis of squamous cell carcinoma. *Int J Cancer,* Vol.125, No.6, (September 2009), pp.1398-406.

Miike, K., Aoki, M., Yamashita, R., Takegawa, Y., Saya, H., Miike, T. & Yamamura, K. (2010). Proteome profiling reveals gender differences in the composition of human serum. *Proteomics,* Vol.10, No.14, (July 2010), pp.2678-91.

Ong, S.E., Blagoev, B., Kratchmarova, I., Kristensen, D.B., Steen, H., Pandey, A. & Mann, M. (2002). Stable isotope labeling by amino acids in cell culture, SILAC, as a simple and accurate approach to expression proteomics. *Molecular and Cellular Proteomics,* Vol.1, No.5, (May 2002), pp.376-86.

Ow, S.Y., Salim, M., Noirel, J., Evans, C., Rehman, I. & Wright, P.C. (2009). iTRAQ underestimation in simple and complex mixtures: "the good, the bad and the ugly". *J Proteome Res,* Vol.8, No.11, (November 2009), pp.5347-55.

Ow, S.Y., Salim, M., Noirel, J., Evans, C. & Wright, P.C. (2011). Minimising iTRAQ ratio compression through understanding LC-MS elution dependence and high-resolution HILIC fractionation. *Proteomics*. Vol.11, No.11, (June 2011), pp. 2341-6.

Pendyala, G., Trauger, S.A., Siuzdak, G. & Fox, H.S. (2010). Quantitative plasma proteomic profiling identifies the vitamin E binding protein afamin as a potential pathogenic factor in SIV induced CNS disease. *J Proteome Res,* Vol.9, No.1, (January 2010), pp.352-8.

Ross, P. L., Huang, Y.N., Marchese, J.N., Williamson, B., Parker, K., Hattan, S., Khainovski, N., Pillai, S., Dey, S., Daniels, S., Purkayastha, S., Juhasz, P., Martin, S., Bartlet-Jones, M., He, F., Jacobson, A. & Pappin, D.J. (2004). Multiplexed Protein Quantitation in Saccharomyces cerevisiae Using Amine-reactive Isobaric Tagging Reagents. *Molecular and Cellular Proteomics,* Vol.3, No.12, (December 2004), pp. 1154-69.

Rudrabhatla, P., Grant, P., Jaffe, H., Strong, M.J. & Pant, H.C. (2010). Quantitative phosphoproteomic analysis of neuronal intermediate filament proteins (NF-M/H) in Alzheimer's disease by iTRAQ. *FASEB J*, Vol.24, No.11, (November 2010), pp.4396-407.

Scheri, R.C., Lee, J., Curtis, L.R. & Barofsky, D.F. (2008). A comparison of relative quantification with isobaric tags on a subset of the murine hepatic proteome using electrospray ionization quadrupole time-of-flight and matrix-assisted laser desorption/ionization tandem time-of-flight. *Rapid Commun Mass Spectrom*, Vol.22, No.20, (October 2008), pp.3137-46.

Shirran, S.L. & Botting, C.H. (2010) A comparison of the accuracy of iTRAQ quantification by nLC-ESI MSMS and nLC-MALDI MSMS methods. *J. Proteomics*, Vol.73, No.7, (May 2010), pp. 1391-403.

Sturgeon, C., Hill, R., Hortin, G.L. & Thompson, D. (2010). Taking a new biomarker into routine use - A perspective from the routine clinical biochemistry laboratory. *Proteomics Clin Appl*, Vol.4, No.12, (December 2010), pp. 892-903.

Tripathi, S.C., Matta, A., Kaur, J., Grigull, J., Chauchan, S.S., Thakar, A., Shukla, N.K., Duggal, R., DattaGupta, S., Ralhan, R., Michael Siu, K.W. (2010) Nuclear S100A7 is associated with poor prognosis in head and neck cancer. *PLos One*, Vol.5, No.8, (August 2010), e11939.

Truscott, R.J., Comte-Walters, S., Ablonczy, Z., Schwacke, J.H., Berry, Y., Korlimbinis, A., Friedrich, M.G. & Schey, K.L. (2010). Tight binding of proteins to membranes from older human cells. *Age (Dordr)*, 2010 Dec 23. [Epub ahead of print].

Wang, M., Wang, X., Ching, C.B. & Chen, W.N. (2010). Proteomic profiling of cellular responses to Carvedilol enantiomers in vascular smooth muscle cells by iTRAQ-coupled 2-D LC-MS/MS. *J Proteomics*,Vol.73, No.8, (June 2010), pp. 1601-11.

Washburn, M.P., Wolters, D. & Yates, J.R. 3rd. (2001). Large-scale analysis of the yeast proteome by multidimensional protein identification technology. *Nature Biotechnology*, Vol.19, No.3, (March 2001), pp.242-7.

Wu, W.W., Wang, G., Baek, S.J., Shen, R-F. (2005). Comparative study of three proteomic quantitative methods, DIGE, cICAT, and iTRAQ, using 2D gel- or LC-MALDI TOF/TOF. *J Proteome Research*, Vol.5, No.3, pp.651-658.

Yang, E.C., Guo, J., Diehl, G., DeSouza, L., Rodrigues, M.J., Romaschin, A.D., Colgan, T.J. & Siu, K.W. (2004). Protein expression profiling of endometrial malignancies reveals a new tumor marker: chaperonin 10. *J Proteome Res*, Vol.3, No.3, (May-June 2004), pp.636-43.

Yang, Y., Zhang, S., Howe, K., Wilson, D.B., Moser, F. & Irwin, D. (2007). Comparison of nLC-ESI-MS/MS and nLC-MALDI-MS/MS for Gel-LC-based protein identification and iTRAQ-based shotgun quantitative proteomics. *J Biomol Tech*, Vol. 18, No.4, (September 2007), pp.226-37.

Zieske, L.R. (2006). A perspective on the use of iTRAQ™ reagent technology for protein complex and profiling studies. *Journal of Experimental Botany*, Vol.57, No.7, (March 2006), pp.1501-08.

Zhang, Y., Ficarro, S.B., Li, S. & Marto, J.A. (2009). Optimized Orbitrap HCD for quantitative analysis of phosphopeptides. *J Am Soc Mass Spectrom*, Vol.20, No.8, (August 2009), pp.1425-34.

19

Functional Proteomics: Mapping Lipid-Protein Interactomes

Clive D'Santos[1] and Aurélia E. Lewis[2]
[1]*Cancer Research UK Cambridge, Cambridge*
[2]*Department of Molecular Biology,*
University of Bergen, Bergen
[1]*United Kingdom*
[2]*Norway*

1. Introduction

Cell function is dependent upon the co-ordinated and dynamic formation of complex interaction networks between molecules of diverse biochemical properties. These networks, or interactomes, are comprised of macromolecular biopolymers; proteins, DNA, RNA and polysaccharides, in addition to non-polymer compounds such as small molecular metabolites. This myriad of interactions is highly regulated and any perturbation or alteration has potential to result in disease.

Profiling protein-protein interactions has been the major focus of interactomics in the past few years (Charbonnier et al. 2008) largely due to the advances in technological platforms that have the capacity to probe globally. Early efforts have included two-hybrid screens to identify binary binding interactions; more recent studies have used a range of mass spectrometry based methods to identify protein complexes that are a better reflection of multi-interactive nature of such complexes. Protein/small molecule interactions are equally important in modulating the function of their target proteins but few studies have analyzed these interactions on a large scale. The field is indeed still in its infancy due to difficulties in identifying metabolites but has recently benefitted from technological advances in mass spectrometry, data analysis software and metabolites database development for the measurement and identification of metabolites. The next step is to integrate metabolomic profiling to functional characterization of metabolic pathways by identifying systematically metabolite-protein interactions.

Research efforts have in general been more focused on lipid-mediated interactions and this chapter reviews the global methods as well as their applications used to map lipid-protein interactomes based on mass spectrometry or arrays. The potential of these studies to deepen our understanding on the biological function of metabolites as protein effectors is also discussed.

2. Metabolomics

Metabolites are defined as small organic molecules produced and modified by a living organism as a result of cellular and physiological metabolism. These molecules constitute an important fraction of the dry weight of a living cell ranging from 17 to 27 % in bacteria and mammalian cells respectively. They consist of a wide variety of small molecules with a vast chemical diversity, including amino acids, nucleotides, sugars and fatty acids that are central to all metabolic pathways existing in the cell. The precise number of metabolites produced in a cell at a certain time point is unknown but is estimated to range from a few hundreds in bacteria to a few thousands in plant and animal cells. Metabolic networks reconstructed from studies in yeast have indicated up to 1494 different metabolic compounds (Herrgard et al. 2008). The human metabolome database (version 2.5) embraces 7982 compounds that have been experimentally confirmed (Wishart et al. 2009). These compounds have been further divided into 52 different classes. The number and diversity of possible metabolites is extensive and this entails that a significant proportion of proteins may form functional but also opportunistic interactions with metabolites.

Overall metabolites constitute the metabolome of a cell, tissue or organism at a specific time and changes in metabolic profiles have enormous potential to understand cellular function and for clinical diagnostics (Vinayavekhin et al. 2010). To this effect, metabolomics has been applied to the general profiling of metabolites in biological samples, the discovery of biomarkers in diseases and the clinical screening of targeted compounds. Metabolomics gives an additional biologically relevant dimension to transcriptomics and proteomics and the integration of these data allows for a deeper understanding of physiological processes in normal and pathological states. While transcriptomics, proteomics and metabolomics allow the cellular inventory of biochemicals, an additional layer of data integration is still necessary to assess the mechanisms of regulation leading to a specific metabolic status. Computational-based metabolic flux analysis provides information on intracellular flux distributions of metabolic processes of a cell or an organism that can be integrated to data generated through transcriptomics, proteomics and metabolomics (Blank & Kuepfer 2010).

The range and diversity of functional metabolites has been highlighted and it is beyond the scope of this review to critique the methods that are currently used for all metabolites. Rather we have chosen to focus on a subset of metabolites, lipids, and in particular inositol-phospholipids, which are key regulators of numerous signalling pathways and which have been the most studied in recent years.

3. Functional lipidomics: From lipidomics to lipid-protein interactomics

The last 5 to 10 years have witnessed an incredible advancement in mass spectrometry and bioinformatics to analyse and identify systematically most lipids existing in biological samples at any one time and under different conditions in a specific entity (Wenk 2010; van Meer & de Kroon 2011). Cells contain thousands of lipids with a large chemical diversity and in light of recent advancement in lipid research, their classification has recently been updated by the LIPID MAPS initiative (Fahy et al. 2009). Lipidomics provide snapshots of the biochemical status of a cell and have the potential to complement transcriptomic and proteomic profiles. Integration of metabolomics, including lipidomics, to transcriptomics and proteomics analyses is expected to improve our understanding of metabolic pathways in health and diseases.

Although metabolite profiling provides important information on the status of a cell or organism, there is still a lack of functional data. Recently a shift has occurred from profiling all existing metabolites in a cell, tissue or an organism (metabolomics) to understanding how they may affect cellular functions (functional metabolomics) by the identification of metabolites-protein interaction networks. In this chapter, we have highlighted the methods as well as their applications used to map lipid-protein interactions in biological systems.

3.1 Methods to identify lipid-protein interactions

Small scale and large scale mapping of lipid-protein interactions methods have been developed using either targeted strategies studying a specific lipid or protein of interest (Figure 1) or large-scale strategies using lipid arrays or protein arrays (Figure 2).
Protein capture using affinity-based pull down has been widely used in combination with mass spectrometry to identify lipid interactomes in particular (Figure 1). In these cases cell extracts are incubated with lipid conjugated to affinity matrices and bound proteins are identified by mass spectrometry (Krugmann et al. 2002; Scholten et al. 2006; Osborne et al. 2007; Pasquali et al. 2007; Catimel et al. 2008; Catimel et al. 2009; Lewis et al. 2011).
Opposite strategies to identify lipids bound to a selection of proteins or a particular protein of interest have also been developed (Tagore et al. 2008; Kim et al. 2011; Li & Snyder 2011) (Figure 1). Recombinant proteins can be purified and immobilised onto a solid support and exposed to a metabolite mixture obtained from cells or tissues where the protein is known to be expressed. Additionally, endogenous proteins or tagged proteins can be immunoprecipitated from a cell or tissue extract (Li et al. 2007; Urs et al. 2007). Metabolites that are bound to the isolated protein are eluted and identified by mass spectrometry.
High-throughput screening strategies of these interactions have also been established using protein and small molecules microarrays (Lueking et al. 2005; Chen & Snyder 2010; Wu et al.

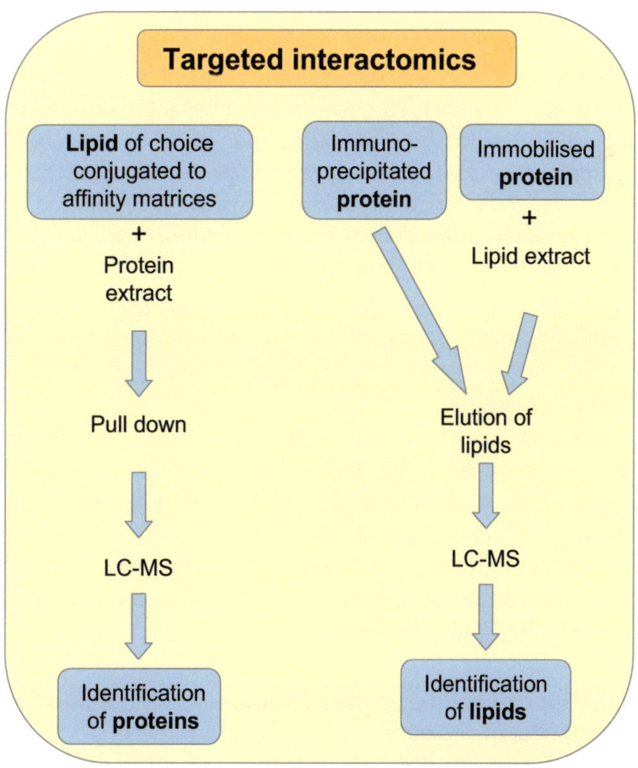

Fig. 1. Methods to identify lipid-protein interactions using targeted methods

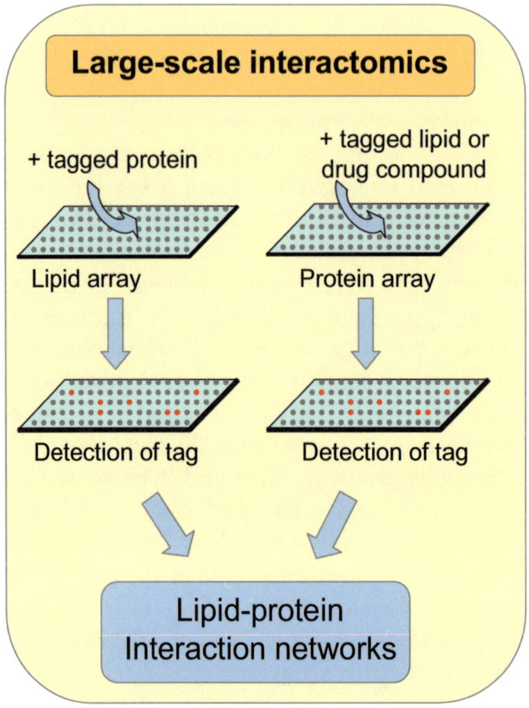

Fig. 2. Methods to identify lipid-protein interactions using large-scale methods

2011) (Figure 2). Microarrays are collections of hundreds to thousands of molecules immobilised on planar surfaces such as glass slides or nitrocellulose coated slides. Protein microarrays consist of individually expressed and purified proteins representing the complete or partial proteome known for a particular organism. Small molecules arrays consist of synthetic or naturally occurring molecules printed or spotted onto solid surfaces. To assess metabolites-protein interactions, protein arrays are exposed to fluorescently labelled metabolites (Zhu et al. 2001). Small molecules arrays are exposed to individual proteins (Rogers et al. 2011) or cellular lysates containing tagged proteins and interactions are detected using antibody recognising the specific tag (Gallego et al. 2010).

3.2 Lipid-protein interactomes

Lipids represent the largest class of metabolites in cells and are involved in a wide variety of cellular functions (van Meer & de Kroon 2011). They are essential structural components of cellular membranes and function as energy stores, cellular signalling molecules and regulators of transcription factor. Recent lipidomics analyses in mammalian cells have highlighted the dynamic remodelling of different lipid molecules (Dennis et al. 2010). These molecules have therefore been the focus of un-biased and systematic interactome studies in an effort to further clarify the functions of these molecules.

3.2.1 Phosphoinositide-protein interactomes mapping using lipid affinity matrices capture combined with MS of proteins

Many studies have focused on the identification of phosphoinositides (PIs)-protein interactomes. Inositol phospholipids, a small subset of the total lipid pool function as key regulators of numerous regulatory pathways (Toker 2002; Janmey & Lindberg 2004; Di Paolo & De Camilli 2006; Poccia & Larijani 2009). They function predominantly but not exclusively as sensors that recruit proteins and protein complexes to sites of synthesis in response to external cues (Lindmo & Stenmark 2006; Lemmon 2008). Target proteins possess well characterised domains within their structure that bind with varying affinity and specificity to the phosphorylated inositol head group. In addition, the hydrolysis of these lipids by phospholipase activities generate further second messengers such as diacylglycerols and polyphosphorylated inositols extending the influence of these lipids on cellular function and highlighting the need for further efforts to understand molecular mechanisms. A first step toward this would be to identify specific effector protein complexes that are regulated directly via binding and from this point of view proteomic methods and their applications are well placed to characterise globally the macromolecular complexes.

A number of studies have focused on the identification of PI-binding proteins using affinity matrices to pull down potential PI interacting proteins from cellular lysates and subsequent mass spectrometry analyses. These studies are summarized in Table 1. In a study using a combination of PI affinity matrices, competitive lipid pull down and protein fractionation

PI interactome analysed	Method	Cell type/subcellular compartment	Reference
PtdIns(3,4,5)P_3	PI conjugated beads and MS	Pig leukocyte cytosolic extract	(Krugmann et al. 2002)
Mostly PtdIns(3,4)P_2	PI conjugated to cleavable S-S bond biotin + streptavidin beads and MS	Primary macrophage cytosolic extract	(Pasquali et al. 2007)
PtdIns(4,5)P_2	Biotinylated PI, streptavidine conjugated beads and MS	Secretory granules from bovine adrenal chromaffin cells	(Osborne et al. 2007)
PtdIns(3,5)P_2 & PtdIns(4,5)P_2	PI conjugated beads or liposomes and MS	LIM1215 colon cancer cell cytosolic extract	(Catimel et al. 2008)
PtdIns(3,4,5)P_3	PI conjugated beads or liposomes and MS	LIM1215 colon cancer cell cytosolic extract	(Catimel et al. 2009)
PtdIns(4,5)P_2	PI conjugated beads and quantitative MS	Neomycin extracted nuclear proteins isolated from murine erythroleukemia (MEL) cells	(Lewis et al. 2011)
PtdIns(3,4)P_2	Stimulation of class I PI3K +/- wortmannin, biotinylated PI coupled to streptavidin beads and SILAC-based quantitative MS	1321N1 astrocytoma membrane fractions	(Dixon et al. 2011)

Table 1. Large-scale proteomics studies for the identification of PI binding proteins by MS

from pig leukocyte cytosol, 16 proteins were identified as PtdIns(3,4,5)P_3 and 5 as PtdIns(3,4)P_2 binding proteins by mass spectrometry (Krugmann et al. 2002). One of these proteins, ARAP3, a GTPase-activating protein, was further characterized as a functional PtdIns(3,4,5)P_3 effector protein (Krugmann et al. 2002). Another study identified 10 known and 11 potentially novel PtdIns(3,4)P_2 interacting proteins using cleavable biotinylated PI baits (Pasquali et al. 2007). None of these proteins overlapped with the ones identified in the previous study.

In a more comprehensive study, Holmes and colleagues have characterized and compared the interactomes of PtdIns(3,5)P_2 and PtdIns(4,5)P_2 (Catimel et al. 2008) as well as PtdIns(3,4,5)P_3 (Catimel et al. 2008; Catimel et al. 2009) determined from the cytosolic extracts of colon cancer cells expressing WT PI3 kinase. PIs immobilized either onto beads or incorporated into liposomes were used for protein capture from the cytosolic extracts. This led to the identification of 388 proteins in complex with PtdIns(3,5)P_2 and/or PtdIns(4,5)P_2 (Catimel et al. 2008) and 282 proteins in complex with PtdIns(3,4,5)P_3 (Catimel et al. 2009). A fraction of these were found to form complexes only with PtdIns(3,5)P_2 (69), PtdIns(4,5)P2 (146) or PtdIns(3,4,5)P_3 (141). In addition significant overlaps were observed for these interactions, consistent with the promiscuous properties of some of these interactions. These studies represent the first comprehensive datasets of potential cytosolic PI-interacting proteins. In addition, the computational analyses of the molecular functions of proteins found in complex with cytosolic PI interactomes have highlighted roles in the regulation of GTPases, in transport/trafficking, cytoskeletal remodelling, phosphorylation-mediated post-translational modifications.

The first organellar PI interactome was deciphered from secretary granules. Secretary granules were isolated from PC12 cells and 5 PtdIns(4,5)P_2 binding proteins were identified by affinity lipid pull down and mass spectrometry. These interactions were all validated by lipid pull down and Western immunoblotting.

PIs are also found in the nucleus (Irvine 2003; Hammond et al. 2004; Ye & Ahn 2008; Keune et al. 2011) and we have established a quantitative and proteomic method to identify PtdIns(4,5)P_2 interacting proteins to gain insight into the PI-mediated nuclear functions (Lewis et al. 2011). The workflow of the method is schematised in Figure 3. The nuclear PtdIns(4,5)P_2 interactome was characterized using PI-conjugated beads incubated in neomycin-extracted nuclear proteins mixtures and quantitative mass spectrometry using isotopic labeling of cells. Neomycin is known to bind to PIs with high affinity (Schacht 1978; Gabev et al. 1989) and we predicted that neomycin would compete for PIs in complex with proteins. Incubation of intact nuclei with neomycin resulted in the specific displacement of 168 nuclear proteins harbouring a PI binding domain. Using neomycin extracts, 34 proteins were shown to interact with PtdIns(4,5)P_2 in quantitative affinity purification using specific lipid conjugated matrices. Neomycin extraction of proteins represented an ideal preparation from which to affinity-purify PI-effector proteins using specific lipid conjugated matrices, avoiding the issues of sample complexity and dynamic range. Functional classification and enrichment analyses of the identified PtdIns(4,5)P_2-interacting proteins pointed to roles in mRNA transcriptional regulation, mRNA splicing and protein folding.

Dixon and colleagues have recently developed a three phase affinity enrichment method to quantitatively identify PtdIns(3,4)P_2 effector proteins targeted to membranes (Dixon et al. 2011). 1321N1 astrocytoma cells labelled with either light or heavy isotope were stimulated with bpV, a vanadate analogue, which induces high levels of PtdIns(3,4)P_2, and in the presence or absence of wortmannin, an inhibitor of the PI3K pathway. After the isolation of

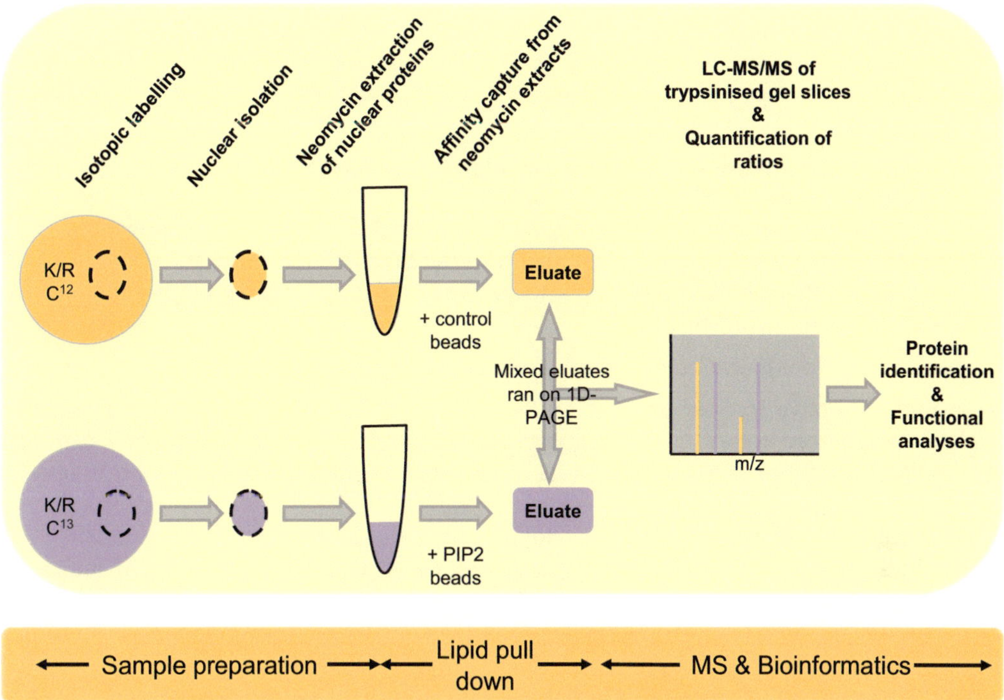

Fig. 3. Quantitative characterisation of nuclear PI interactome by combining isotopic labelling of cells, affinity capture of proteins using PI matrices and mass spectrometry (Lewis et al. 2011): C^{13} K/R-labelled and C^{12} K/R-labelled nuclei were incubated with 5 mM neomycin. Displaced proteins were pulled down at equal concentration with control beads or PtdIns(4,5)P_2 (PIP2)-conjugated beads. Proteins in mixed eluates were resolved by SDS-PAGE, Coomassie stained and trypsin digested. Peptides were analysed by LC-MS/MS and 13C/12C ratios were quantified using MSQuant (http://msquant.alwaysdata.net/msq/) and statistics were determined with StatQuant (van Breukelen et al. 2009).

membranes, proteins specifically recruited to membrane fractions following bpV stimulation, were eluted with Ins(1,3,4)P3. Eluted proteins were subjected to ion-exchange chromatography, affinity capture with streptavidin beads pre-coupled to PtdIns(3,4)P_2, followed by SDS-PAGE, LC-MS/MS and quantitative assessment of PtdIns(3,4)P_2 effector proteins. Previously established PtdIns(3,4)P_2-binding proteins, such as TAPP1 and Akt1-3, were identified, providing a strong proof of principle of the method. Overall 80-85 potential proteins were identified and this study provided the first quantitative MS-based identification of PtdIns(3,4)P_2 effector proteins. Many but not all proteins harboured lipid binding domains. The binding characteristics of a novel binding protein, IQGAP1, to PtdIns(3,4)P_2 were determined, demonstrating the existence of an atypical PI binding domain.

Overall, studies based on affinity capture combined with mass spectrometry serve as useful resources and have the advantage to give a global view of the biological functions of proteins regulated by PIs in different cellular compartments. However a main drawback remains in the inability to discriminate between direct and indirect interactions through associated proteins. Such analyses should be complemented by biochemical approaches analysing direct interactions for individual proteins. In addition, the potential existence of indirect protein complex networks can be assessed using known data for protein-protein interaction networks for the corresponding cell line or tissue explored.

3.2.2 Lipid-protein interactomes mapping by protein immobilization or affinity purification and identification of lipids by MS

Mass spectrometry has allowed the identification of lipids interacting with proteins both in targeted and large-scale systematic analyses. Different methods have been developed to affinity capture proteins followed by the extraction of bound lipids and their identification by tandem mass spectrometry.

3.2.2.1 Targeted identification of ligands for nuclear receptors

Several studies have focused on developing methods to identify physiological ligands for orphan nuclear receptors. Nuclear receptors represent a family of transcription factors that are activated by binding to specific small molecules to regulate the expression of specific genes.

Saghatelian and colleagues have developed methods to identify indiscriminately the metabolites bound to recombinant proteins (Tagore et al. 2008). A protein of interest is purified from bacteria, immobilized on a solid support via a 6xHis or GST tag and incubated with a lipid extract obtained from cells known to express the corresponding protein. Eluted metabolites are analysed by LC-MS and the metabolite chromatogram profiles are compared computationally to control samples obtained from solid support alone. This strategy was applied for the nuclear receptors, peroxisome proliferator-activated receptors (PPAR)α and γ (Kim et al. 2011) involved in lipid metabolism. Free fatty acids (FFA) such as arachidonic (C20:4), linoleic (C18:2) and oleic (C18:1) acids were identified as endogenous ligands for both nuclear receptors. Palmitoleic acid (C16:1) was also identified as a ligand for PPARγ.

In an alternative method, a physiological ligand was discovered for PPPARα by isolating the receptor from liver nuclear extracts obtained from mice either WT or lacking fatty acid synthase (FAS) (Chakravarthy et al. 2009). FAS is an enzyme that synthesize saturated FA which was previously shown to synthesize *de novo* a potential ligand for PPARα in liver cells (Chakravarthy et al. 2005). After elution of the receptor, lipids were extracted and subjected to tandem MS, which identified the phospholipid, 1-palmitoyl-2-oleyl-sn-glycerol-3-phosphocholine as a FAS-dependent ligand of PPARα (Chakravarthy et al. 2009). This is a compelling approach to decipher endogenous ligand occupancy of orphan nuclear receptor in an *in vivo* setting.

Sewer and colleagues were able to identify several phospholipids bound to another orphan nuclear receptor, steroidogenic factor 1 (SF-1) (Li et al. 2007) with roles in the regulation of steroidogenic hormones expression. SF-1 was immunoprecipitated from adrenocortical cells which express the receptor endogenously and phospholipids were analysed by LC-MS. Phosphatidic acid was found to be a major lipid bound to SF-1 and to activate the transcriptional activity of the receptor. A similar approach allowed the identification of linoleic acid (C18:2) as a ligand for another orphan nuclear receptor hepatocyte nuclear factor 4 (HNF4) α, affinity purified from mammalian cells (Yuan et al. 2009). Importantly the occupancy of the ligand was dependent upon the physiological condition studied: HNF4α was bound to linoleic acid when the receptor was isolated from livers of fed mice but not of fasted mice. Additionally the ligand did not have any effect on the transcriptional activity of HNF4α.

3.2.2.2 Large scale identification of metabolite-protein interactions

An affinity purification protocol in yeast was recently established by Snyder and colleagues to identify hydrophobic metabolites bound to 103 protein kinases as well as to a selection of proteins including 21 enzymes involved in the ergosterol biosynthetic pathways (yeast molecular analogue of cholesterol) (Li et al. 2010). In this case proteins were fused to an immunoglobulin binding domain and isolated from yeast extracts by affinity pull down. Metabolites interacting with the affinity purified proteins were extracted and identified by LC-MS. Control samples consisted of a yeast strain extract devoid of the corresponding fused protein. Such systematic analysis revealed that about 70% of the ergosterol biosynthetic enzymes and 20% of all protein kinases analyzed were bound to hydrophobic molecules. Known protein-metabolites interactions were observed but a majority of new interactions were also uncovered. Some interactions were unexpected and suggested important roles for ergosterol in the regulation of not only lipid biosynthetic pathways but also of many kinases, amongst which Ypk1 yeast kinase homologue to the mammalian kinase Akt.

3.2.3 Large scale interactomics to identify lipid-protein interactions

The magnitude of metabolites-protein interactions has been highlighted in large-scale screens in budding yeasts using different approaches using protein or lipid arrays (Zhu et al. 2001; Gallego et al. 2010). Using systematic approaches such as these, a comprehensive set of proteins can be simultaneously assessed for their potential interactions with lipids but also with other small molecules.

Firstly, Snyder and colleagues developed protein chips for the yeast proteome of *Saccharomyces cerevisiae*. These were the first protein arrays for any organism to be engineered (Zhu et al. 2001). The yeast proteome array contained 5800 proteins fused to GST-6xHis and was screened for PI interactions using PIs assembled in liposomes containing phosphatidylcholine (PC) and an additional biotinylated lipid. PtdIns(3)*P*, PtdIns(4)*P*, PtdIns(3,4)P_2, PtdIns(4,5)P_2 or PtdIns(3,4,5)P_3, containing liposomes were applied to the arrays, followed by an incubation with fluorescently-labeled streptavidin. Following the fluorescence detection of the arrays, 49 proteins were found to interact significantly with PIs compared to PC liposomes, with different affinities and specificities for the different PI molecules. Conventional methods were applied to confirm protein-PI interactions for 3 proteins involved in glucose metabolism that were not previously expected to bind to PIs.

The second screen reported is a large-scale analysis of yeast lipid-proteins interactions that was recently performed by Gavin and colleagues (Gallego et al. 2010). An opposite approach was used and lipid arrays were generated using 56 different lipids spanning the main classes of lipids found in yeast applied on nitrocellulose membranes. These arrays were incubated with cell extracts expressing single tandem affinity purification (TAP)-tagged proteins in *S. cerevisiae*. Interactions with 172 single TAP-tagged protein containing extracts were assessed for their potential interaction with the lipid array and gave rise to 530 interactions involving 124 proteins and 30 lipids. Amongst the 56 lipids studied, PIs were represented by PtdIns(3)*P*, PtdIns(4)*P*, PtdIns(4,5)P_2 and PtdIns(3,4,5)P_3, and 86 proteins were found to bind PIs of which 77% harboured a lipid binding domain. PIs represented indeed the lipid category that interacted with most proteins, which is consistent with the wide range of cellular functions they are reported to take part into. Importantly, this study also assessed the quality of the data generated by the lipid array by comparing data retrieved from the literature and from genetic interactions. In addition the identified interactions by lipid overlay were validated for 8 proteins chosen amongst the obtained dataset. Overall, Gavin and colleagues reported that 54% of the identified interactions were validated by additional genetic evidence, making this interactome dataset the most comprehensive resource for lipid biology.

Unfortunately, very little overlap, accounting for about 5%, could be observed for PI-protein interactions between the 2 previously described studies despite an overlap of 88% of the proteins analysed (Zhu et al. 2001; Gallego et al. 2010). These contrasting datasets may be explained by different interaction properties being measured. For example, the protein chips may not allow the access of potential binding domain of all proteins to the phospholipids and therefore prevent potential interactions. In addition not all the proteins are overexpressed and purified as full length proteins or at sufficiently high level for the assay. As for the lipid array experiments, indirect interactions are also possible.

Although extensive datasets are now available from PI interactomes in yeast and mammalian cell lines, the overlap between the PI interactomes determined in these different species is unknown. We have therefore attempted to compare the dataset obtained from PtdIns(4,5)P_2 nuclear interactome studies (Lewis et al. 2011) to the datasets obtained in *S. cerevisiae*. Using InParanoid 7 (Ostlund et al. 2010), 18 yeast orthologues were recovered from the 34 murine proteins reported in the PtdIns(4,5)P_2 nuclear interactome dataset. Out of the 18 yeast orthologues, 3 proteins were in common with PI binding proteins identified by protein chip lipid overlay by Snyder and colleagues (Zhu et al. 2001). These proteins are listed in Table 2. Cdc19 and Cct8, were found to interact with PtdIns(4)P in the yeast PI interactomics study. Cam1 was identified as a PtdIns(4,5)P_2-interacting protein in the same study, which was consistent with our study. These findings certainly warrant further characterization of these proteins. In contrast, none of the 18 orthologues were found to be common to the dataset obtained from the lipid array screen by Gavin and colleagues (Gallego et al. 2010) and this may be explained by the following points. Firstly, the majority of proteins chosen in the lipid array screen were known to harbour at least one lipid binding domain (LBD) as defined by the online tools for protein domain assignment SMART, Pfam or SuperFamily, whereas none of the identified nuclear proteins harboured such domains but rather simple basic amino acid rich patches (Lewis et al. 2011). Secondly, the proportion of proteins annotated to the nucleus compartment analysed in the lipid array screen is not known.

ORF (yeast)	Gene name (yeast)	Uniprot ID (mouse)	Gene name (mouse)	Protein description (mammalian)
YAL038W	Cdc19	P52480	Pkm2	Isoform M2 of Pyruvate kinase isozymes M1/M2
YJL008C	Cct8	P42932	Cct8	T-complex protein 1 subunit theta
YPL048W	Cam1	Q9D8N0	Eef1g	Elongation factor 1-gamma

Table 2. List of genes found in common between the datasets obtained from the mammalian PtdIns(4,5)P_2 nuclear interactome (Lewis et al. 2011) and the yeast PI interactome (Zhu et al. 2001). Datasets that were compared include all yeast proteins found to bind PIs in the study by Zhu *et al* and 18 yeast orthologues retrieved from the 34 murine proteins identified in the PtdIns(4,5)P_2 nuclear interactome.

4. Chemical proteomics for lipid pathway drug specificity validation

Chemical proteomics has recently been used to explore the specificity of known drugs. In the case of drugs targeting lipid pathways, this method was used to identify all possible protein targets of a class I PI3K inhibitor, LY294002 (Gharbi et al. 2007). An analogue of the PI3K class I inhibitor LY294002, PI828, was immobilized onto epoxy-activated sepharose beads and used to pull out protein targets from whole cell extracts obtained from a human epithelial cell line (HeLa) and a mouse lymphoma B-cell line (WEHI231). Protein targets were eluted and identified by LC-MS/MS. This study demonstrated that this compound, bound not only to class I PI3Ks and other PI3K-related kinases, but also non-lipid kinases which was consistent with the inhibitory profile previously known for this compound. However novel targets were also identified which were reported to possibly explain some of the off-target cellular effects of this compound. The use of such proteomic approach has the potential to determine the specificity of known or new drugs at the cellular level as well as the potential cellular functions altered by the compound studied.

5. Comparison of PI interactomics data to whole genome genetic screens

Chemical genomics preceded the advent of chemical proteomics due to the completion of the *S. cerevisiae* Deletion Project which allowed whole genome genetic screens of different drugs. Such screen was performed to reveal new functions of PI metabolism by using wortmannin to identify genes which could confer altered sensitivity to the drug (Zewail et al. 2003). In yeast, wortmannin inhibits PtdIns(4)P kinase, Stt4p, and its inhibitory effects have been reported to be due to the depletion of PtdIns(4,5)P_2 (Cutler et al. 1997). This screen allowed the identification of 591 genetic interactions due to wortmannin resistance and provided an overview of the actions of the PI pathway. New functions that were not previously attributed to the PI metabolic pathway were uncovered, namely DNA replication and DNA damage checkpoint, chromatin remodelling and proteasome-mediated protein degradation.

A fraction of protein-protein interaction networks can be correlated to genetic interaction networks in yeast. Since wortmannin has been reported to affect the pool of PtdIns(4,5)P_2, we assumed that a fraction of PtdIns(4,5)P_2 effector proteins identified in physical screens would coincide with a fraction of the wortmannin genetic interaction screen. We have therefore compared our PtdIns(4,5)P_2 interaction networks obtained from mammalian nuclei to the wortmannin genetic screen performed in yeast. We were able to identify 4 genes in common between the physical and the genetic interaction datasets and these are listed in Table 3 and shown in the Venn diagram in Figure 4. In addition, 1 of these genes, Cam1, corresponding to the mammalian orthologue Eef1g (Elongation factor-1 γ), was also found to be common to the PI-protein interactomics study from Snyder and colleagues (Zhu et al. 2001). Interestingly, Cam1 was initially characterised as a possible phospholipid binding protein (Creutz et al. 1991; Kambouris et al. 1993), which would therefore be consistent with both physical and genetic studies.

Moreover, comparing the datasets obtained from the PI interactome study from Zhu *et al* to the wortmannin genetic screen from Zewail *et al* identified 18 proteins in common (14%). The overlapping data is presented as a Venn diagram in Figure 4.

ORF (yeast)	Gene name (yeast)	Uniprot ID (mouse)	Gene name (mouse)	Protein description (mammalian)
YER081W	Cam1	Q9D8N0	Eef1g	Elongation factor 1-gamma
YDR518W	Eug1	P27773	Pdia3	Protein disulfide-isomerase A3 precursor
YMR072W	Abf2	P63158 P30681	Hmgb1 Hmgb2	High mobility group protein B1 High mobility group protein B2

Table 3. List of proteins found in common in the PtdIns(4,5)P2 nuclear interactome study (Lewis et al. 2011) and in the wortmannin genetic screen in *S. cerevisiae* (Zewail et al. 2003). Datasets included 18 yeast orthologues retrieved from the 34 murine proteins identified in the PtdIns(4,5)P_2 nuclear interactome and all of the genes that conferred wortmannin resistance when deleted individually in yeast.

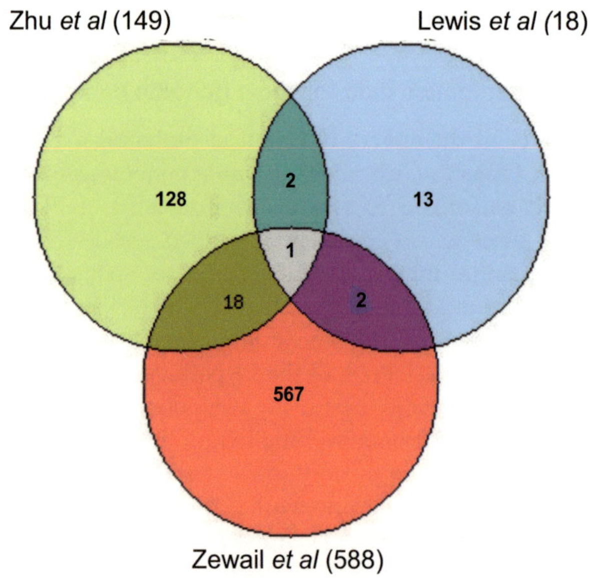

Fig. 4. Venn diagram representation of common proteins identified in PI interactome studies (Zhu et al. 2001; Lewis et al. 2011) and chemical genomics screen using wortmannin (Zewail et al. 2003). Datasets that were compared included all yeast proteins found to bind PIs in the study by Zhu *et al*, the 18 yeast orthologues retrieved from the 34 murine proteins identified in the PtdIns(4,5)P_2 nuclear interactome (Lewis *et al*) and all of the genes that conferred wortmannin resistance when deleted individually in yeast (Zewail *et al*).

6. Conclusion: Challenges and future direction in lipid-protein interactomics

Systematic and unbiased proteomics studies have answered some of the questions regarding lipid-mediated pathway functions. Most studies have focused on mapping PI interactomes from separate cellular compartments while more recent studies have expanded our knowledge to other lipid subclass interaction networks. In addition, the availability of genome wide genetic screen in yeast allows the potential discovery of overlaps with physical interactions datasets, thereby strengthening interaction data. Identifying lipid binding proteins has indeed provided some insights into the possible biological functions of the corresponding lipids mainly by inference of protein function. The next challenge is to assign biological functions to each of these interactions.

Several large scale interactome studies have shown little overlap in findings and both false positive and false negative are likely to be generated by these types of methods. Overall a great body of data is now indeed available and studies will still be required to further validate these interactions at the biochemical and cellular level, *in vitro* and *in vivo*.

The reported interactomes are at present synonymous with a static view of molecular complexes and this raises therefore a number of questions and challenges worthy of further scrutiny. What are the lipid-protein interactomes at the sub-cellular level? How are these interactions regulated in time and space? What are the mechanisms of regulation? What are the modes of interactions? Moreover do these interactions affect other types of interactions mediated by other macromolecules? This last question entails probably a new challenge in systems biology, i.e. the integration of data obtained from lipid-protein interactomes to those obtained from other protein-macromolecule interactomes.

Finally, lipids, in particular signalling lipids such as PIs, but also eicosanoids, sphingolipids and fatty acids are known to control critical cellular functions and the alteration of lipid-mediated pathways is known to contribute to the development of pathologies, such as chronic inflammation, cancer, neurodegenerative and metabolic diseases (Pendaries et al. 2003; Wymann & Schneiter 2008; Skwarek & Boulianne 2009). Newly acquired knowledge on lipid-protein interactions may pinpoint potential lipid effectors that may become targets for drug therapies. This may become even more relevant if changes in lipid-protein networks are identified in pathological states.

7. References

Blank, L. M. & Kuepfer, L. (2010). "Metabolic flux distributions: genetic information, computational predictions, and experimental validation." *Appl Microbiol Biotechnol* 86, 5, (May), 1243-1255.

Catimel, B., Schieber, C., Condron, M., Patsiouras, H., Connolly, L., Catimel, J., Nice, E. C., Burgess, A. W. & Holmes, A. B. (2008). "The PI(3,5)P2 and PI(4,5)P2 interactomes." *J Proteome Res* 7, 12, (Dec), 5295-5313.

Catimel, B., Yin, M. X., Schieber, C., Condron, M., Patsiouras, H., Catimel, J., Robinson, D. E., Wong, L. S., Nice, E. C., Holmes, A. B. & Burgess, A. W. (2009). "PI(3,4,5)P3 Interactome." *J Proteome Res* 8, 7, (Jul), 3712-3726.

Chakravarthy, M. V., Lodhi, I. J., Yin, L., Malapaka, R. R., Xu, H. E., Turk, J. & Semenkovich, C. F. (2009). "Identification of a physiologically relevant endogenous ligand for PPARalpha in liver." *Cell* 138, 3, (Aug 7), 476-488.

Chakravarthy, M. V., Pan, Z., Zhu, Y., Tordjman, K., Schneider, J. G., Coleman, T., Turk, J. & Semenkovich, C. F. (2005). ""New" hepatic fat activates PPARalpha to maintain glucose, lipid, and cholesterol homeostasis." *Cell Metab* 1, 5, (May), 309-322.

Charbonnier, S., Gallego, O. & Gavin, A. C. (2008). "The social network of a cell: recent advances in interactome mapping." *Biotechnol Annu Rev* 14, 1-28.

Chen, R. & Snyder, M. (2010). "Yeast proteomics and protein microarrays." *J Proteomics* 73, 11, (Oct 10), 2147-2157.

Creutz, C. E., Snyder, S. L. & Kambouris, N. G. (1991). "Calcium-dependent secretory vesicle-binding and lipid-binding proteins of Saccharomyces cerevisiae." *Yeast* 7, 3, (Apr), 229-244.

Cutler, N. S., Heitman, J. & Cardenas, M. E. (1997). "STT4 is an essential phosphatidylinositol 4-kinase that is a target of wortmannin in Saccharomyces cerevisiae." *J Biol Chem* 272, 44, (Oct 31), 27671-27677.

Dennis, E. A., Deems, R. A., Harkewicz, R., Quehenberger, O., Brown, H. A., Milne, S. B., Myers, D. S., Glass, C. K., Hardiman, G., Reichart, D., Merrill, A. H., Jr., Sullards, M. C., Wang, E., Murphy, R. C., Raetz, C. R., Garrett, T. A., Guan, Z., Ryan, A. C., Russell, D. W., McDonald, J. G., Thompson, B. M., Shaw, W. A., Sud, M., Zhao, Y., Gupta, S., Maurya, M. R., Fahy, E. & Subramaniam, S. (2010). "A mouse macrophage lipidome." *J Biol Chem* 285, 51, (Dec 17), 39976-39985.

Di Paolo, G. & De Camilli, P. (2006). "Phosphoinositides in cell regulation and membrane dynamics." *Nature* 443, 7112, (Oct 12), 651-657.

Dixon, M. J., Gray, A., Boisvert, F. M., Agacan, M., Morrice, N. A., Gourlay, R., Leslie, N. R., Downes, C. P. & Batty, I. H. (2011). "A screen for novel phosphoinositide 3-kinase effector proteins." *Mol Cell Proteomics* 10, 4, (Apr), M110 003178.

Fahy, E., Subramaniam, S., Murphy, R. C., Nishijima, M., Raetz, C. R., Shimizu, T., Spener, F., van Meer, G., Wakelam, M. J. & Dennis, E. A. (2009). "Update of the LIPID MAPS comprehensive classification system for lipids." *J Lipid Res* 50 Suppl, (Apr), S9-14.

Gabev, E., Kasianowicz, J., Abbott, T. & McLaughlin, S. (1989). "Binding of neomycin to phosphatidylinositol 4,5-bisphosphate (PIP2)." *Biochim Biophys Acta* 979, 1, (Feb 13), 105-112.

Gallego, O., Betts, M. J., Gvozdenovic-Jeremic, J., Maeda, K., Matetzki, C., Aguilar-Gurrieri, C., Beltran-Alvarez, P., Bonn, S., Fernandez-Tornero, C., Jensen, L. J., Kuhn, M., Trott, J., Rybin, V., Muller, C. W., Bork, P., Kaksonen, M., Russell, R. B. & Gavin, A. C. (2010). "A systematic screen for protein-lipid interactions in Saccharomyces cerevisiae." *Mol Syst Biol* 6, (Nov 30), 430.

Gharbi, S. I., Zvelebil, M. J., Shuttleworth, S. J., Hancox, T., Saghir, N., Timms, J. F. & Waterfield, M. D. (2007). "Exploring the specificity of the PI3K family inhibitor LY294002." *Biochem J* 404, 1, (May 15), 15-21.

Hammond, G., Thomas, C. L. & Schiavo, G. (2004). "Nuclear phosphoinositides and their functions." *Curr Top Microbiol Immunol* 282, 177-206.

Herrgard, M. J., Swainston, N., Dobson, P., Dunn, W. B., Arga, K. Y., Arvas, M., Bluthgen, N., Borger, S., Costenoble, R., Heinemann, M., Hucka, M., Le Novere, N., Li, P., Liebermeister, W., Mo, M. L., Oliveira, A. P., Petranovic, D., Pettifer, S., Simeonidis, E., Smallbone, K., Spasic, I., Weichart, D., Brent, R., Broomhead, D. S., Westerhoff, H. V., Kirdar, B., Penttila, M., Klipp, E., Palsson, B. O., Sauer, U., Oliver, S. G., Mendes, P., Nielsen, J. & Kell, D. B. (2008). "A consensus yeast metabolic network reconstruction obtained from a community approach to systems biology." *Nat Biotechnol* 26, 10, (Oct), 1155-1160.

Irvine, R. F. (2003). "Nuclear lipid signalling." *Nat Rev Mol Cell Biol* 4, 5, (May), 349-360.

Janmey, P. A. & Lindberg, U. (2004). "Cytoskeletal regulation: rich in lipids." *Nat Rev Mol Cell Biol* 5, 8, (Aug), 658-666.

Kambouris, N. G., Burke, D. J. & Creutz, C. E. (1993). "Cloning and genetic characterization of a calcium- and phospholipid-binding protein from Saccharomyces cerevisiae that is homologous to translation elongation factor-1 gamma." *Yeast* 9, 2, (Feb), 151-163.

Keune, W., Bultsma, Y., Sommer, L., Jones, D. & Divecha, N. (2011). "Phosphoinositide signalling in the nucleus." *Adv Enzyme Regul* 51, 1, 91-99.

Kim, Y. G., Lou, A. C. & Saghatelian, A. (2011). "A metabolomics strategy for detecting protein-metabolite interactions to identify natural nuclear receptor ligands." *Mol Biosyst* 7, 4, (Apr), 1046-1049.

Krugmann, S., Anderson, K. E., Ridley, S. H., Risso, N., McGregor, A., Coadwell, J., Davidson, K., Eguinoa, A., Ellson, C. D., Lipp, P., Manifava, M., Ktistakis, N., Painter, G., Thuring, J. W., Cooper, M. A., Lim, Z. Y., Holmes, A. B., Dove, S. K., Michell, R. H., Grewal, A., Nazarian, A., Erdjument-Bromage, H., Tempst, P., Stephens, L. R. & Hawkins, P. T. (2002). "Identification of ARAP3, a novel PI3K effector regulating both Arf and Rho GTPases, by selective capture on phosphoinositide affinity matrices." *Mol Cell* 9, 1, (Jan), 95-108.

Lemmon, M. A. (2008). "Membrane recognition by phospholipid-binding domains." *Nat Rev Mol Cell Biol* 9, 2, (Feb), 99-111.

Lewis, A. E., Sommer, L., Arntzen, M. O., Strahm, Y., Morrice, N. A., Divecha, N. & D'Santos, C. S. (2011). "Identification of nuclear phosphatidylinositol 4,5-bisphosphate-interacting proteins by neomycin extraction." *Mol Cell Proteomics* 10, 2, (Feb), M110 003376.

Li, D., Urs, A. N., Allegood, J., Leon, A., Merrill, A. H., Jr. & Sewer, M. B. (2007). "Cyclic AMP-stimulated interaction between steroidogenic factor 1 and diacylglycerol kinase theta facilitates induction of CYP17." *Mol Cell Biol* 27, 19, (Oct), 6669-6685.

Li, X., Gianoulis, T. A., Yip, K. Y., Gerstein, M. & Snyder, M. (2010). "Extensive in vivo metabolite-protein interactions revealed by large-scale systematic analyses." *Cell* 143, 4, (Nov 12), 639-650.

Li, X. & Snyder, M. (2011). "Metabolites as global regulators: A new view of protein regulation: Systematic investigation of metabolite-protein interactions may help bridge the gap between genome-wide association studies and small molecule screening studies." *Bioessays* 33, 7, (Jul), 485-489.

Lindmo, K. & Stenmark, H. (2006). "Regulation of membrane traffic by phosphoinositide 3-kinases." *J Cell Sci* 119, Pt 4, (Feb 15), 605-614.

Lueking, A., Cahill, D. J. & Mullner, S. (2005). "Protein biochips: A new and versatile platform technology for molecular medicine." *Drug Discov Today* 10, 11, (Jun 1), 789-794.

Osborne, S. L., Wallis, T. P., Jimenez, J. L., Gorman, J. J. & Meunier, F. A. (2007). "Identification of secretory granule phosphatidylinositol 4,5-bisphosphate-interacting proteins using an affinity pulldown strategy." *Mol Cell Proteomics* 6, 7, (Jul), 1158-1169.

Ostlund, G., Schmitt, T., Forslund, K., Kostler, T., Messina, D. N., Roopra, S., Frings, O. & Sonnhammer, E. L. (2010). "InParanoid 7: new algorithms and tools for eukaryotic orthology analysis." *Nucleic Acids Res* 38, Database issue, (Jan), D196-203.

Pasquali, C., Bertschy-Meier, D., Chabert, C., Curchod, M. L., Arod, C., Booth, R., Mechtler, K., Vilbois, F., Xenarios, I., Ferguson, C. G., Prestwich, G. D., Camps, M. & Rommel, C. (2007). "A chemical proteomics approach to phosphatidylinositol 3-kinase signaling in macrophages." *Mol Cell Proteomics* 6, 11, (Nov), 1829-1841.

Pendaries, C., Tronchere, H., Plantavid, M. & Payrastre, B. (2003). "Phosphoinositide signaling disorders in human diseases." *FEBS Lett* 546, 1, (Jul 3), 25-31.

Poccia, D. & Larijani, B. (2009). "Phosphatidylinositol metabolism and membrane fusion." *Biochem J* 418, 2, (Mar 1), 233-246.

Rogers, C. J., Clark, P. M., Tully, S. E., Abrol, R., Garcia, K. C., Goddard, W. A., 3rd & Hsieh-Wilson, L. C. (2011). "Elucidating glycosaminoglycan-protein-protein interactions using carbohydrate microarray and computational approaches." *Proc Natl Acad Sci U S A* 108, 24, (Jun 14), 9747-9752.

Schacht, J. (1978). "Purification of polyphosphoinositides by chromatography on immobilized neomycin." *J Lipid Res* 19, 8, (Nov), 1063-1067.

Scholten, A., Poh, M. K., van Veen, T. A., van Breukelen, B., Vos, M. A. & Heck, A. J. (2006). "Analysis of the cGMP/cAMP interactome using a chemical proteomics approach in mammalian heart tissue validates sphingosine kinase type 1-interacting protein as a genuine and highly abundant AKAP." *J Proteome Res* 5, 6, (Jun), 1435-1447.

Skwarek, L. C. & Boulianne, G. L. (2009). "Great expectations for PIP: phosphoinositides as regulators of signaling during development and disease." *Dev Cell* 16, 1, (Jan), 12-20.

Tagore, R., Thomas, H. R., Homan, E. A., Munawar, A. & Saghatelian, A. (2008). "A global metabolite profiling approach to identify protein-metabolite interactions." *J Am Chem Soc* 130, 43, (Oct 29), 14111-14113.

Toker, A. (2002). "Phosphoinositides and signal transduction." *Cell Mol Life Sci* 59, 5, (May), 761-779.

Urs, A. N., Dammer, E., Kelly, S., Wang, E., Merrill, A. H., Jr. & Sewer, M. B. (2007). "Steroidogenic factor-1 is a sphingolipid binding protein." *Mol Cell Endocrinol* 265-266, (Feb), 174-178.

van Breukelen, B., van den Toorn, H. W., Drugan, M. M. & Heck, A. J. (2009). "StatQuant: a post-quantification analysis toolbox for improving quantitative mass spectrometry." *Bioinformatics* 25, 11, (Jun 1), 1472-1473.

van Meer, G. & de Kroon, A. I. (2011). "Lipid map of the mammalian cell." *J Cell Sci* 124, Pt 1, (Jan 1), 5-8.

Vinayavekhin, N., Homan, E. A. & Saghatelian, A. (2010). "Exploring disease through metabolomics." *ACS Chem Biol* 5, 1, (Jan 15), 91-103.

Wenk, M. R. (2010). "Lipidomics: new tools and applications." *Cell* 143, 6, (Dec 10), 888-895.

Wishart, D. S., Knox, C., Guo, A. C., Eisner, R., Young, N., Gautam, B., Hau, D. D., Psychogios, N., Dong, E., Bouatra, S., Mandal, R., Sinelnikov, I., Xia, J., Jia, L., Cruz, J. A., Lim, E., Sobsey, C. A., Shrivastava, S., Huang, P., Liu, P., Fang, L., Peng, J., Fradette, R., Cheng, D., Tzur, D., Clements, M., Lewis, A., De Souza, A., Zuniga, A., Dawe, M., Xiong, Y., Clive, D., Greiner, R., Nazyrova, A., Shaykhutdinov, R., Li, L., Vogel, H. J. & Forsythe, I. (2009). "HMDB: a knowledgebase for the human metabolome." *Nucleic Acids Res* 37, Database issue, (Jan), D603-610.

Wu, H., Ge, J., Uttamchandani, M. & Yao, S. Q. (2011). "Small molecule microarrays: the first decade and beyond." *Chem Commun (Camb)* 47, 20, (May 28), 5664-5670.

Wymann, M. P. & Schneiter, R. (2008). "Lipid signalling in disease." *Nat Rev Mol Cell Biol* 9, 2, (Feb), 162-176.

Ye, K. & Ahn, J. Y. (2008). "Nuclear phosphoinositide signaling." *Front Biosci* 13, 540-548.

Yuan, X., Ta, T. C., Lin, M., Evans, J. R., Dong, Y., Bolotin, E., Sherman, M. A., Forman, B. M. & Sladek, F. M. (2009). "Identification of an endogenous ligand bound to a native orphan nuclear receptor." *PLoS One* 4, 5, e5609.

Zewail, A., Xie, M. W., Xing, Y., Lin, L., Zhang, P. F., Zou, W., Saxe, J. P. & Huang, J. (2003). "Novel functions of the phosphatidylinositol metabolic pathway discovered by a chemical genomics screen with wortmannin." *Proc Natl Acad Sci U S A* 100, 6, (Mar 18), 3345-3350.

Zhu, H., Bilgin, M., Bangham, R., Hall, D., Casamayor, A., Bertone, P., Lan, N., Jansen, R., Bidlingmaier, S., Houfek, T., Mitchell, T., Miller, P., Dean, R. A., Gerstein, M. & Snyder, M. (2001). "Global analysis of protein activities using proteome chips." *Science* 293, 5537, (Sep 14), 2101-2105.

20

Protein Thiol Modification and Thiol Proteomics

Yingxian Li, Xiaogang Wang and Qi Li
State Key Lab of Space Medicine Fundamentals and Application,
China Astronaut Research and Training Centre, Beijing
China

1. Introduction

Cysteine plays an important role in the regulation of redox chemistry and gene expression and is essential in the structural and macromolecular organisation of proteins. Thiol oxidation leads to misfolding and the influencing of the protein function (Buczek et al., 2007). In our experiments, we have identified an oxidised cytoskeletal protein actin involved in the rearrangement of filament in the cells, leading to cellular apoptosis (Wang et al., 2010). Redox signalling can be relayed through intramolecular or intermolecular disulphide formation (Li et al., 2005).

Redox proteomics is an emerging branch of proteomics aimed at detecting and analysing redox-based changes within the proteome in different redox statuses (D'Alessandro et al., 2011). For this reason, several experimental approaches have been developed for the systematic characterisation of thiol proteome. One major limit in such an analysis is the chemical labile nature of Cys redox modifications; thus - basically - two critical steps are needed in analysing the thiol proteome, which consists of a temporary trapping of free thiols and their subsequent reduction (Avellini et al., 2007; Butterfield and Sultana, 2007).

2. Analysis of redox-sensing and signalling thiols

The cysteine proteome includes 214000 Cys with thiols and other forms. A relatively small subset functions in cell signalling, while a large number functions in response to the redox state. The former are redox-signalling thiols and the latter are defined as redox-sensing thiols (Dietz, 2003; Jones and Go, 2011; Sen, 1998; Sen, 2000). Some proteins contain Cys residues that are regulatory: their oxidation leads to misfolding and the influencing of protein activity. Several cytoskeletal proteins have been identified to be oxidative sensitive. Specific Cys residues' oxidation in these proteins has been identified. Among them, actin is the main component of the microfilament cytoskeleton and exists as monomeric G-actin, which can polymerise into filamentous F-actin upon extracellular stimuli. The constant and rapid reorganisation of the actin microfilament system is highly regulated (Carlier, 1991). A growing body of evidence indicates that the actin system is one of the most sensitive constituents of the cytoskeleton to oxidant attack. Recent redox proteomics studies have detected actin as the most prominent protein oxidised in response to the exposure of cells to oxidants (Fiaschi et al., 2006). The direct redox regulation of actin *in vivo* is one of the most important processes regulating the dynamics of the microfilament system. Trx1 was identified as interacting with actin and protecting the actin cytoskeleton from oxidative stress. Moreover, actin can be kept at a reduced status, even at a higher concentration of H_2O_2 stimulation, under the protection of Trx1. Trx1 is expressed ubiquitously in mammalian cells and contains a conserved Cys-Gly-Pro-Cys active site (Cys 32 and Cys 35) that is essential for the redox regulatory function (Carlier, 1991). In addition to the conserved cysteine residues in the active site, three additional structural cysteine residues (Cys 62, Cys 69, and Cys 73) are present in the structure of the human Trx1 (Nishiyama et al., 2001). Trx1 is S-nitrosylated on Cys 69, which is required for scavenging ROS and preserving the redox regulatory activity, and contribute to the protein's anti-apoptotic functions (Haendeler et al., 2002). Cys 73 residue is involved in dimerisation of Trx1 via an intermolecular disulphide bond formation between Cys 73 of each monomer in the oxidised state. The biological function of the Cys 62 and Cys 69 residues in the non-active domain remains to be fully elucidated. Some studies suggest that the formation of a disulphide bond between Cys 62 and Cys 69 created a way to transiently inhibit Trx1 activity for redox signalling under oxidative stress (Watson et al., 2003). A new role for Cys 62 - although it is not a key site that is involved in cellular redox regulation - plays an important role in mediating its interaction with actin. This interaction disappeared with the increasing concentration of H_2O_2 stimulation. One possible reason is that the intramolecular disulphide bond formation inhibits the activity of Trx1. Different H_2O_2 concentrations have different oxidative effects of functional relevance, leading to dimer formation, glutathionylation and depolymerisation of the actin system, depending on the location of the actin molecules, the source of the oxidant and the availability of surrounding reducing systems (Lassing et al., 2007). Many studies on oxidative stress have shown that both Cys 374 and Cys 272 of β-actin are highly reactive to oxidising agents. Chemical modification of Cys 374 affects polymerization ability and profilin binding (Dalle-Donne et al., 2007). The intracellular thiol homeostasis is maintained by the thioredoxin and glutaredoxin systems, which utilise NADPH as reducing equivalents in order to reduce proteins (Kalinina et al., 2008). Thus, oxidative modification may be restored by these redoxins and glutaredoxins. *In vivo*, the direct redox control of actin by Trx1 could be one of the most important processes regulating the dynamics of the microfilament system. It has been demonstrated that Trx1 could protect cells from apoptosis by the thiol oxidoreductase activity (Damdimopoulos et al., 2002; Poerschke and Moos, 2011; Smeets et al., 2005). Moreover, reduced Trx1 forms a complex with the apoptosis signalling regulating kinase-1 (ASK1) and protects cells from apoptosis by inhibiting ASK1 (Saitoh et al., 1998). Cys 62 in Trx1 plays an important role in protecting cells from apoptosis, independently of its role in the enzyme active site. Trx1, by binding to actin and regulating its dynamics, could protect cells from apoptosis(Wang et al., 2010). The results of oxidative stress on protein thiols and disulphides in *Mytilus edulis* revealed by proteomics also suggest that actin and protein disulphide isomerase are redox targets (McDonagh and Sheehan, 2008). Actin was also identified by affinity chromatography assay to be a Trx1 target in eukaryotic unicellular green algae (Saitoh et al., 1998). Both actin and Trx1 are evolutionarily conserved proteins, the protection of actin from oxidative insult by the TRX system could be a universal regulatory mechanism.

Many crucial signalling pathways utilise the reversible oxidation and reduction of cysteine thiols as a molecular switch. Redox-based regulation of gene expression has emerged as a fundamental regulatory mechanism in cell biology. Some proteins have apparent redox-sensing activity: electron flow through side-chain functional CH2-SH groups of conserved cysteinyl residues in these proteins accounts for their redox-sensing properties. Protein thiol groups with high thiol-disulphide oxidation potentials are likely to be redox-sensitive.

A class of signalling factors has been identified which uses cysteine residues in the conserved motifs (such as CXXC or CXXS) as redox-sensitive sulphydryl switches to modulate specific signal transduction cascades that have similar redox-sensitive sites (Lemaire et al., 2005). The identity of the amino acids separating the two cysteines in the CXXC motif and protein location influences the redox properties of CXXC-containing proteins - these proteins may serve as reductants or oxidants. TRX is a key molecule in the maintenance of the cellular redox balance. In addition to the cytoprotective action against oxidative stresses, it is involved in various cellular processes, including gene expression, signal transduction, proliferation and apoptosis (Klemke et al., 2008). Hepatopoietin (HPO) is a novel hepatotrophic growth factor, which is involved in the process of liver regeneration in rats, mice and humans (Francavilla et al., 1994; Hagiya et al., 1994). It belongs to the family essential for respiration and vegetative growth (ERV) 1/augmenter of liver regeneration (ALR). Both HPO and TRX have conserved CXXC motifs as their enzymatic active site. These cysteines in the redox regulatory domain are reactive and can be covalently linked to other proteins by forming disulphide bridges. It has been known that the family of FAD-dependent sulphydryl oxidase/quiescin-Q6-related genes contains thioredoxin (TRX) and yeast ERV1 domains (Hoober et al., 1999). If a composite protein is uniquely similar to two component proteins, no matter whether they are in the same species or not, the component proteins are most likely to interact or be involved in the same signal transduction (Marcotte et al., 1999a; Marcotte et al., 1999b). HPO is identified as functioning in conjunction with TRX by which it plays an important role in sensing the extracellular redox signals.

Homologs of this family have been found in a large number of lower and higher eukaryotes and in some viruses. Recently, ALR was identified as a sulphydryl oxidase by its ability to oxidise thiol groups of protein substrates and the presence of an FAD moiety in the carboxyl-terminal domain and the formation of dimer *in vivo* (Hofhaus et al., 2003; Lisowsky et al., 2001). It has also been shown that the effect of HPO on the activator protein-1 (AP-1) is dependent on its sulphydryl oxidase activity. ERV2, a member of ERV1/ALR in yeast, is an essential element of the pathway for the formation of disulphide bonds within the endoplasmic reticulum. E10R, a viral member of the ERV1/ALR protein family, participates in a cytoplasmic pathway of disulphide bond formation (Senkevich et al., 2000) and is responsible for the oxidation of the viral G4L gene product, which is homologous to glutaredoxin. A common characteristic of the proteins in this family is that they are involved in the redox reaction by the regulation of disulphide bond formation.

Two major mechanisms involving the reversible modification of amino acid side chains to modulate protein activity are phosphorylation/dephosphorylation by kinase and phosphatase systems and reduction/oxidation by thiol-dependent enzymes (Nakashima et al., 2002). Whereas many signalling processes involving phosphorylation are well-understood in terms of the mechanisms and identities of participating enzymes, redox regulation of cellular processes remains a poorly characterised area. HPO directly interacted with TRX, by which the redox state of TRX was changed, and then its effects on the activity of AP-1 and NF-κB were potentiated. The transcription factors NF-κB and AP-1 have been implicated in the inducible expression of a variety of genes involved in responses to oxidative stress and cellular defence mechanisms (Xanthoudakis et al., 1992). A feature of NF-κB is that both oxidants as well as reductants are known to activate it (Byun et al., 2002; Jeon et al., 2003). Activation of NF-κB by TRX could be attributed to its reduced form for the overexpression of TRX caused activation of NF-κB and the degradation of IκB in the cytosol;

at the same time, the c-Jun NH2-terminal kinase (JNK) signalling cascade was also activated. However, some investigations have shown that the transient expression of TRX resulted in a pronounced inhibition of NF-κB-dependent transactivation in CAT assays. Our studies showed that when the COS7 cells were cotransfected with HPO and TRX, a part of TRX is oxidised in cells expressing TRX, but the DNA binding activity of NF-κB and its transactivation were increased. The results of yeast two-hybrid analysis demonstrate that the binding ability of HPO with HPO is higher than that of HPO with TRX, implying that under the stimulation of an oxidative signal HPO tended to be assembled into dimers. The direct transfer of oxidising equivalents by dithiol/disulphide exchange reactions can be demonstrated by the oxidisation of TRX by HPO *in vivo* and *in vitro*. We could infer that the oxidising equivalents' flow might be from HPO to TRX and then to substrate proteins, leading to the change of the redox state of substrate protein and finally to affecting the activity of AP-1 or NF-κB. Recently, we have demonstrated that HPO can exist as a homodimer via disulphide bonds and that HPO has the capacity to form both homodimers and heterodimers with its alternatively spliced forms, which might contribute to the existence of various HPO compounds in hepatic cells. The HPO dimer still has sulphydryl oxidase activity and serves as an oxidant under oxidative conditions. These results imply that under oxidative stress conditions, intermolecular disulphide bonds formed within HPO could be transferred by a dithiol/disulphide exchange reaction to the active site of TRX and then to substrate proteins. In this sense, HPO links the redox chemistry of the cell to the formation of disulphide bonds within cytoplasm, while TRX acts as a mobile carrier of oxidising equivalents inducing the latter into nucleus to activate the expression of related genes. Two members of the ERV/ALR protein family (ERV2 and E10R) could use molecular oxygen directly to contribute oxidising equivalents for disulphide bond formation. Here, we found that another member of this family (HPO, as a FAD-linked sulphydryl oxidase) could also use reactive oxygen to generate disulphide bridges in protein substrate.

HPO is assembled into a dimer under the stimulation of oxidants, such as H_2O_2 and diamide, and the HPO dimer could be dissociated into monomer by DTT both *in vivo* and *in vitro*. TRX, another component protein in Q6, is also sensitive to the alteration of the redox state by the change of its free thiols. These results support the assumption that both HPO and TRX are sensitive to the cellular redox state and involved in the modulation to it.

TRX has been shown to interact directly with the apoptosis signal-regulating kinase 1 (ASK1) by forming disulphide bridges and acting as a physiological inhibitor of ASK1 in stress-free cells (Saitoh et al., 1998). The interaction is dependent on the redox status of TRX and can be regulated by intracellular reactive oxygen species (ROS) levels. In particular, an increase in ROS concentration causes the dissociation of TRX from ASK. As a result, ASK1 can undergo polymerisation, which corresponds to the active form of the enzyme. ASK1 has been suggested to activate the p38 and JNK upstream kinases, MKK3/6 and MKK4/7, respectively. We therefore speculate that the interaction between HPO and TRX might disrupt the interaction between TRX and ASK1. It is reduced but not oxidised thioredoxin that acts as a high affinity inhibitor of ASK1. When thioredoxin is oxidised by HPO in cytoplasm, it leads to the dissociation of TRX from ASK1 and polymerisation of ASK, it activates the stress-activated protein kinase pathway, and it promotes JNK activation and increases the activity of AP-1 and NF-κB.

A comparison of HPO-specific redox components with those of other known pathways for disulphide bond formation suggests some interesting analogies. The upstream components of the three known pathways - namely E. coli DsbB, yeast ERO1p and ERV2p - are proteins having two pairs of active cysteines each. In each case, the catalytic pair of cysteines interacts with ubiquinone, oxygen or another nonthiol electron acceptor forming a CXXC motif. The oxidative equivalents are then transferred to the second pair of cysteines on the same polypeptide chain, in the case of ERV2p, to the second subunit of a homodimer, and so is HPO. The second protein in the cascade of disulphide bond formation invariably is a thioredoxin-like protein - namely DsbA in E. coli, protein disulphide isomerase or its homologs in the yeast ER and the G4L thioredoxin-like protein in poxviruses. The HPO pathway therefore represents the first eukaryotic pathway for disulphide transmission in cytoplasm.

The importance of these findings is that the redox signalling transduction is conducted by the thiol-disulphide cascade in cytoplasm of mammalian cells. Thus, the pathways of disulphide bond formation in such diverse systems appear to use the same general principles of thiol-disulphide transfer between protein components. In this sense, HPO serves as a signal factor in the regulation of AP-1 and NF-κB activity via its cysteine. Early in the course of liver regeneration initiation, the expression of HPO increases quickly so that the cellular milieu becomes highly oxidising and these conditions shift the thiol-disulphide equilibrium of cellular proteins, which may play an important role in the stimulation of signalling transduction for promoting hepatocyte proliferation. This issue also explains the important role of HPO in liver regeneration and the mechanisms found in the calcium-dependent oxidation of TRX during cellular growth initiation. The rise in intracellular calcium induced by a growth factor binding to their receptor resulted in a marked conversion of reduced thioredoxin to the oxidised disulphide form. This apparent inhibition of thioredoxin reductase, coupled with the burst of H_2O_2 formation, leads to transient redox changes in cellular thiol proteins that may play an essential role in mitogen signal transduction. Thus, the relationship between HPO and TRX demonstrated by our results might shed new light on the signal transduction that oxidoreductase is involved in the processes of cell proliferation, apoptosis and organogenesis (Li et al., 2005).

3. Proteomics studies to analyse the oxidation state of proteins

3.1 Covalent modification to identify oxidation/nitrosylation of cysteine thiol groups

To date, 2DE coupled with mass spectrometry (MS) is still the best separation tool for analysing redox-based protein changes. ROS/RNS caused covalent modifications to proteins, which makes it possible to reveal these changes by applying specific labelling. Among the many kinds of amino acid residues susceptible to oxidative stress, cysteine is one of the most sensitive. Its free thiol groups play an important role in regulating protein functions and are often the target of oxidative stress. So far, several approaches have been developed to analyse the thiol proteome. The main limit is the chemical labile nature of Cys redox modifications. So, there are two critical steps needed in analysing the thiol proteome, which consists in trapping and reducing of free thiols. TCA (trichloroacetic acid)-based acidification was often used to quench the thiol groups, and then cell permeable Cys-specific reagents such as the alkylating agents iodoacetamide (IAA) or N-ethylmaleimide (NEM) were used to label the free thiols. Some specific reducing agents can be used to detect specific forms of oxidation. For example, cysteine residues in the form of sulphenic acid are difficult to identify because of their unstable chemical nature; however, this has been achieved by the exclusive reduction of sulphenic acid by sodium arsenite or through its reaction with specific chemicals, such as dimedone. S-nitrosothiols are reduced by ascorbate, whereas stronger reductants such as DTT reduce both nitrosothiols and disulphides (Jones and Go, 2011).

3.2 Quantification of redox proteomics

Several thiol-reactive reagents have been used to reveal the extent of Cys oxidation by 2DE gels, which include the IAM-derivatives 5-iodoacetamidofluorescein and Cys-specific fluorescent reagent monobromobimane. Differentials in the gel electrophoresis (DIGE) technique have been used to analyse the "redoxome" (Sethuraman et al., 2004). In this method, a set of fluorophores of similar molecular weights and chemical structures that differ according to their spectral features (absorption and emission wavelengths) were applied. NEM or IAM derivatives of cyanine (Cy3, Cy5) and DY-dyes were used in Redox-DIGE. The limitation of the above method is that only abundant proteins are detected, often missing low amplitude proteins such as transcription proteins and regulatory proteins. One way solving this problem is to perform an upstream enrichment step for the oxidised protein-thiol fraction of the proteome using the biotin-switch method originally developed by Jaffrey et al. (Salsbury et al., 2008). Biotin-based strategies are largely used for the detection of S-glutathionylation and S-nitrosylation - two Cys modifications which occur extensively in diseases characterised by oxidative stress.

Sethuraman et al. described the shotgun proteomic approach: isotope-coded affinity tag (ICAT) reagents were applied to quantify oxidant-sensitive protein thiols. This technique uses a certain type of marker which consists of three different parts: (i) a thiol-reactive compound (an iodoacetamide analogue), (ii) a linker containing either heavy or light isotopes, and (iii) a biotin tag for separation by avidin-coupled affinity chromatography. The principle of the ICAT approach in redox proteomics is that only free thiols are modified by the IAM moiety of the ICAT reagent. After equivalent samples were exposed to either control or oxidant conditions in a non-reducing environment, they are differentially labelled with the heavy or light form of the ICAT. The protein samples are mixed and then, with tryptic digestion, the labelled peptides are separated by affinity chromatography. Finally, the captured peptides are analysed by LC-MS/MS for the identification of the oxidant-sensitive cysteine thiols.

3.3 Shotgun proteomics

At present, 2DE-based methods are gradually substituted by gel-free technologies, such as shotgun-proteomics strategies. Shotgun-proteomics refers to the direct analysis by MS/MS of proteolysed protein mixtures so as to rapidly generate a global profile of the protein complement within the mixture itself. This mixture is highly complex. A solution to overcome this is represented by alternative sample preparation strategies, which could be suitable for performing a preliminary enrichment of peptides containing redox-modified cysteines. Several methods have been developed for isolating peptides containing oxidised cysteines. One of these approaches designed for the specific enrichment of sulpho peptides in tryptic digests is based on anionic affinity capture using poly-arginine-coated nanodiamonds as high affinity probes (Aggarwal et al., 2006; Barrios-Llerena et al., 2006; Haas et al., 2006).

4. Human diseases and early hints from redox proteomics

Redox biology is key to the life sciences because an increasing number of cellular functions and impairments are found to be linked to redox processes. Accumulating evidence suggests that a large number of diseases are closely related to oxidative stress. There is a growing need for the assessment of metabolic/oxidative stress and its modulation by the administration of pharmaceutical products. From a therapeutic point of view, drugs need to be designed to target oxidative stress sensitive biomarkers. In this context, redox proteomics might be pivotal in highlighting the main targets of protein oxidations and the biological pathways involved or compromised by these phenomena. Although the application of proteomics to drug design and development is in its earliest phase, preliminary redox proteomics results help to pave the way for further research in this field (D'Alessandro et al., 2011).

4.1 Neurodegenerative diseases

Neurodegenerative diseases such as AD, PD and HD each have distinct clinical symptoms and pathologies: they all share common mechanisms, such as protein aggregation, oxidative injury, inflammation, apoptosis, and mitochondrial injury which all contribute to neuronal loss. In neurodegenerative diseases, ROS generated by dysfunctional mitochondria are known to have a strong impact on the cellular proteome. Redox proteomic analysis of the post-mortem brains of AD patients revealed the presence of oxidative modifications of various protein substrates. Among them, some are relevant mitochondrial proteins, such as ATP synthase α- and β-chain and VDAC(Robinson et al., 2011). These mitochondrial resident proteins were found to be oxidised in the hippocampus and the observed modifications could play a role in the mitochondrial dysfunction and cell death observed in AD. Extensive oxidative stress has also been detected in PD. It has been reported that α-SYN was oxidised in the substantial nigra at the early stages of the disease. In addition, DJ-1 has been found to be modified by carbonylation and parkin to be S-nitrosylated, which results in a decrease of its E3 ligase activity. Several subunits of the respiratory complex I are subjected to oxidative damage, resulting in misassembling and the functional impairment of the complex. Redox proteomic analysis of HD R6/2 transgenic mice striatum revealed increased carbonyl levels in six proteins, including aconitase, creatine kinase and VDAC. Aconitase is an iron-sulphur protein that catalyses the isomerisation of citrate to isocitrate via cis-aconitate, and its inactivation may lead to an accumulation of reduced metabolites, such as NADH. The increased carbonyl levels, associated with the decreased activity of creatine kinase, could be relevant to the energetic impairments observed in HD (Sorolla et al., 2008). The role played by oxidative stress in ALS pathogenesis seems to be more relevant than in other neurodegenerative diseases. Oxidative damage induced oxidative modification of SOD1, UCHL-1 and Hsp70 proteins, which leads to the formation of SOD1 aggregation (Sorolla et al., 2008; Sussmuth et al., 2008).

Therefore, in the treatment of neurodegenerative diseases, neuroprotective agents which target ROS sources and aim at preventing their generation represent one class of drug therapeutics of great interest in pharmaceutical endeavours. Among them, the inhibitors of type B monoamine oxidase (such as selegiline and rasagiline) are the most promising neuroprotective agents to date, in that they prevent ROS generation. These inhibitors protect neuronal cells against cell death induced in cellular and animal models. The neuroprotective functions are ascribed to the stabilisation of mitochondria, the prevention of the death signalling process and the induction of the pro-survival anti-apoptotic Bcl-2 protein family and neurotrophic factors, thus counteracting mitochondria-mediated apoptotic pathways (Jones and Go, 2011).

4.2 Cardiovascular aging under oxidative stress

Reactive oxygen species (ROS) play an important role in the pathologic genesis of cardiovascular disease. Vascular enzymes such as NADPH oxidases, xanthine oxidase and uncoupled endothelial nitric oxide synthase, are involved in the production of ROS. NO · is produced in endothelial cells by the activation of eNOS during the normal functioning of the vessel wall. Vasodilator hormones raise intracellular Ca_2, leading to an increase in eNOS activity and NO · release. Physical forces such as fluid shear stress activate eNOS via protein kinase A- or Akt-dependent phosphorylation. The pathophysiological expression of inducible NOS in both macrophages and VSMCs elevates cytokine levels, resulting in localised inflammation. This, in turn, results in the production of NO · in the absence of further stimuli. Moreover, under some circumstances, eNOS becomes uncoupled and O_2 and is made instead of NO. The NOS enzymes are thus potentially important sources of both NO and O_2, depending on the surrounding environment. Virtually all types of vascular cells produce O_2 and H_2O_2. In addition to mitochondrial sources of ROS, O_2 and/or H_2O_2 can be made by many enzymes. Two of the most important sources in the normal vessel are thought to be cytochrome P450 and the membrane-associated NAD(P)H oxidase(s). A cytochrome P450 isozyme homologous to CYP 2C9 has been identified in coronary arteries and has been shown to produce O_2 in response to bradykinin. NAD(P)H oxidases that are similar in structure to the neutrophil respiratory burst NADPH oxidase, but which produce less O_2 for a longer time, have been identified in vascular cells. The endothelial, VSMC and fibroblast enzymes are not identical but have unique subunit structures and mechanisms of regulation. One important aspect of ROS production by at least the VSMC NAD(P)H oxidase is that it occurs largely intracellularly, making it ideally suited to modify signalling pathways and gene expression. The activity of the NAD(P)H oxidases can be modulated by vasoactive hormones and the small molecular weight G-protein rac-1. Angiotensin II, tumour necrosis factor-α, thrombin and platelet-derived growth factor all increase oxidase activity and raise intracellular levels of O_2 and H_2O_2 in VSMCs. Angiotensin II and lactosylceramide activate the endothelial cell enzyme, whereas fibroblasts increase O_2 production in response to angiotensin II, tumour necrosis factor-α, interleukin-1 and the platelet-activating factor. Physical forces, including cell stretch, laminar shear stress and the disturbed oscillatory flow that occurs at branch points, are also potent activators of O_2 production in endothelial cells. There are two major mechanisms by which hormones and physical forces activate the NAD(P)H oxidase: (1) acutely, whereby the expressed enzyme is activated by phosphorylation, GTPase activity and production of the relevant lipid second messenger, and (2) chronically, when the expression of rate-limiting subunits of the enzyme is induced, thereby providing higher levels of enzyme susceptible to activation. Macrophages are perhaps the major vascular source of O_2 in disease states. They oxidise LDL via the activation of diverse enzymes. Neutrophils and monocytes may also secrete myeloperoxidase, which appears to initiate lipid peroxidation. Two potential diffusible candidates to initiate myeloperoxidase dependent lipid peroxidation are the tyrosyl radical and nitrogen dioxide (NO2) (Elahi et al., 2009; Fearon and Faux, 2009; Lakshmi et al., 2009; Strobel et al., 2011).

4.3 Aging and metabolism

Early attempts at antioxidant intervention as a means to delay aging were initiated soon after the free radical theory of aging was proposed. These attempts stemmed from the postulation of the free radical theory of ageing which posits that the accumulation of

oxidative damage underlies the increased cellular, tissue and organ dysfunction and failure associated with advanced age.
However, these antioxidant interventions have so far failed to extend life spans in most cases. At present, a series of encouraging - albeit preliminary - results have been reported in *C. elegans* and drosophila through the use of enzymatic synthetic drugs miming SOD and CAT activities, such as EUK-8 and EUK-134. However, while increasing antioxidant defences in these organisms, the drugs have not produced any significant increase in lifespan. Transgenic mice that constitutively over-express human CuZn-SOD did not live longer than control animals, while heterozygous mice with reduced MnSOD activity have a life expectancy that is similar to wild-type mice (although these animals have increased oxidative damage to their DNA). If free radicals are actually correlated to aging, a winning strategy should be targeted at preventing their production rather than increasing defences and repairing mechanisms against ROS-induced damages. The Mitochondrial Free Radical Theory of Aging (MFRTA) proposes that mitochondrial free radicals are the major source of oxidative damage. According to MFRTA, the accumulation of these oxidative phenomena is the main driving force in the aging process (Sanz and Stefanatos, 2008).
Recent findings shed further light on the strong linkage between aging and metabolism and have opened brand new scenarios in the field of drug discovery. Insulin-like signalling in *C. elegans* activates the transcription factor SNK-1, which is known to defend against oxidative stress by mobilising the conserved phase 2 detoxification responses and it is thus referred to as the longevity-promoting factor. While aging remains a controversial issue, good results have been obtained in the field of cosmesis, as far as skin-aging is concerned. Antioxidant drug developments against skin aging have been extensively developed. A role has been proposed for ascorbic acid, alpha-tocopherol, carotenoids, polyphenols and other substances, such as ergothioneine, Zn(II)-glycine and CoQ10 in the treatment of skin-aging. In particular, the topical application of CoQ10 and antioxidants like alpha-glucosylrutin diminished resistance in the keratinocytes of old donors against UV irradiation, both in *in vitro* and in *in vivo* studies (D'Alessandro et al., 2011).

5. Challenges to mapping the thiol proteome

At present, mass spectrometry based proteomics makes rapid progress in mapping the Cys proteome (Chiappetta et al., 2011). These methods were also used to develop quantitative Cys proteomic databases and maps of redox systems biology. The full spectrum of Cys reactivity, such as glutathionylation, nitrosylation and other Cys modifications, needs to be analysed in order to address multiple modifications of the same Cys. Multiple modifications (e.g., products of benzene or acetaminophen oxidation) of a single Cys (e.g., C34 in albumin or Cb93 in haemoglobin) are used to identify chemical exposures. Links to chemical reactivity data, such as that provided by systematic comparisons of maleimide and iodoacetamide reactivity, would support an important chemical-biology interface which is currently lacking (Marino and Gladyshev, 2011).
To address the entirety of the Cys proteome, there is a need to understand the fractional contribution of Cys with high and low reactivities. Considerable attention has been given to oxidation of the Cys proteome by H_2O_2. However, protein thiols can be oxidised by many other chemicals, including hydroperoxides, endoperoxides and quinones. CySS reacts slowly with GSH, but many protein Cys residues are much more reactive. Cys/CySS shuttle functions in the regulation of extracellular thiol/disulphide pools (Mannery et al., 2011).

An alternative possibility explaining the maintenance of cellular proteins under a non-equilibrium, kinetically-controlled steady-state oxidation involves the pseudo-oxidase and/or pseudo-peroxidase activities of proteins. Very slow oxidative and peroxidative activities can be considered to be pseudo-oxidase and pseudo-peroxidase activities because the reaction rates and specificity for reactants are more similar to chemical reactions than to enzyme-catalysed reactions. Such reactions can depend upon low levels of associated metals, such as Cu2+ and Fe3+. For instance, Cu2+ can catalyse the oxidation of thiols in the presence of O_2, resulting in thiol oxidation to a sulphenic acid or disulphide. Reduction back to a thiol by TRX or GSH would complete a pseudo-oxidase cycle. At low rates of oxidation of the Cys proteome where ongoing cellular H_2O_2 generation occurs by other mechanisms, such a reaction sequence is difficult to verify. Earlier studies showed that H_2O_2 production in cellular fractions increases in proportion to O_2 partial pressure, and that protein oxidation occurs as a function of cellular iron and copper. Consequently, for the development of redox maps of the Cys proteome, additional information will be needed regarding the contribution of reactions of Cys at relevant, slow biologic reaction rates so that the system descriptions will adequately interpret reaction rates in systems biology models (Jones and Go, 2011).

6. Conclusion

Redox regulation is a fundamental physiological process which plays an important role in pathophysiological events. It is via reversible thiol modification that transcriptional and posttranslational responses are triggered. At present, an increasing number of techniques have been developed that make it possible to investigate, either qualitatively or quantitatively, modifications to specific amino acids (cysteines, tyrosines, etc.) or specific groups (carbonylations, nitrosylations, etc.). Redox proteomics is a powerful tool for monitoring physiological changes under oxidative stress. The identification of redox regulated proteins will provide great help in directing drug design and administration, new therapeutic targets and their validation. Currently, an accurate quantification of oxidised proteins remains difficult. A major task for future proteomics studies will be to develop tools to identify the different types of oxidation forms and establish the means to quantify the extent of such modification.

7. Acknowledgements

This work was supported by the National Basic Research Program of China (973 program, 2011CB711003), State Key Lab of Space Medicine Fundamentals and Application grants (SMFA1002), National Natural Science Foundation of China Project (31170811，31000386).

8. References

Aggarwal, K., Choe, L.H. and Lee, K.H. (2006) Shotgun proteomics using the iTRAQ isobaric tags. *Brief Funct Genomic Proteomic*, 5, 112-120.

Avellini, C., Baccarani, U., Trevisan, G., Cesaratto, L., Vascotto, C., D'Aurizio, F., Pandolfi, M., Adani, G.L. and Tell, G. (2007) Redox proteomics and immunohistology to study molecular events during ischemia-reperfusion in human liver. *Transplant Proc*, 39, 1755-1760.

Barrios-Llerena, M.E., Chong, P.K., Gan, C.S., Snijders, A.P., Reardon, K.F. and Wright, P.C. (2006) Shotgun proteomics of cyanobacteria--applications of experimental and data-mining techniques. *Brief Funct Genomic Proteomic*, 5, 121-132.

Buczek, O., Green, B.R. and Bulaj, G. (2007) Albumin is a redox-active crowding agent that promotes oxidative folding of cysteine-rich peptides. *Biopolymers*, 88, 8-19.

Butterfield, D.A. and Sultana, R. (2007) Redox proteomics identification of oxidatively modified brain proteins in Alzheimer's disease and mild cognitive impairment: insights into the progression of this dementing disorder. *J Alzheimers Dis*, 12, 61-72.

Byun, M.S., Jeon, K.I., Choi, J.W., Shim, J.Y. and Jue, D.M. (2002) Dual effect of oxidative stress on NF-kappakB activation in HeLa cells. *Exp Mol Med*, 34, 332-339.

Carlier, M.F. (1991) Actin: protein structure and filament dynamics. *J Biol Chem*, 266, 1-4.

Chiappetta, G., Ndiaye, S., Igbaria, A., Kumar, C., Vinh, J. and Toledano, M.B. (2011) Proteome screens for Cys residues oxidation: the redoxome. *Methods Enzymol*, 473, 199-216.

D'Alessandro, A., Rinalducci, S. and Zolla, L. (2011) Redox proteomics and drug development. *J Proteomics*, [Epub ahead of print].

Dalle-Donne, I., Carini, M., Vistoli, G., Gamberoni, L., Giustarini, D., Colombo, R., Maffei Facino, R., Rossi, R., Milzani, A. and Aldini, G. (2007) Actin Cys374 as a nucleophilic target of alpha,beta-unsaturated aldehydes. *Free Radic Biol Med*, 42, 583-598.

Damdimopoulos, A.E., Miranda-Vizuete, A., Pelto-Huikko, M., Gustafsson, J.A. and Spyrou, G. (2002) Human mitochondrial thioredoxin. Involvement in mitochondrial membrane potential and cell death. *J Biol Chem*, 277, 33249-33257.

Dietz, K.J. (2003) Redox control, redox signaling, and redox homeostasis in plant cells. *Int Rev Cytol*, 228, 141-193.

Elahi, M.M., Kong, Y.X. and Matata, B.M. (2009) Oxidative stress as a mediator of cardiovascular disease. *Oxid Med Cell Longev*, 2, 259-269.

Fearon, I.M. and Faux, S.P. (2009) Oxidative stress and cardiovascular disease: novel tools give (free) radical insight. *J Mol Cell Cardiol*, 47, 372-381.

Fiaschi, T., Cozzi, G., Raugei, G., Formigli, L., Ramponi, G. and Chiarugi, P. (2006) Redox regulation of beta-actin during integrin-mediated cell adhesion. *J Biol Chem*, 281, 22983-22991.

Francavilla, A., Hagiya, M., Porter, K.A., Polimeno, L., Ihara, I. and Starzl, T.E. (1994) Augmenter of liver regeneration: its place in the universe of hepatic growth factors. *Hepatology*, 20, 747-757.

Haas, W., Faherty, B.K., Gerber, S.A., Elias, J.E., Beausoleil, S.A., Bakalarski, C.E., Li, X., Villen, J. and Gygi, S.P. (2006) Optimization and use of peptide mass measurement accuracy in shotgun proteomics. *Mol Cell Proteomics*, 5, 1326-1337.

Haendeler, J., Hoffmann, J., Tischler, V., Berk, B.C., Zeiher, A.M. and Dimmeler, S. (2002) Redox regulatory and anti-apoptotic functions of thioredoxin depend on S-nitrosylation at cysteine 69. *Nat Cell Biol*, 4, 743-749.

Hagiya, M., Francavilla, A., Polimeno, L., Ihara, I., Sakai, H., Seki, T., Shimonishi, M., Porter, K.A. and Starzl, T.E. (1994) Cloning and sequence analysis of the rat augmenter of liver regeneration (ALR) gene: expression of biologically active recombinant ALR and demonstration of tissue distribution. *Proc Natl Acad Sci U S A*, 91, 8142-8146.

Hofhaus, G., Lee, J.E., Tews, I., Rosenberg, B. and Lisowsky, T. (2003) The N-terminal cysteine pair of yeast sulfhydryl oxidase Erv1p is essential for in vivo activity and interacts with the primary redox centre. *Eur J Biochem*, 270, 1528-1535.

Hoober, K.L., Sheasley, S.L., Gilbert, H.F. and Thorpe, C. (1999) Sulfhydryl oxidase from egg white. A facile catalyst for disulfide bond formation in proteins and peptides. *J Biol Chem*, 274, 22147-22150.

Jeon, K.I., Byun, M.S. and Jue, D.M. (2003) Gold compound auranofin inhibits IkappaB kinase (IKK) by modifying Cys-179 of IKKbeta subunit. *Exp Mol Med*, 35, 61-66.

Jones, D.P. and Go, Y.M. (2011) Mapping the cysteine proteome: analysis of redox-sensing thiols. *Curr Opin Chem Biol*, 15, 103-112.

Kalinina, E.V., Chernov, N.N. and Saprin, A.N. (2008) Involvement of thio-, peroxi-, and glutaredoxins in cellular redox-dependent processes. *Biochemistry (Mosc)*, 73, 1493-1510.

Klemke, M., Wabnitz, G.H., Funke, F., Funk, B., Kirchgessner, H. and Samstag, Y. (2008) Oxidation of cofilin mediates T cell hyporesponsiveness under oxidative stress conditions. *Immunity*, 29, 404-413.

Lakshmi, S.V., Padmaja, G., Kuppusamy, P. and Kutala, V.K. (2009) Oxidative stress in cardiovascular disease. *Indian J Biochem Biophys*, 46, 421-440.

Lassing, I., Schmitzberger, F., Bjornstedt, M., Holmgren, A., Nordlund, P., Schutt, C.E. and Lindberg, U. (2007) Molecular and structural basis for redox regulation of beta-actin. *J Mol Biol*, 370, 331-348.

Lemaire, S.D., Quesada, A., Merchan, F., Corral, J.M., Igeno, M.I., Keryer, E., Issakidis-Bourguet, E., Hirasawa, M., Knaff, D.B. and Miginiac-Maslow, M. (2005) NADP-malate dehydrogenase from unicellular green alga Chlamydomonas reinhardtii. A first step toward redox regulation? *Plant Physiol*, 137, 514-521.

Li, Y., Liu, W., Xing, G., Tian, C., Zhu, Y. and He, F. (2005) Direct association of hepatopoietin with thioredoxin constitutes a redox signal transduction in activation of AP-1/NF-kappaB. *Cell Signal*, 17, 985-996.

Lisowsky, T., Lee, J.E., Polimeno, L., Francavilla, A. and Hofhaus, G. (2001) Mammalian augmenter of liver regeneration protein is a sulfhydryl oxidase. *Dig Liver Dis*, 33, 173-180.

Mannery, Y.O., Ziegler, T.R., Hao, L., Shyntum, Y. and Jones, D.P. (2011) Characterization of apical and basal thiol-disulfide redox regulation in human colonic epithelial cells. *Am J Physiol Gastrointest Liver Physiol*, 299, G523-530.

Marcotte, E.M., Pellegrini, M., Ng, H.L., Rice, D.W., Yeates, T.O. and Eisenberg, D. (1999a) Detecting protein function and protein-protein interactions from genome sequences. *Science*, 285, 751-753.

Marcotte, E.M., Pellegrini, M., Thompson, M.J., Yeates, T.O. and Eisenberg, D. (1999b) A combined algorithm for genome-wide prediction of protein function. *Nature*, 402, 83-86.

Marino, S.M. and Gladyshev, V.N. (2011) Proteomics: mapping reactive cysteines. *Nat Chem Biol*, 7, 72-73.

McDonagh, B. and Sheehan, D. (2008) Effects of oxidative stress on protein thiols and disulphides in Mytilus edulis revealed by proteomics: actin and protein disulphide isomerase are redox targets. *Mar Environ Res*, 66, 193-195.

Nakashima, I., Kato, M., Akhand, A.A., Suzuki, H., Takeda, K., Hossain, K. and Kawamoto, Y. (2002) Redox-linked signal transduction pathways for protein tyrosine kinase activation. *Antioxid Redox Signal*, 4, 517-531.

Nishiyama, A., Masutani, H., Nakamura, H., Nishinaka, Y. and Yodoi, J. (2001) Redox regulation by thioredoxin and thioredoxin-binding proteins. *IUBMB Life*, 52, 29 33.

Poerschke, R.L. and Moos, P.J. (2011) Thioredoxin reductase 1 knockdown enhances selenazolidine cytotoxicity in human lung cancer cells via mitochondrial dysfunction. *Biochem Pharmacol*, 81, 211-221.

Robinson, R.A., Lange, M.B., Sultana, R., Galvan, V., Fombonne, J., Gorostiza, O., Zhang, J., Warrier, G., Cai, J., Pierce, W.M., Bredesen, D.E. and Butterfield, D.A. (2011) Differential expression and redox proteomics analyses of an Alzheimer disease transgenic mouse model: effects of the amyloid-beta peptide of amyloid precursor protein(Xi). *Neuroscience*, 177, 207-222.

Saitoh, M., Nishitoh, H., Fujii, M., Takeda, K., Tobiume, K., Sawada, Y., Kawabata, M., Miyazono, K. and Ichijo, H. (1998) Mammalian thioredoxin is a direct inhibitor of apoptosis signal-regulating kinase (ASK) 1. *Embo J*, 17, 2596-2606.

Salsbury, F.R., Jr., Knutson, S.T., Poole, L.B. and Fetrow, J.S. (2008) Functional site profiling and electrostatic analysis of cysteines modifiable to cysteine sulfenic acid. *Protein Sci*, 17, 299-312.

Sanz, A. and Stefanatos, R.K. (2008) The mitochondrial free radical theory of aging: a critical view. *Curr Aging Sci*, 1, 10-21.

Sen, C.K. (1998) Redox signaling and the emerging therapeutic potential of thiol antioxidants. *Biochem Pharmacol*, 55, 1747-1758.

Sen, C.K. (2000) Cellular thiols and redox-regulated signal transduction. *Curr Top Cell Regul*, 36, 1-30.

Senkevich, T.G., White, C.L., Koonin, E.V. and Moss, B. (2000) A viral member of the ERV1/ALR protein family participates in a cytoplasmic pathway of disulfide bond formation. *Proc Natl Acad Sci U S A*, 97, 12068-12073.

Sethuraman, M., McComb, M.E., Huang, H., Huang, S., Heibeck, T., Costello, C.E. and Cohen, R.A. (2004) Isotope-coded affinity tag (ICAT) approach to redox proteomics: identification and quantitation of oxidant-sensitive cysteine thiols in complex protein mixtures. *J Proteome Res*, 3, 1228-1233.

Smeets, A., Evrard, C., Landtmeters, M., Marchand, C., Knoops, B. and Declercq, J.P. (2005) Crystal structures of oxidized and reduced forms of human mitochondrial thioredoxin 2. *Protein Sci*, 14, 2610-2621.

Sorolla, M.A., Reverter-Branchat, G., Tamarit, J., Ferrer, I., Ros, J. and Cabiscol, E. (2008) Proteomic and oxidative stress analysis in human brain samples of Huntington disease. *Free Radic Biol Med*, 45, 667-678.

Strobel, N.A., Fassett, R.G., Marsh, S.A. and Coombes, J.S. (2011) Oxidative stress biomarkers as predictors of cardiovascular disease. *Int J Cardiol*, 147, 191-201.

Sussmuth, S.D., Brettschneider, J., Ludolph, A.C. and Tumani, H. (2008) Biochemical markers in CSF of ALS patients. *Curr Med Chem*, 15, 1788-1801.

Wang, X., Ling, S., Zhao, D., Sun, Q., Li, Q., Wu, F., Nie, J., Qu, L., Wang, B., Shen, X., Bai, Y., Li, Y. and Li, Y. (2010) Redox regulation of actin by thioredoxin-1 is mediated by the interaction of the proteins via cysteine 62. *Antioxid Redox Signal*, 13, 565-573.

Watson, W.H., Pohl, J., Montfort, W.R., Stuchlik, O., Reed, M.S., Powis, G. and Jones, D.P. (2003) Redox potential of human thioredoxin 1 and identification of a second dithiol/disulfide motif. *J Biol Chem*, 278, 33408-33415.

Xanthoudakis, S., Miao, G., Wang, F., Pan, Y.C. and Curran, T. (1992) Redox activation of Fos-Jun DNA binding activity is mediated by a DNA repair enzyme. *Embo J*, 11, 3323-3335.

Part 5

Structural Proteomics

21

The Utility of Mass Spectrometry Based Structural Proteomics in Biopharmaceutical Biologics Development

Parminder Kaur and Mark R. Chance
Center for Proteomics and Bioinformatics,
Case Western Reserve University, Cleveland, OH
NeoProteomics, Inc., Cleveland, OH
USA

1. Introduction

Proteins are essential components of living organisms and participate in nearly every biochemical process within cells. Examples of the processes include enzyme catalysis, cell signalling, host defense, metabolism, etc. These large, complex bio-molecules are connected by long chains of amino acids, that fold in very intricate patterns giving rise to a unique three-dimensional conformation. The biological function and physicochemical properties of a protein are determined by this higher order structure Ecroyd & Carver (2008); Hegyi & Gerstein (1999); Sadowski & Jones (2009). Most proteins tend to achieve the lowest possible free energy of the polypeptide chain and the surrounding solvent forming a native structure under physiological conditions Anfinsen (1973). This tightly folded conformation typically represents the biologically active state necessary for performing the required biochemical task. However, these macromolecules also have a temporal behavior leading to significant flexibility and dynamic motion because of the fluctuations in the surrounding electrostatic forces and hydrogen bonds that are important for maintaining conformations Henzler-Wildman & Kern (2007); Teilum et al. (2009). These temporal variations are important for certain functions such as protein-protein interactions and protein stability Kamerzell & Middaugh (2008); Travaglini-Allocatelli et al. (2009); van der Kamp et al. (2010). Thus, it is the presence of both spatial and temporal characteristics that allows for an ensemble of various molecular conformations to exist in solution. Change in environment of proteins, such as solvent acidity, urea concentration, temperature fluctuations, can change the folding pattern of the protein. Studying these partially or fully denatured states provides insights for understanding a variety of *in vivo* processes such as structural changes associated with aggregation, signal transduction, and transportation across membranes. Certain biological conditions can cause misfolding and aggregation of proteins, often causing severe disorders such as Alzheimer's disease, spongiform encephalopathies, and certain forms of diabetes Dobson (2003). Many genetic diseases are caused by protein-folding disorders, because an altered gene results in a modified protein sequence which is not able to undergo native folding and results in the disease phenotype Dobson (2001). Proteins have the ability to interact with one another, and can also bind to smaller ligands, which forms the basis of signaling and regulatory processes, playing a critical role in the mechanisms of drug activity. Owing to the

critical importance of structure-function paradigm, the pioneering scientists, Anfinsen and Stein and Moore, who established the relationship between protein structure and function, were awarded the Nobel Prize in Chemistry in 1972 Anfinsen (1973); Moore & Stein (1973).
The static protein structures commonly seen in X-ray images from the literature depict only one of the many possible conformations that the protein assumes at a particular instant in time. In fact, conformational dynamics are essential for mediating multifaceted functional roles performed by many proteins Fenimore et al. (2002); Frauenfelder et al. (1991); Huang & Montelione (2005). One particular example is the case of enzymes that require induced-fit binding behavior for proper operation Falke (2002); Schulz (1992). This suggests that the structural rearrangements of of such proteins represents a well balanced compromise between a highly ordered core conformation to ensure specificity, and a relatively flexible and dynamic state that maintains diverse functionality. However, there are certain proteins that remain disordered, and without any associated characteristic structure under physiological conditions, that fold specifically only while binding to another target Gunasekaran et al. (2003); Sugase et al. (2007); Wright & Dyson (1999); Yi et al. (2007). Nevertheless, most proteins behave according to the function-dictated-by-structure principle.
The above discussion illustrates the close interplay between the processes of protein folding, dynamics, conformation, intermolecular interaction, and function. In this Chapter, we will explore these structure function concepts as they relate to the design of novel pharmaceutical products. For example, these structure-dynamics-function aspects outlined above dictate action mechanisms of protein drugs (also called therapeutic protein, protein pharmaceutical, protein biopharmaceutical, or just biopharmaceutical) including their appropriate design and development to treat disease. Second, technologies like x-ray crystallography have been more difficult to apply to membrane proteins, likely the most important targets for small molecule drugs, and where information on the structural consequence of ligand binding are critical to drug development. This provides a strong interest in the development and application of reliable, sophisticated analytical techniques for thorough structural examination of therapeutic and membrane proteins in order to ensure and/or understand appropriate functionality and safety of the both small and large molecule drug development. This chapter introduces effective techniques that help realize this goal and demonstrates their application across monoclonal antibodies and membrane proteins. Monoclonal antibodies are designed to bind to specific protein targets in the cell blocking function of the targets; while membrane proteins perform essential processes in the cell, such as controlling the flow of information and materials between cells and mediating activities like nerve impulses and hormone action.

2. Significance

The function and efficacy of protein drug and biologic therapies is determined by the structure of the protein and its ability to interact with the surrounding partners. The interrogation and verification of the three dimensional conformation becomes critical in order to demonstrate the consistency of structure and function in biologics development. This necessitates the deployment of reliable, sensitive, and high-resolution techniques capable of examining higher order structure of such biomolecules in detail.
Biopharmaceutical manufacturers are required to demonstrate the consistency of the conformational complexity to the regulatory agencies. Traditional biophysical techniques used for this purpose include circular dichroism (CD), fluorescence, ultraviolet (UV), differential scanning calorimetry (DSC), isothermal titration calorimetry (ITC), analytical ultracentrifugation (AUC), and Fourier transform infrared spectroscopy (FTIR) Pain (2000).

These techniques provide information on a global state of the molecule, and typically output the average of overall information across the whole protein. These methods can in some cases successfully detect small changes between highly similar proteins, but the differential signal will not specifically identify the defect across the global readout. Some methods may reveal signal from only a limited number of residues (e.g., aromatic amino acid residues) in the protein and the absence of a such residues in the area of interest may lead to loss of information. Alternative sensitive, sophisticated tools for structure determination, nuclear magnetic resonance spectroscopy (NMR) and X-ray crystallography, allow both sensitive and specific probes of small structural changes Drenth (1999); Ramsey & Purcell (1952). NMR spectroscopy provides detailed structural information on proteins in solution, which is based upon distance constraints obtained from nuclear Overhauser effect Wuthrich (1990). However, these methods require high protein concentrations and can face significant challenges in the case of large proteins (in the case of NMR), and some proteins are not amenable to X-ray crystallography due to their inability to form crystals. In addition, these techniques tend to be fairly complex, and the conformation of the protein observed in the crystal is just a particular (although high resolution) conformation pulsed by the crystal lattice. This provides a strong interest to employ alternative practical methods to detect small, local differences within proteins in solution for both large and small proteins and at a range of protein concentrations. Mass spectrometry (MS) has established itself as a crucial technique in the biochemist's repository of tools over the past two decades, and many different flavors of protein MS are available with a variety of choices for sample preparation, molecular ionization, detection, and instrumentation. Structural proteomics techniques such as covalent labeling Maleknia et al. (2001); Suckau et al. (1992), hydrogen/deuterium exchange (H/DX) Wales & Engen (2006), and chemical cross-linking Back et al. (2003); when coupled with highly sensitive mass spectrometry instruments; alleviate many of the above limitations and have shown promising results in the past decade. This chapter focuses on the fundamentals of these techniques, discusses challenges and limitations experienced by each method, and concludes with the successful application examples of these analytical tools.

3. Techniques

The following techniques can be used to determine conformational change, binding stoichiometry, and affinity for protein-ligand interactions.

3.1 Hydrogen/deuterium exchange mass spectrometry(H/DX-MS)

H/DX methods were introduced in 1990s and have now powerfully established themselves for probing the biomolecular structure Bai et al. (1995); Englander & Kallenbach (1983); Hvidt & Nielsen (1966); Krishna et al. (2004); Wales & Engen (2006); Woodward et al. (1982). The principle behind the technique is that protein backbone amide hydrogens are exchangeable with deuterium atoms from the solvent surrounding the protein at specific exchange rates that can be measured experimentally. The amide hydrogens at the surface exchange very rapidly, while those buried in the core have much slower exchange rates. The backbone amide hydrogens participating in the formation of hydrogen bonds will also have relatively slower exchange rates. Hence, the rate of exchange of hydrogens provides valuable insights into the bio-molecular backbone, secondary structure and structural stability.

Fig 1 shows the overall schematics of an H/DX experiment Wales & Engen (2006). The protein under consideration is subjected to a deuterium rich environment that labels surface accessible residues, followed by quenching of the reaction. The incorporation of deuterium into the biomolecule under consideration results from the natural process of hydrogen exchange with deuterium from the surrounding environment. There are a variety of methods available for the introduction of deuterium into a peptide or protein; and various experimental strategies are used for investigating the biomolecular exchange as seen in Fig 1 . The protein can either be studied intact (for global exchange analysis), or can be digested by proteolysis (for local exchange analysis), and intact proteins or peptide fragments can be analyzed using mass spectrometry, which is able to measure the increase in mass as hydrogen atoms are exchanged for deuterium. Specific solvent accessible residues in the protein show an increased mass in the mass spectrometer readout. The experiment is repeated multiple times, each time increasing the duration of the deuterium pulse exposure to the protein, allowing for the study of deuterium exchange rate kinetics.

Although H/DX technique has shown promise for utility in biopharmaceutical studies, greater automation, seamless coupling of high performance separation, and sophisticated software for automated data interpretation is required for a more routine implementation of H/DX-MS into commercial experiments Houde et al. (2011); Wales et al. (2008). Various approaches have been developed recently towards the automated data analysis for data acquisition and post-processing Chalmers et al. (2006); Kazazic et al. (2010); Pascal et al. (2009). The H/DX-MS method has been successfully applied to study structural changes introduced by kinase activation, compare isoform-specific differences in binding to a common ligand, and to map the epitopes of monoclonal antibodies Houde et al. (2009); Lee et al. (2004); Stokasimov & Rubenstein (2009).

3.2 Hydroxyl-radical mediated covalent labeling mass spectrometry or protein footprinting

Another popular method for investigating macromolecular conformation in solution is protein footprinting, also called Covalent Labeling, which was invented initially to characterize the sites of DNA-protein interaction Brenowitz et al. (1986); Galas & Schmitz (1978); Humayun et al. (1977); Schmitz & Galas (1980). The technique was further extended to protein structure examination by subjecting them to limited proteolysis in conjunction with separation using SDS polyacrylamide gel electrophoresis, with the first report of protein footprinting appearing in 1988 Sheshberadaran & Payne (1988). The advent of sophisticated analytical tools such as mass spectrometry for examining cleaved fragments of proteins, significantly improved the spatial resolution of the technique. A key distinction from H/DX-MS is that most labeling reagents target side-chains, while HDX-MS specifically examines the bio-molecular backbone and protein secondary structure.

The basic principle behind hydroxyl radical mediated protein footprinting approaches for probing solvent accessible residues is similar to that of H/DX-MS technique. The overall schematic for experimental setup for a covalent labeling experiment is shown in Fig 2. The protein solution is exposed to hydroxyl radicals, generated by multiple methods, which leads to stable, covalent oxidative modifications on the surface accessible residues Hambly & Gross (2005); Maleknia et al. (2001); Sharp et al. (2003; 2004); Takamoto & Chance (2006). The chemistry of amino acid and peptide oxidation using MS revealed that in dilute aqueous solution, oxidative modification of side chains is observed in a much more predominant form as compared with backbone cleavage or cross-linking. These stable side chain modifications result in mass shifts, which can be easily revealed by isolating protein fragment and then comparing the masses to unmodified forms of the protein. Thus, labeling is followed by subjecting the protein to proteolysis and high pressure liquid chromatography coupled with mass spectrometry as in the case of H/DX-MS method. Tandem mass spectrometry (MS/MS) methods have been found to be particularly suited for further identifying and localizing the specific sites of oxidation Chance (2001); Kiselar et al. (2002); Maleknia et al. (1999). Thus, the structural resolution of covalent labeling is very high, and at the single side chain level. In the

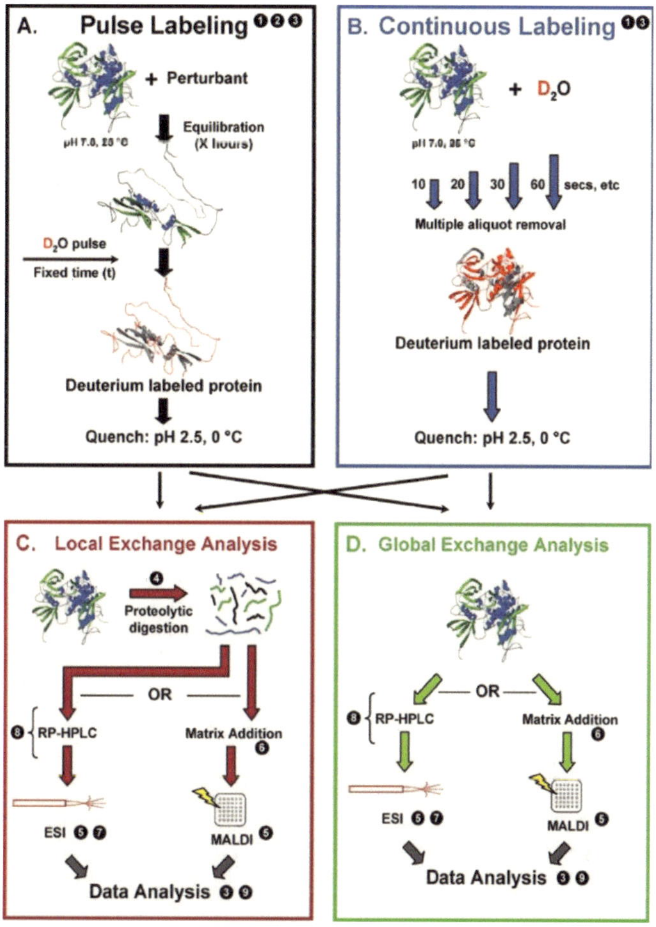

Fig. 1. Overall scheme for hydrogen exchange mass spectrometry experiments. A: Pulse labeling. After a protein has been exposed to a perturbant (chemical denaturant, heat, pH, binding, complex formation, pressure, etc.), unfolded regions (gray) become labeled with deuterium (red) during a quick pulse of D2O (typically 10 s). Deuterium exchange is quenched by reducing the pH and temperature. B: Continuous labeling. D2O buffer is added to a protein (in H2O buffer) such that the final D concentration is >95%. After a set period of time, an aliquot of the labeled protein is removed from the original tube and mixed with quench buffer to reduce the pH and temperature. Aliquot removal is repeated for subsequent labeling times. The protein concentration and solution volume are controlled such that all the aliquots are identical upon quench except for the amount of time the protein was exposed to D2O. C:. Localized exchange information. Quenched samples (from part A, part B, or both) are digested with pepsin or another acid protease. The resulting peptides are analyzed with online HPLC-ESI-MS or with MALDI-MS. The resulting data analysis provides information on deuterium exchange in short fragments of the peptide backbone. D: Global exchange information. Quenched samples (from part A, part B, or both) are directly analyzed with HPLC-ESI-MS or MALDI-MS. The data provide a global picture of how the protein behaves in D2O. Reprinted with permissions from Wales & Engen (2006) Copyright 2006 John Wiley and Sons.

typical workflow, series of samples are exposed to variable doses, and a dose-response curve is generated for observed peptides individually in order to provide relative quantitation of oxidation as a function of hydroxyl radical exposure time. The generation of stable, covalent modifications allows a wide range of samples and proteases to be employed under broad solution conditions and pH values. The reactivity of side chains to hydroxyl radical attack initially and the attenuation of this reactivity as a result of structural perturbation such as ligand binding, unfolding, or macromolecular interactions provides insights into the change in surface accessibility at particular sites under consideration. Since the side chains get modified during the procedure, specific probe sites can be investigated using tandem mass spectrometry, while, in the case of H/DX-MS, the conformational changes may be attributed only to a specific peptide fragment. However, the two approaches are complementary to each other since H/DX-MS characterizes backbone secondary structure and stability while protein footprinting probes the side chains of residues.

The interpretation of high volumes of data resulting from covalent labeling experiments used to pose as the biggest bottleneck for the overall experiment, thus, limiting their potential. A typical hydroxyl radical-mediated covalent labeling experiment leads to multiple oxidation states of various amino acid side chains Takamoto & Chance (2006); Xu & Chance (2007), leading to a challenging task for data analysis. This bottleneck has now been eliminated with the advent of ProtMapMS, a computational analytical tool, that is specifically tailored to meet the needs of covalent labelling experiments Kaur et al. (2009). Figure 3 illustrates typical liquid chromatographic elution profile results automatedly generated in a covalent labeling experiment using ProtMapMS. The four plots in Fig 3 represent the chromatographic elution plots from a doubly charged insulin B-chain peptide 23-29 for an X-ray exposure time of 0, 8, 15, and 20 ms successively. The unoxidized form of the peptide is indicated by cyan (m/z = 430.22) color, while the different oxidative forms are blue, green, and red. Interestingly, five green peaks labeled A-E in Fig 3 represent the oxidatively labeled products for the peptide incorporating one oxygen atom represent five unique isomeric forms of the peptide molecule, differing in the position of the attached oxygen atom within the same peptide. Fig 3 shows that the relative intensities of the modified forms increase as the amount of X-ray exposure time to the protein increases. This behavior is expected since the protein molecules have increased opportunity to react with hydroxyl radicals. The oxidative forms of peptide are seen to elute at a slightly different time (although in close proximity) than their unoxidized counterpart.

Improvements in two specific areas will help in making the covalent labeling experiments more routine. More accurate quantitative relationship of solvent accessibility and the side chain reactivity is highly desirable to add quantitative rigor to the structural characterization. This will provide specific constraints that can be used in computational modeling approaches for a more comprehensive analysis. Flexible computational modeling approaches should be developed that allow for including surface accessibility constraints in a quantitative manner for more accurate results. Such improvements will lead the way for oxidative footprinting and other covalent labeling approaches using MS to be utilized for understanding the conformation and dynamics of very complex macromolecular assemblies. Footprinting technique has been very successfully applied in the past for RNA structure analysis Sclavi et al. (1998). There is a great potential for similar progress for protein structure prediction by incorporating the technique into a wider utilization.

3.3 Chemical cross-linking

Covalent cross-linking is another important technique for characterizing the connectivity of solution-phase complexes, and for obtaining new intramolecular or intermolecular distance constraints between biomolecules Sinz (2003); Vasilescu et al. (2004); Wine et al. (2002).

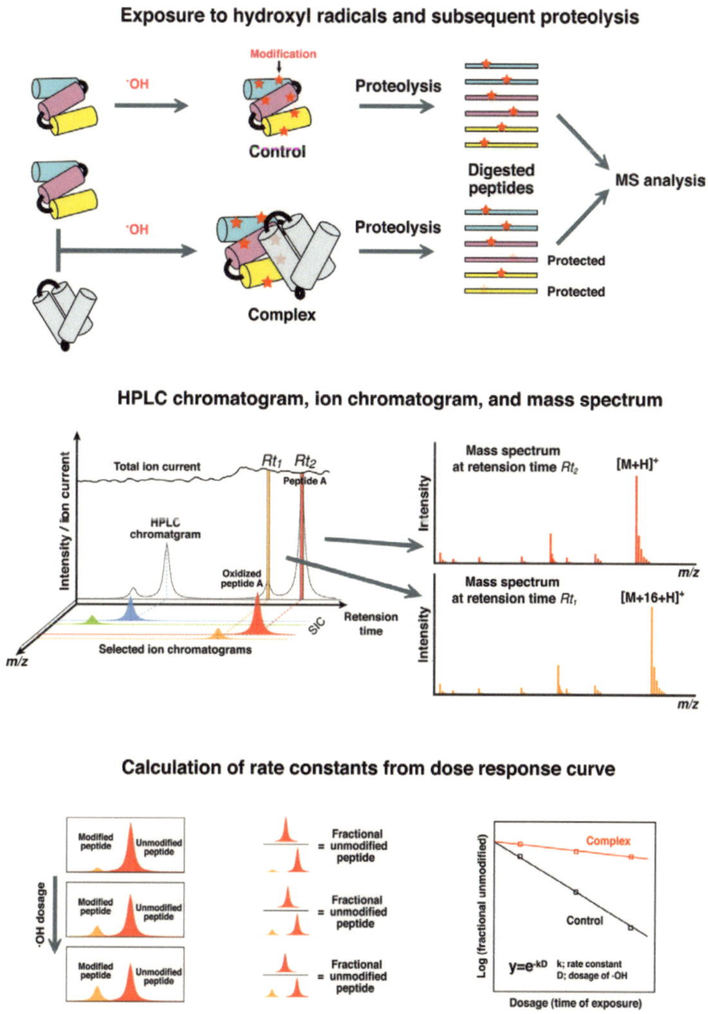

Takamoto K, Chance MR. 2006.
Annu. Rev. Biophys. Biomol. Struct. 35:251–76

Fig. 2. Hydroxyl radical footprinting: data collection and data analysis. Top panel: Protein is exposed to hydroxyl radical and modified covalently. The resulting protein sample is then digested by protease or chemical cleavage to fragments that are suitable in size for mass spectrometry. The experiment is carried out for each individual protein and for the protein complex. In a tight binding interface, some regions are protected from hydroxyl radical attack. Middle panel: Peptides are separated by liquid chromatography and introduced into a mass analyzer. The selected ion chromatograms (SIC) are constructed for each ion (with particular mass) as a function of retention time. By monitoring the mass and time, we know what species appears at what retention time. By integrating peak areas in SIC, we can calculate the total indicated ion abundance. Bottom panel: The determinations of modification rates are performed by calculating the loss of intact peptide in order to maximize the interrogation of intact material. Reprinted with permissions from Takamoto & Chance (2006) Copyright 2006 Anuual Reviews.

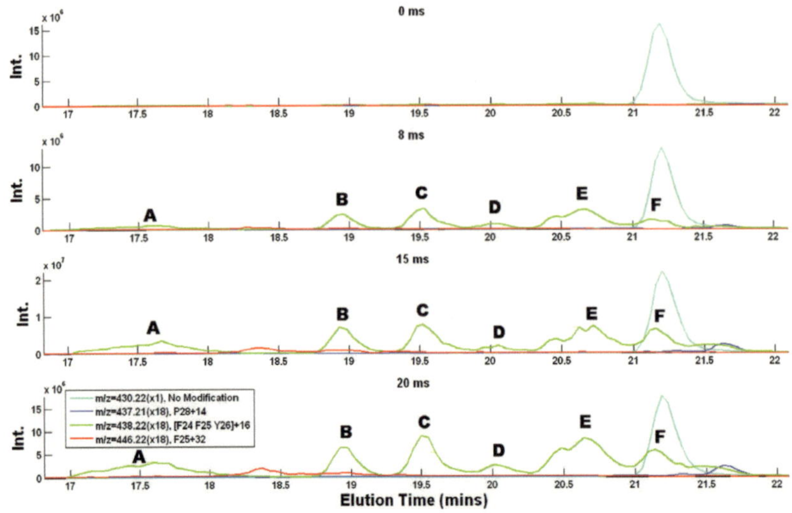

Fig. 3. Chromatographic elution plots for doubly charged human insulin B-chain peptide 23-29. The unmodified form is shown in cyan, while the modified forms (magnified by a factor of 18) are shown in blue (P28 + 14), green (mixture of F24 + 16, F25 + 16, and Y26 + 16), and red (F25 + 32). Reprinted with permission from Kaur et al. (2009). Copyright 2009 American Chemical Society.

It involves covalently attaching two specific functional groups of the protein(s) under investigation by means of a special reagent called cross-linker. The location and the identity of the created cross-links imposes a distance constraint on the location of the respective side chain sites and provides important clues on the three-dimensional conformation of the protein or a protein complex. Coupling chemical cross-linking with sensitive mass spectrometric analysis allows to characterize the position(s) of the introduced cross links for generating distance constraints. The wide variety of crosslinking reagents allow for varied specificities towards numerous functional groups such as primary amines, sulfhydryls, or carboxylic acids; and the wide range of spacer lengths offered by different cross-linking reagents allow the possibility to address a broad range of scientific questions. However, owing to the inherent complexity of the reaction mixtures, the identification of the cross-linked products can be quite tedious. The greatest challenge of utilizing chemical cross-linking and MS analysis is the lack of computational tools that can effectively interpret the enormous complexity of the reaction mixtures. There is a significant overhead of labor intensive manual processing involved in the data analysis since all the existing programs exhibit their specific limitations Sinz (2006). Some of the limiting bottlenecks have been eliminated with the advent of specialized search programs such as GPMAW, xQuest, searchXlinks, VIRTUALMSLAB, and ASAP de Koning et al. (2006); El-Shafey et al. (2006); Peri et al. (2001); Rinner et al. (2008); Wefing et al. (2006). Further progress into an integrated suite of algorithms addressing comprehensive needs for chemical cross-linking combined with mass spectrometry would greatly facilitate making it a generally applicable technique for rapid protein structure characterization for biopharmaceutical experiments.

Cross-linking and covalent labeling methods are complementary to each other - cross-linking methods provide distance constraints for two amino acids whereas covalent labeling approaches derive information about protein surface mapping since the reactions are typically controlled by the accessibility of surface amino acids to the covalent labeling reagent.

3.4 Future trends

Experimental data from structural proteomics experiments can be used together with computational structure modeling techniques such as comparative modeling and threading. The stand alone theoretical models without support from experimental data lack reliability, especially in the case of ab-initio modeling where suitable templates may not be available. Hybrid approaches resulting from a combination of theoretical modeling and experimental methods such as hydrogen-deuterium exchange and covalent labelling are gaining increasing popularity, allowing to combine the merits from both the methodsPantazatos et al. (2004); Zhu et al. (2003). The results from experimental analysis specifically provide explicit constraints such as distance constraints in the case of chemical cross-linking, that reflect the surface accessibility or burial of particular sites, which can be included for refining computational structure prediction models, hence greatly reducing the model space to be considered while increasing the reliability from complementary approaches.

4. Applications

4.1 Study of membrane proteins using protein footprinting

Proteins embedded in membranes assist in water or ion transport in signaling processes across the biological membrane. Typically, transmembrane proteins are comprised of hydrophobic cores with ionizable or charged residues at specific locations that are crucial for their appropriate functionality Muller et al. (2008). G protein-coupled receptors (GPCRs) comprise a large protein family of transmembrane receptors that sense molecules outside the cell and activate the signal transduction pathways inside and regulate cellular responses Rosenbaum et al. (2009). The presence of ordered, structural waters are likely to be important factors to impart structural plasticity required for agonist-induced signal transmission for allosteric activation of the G protein-coupled receptors (GPCRs) Rosenbaum et al. (2007). The functionality of these ordered water molecules is not clearly known. They may provide structural stabilization, mediate conformational changes in signaling, neutralize charged residues, or carry out a combination of all these functions. Structural investigation of GPCR superfamily members using radiolytic footprinting revealed the presence of conserved embedded water molecules likely to be important for GPCR function Angel, Gupta, Jastrzebska, Palczewski & Chance (2009).

The behavior of soluble proteins with hydroxyl radical footprinting is well-characterized, such that the intrinsic reactivity and the solvent accessibility of the side chains govern their observed reactivity Chance et al. (1997); Kiselar et al. (2002); Takamoto & Chance (2006). However, these approaches have not been investigated for membrane proteins, factors influencing labeling or the overall scavenging effects of detergents or lipids have not been well understood. Recently, in order to gain insights into membrane proteins, radiolytic protein footprinting was used to interrogate the structural dynamics of ground state (rhodopsin), photoactivated (Meta II), and inactive ligand-free receptor (opsin) and native membranes Angel, Gupta, Jastrzebska, Palczewski & Chance (2009). In contrast to the previous literature on soluble proteins, oxidative modifications were found on residues located in both solvent-accessible and solvent-inaccessible regions. The oxidized residues within the transmembrane domain were labeled, and their reactivity was found to be varying as a function of rhodopsin activation state. Using radiolytic hydroxyl radical labeling in conjunction with H_2O^{18} solvent mixing, it was discovered that labeling within the transmembrane region is highly influenced by the tightly bound waters and that regions undergoing local conformational alterations and water reorganization experience changes in the oxidation status.

Fig 4 shows the findings of the study, indicating alterations in rates of oxidation introduced going from ground state to activated receptor, reflecting local structural changes upon the formation of both Meta II and opsin. No exchange of the structural waters was observed with the surrounding solvent in either the ground state or for the Meta II or opsin states. However, oxidative labeling of selected side chain residues within the transmembrane helices was observed and activation-induced changes in local structural constraints likely mediated by dynamics of both water and protein were revealed. This work suggests a possible general mechanism for water-dependent communication in family A GPCRs, and illustrates the role of radiolytic footprinting for characterizing the structure and dynamics of the transmembrane region, including dynamics of water in membrane proteins, and has the potential to define allosteric channels for other transmembrane signalling proteins, and ion channels.

The implications for these results are considerable for the design of drugs to target important Type A GPCRs, e.g. serotonin, adenosine, or β_2-adrenergic receptor. If this model for rhodopsin functional activation is correct, it means that effective drugs mediate specific and local changes in water/side-chain interactions within the transmembrane region and that this rearrangement mediates the correct and efficient signaling across the membrane. In addition it focuses attention on water molecules and their potential rearrangements in recent crystallographic data of Type A family GPCRs Angel, Chance & Palczewski (2009). Recent extensions of the covalent labeling methodology to ion channel structure-function studies Gupta et al. (2010), where the movement of water coupled to the rearrangement of specific side chains was specifically tracked in the mechanism of channel opening, show the emerging power of the method to reveal important structure function considerations relevant for drug development.

4.2 Characterizing monoclonal antibody IgG1 using hydrogen-deuterium exchange

Monoclonal antibodies (mAb) are used both in fundamental research and in clinical settings as highly specific therapeutic agents for treating an array of different diseases. Currently, recombinant immunoglobulin gamma (IgG) mAbs comprise the largest percentage of molecules in the biopharmaceutical development pipeline. This provides a strong motivation for utilizing new or improved analytical methods and tools for mAb characterization. Presently, there is very limited information available on crystal structures of entire IgGs. Moreover, such cases provide information only of a very stable structure sampled by the protein; while lacking information on the conformational dynamics or in-solution motion.

In a recent study, H/DX-MS has been used to study both global and local conformational behavior of a recombinant monoclonal IgG1 antibody to obtain detailed conformational dynamics Houde et al. (2009). It demonstrates the capabilities of H/DX-MS as a powerful analytical tool to study large protein biopharmaceuticals such as mAbs. The conformational features of an intact glycosylated IgG1 are compared against its deglycosylated form. This assists in drawing conclusions to determine how glycosylation affects the IgG1 conformation. First, deuterium exchange into the intact form of IgG1 was measured. This is useful for providing information about the overall solvent accessibility of the protein and, also indicates whether the protein is amenable to H/DX-MS experiments so that further investigation can be performed. The next step performed the analysis of exchange into isolated Fab/Fc fragments. Next, the intact protein was labeled and digested (after quenching the deuterium labeling reaction) using pepsin as a protease. Digestion protocols were performed for both glycosylated and deglycosylated versions of the IgG1, followed by analysis using mass spectrometry. Five independent experiments were performed, each containing an undeuterated sample and five different labeling times

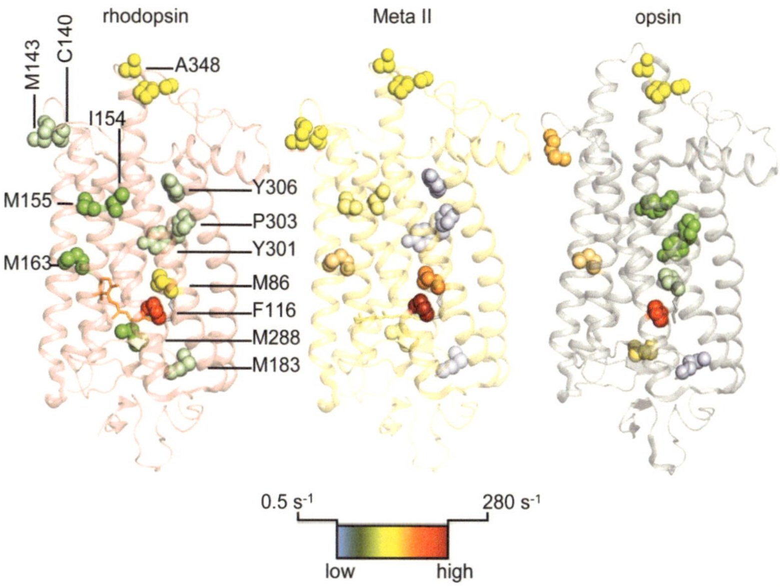

Fig. 4. Pictorial summary of modification rate constants. Radiolytic modification rate constants were determined for many residues in rhodopsin (Left), Meta II (Center), and opsin (Right). Residues with rate constants >0.1 s^{-1} are rendered as spheres colored by rate constant ranges: 0.5-1.2 s^{-1}, light blue; 1.3-3.9 s^{-1}, light green; 4.0-5.9 s^{-1}, green; 6.0-7.9 s^{-1}, light-yellow; 8.0-9.9 s^{-1}, yellow; 10-14.9 s^{-1}, light-orange; 15-25 s^{-1}, orange; >200 s^{-1}, red. Following photoactivation, modification rates increased for M86, C140, M143, the pair of residues in helix IV I154 and M155, M163, and M288. Residues exhibiting decreased modification rates were Y301, P303, and Y306 in helix VII. There also was a reduced modification rate of M86 and F116 in opsin as compared with the two other states. The mixed modification of peptide 137-146, comprising part of the C-II loop, showed a large increase in the rates of detectable modification for opsin relative to ground state and activated rhodopsin, whereas M183 in the E-II loop exhibited no change in modification rate as a function of receptor activation state. The carboxyl terminal peptide did not show a marked difference in modification rates between the three states of the receptor. Changes in rates of oxidation observed when comparing ground state and activated receptor reflect local structural changes upon formation of both Meta II and opsin. Reprinted with permissions from Angel, Gupta, Jastrzebska, Palczewski & Chance (2009) Copyright 2009 National Academy of Sciences.

Figure 5 depicts the oxidative changes detected in IgG1 in the presence and absence of glycosylation. Analysis of H/D exchange pattern into the intact, glycosylated IgG1 indicated that the molecule was folded, very stable, and could be analyzed with very high sensitivity. Since the approach can detect subtle, localized changes within the protein, H/D exchange could be localized to very specific regions of the antibody. Degylosylation resulted in changes in the IgG1 conformation, and were characterized by comparing H/D exchange rates of the glycosylated and deglycosylated forms of the antibody. Two specific regions of the IgG1 (residue positions 236-253 and 292-308) were found to have experienced change in H/D exchange properties upon deglycosylation. These results are consistent with previous findings using X-Ray crystallography and NMR techniques associating the role of glycosylation in the interaction of IgG1 with Fc receptors. Overall, H/DX-MS showed that changes in

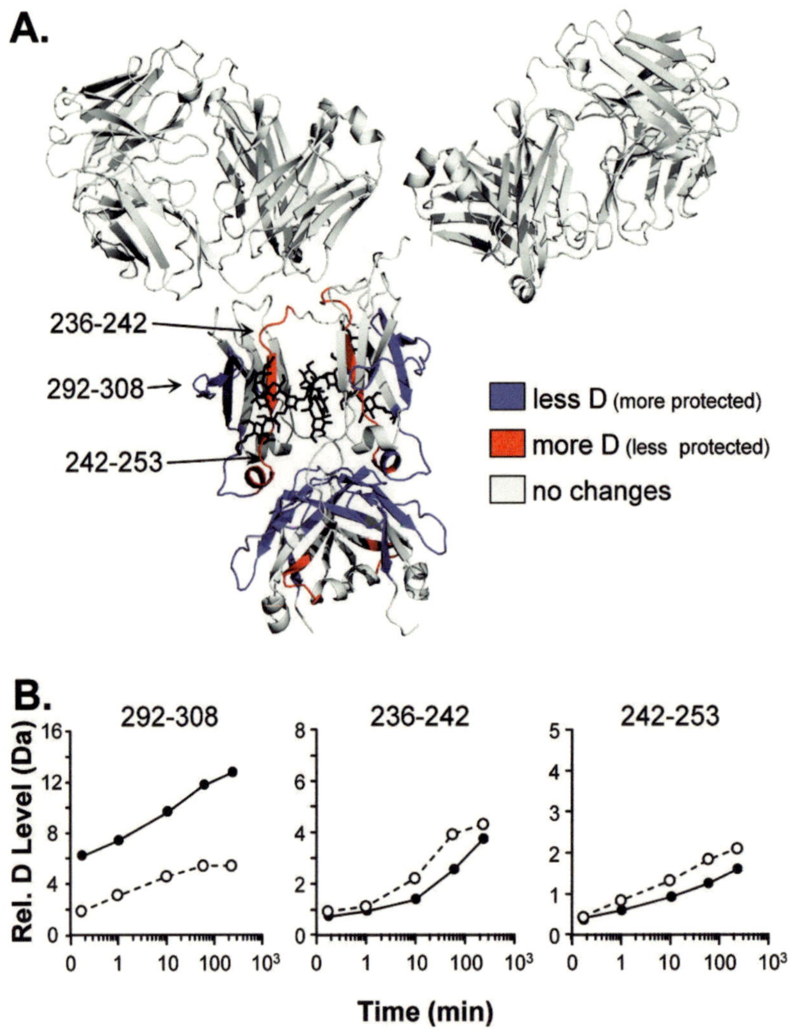

Fig. 5. Comparison of deuterium levels in IgG1 with and without glycosylation. (A) The model structure of IgG1, with the glycosylation indicated in black sticks. Parts colored blue indicate regions where the deglycosylated form had, over all time points, less deuterium (more protection from exchange). Parts colored red indicate regions where the deglycosylated form had, over all time points, more deuterium (less protection from exchange). Note that although blue regions appear to be more surface accessible than the red regions, the conformational distrubances introduced during the deglycosylation process result in greater protection in blue areas than their red counterparts (B) Representative deuterium incorporation profiles comparing exchange in heavy chain residues 292-308 (PREEQYNSTYRVVSVLT), 236-242 (LGGPSVF), and 242-253 (FLFPPKPKDTLM). The solid, black line represents data from the glycosylated form, and the dotted line represents data from the deglycosylated form. Reprinted with permission from Houde et al. (2009). Copyright 2009 American Chemical Society.

conformation as a result of deglycosylation were in areas critical for Fc receptor binding. The data illustrate the utility of H/DX-MS to provide valuable information on the higher order structure of antibodies and characterizing conformational changes that these molecules may experience upon modifications such as glycosylation, possibly affecting the functionality of the protein.

Recently, the application of covalent labeling to the analysis of glycoprotein structure was demonstrated Wang et al. (2010). This shows that covalent labeling can also be successfully applied to monoclonal antibodies to determine the structure-function relationships relevant to biopharmaceutical development.

5. Conclusions

There is a close inter-relation between the functionality of a protein and the processes of protein folding, dynamics, three-dimensional structure, and intermolecular interaction. These aspects further dictate action mechanisms of a protein drug and its effectiveness in the treatment of a disease. The examination of primary sequence, specific protein modifications, and the three dimensional conformation are important variables used to demonstrate structural equivalency of clinically relevant formulations for biologics development. An evolving regulatory climate is likely to require highly accurate data on "higher-order structure" of biologics for the approval of biosimilars in the US. This provides strong impetus for developing reliable, sensitive, and high-resolution analytical technologies along with associated computational methods for detailed primary and secondary structural interrogation of biomolecules. MS based techniques are proving to be indispensable tools for monitoring protein folding, structure, and dynamics. Rapid progress in this technology has allowed the possibility of highly sophisticated experiments for addressing complex biological questions.

Many different configurations of protein MS are now available with a wide range of choices for sample preparation, molecular ionization, detection, and instrumentation. This chapter introduced MS based structural proteomics techniques such as H/D exchange, oxidative covalent labeling, and chemical cross-linking. These techniques, when used in conjunction with highly sensitive mass spectrometry instruments; overcome many of the limitations experienced by traditional biophysical methods, and have gained wide popularity with promising results in the past decade. In the H/D exchange process, a covalently bonded hydrogen atom from the protein backbone is replaced by a deuterium atom from the surrounding environment, or vice versa. It provides information about the solvent accessibility of various parts of the molecule. The rate of H/D exchange imparts understanding of protein backbone and the secondary structure. Covalent labeling is based on a similar principle, except that specific labeling reagents can be used that lead to stable, covalent modifications on the solvent accessible residues. Protein footprinting is a popular covalent technique in which hydroxyl radicals are generated that create specific oxidative modifications on the surface accessible residues, which helps in mapping protein surfaces. Most covalent labeling reagents target side chains, while H/DX-MS specifically probes protein backbone and tertiary structure. Chemical cross linking involves covalently attaching two specific inter- or intra-molecular functional groups of side chains by means of a cross-linking reagent. The cross-linking agent imposes a distance constraint on the respective functional groups, providing valuable information on the three dimensional structure of a macromolecule or macromolecular assembly.

Greater automation of experimental workflow along with reliable and sophisticated computer software will pave the way for a more routine incorporation of structural mass spectrometry into commercial experiments. The future trends lie in the use of structural proteomics experimental data coupled with computational modeling techniques such as comparative modeling and threading. Hybrid approaches combining the theoretical models and experimental methods will allow to combine the merits from both approaches. Experimental results can provide specific, explicit constraints such as distance limitations or surface accessibility information, which can be included for refining computational structure prediction models for more robust and reliable results.

The chapter concludes with the discussion of successful application examples of MS based structural proteomics tools in the context of the design of novel pharmaceutical products. Structural examination of GPCR superfamily members using hydroxyl radical mediated footprinting coupled with H_2O^{18} labeling showed the presence of conserved embedded water molecules that are likely to be important for GPCR function. These structural waters were not found to exchange with the surrounding solvent in either the ground state or for the Meta II or opsin states. On the other hand, oxidative modification of selected side chain residues within the transmembrane helices was detected and activation-induced changes in local structural constraints were revealed, likely to be mediated by the dynamics of both water and protein. The results suggest the possibility of a general mechanism for water-dependent communication in family A GPCRs. This example illustrates the importance and potential of radiolytic footprinting for characterizing the structure and dynamics of the transmembrane region, including dynamics of water in membrane proteins.

H/DX-MS has been used to characterize both global and local structural behavior of a recombinant monoclonal IgG1 antibody to study detailed conformational dynamics. The intact, glycosylated form of IgG1 was found to be folded and stable, while the degylosylation resulted in changes in the IgG1 conformation. This was evident by comparing H/D exchange rates of the glycosylated and deglycosylated forms of the antibody. Two specific regions of the IgG1 (residue positions 236-253 and 292-308) were found to have experienced change in exchange properties upon deglycosylation. The data illustrate the utility of H/DX-MS to provide insights into the higher order structure of antibodies and characterizing conformational changes that these molecules may experience upon modifications such as glycosylation, possibly affecting the functionality of the protein.

Covalent labeling has also shown successful deployment towards monoclonal antibodies characterization to establish the structure-function relationships in context of biopharmaceutical development. Such studies are currently ongoing at several pharmaceutical companies and are likely to have a significant impact on the development of both new drugs and biosimilars in the near future.

6. References

Anfinsen, C. B. (1973). Principles that govern the folding of protein chains, *Science* 181: 223–230.

Angel, T. E., Chance, M. R. & Palczewski, K. (2009). Conserved waters mediate structural and functional activation of family A (rhodopsin-like) G protein-coupled receptors, *Proc. Natl. Acad. Sci. U.S.A.* 106: 8555–8560.

Angel, T. E., Gupta, S., Jastrzebska, B., Palczewski, K. & Chance, M. R. (2009). Structural waters define a functional channel mediating activation of the GPCR, rhodopsin, *Proc. Natl. Acad. Sci. U.S.A.* 106: 14367–14372.

Back, J. W., de Jong, L., Muijsers, A. O. & de Koster, C. G. (2003). Chemical cross-linking and mass spectrometry for protein structural modeling., *J Mol Biol* 331(2): 303–313.

Bai, Y., Sosnick, T. R., Mayne, L. & Englander, S. W. (1995). Protein folding intermediates: native-state hydrogen exchange, *Science* 269: 192–197.

Brenowitz, M., Senear, D. F., Shea, M. A. & Ackers, G. K. (1986). "Footprint" titrations yield valid thermodynamic isotherms, *Proc. Natl. Acad. Sci. U.S.A.* 83: 8462–8466.

Chalmers, M. J., Busby, S. A., Pascal, B. D., He, Y., Hendrickson, C. L., Marshall, A. G. & Griffin, P. R. (2006). Probing protein ligand interactions by automated hydrogen/deuterium exchange mass spectrometry, *Anal. Chem.* 78: 1005–1014.

Chance, M. R. (2001). Unfolding of apomyoglobin examined by synchrotron footprinting, *Biochem. Biophys. Res. Commun.* 287: 614–621.

Chance, M. R., Sclavi, B., Woodson, S. A. & Brenowitz, M. (1997). Examining the conformational dynamics of macromolecules with time-resolved synchrotron x-ray 'footprinting'., *Structure* 5(7): 865–869.

de Koning, L. J., Kasper, P. T., Back, J. W., Nessen, M. A., Vanrobaeys, F., Van Beeumen, J., Gherardi, E., de Koster, C. G. & de Jong, L. (2006). Computer-assisted mass spectrometric analysis of naturally occurring and artificially introduced cross-links in proteins and protein complexes, *FEBS J.* 273: 281–291.

Dobson, C. M. (2001). The structural basis of protein folding and its links with human disease, *Philos. Trans. R. Soc. Lond., B, Biol. Sci.* 356: 133–145.

Dobson, C. M. (2003). Protein folding and misfolding, *Nature* 426: 884–890.

Drenth, J. (1999). *Principles of protein x-ray crystallography*, Springer-Verlag.

Ecroyd, H. & Carver, J. A. (2008). Unraveling the mysteries of protein folding and misfolding, *IUBMB Life* 60: 769–774.

El-Shafey, A., Tolic, N., Young, M. M., Sale, K., Smith, R. D. & Kery, V. (2006). "Zero-length" cross-linking in solid state as an approach for analysis of protein-protein interactions, *Protein Sci.* 15: 429–440.

Englander, S. W. & Kallenbach, N. R. (1983). Hydrogen exchange and structural dynamics of proteins and nucleic acids, *Q. Rev. Biophys.* 16: 521–655.

Falke, J. J. (2002). Enzymology. A moving story, *Science* 295: 1480–1481.

Fenimore, P. W., Frauenfelder, H., McMahon, B. H. & Parak, F. G. (2002). Slaving: solvent fluctuations dominate protein dynamics and functions, *Proc. Natl. Acad. Sci. U.S.A.* 99: 16047–16051.

Frauenfelder, H., Sligar, S. G. & Wolynes, P. G. (1991). The energy landscapes and motions of proteins, *Science* 254: 1598–1603.

Galas, D. J. & Schmitz, A. (1978). DNAse footprinting: a simple method for the detection of protein-DNA binding specificity, *Nucleic Acids Res.* 5: 3157–3170.

Gunasekaran, K., Tsai, C. J., Kumar, S., Zanuy, D. & Nussinov, R. (2003). Extended disordered proteins: targeting function with less scaffold, *Trends Biochem. Sci.* 28: 81–85.

Gupta, S., Bavro, V. N., D'Mello, R., Tucker, S. J., Venien-Bryan, C. & Chance, M. R. (2010). Conformational changes during the gating of a potassium channel revealed by structural mass spectrometry, *Structure* 18: 839–846.

Hambly, D. M. & Gross, M. L. (2005). Laser flash photolysis of hydrogen peroxide to oxidize protein solvent-accessible residues on the microsecond timescale, *Journal of the American Society for Mass Spectrometry* 16(12): 2057 – 2063.

Hegyi, H. & Gerstein, M. (1999). The relationship between protein structure and function: a comprehensive survey with application to the yeast genome, *J. Mol. Biol.* 288: 147–164.

Henzler-Wildman, K. & Kern, D. (2007). Dynamic personalities of proteins, *Nature* 450: 964–972.

Houde, D., Arndt, J., Domeier, W., Berkowitz, S. & Engen, J. R. (2009). Characterization of IgG1 Conformation and Conformational Dynamics by Hydrogen/Deuterium Exchange Mass Spectrometry, *Anal. Chem.* 81: 5966.

Houde, D., Berkowitz, S. A. & Engen, J. R. (2011). The utility of hydrogen/deuterium exchange mass spectrometry in biopharmaceutical comparability studies, *J Pharm Sci* 100: 2071–2086.

Huang, Y. J. & Montelione, G. T. (2005). Structural biology: proteins flex to function, *Nature* 438: 36–37.

Humayun, Z., Kleid, D. & Ptashne, M. (1977). Sites of contact between lambda operators and lambda repressor, *Nucleic Acids Res.* 4: 1595–1607.

Hvidt, A. & Nielsen, S. O. (1966). Hydrogen exchange in proteins, *Adv. Protein Chem.* 21: 287–386.

Kamerzell, T. J. & Middaugh, C. R. (2008). The complex inter-relationships between protein flexibility and stability, *J Pharm Sci* 97: 3494–3517.

Kaur, P., Kiselar, J. G. & Chance, M. R. (2009). Integrated algorithms for high-throughput examination of covalently labeled biomolecules by structural mass spectrometry, *Anal. Chem.* 81: 8141–8149.

Kazazic, S., Zhang, H. M., Schaub, T. M., Emmett, M. R., Hendrickson, C. L., Blakney, G. T. & Marshall, A. G. (2010). Automated data reduction for hydrogen/deuterium exchange experiments, enabled by high-resolution Fourier transform ion cyclotron resonance mass spectrometry, *J. Am. Soc. Mass Spectrom.* 21: 550–558.

Kiselar, J. G., Maleknia, S. D., Sullivan, M., Downard, K. M. & Chance, M. R. (2002). Hydroxyl radical probe of protein surfaces using synchrotron x-ray radiolysis and mass spectrometry., *Int J Radiat Biol* 78(2): 101–114.
URL: *http://dx.doi.org/10.1080/09553000110094805*

Krishna, M. M., Hoang, L., Lin, Y. & Englander, S. W. (2004). Hydrogen exchange methods to study protein folding, *Methods* 34: 51–64.

Lee, T., Hoofnagle, A. N., Kabuyama, Y., Stroud, J., Min, X., Goldsmith, E. J., Chen, L., Resing, K. A. & Ahn, N. G. (2004). Docking motif interactions in MAP kinases revealed by hydrogen exchange mass spectrometry, *Mol. Cell* 14: 43–55.

Maleknia, S. D., Brenowitz, M. & Chance, M. R. (1999). Millisecond radiolytic modification of peptides by synchrotron x-rays identified by mass spectrometry., *Anal Chem* 71(18): 3965–3973.

Maleknia, S. D., Ralston, C. Y., Brenowitz, M. D., Downard, K. M. & Chance, M. R. (2001). Determination of macromolecular folding and structure by synchrotron x-ray radiolysis techniques., *Anal Biochem* 289(2): 103–115.
URL: *http://dx.doi.org/10.1006/abio.2000.4910*

Moore, S. & Stein, W. H. (1973). Chemical structures of pancreatic ribonuclease and deoxyribonuclease, *Science* 180: 458–464.

Muller, D. J., Wu, N. & Palczewski, K. (2008). Vertebrate membrane proteins: structure, function, and insights from biophysical approaches, *Pharmacol. Rev.* 60: 43–78.

Pain, R. H. (2000). *Mechanisms of protein folding*, Oxford University Press.

Pantazatos, D., Kim, J. S., Klock, H. E., Stevens, R. C., Wilson, I. A., Lesley, S. A. & Woods, V. L. (2004). Rapid refinement of crystallographic protein construct definition employing enhanced hydrogen/deuterium exchange MS, *Proceedings of the National Academy of Sciences of the United States of America* 101(3): 751–756.
URL: *http://www.pnas.org/content/101/3/751.abstract*

Pascal, B. D., Chalmers, M. J., Busby, S. A. & Griffin, P. R. (2009). HD desktop: an integrated platform for the analysis and visualization of H/D exchange data, *J. Am. Soc. Mass Spectrom.* 20: 601–610.

Peri, S., Steen, H. & Pandey, A. (2001). GPMAW–a software tool for analyzing proteins and peptides, *Trends Biochem. Sci.* 26: 687–689.

Ramsey, N. F. & Purcell, E. M. (1952). Interactions between nuclear spins in molecules, *Physical Review* 85(1): 143–144.

Rinner, O., Seebacher, J., Walzthoeni, T., Mueller, L. N., Beck, M., Schmidt, A., Mueller, M. & Aebersold, R. (2008). Identification of cross-linked peptides from large sequence databases, *Nat. Methods* 5: 315–318.

Rosenbaum, D. M., Cherezov, V., Hanson, M. A., Rasmussen, S. G., Thian, F. S., Kobilka, T. S., Choi, H. J., Yao, X. J., Weis, W. I., Stevens, R. C. & Kobilka, B. K. (2007). GPCR engineering yields high-resolution structural insights into beta2-adrenergic receptor function, *Science* 318: 1266–1273.

Rosenbaum, D. M., Rasmussen, S. G. & Kobilka, B. K. (2009). The structure and function of G-protein-coupled receptors, *Nature* 459: 356–363.

Sadowski, M. I. & Jones, D. T. (2009). The sequence structure relationship and protein function prediction, *Curr. Opin. Struct. Biol.* 19: 357–362.

Schmitz, A. & Galas, D. J. (1980). Sequence-specific interactions of the tight-binding I12-X86 lac repressor with non-operator DNA, *Nucleic Acids Res.* 8: 487–506.

Schulz, G. E. (1992). Induced-fit movements in adenylate kinases, *Faraday Discuss.* pp. 85–93.

Sclavi, B., Sullivan, M., Chance, M. R., Brenowitz, M. & Woodson, S. A. (1998). RNA Folding at Millisecond Intervals by Synchrotron Hydroxyl Radical Footprinting, *Science* 279(5358): 1940–1943.
URL: *http://www.sciencemag.org/cgi/content/abstract/279/5358/1940*

Sharp, J. S., Becker, J. M. & Hettich, R. L. (2003). Protein surface mapping by chemical oxidation: structural analysis by mass spectrometry., *Anal Biochem* 313(2): 216–225.

Sharp, J. S., Becker, J. M. & Hettich, R. L. (2004). Analysis of protein solvent accessible surfaces by photochemical oxidation and mass spectrometry., *Anal Chem* 76(3): 672–683.
URL: *http://dx.doi.org/10.1021/ac0302004*

Sheshberadaran, H. & Payne, L. G. (1988). Protein antigen-monoclonal antibody contact sites investigated by limited proteolysis of monoclonal antibody-bound antigen: protein "footprinting"., *Proc Natl Acad Sci USA* 85(1): 1–5.

Sinz, A. (2003). Chemical cross-linking and mass spectrometry for mapping three-dimensional structures of proteins and protein complexes, *J Mass Spectrom* 38: 1225–1237.

Sinz, A. (2006). Chemical cross-linking and mass spectrometry to map three-dimensional protein structures and protein-protein interactions, *Mass Spectrom Rev* 25: 663–682.

Stokasimov, E. & Rubenstein, P. A. (2009). Actin isoform-specific conformational differences observed with hydrogen/deuterium exchange and mass spectrometry, *J. Biol. Chem.* 284: 25421–25430.

Suckau, D., Mak, M. & Przybylski, M. (1992). Protein surface topology-probing by selective chemical modification and mass spectrometric peptide mapping., *Proc Natl Acad Sci USA* 89(12): 5630–5634.

Sugase, K., Dyson, H. J. & Wright, P. E. (2007). Mechanism of coupled folding and binding of an intrinsically disordered protein, *Nature* 447: 1021–1025.

Takamoto, K. & Chance, M. R. (2006). Radiolytic protein footprinting with mass spectrometry to probe the structure of macromolecular complexes., *Annu Rev Biophys Biomol Struct* 35: 251–276.
URL: *http://dx.doi.org/10.1146/annurev.biophys.35.040405.102050*

Teilum, K., Olsen, J. G. & Kragelund, B. B. (2009). Functional aspects of protein flexibility, *Cell. Mol. Life Sci.* 66: 2231–2247.

Travaglini-Allocatelli, C., Ivarsson, Y., Jemth, P. & Gianni, S. (2009). Folding and stability of globular proteins and implications for function, *Curr. Opin. Struct. Biol.* 19: 3–7.

van der Kamp, M. W., Schaeffer, R. D., Jonsson, A. L., Scouras, A. D., Simms, A. M., Toofanny, R. D., Benson, N. C., Anderson, P. C., Merkley, E. D., Rysavy, S., Bromley, D., Beck, D. A. & Daggett, V. (2010). Dynameomics: a comprehensive database of protein dynamics, *Structure* 18: 423–435.

Vasilescu, J., Guo, X. & Kast, J. (2004). Identification of protein-protein interactions using in vivo cross-linking and mass spectrometry, *Proteomics* 4: 3845–3854.

Wales, T. E. & Engen, J. R. (2006). Hydrogen exchange mass spectrometry for the analysis of protein dynamics, *Mass Spectrom. Rev.* 25: 158–70.

Wales, T. E., Fadgen, K. E., Gerhardt, G. C. & Engen, J. R. (2008). High-speed and high-resolution UPLC separation at zero degrees Celsius, *Anal. Chem.* 80: 6815–6820.

Wang, L., Qin, Y., Ilchenko, S., Bohon, J., Shi, W., Cho, M. W., Takamoto, K. & Chance, M. R. (2010). Structural analysis of a highly glycosylated and unliganded gp120-based antigen using mass spectrometry, *Biochemistry* 49: 9032–9045.

Wefing, S., Schnaible, V. & Hoffmann, D. (2006). SearchXLinks. A program for the identification of disulfide bonds in proteins from mass spectra, *Anal. Chem.* 78: 1235–1241.

Wine, R. N., Dial, J. M., Tomer, K. B. & Borchers, C. H. (2002). Identification of components of protein complexes using a fluorescent photo-cross-linker and mass spectrometry, *Anal. Chem.* 74: 1939–1945.

Woodward, C., Simon, I. & Tuchsen, E. (1982). Hydrogen exchange and the dynamic structure of proteins, *Mol. Cell. Biochem.* 48: 135–160.

Wright, P. E. & Dyson, H. J. (1999). Intrinsically unstructured proteins: re-assessing the protein structure-function paradigm, *J. Mol. Biol.* 293: 321–331.

Wuthrich, K. (1990). Protein structure determination in solution by NMR spectroscopy, *The Journal of biological chemistry* 265(36): 22059–22062.
URL: *http://www.jbc.org/cgi/content/abstract/265/36/22059*

Xu, G. & Chance, M. R. (2007). Hydroxyl radical-mediated modification of proteins as probes for structural proteomics, *Chemical Reviews* 107(8): 3514–3543.
URL: *http://pubs.acs.org/doi/abs/10.1021/cr0682047*

Yi, S., Boys, B. L., Brickenden, A., Konermann, L. & Choy, W. Y. (2007). Effects of zinc binding on the structure and dynamics of the intrinsically disordered protein prothymosin alpha: evidence for metalation as an entropic switch, *Biochemistry* 46: 13120–13130.

Zhu, M. M., Rempel, D. L., Du, Z. & Gross, M. L. (2003). Quantification of protein-ligand interactions by mass spectrometry, titration, and H/D exchange: PLIMSTEX, *Journal of the American Chemical Society* 125(18): 5252–5253.
URL: *http://pubs.acs.org/doi/abs/10.1021/ja029460d*

Part 6

Bioinformatics Tools

22

nwCompare and AutoCompare Softwares for Proteomics and Transcriptomics Data Mining – Application to the Exploration of Gene Expression Profiles of Aggressive Lymphomas

Fréderic Pont[1,2,3], Marie Tosolini[1,2,3],
Bernard Ycart[4] and Jean-Jacques Fournié[1,2,3]
[1]*INSERM UMR1037-Cancer Research Center of Toulouse*
[2]*ERL 5294 CNRS, BP3028, CHU Purpan, Toulouse,*
[3]*Université Toulouse III Paul-Sabatier, Toulouse,*
[4]*Laboratoire Jean Kuntzmann, CNRS UMR 5224,*
Université Joseph Fourier, Grenoble
France

1. Introduction

The global protein and gene expression profiling technologies have revolutionized the study of normal and malignant cells. Transcriptomes permitted to delineate subtypes of B-cell lymphomas which were otherwise histologically and clinically undistinguishable. Although the data mining of proteomes or transcriptomes from these malignant cells can unveil new aspects of their biology, tools to simultaneously compare several samples are scarce. Here we depict nwCompare and Autocompare, two new freewares we developed with this aim, and examplify their use for the comparative data mining of transcriptomes from normal human B cells and B cell lymphomas such as follicular lymphomas (FL) and diffuse large B-cell lymphomas (DLBCL).

2. nwCompare software

Proteomics, transcriptomics and metabolomics implies the handling of a huge amount of data. Nano liquid chromatography combined with electrospray mass spectrometry enables the identification of hundreds of proteins in one complex sample whereas transcriptomics analyzes the expression level of about twenty thousand genes. It is very useful to be able to quickly compare lists of proteins, genes or molecules obtained from different patients, different pathological situations.

We designed nwCompare (Pont & Fournié, 2010), a software for n-way comparison of text files. nwCompare performs a line by line comparison of characters, thus, it can be quite useful to compare proteins names, gene names, molecules names, biological pathways names etc.

nwCompare has proven efficacy in proteomics to compare pathological situations (Pont & Fournié, 2010) or large-scale protein analysis (Pottiez & al., 2010).

The first versions of nwCompare were limited to analyse a maximum of 300 files, but, starting from version 3.20, this software is now only limited by the amount of memory of the computer. Moreover, a new feature has been introduced recently, by allowing the computation of a repartition table. It is thus possible to classify each file entry depending of its occurrence. nwCompare is light, very easy to use and enables users to run very complex comparisons just by selecting radio buttons, without learning any comparison syntax (Fig 1). nwCompare is a free software that can be download at:
https://sites.google.com/site/fredsoftwares/home

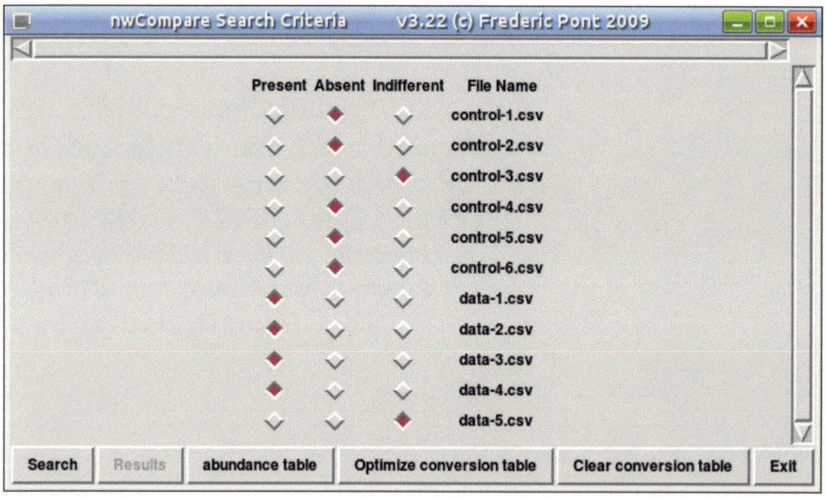

Fig. 1. Screenshot of nwCompare version 3.22 with the simultaneous comparison of eleven protein lists. This example shows the computation of the proteins present in four samples and absent in five controls with two files indifferents. The list of proteins matching those criteria is typically obtained in one second.

3. AutoCompare software

Autocompare freeware was developed as an evolution of nwCompare program to understand the biological significance of large lists of genes or proteins. This software takes as input any data text file and performs string comparisons by line of this file, with a collection of reference files (Fig 2). Then, for each of them, it computes the p-value of the comparison test from hypergeometric distribution tails, then corrects the raw p-values to account for multiple testing, using Bonferroni and Benjamini-Yekutieli methods (Fig 3). We provide AutoCompare with a starting collection of about 5000 genes reference lists based on GSEA (http://www.broadinstitute.org/gsea/) version 3.0 pathways and 162 protein lists based on PANTHER pathways (http://www.pantherdb.org/pathway/) (Mi & al., 2005). Indeed users can also implement in a very straightforward fashion any additional reference list (as .txt format) of their choice.
Autocompare was developed using the Perl programming language (Perl v5.10.1, http://www.perl.org/) and the R statistical programming language under the Linux operating system (ubuntu 10.04, http://www.ubuntu.com/). Autocompare is available for Linux and Windows (https://sites.google.com/site/fredsoftwares/home), and runs on any operating system with Perl, either as a command line tool or with a graphical interface.

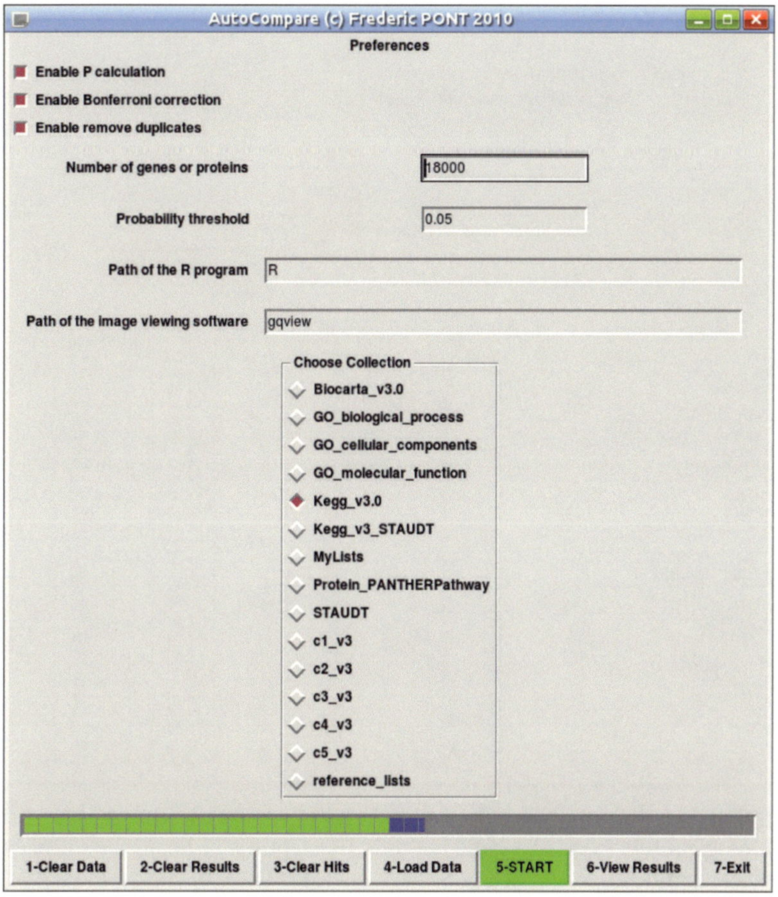

Fig. 2. Screenshot of AutoCompare version 2.31. The software is very easy to use and provides users with biological significance of a large list of genes or proteins.

The main advantages of AutoCompare are that it is very easy to use, rapid and fully automated, it works off line, the number of data files and reference files is only limited by the available disk space, it is very easy to add personalized reference files. In addition, the memory consumption of AutoCompare is very low because only two files are analyzed simultaneously. As nwCompare, AutoCompare performs text file comparisons, so, any kind of data files can potentially be analyzed with it, provided that reference files of the same kind are used.

3.1 AutoCompare false discovery rate (FDR)

To calculate the FDR of AutoCompare, we randomized the genes comprised in the 186 genes lists from the Kegg library on the one hand and the genes comprised in the 3272 gene lists from the Broad institute's GSEA C2' curated library. We then compared the AutoCompare results obtained by querying the same experimental genes list with both the randomized and the correct libraries. With the C2 library, the first false positive was associated with a probability 44 times higher than the Bonferroni threshold. With the Kegg library, the first false positive was associated with a probability 212 times higher than the

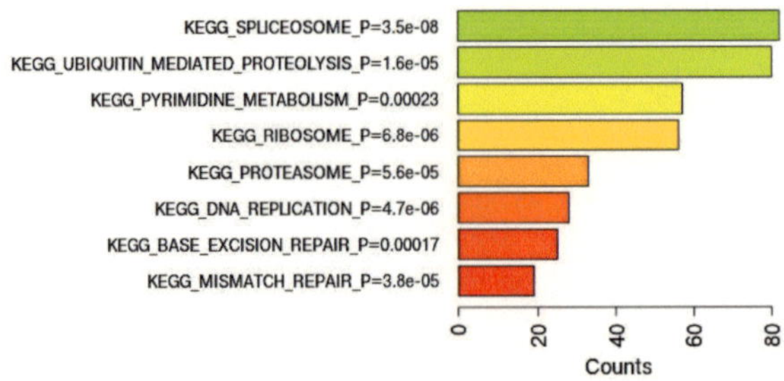

Kegg_v3.0	Score	P value	Counts / Total	(%)	Bonferroni	Benjamini/Yekutieli
KEGG_SPLICEOSOME	7	3.5e-08	82/128	(64 %)	6.51e-06	3.78e-05
KEGG_DNA_REPLICATION	5	4.7e-06	28/36	(78 %)	8.74e-04	2.45e-03
KEGG_RIBOSOME	5	6.8e-06	56/88	(64 %)	1.26e-03	2.45e-03
KEGG_UBIQUITIN_MEDIATED_PROTEOLYSIS	5	1.6e-05	80/138	(58 %)	2.98e-03	4.32e-03
KEGG_MISMATCH_REPAIR	4	3.8e-05	19/23	(83 %)	7.07e-03	8.21e-03
KEGG_PROTEASOME	4	5.6e-05	33/48	(69 %)	1.04e-02	1.01e-02
KEGG_BASE_EXCISION_REPAIR	4	0.00017	25/35	(71 %)	3.16e-02	2.62e-02
KEGG_PYRIMIDINE_METABOLISM	4	0.00023	57/98	(58 %)	4.28e-02	3.10e-02
KEGG_RNA_DEGRADATION	3	0.00037	37/59	(63 %)	6.88e-02	4.44e-02
KEGG_RNA_POLYMERASE	3	0.00043	21/29	(72 %)	8.00e-02	4.64e-02
KEGG_NUCLEOTIDE_EXCISION_REPAIR	3	0.00048	29/44	(66 %)	8.93e-02	4.71e-02
KEGG_CELL_CYCLE	3	0.0019	68/128	(53 %)	3.53e-01	1.71e-01
KEGG_AMINOACYL_TRNA_BIOSYNTHESIS	3	0.0022	26/41	(63 %)	4.09e-01	1.83e-01
KEGG_CHRONIC_MYELOID_LEUKEMIA	2	0.004	41/73	(56 %)	7.44e-01	3.09e-01
KEGG_APOPTOSIS	2	0.0043	48/88	(55 %)	8.00e-01	3.10e-01
KEGG_PURINE_METABOLISM	2	0.0057	80/159	(50 %)	1.00e+00	3.85e-01
KEGG_COLORECTAL_CANCER	2	0.013	34/62	(55 %)	1.00e+00	8.26e-01
KEGG_PANCREATIC_CANCER	2	0.021	37/70	(53 %)	1.00e+00	1.00e+00
KEGG_LYSOSOME	2	0.021	60/121	(50 %)	1.00e+00	1.00e+00
KEGG_HOMOLOGOUS_RECOMBINATION	2	0.022	17/28	(61 %)	1.00e+00	1.00e+00
KEGG_ENDOMETRIAL_CANCER	2	0.031	28/52	(54 %)	1.00e+00	1.00e+00
KEGG_N_GLYCAN_BIOSYNTHESIS	1	0.035	25/46	(54 %)	1.00e+00	1.00e+00
KEGG_ENDOCYTOSIS	1	0.047	85/183	(46 %)	1.00e+00	1.00e+00
KEGG_PROTEIN_EXPORT	1	0.055	14/24	(58 %)	1.00e+00	1.00e+00
KEGG_PROSTATE_CANCER	1	0.071	43/89	(48 %)	1.00e+00	1.00e+00
KEGG_P53_SIGNALING_PATHWAY	1	0.077	34/69	(49 %)	1.00e+00	1.00e+00

Fig. 3. Example of results obtained with AutoCompare version 2.31. Top : histogram of the significant biological functions of a gene list data file. Counts indicate the number of genes found in the corresponding reference gene sets. Bottom: Table of the biological functions identified in a data file, sorted by statistical significance.

Bonferroni threshold. We thus implemented the FDR method of Benjamini and Yekutieli (Benjamini, Y. & Yekutieli, 2005, 2001) to adjust the P values in AutoCompare. This method controls the false discovery rate, the expected proportion of false discoveries among the rejected hypotheses. With the C2 library, the first false positive was associated with a probability 2.3 times higher than the probability of the first false positive. With the Kegg library by contrast, the first false positive was associated with a 119 times higher probability than for the first false positive. Hence, AutoCompare hits that are above the Bonferroni threshold are highly significant and without any false positive result. Furthermore, a good estimate of the FDR is thus given by the correction of Benjamini and Yekutieli.

4. Application of nwCompare and AutoCompare to explore the functional significance of gene expression profiles from normal B-cell subsets and of aggressive lymphomas

Normal differentiation of mature B lymphocytes comprises successive stages of maturation, in which naïve B cells reach germinal centers (GC) in lymph nodes and are activated by antigen to form centroblasts. These highly dividing GC centroblasts may further differentiate into centrocytes which, in turn, mature into either quiescent memory B cells or Ig-secreting plasmablasts which leave lymph nodes to home in bone arrow. Hence the normal B cell maturation in lymph nodes comprises the following sequence: Naïve > GC centroblasts > GC centrocytes > Memory cells, so we searched for the functions associated with the corresponding switches of gene expression signatures.

The transcriptome datasets (Affymetrix CEL files) GSE12195 (Compagno et al., 2009) and GSE15271 (Caron et al., 2009) produced with HG U133-Plus 2.0 platform were downloaded from the NCBI repository GEO database. Together, these comprised 27 normal B cell samples, including 4 naïve B cells, 9 tonsillar germinal center-derived centroblastic cells, 9 tonsillar germinal center-derived centrocytic and 5 memory B cells, 5 lymphoblastoid B-cell lines (B-LCL), 39 follicular lymphomas (FL) and 73 diffuse large B-cell lymphomas (DLBCL) (Caron et al., 2009; Compagno et al., 2009). The raw data from these 144 samples were log (base 2) transformed, normalized in batch by the RMA software and the 54676 probe sets were then reduced to a total of 20606 genes (HUGO symbols) by using the GSEA collapse function set on maximal probe mode (GSEA, http://www.broadinstitute.org/gsea), 18236 of which were fully annotated and thus kept for further study. The genes differentially expressed between two sample groups were defined using two-way Student's tests and $P<0.05$. These gene lists with one gene name per line were converted to text files and then uploaded in Autocompare. More than 4609 genes reference lists based on GSEA (http://www.broadinstitute.org/gsea/) version 3.0 pathways and 162 protein lists based on PANTHER pathways (http://www.pantherdb.org/pathway/) were collected. The differentially-expressed gene subsets were analyzed for enrichment in functionally-related genes among lists downloaded from the gene sets collection. Selective enrichment analysis was then computed with Autocompare using one-sided hypergeometric comparison tests, and False Discovery Rate corrections.

By using this approach, the genes that appeared differentially expressed (P-value <0.05) between respectively, naïve-and GC centroblast, GC centroblasts and GC centrocytes and between GC centrocytes and memory B cells were thus analyzed for functional significance by Autocompare using the KEGG library (V 3.0) of functional genesets in *H. sapiens*. In this example, Autocompare performed the corresponding 1970 comparisons within 529 seconds. The GEP of the naïve-to-GC centroblast transition, the so-called "GC GEP signature", comprised 5516 differentially expressed genes. These latter witnessed of a significant increase of cell cycle ($P<10^{-20}$), DNA replication ($P<10^{-11}$), DNA damage and mismatch repair response, STAT3 signaling pathways together with reduced expression of genes for Krebs cycle metabolism and of IRF4-dependent plasmacytic differentiation genes (all with $P<10^{-5}$). Overall, this pattern reflected the unique differentiation program of B cells at the germinal center stage: a strong proliferation and high mutational activities which are both controlled by the Bcl-6 repressor, a program necessary for the clonal expansion of B-cells expressing mutated Ig. The profile of the centroblast-to-centrocyte GC transition comprised fewer differences (1966 differentially expressed genes) which corresponded to up-regulation of genes normally repressed by Bcl-6 ($P<10^{-8}$), hence reflecting the progressive disappearance of this transcriptional repressor. Finally, the GC centrocyte-to-memory B cell transition (5602 differentially expressed genes) showed a significant up-regulation of genes usually repressed by BLIMP-1, together with down-regulation of both cell cycle ($P<10^{-15}$), DNA replication ($P<10^{-11}$), DNA damage and mismatch repair response. This maturation profile, almost reverse to that of the N-to-GC centroblastic transition genes, indicated not only termination of the Bcl-6 dependent GC reaction but also a switch-off of the Blimp-1-dependent plasmacytic differentiation which, together, characterize quiescent memory cells. Hence the main physiological significance emerging from these comparisons is a signature of the germinal center reaction occurring in centroblasts: a unique combination of rapid proliferation and DNA remodeling (somatic hypermutations) without cell death.

4.1 Significance of gene signatures of non-Hodgkin's B cell lymphoma

Most non-Hodgkin's B cell lymphomas emerge from B cells in the germinal center (GC) stage however, by juxtaposing on their normal development the additional programs triggered by their genetic alterations. Accordingly, follicular lymphoma and diffuse large B-cell lymphoma are known to arise in normal GC B cells through genome alterations and mutations targeting genes controlling apoptotic cell death (BCL2, NFKB), differentiation (BCL6, MYC, BLIMP1, IRF4, CREBBP, EP300) or proliferation (BCR, CARD11, MYD88, NFKB, A20, STAT3) (for review, see Lenz & Staudt, 2010). The spectrum of oncogenes over-expressed and tumor suppressor genes under-expressed in each lymphoma translates to a corresponding profile which now defines the clinical subtype and contributes to predict outcome. We asked which of the 457 known human cancer genes (downloaded from the human cancer gene census (http:// www.sanger.ac.uk/ genetics/CGP/Census) were significantly deregulated in terms of either over-expressed oncogenes or down-regulated tumor suppressor genes in each lymphoma sample. Autocompare yielded such a list for each sample, and we then asked which were present in only one, in just two, in several, or in all of the these samples. Using corresponding requests, the 112 individual cancer gene lists were thus compared using nwCompare. This approach revealed that a total of 221 cancer genes were significantly deregulated in FL and DLBCL, among which 23 oncogenes were consistently upregulated in most (>75%) of the samples, like MAFB, ETV6 and COL1A1 which are strongly up-regulated in all (100%) of the samples (Figure 4). On the other hand 49 cancer genes were deregulated in only one or two patients, indicating these cancer genes are probably not driver cancer genes in the B-cell lymphomagenesis.

We then determined the complete set of genes which were differentially expressed by each individual lymphoma relative to the normal GC B cells ($P<0.05$). On average, 6735 genes were differentially expressed by each follicular lymphoma, 601 of which were shared by all FL. Although these comprised the hallmark over-expression (on average 20-fold) of the anti-apoptotic BCL2 gene, this 601 FL gene set also comprised other deregulated pathways. Using Autocompare, we found that these FL-deregulated pathways were significantly enriched for cytokine-cytokine receptor interactions (37/267 genes, P= 6.7e-12), complement and coagulation cascades (18/69 genes, P=5.1e-11), chemokine signalling (23/190 genes, *P*= 7.9e-07), ECM receptor interactions (14/84 genes, *P*=2.7e-06), focal adhesion (22/201 genes, P=7.2e-06), cell adhesion molecules (15/134 genes, *P*=0.0001), targets of BCL6 (7/19 genes, P= 3.5 e-06), targets of HIF-α (17/164 genes, P=0.0001). In addition, the FL GEP was significantly enriched in the previously depicted FL-type 1 (favourable outcome -associated) and type 2 (poor outcome-associated) immune response genes (respectively 10/40 genes, P=1.0e-06 and 6/23 genes, *P*=0.0001).

Within GC-type DLBCLs on average, 7365 genes were differentially expressed by each lymphoma, 376 of which were shared by all GC-type DLBCL. With ABC type DLBCL on average 7184 genes were differentially expressed by each ABC type DLBCL, 618 of which were shared by all ABC-type DLBCL.This suggested that DLBCL are more heterogeneous than FL, and that ABC-type DLBCL harbour the most genetically diversified profiles. The functional significance of both GC-type and ABC-type DLBCL gene sets comprised the same pathways as for FL plus the lysosome pathway (12/121 genes, P=5.1e-5).

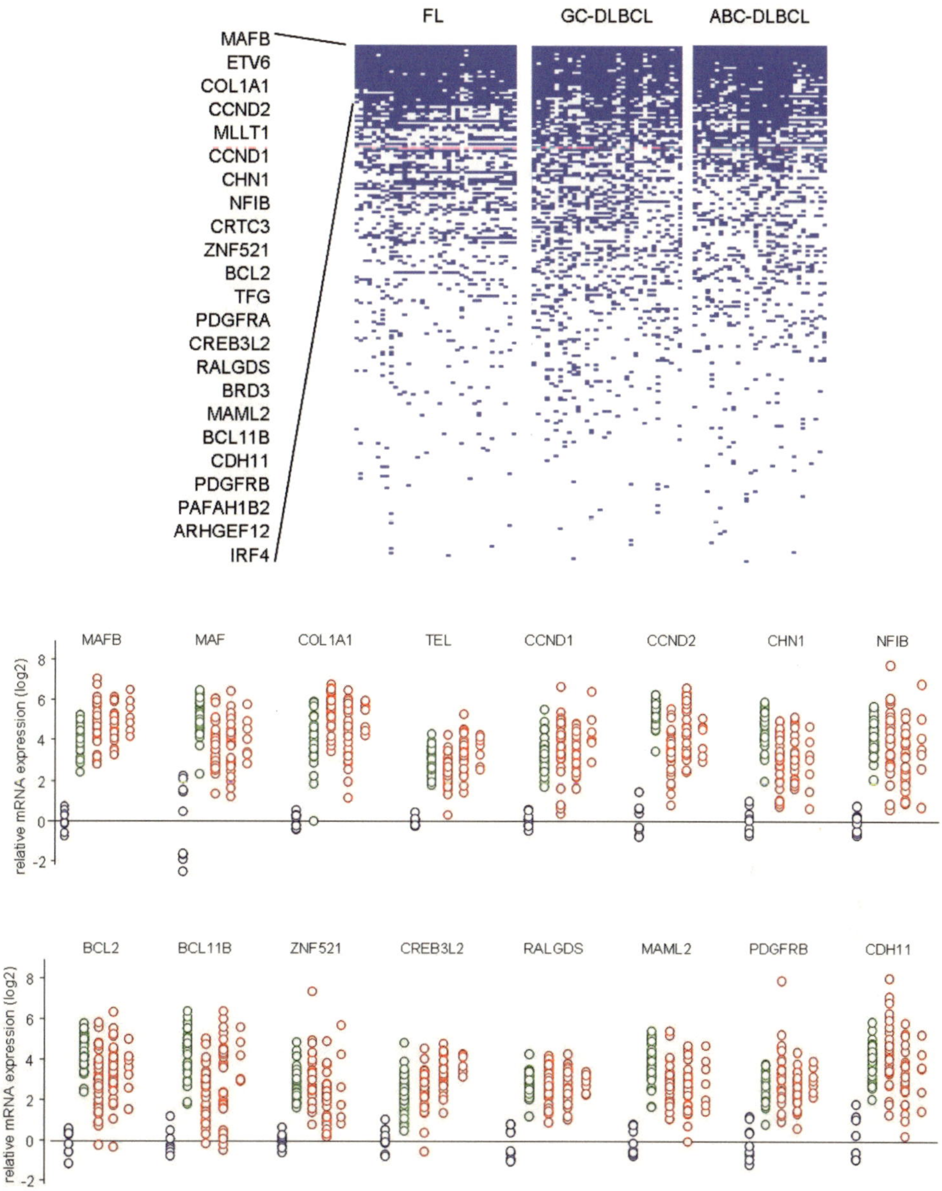

Top: Patterns of cancer genes differentially expressed (oncogenes over-expressed and tumor suppressor genes under-expressed) in follicular lymphomas and diffuse large B cell lymphomas shows that Follicular lymphomas are more homogeneous than DLBCL. Each column corresponds to a patient sample and genes are lines. A blue dot means that the expression of the gene was deregulated for the corresponding patient, a white dot means that the expression of the gene was similar to normal individuals. MAFB, for example, is represented by an horizontal blue line, which mean that this gene was deregulated in all patients. Bottom: Most significantly up-regulated oncogenes in aggressive lymphomas. mRNA expression was normalized to the mean of normal samples (blue), compared to patient's samples: follicular lymphomas (green), DLBCL (red circles), successively grouped as GC, ABC and unclassified DLBCL subtypes, respectively).

Fig. 4. Pattern of oncogenes overexpressed in aggressive lymphomas.

5. Using AutoCompare with proteomics datasets

Proteomic scientists have two options to take benefit of AutoCompare with their proteomic datasets. The first option is straightforward: it is to use directly the starting collection of PANTHER protein pathways provided with AutoCompare. PANTHER protein pathways are built with Uniprot (http://www.uniprot.org/) protein accession numbers. If another protein database is to be used, the Protein Identifier Cross-Reference Service (PICR, http://www.ebi.ac.uk/Tools/picr/) can be applied to convert the data. Further, we present below two examples illustrating how AutoCompare can help data mining proteomes

Example 1: A virtual follicular lymphoma's proteome was created by converting genes up-regulated in follicular lymphoma (as depicted in §4) into protein accession numbers with PANTHER protein pathways. By using AutoCompare, this virtual proteome was then conveniently compared to a series of other proteomes, namely the whole PANTHER pathways proteome collection. Table 1 shows that the top ranking matches concerned proteins of apoptotic cell death, differentiation or proliferation (apoptosis, p53, p38 MAPK and Wnt pathways), focal adhesion (integrin and cadherin pathways), coagulation pathways, cytokines and chemokines signaling and immune response were differentially expressed in FL. In addition, angiogenesis (118/1231 proteins) and various growth factor signaling pathways (PDGF 65/938 proteins; VEGF 39/416 proteins; EGF 61/1071 proteins; IGF 32/276 proteins; FGF 53/978 proteins) were also enriched. Indeed in this example, these proteome comparisons matched with the results from transcriptome comparisons depicted in §4. Of note, the reverse strategy: converting protein accession numbers into gene names is also possible via the Protein Information Resource (PIR) (http://pir.georgetown.edu/pirwww/search/idmapping.shtml). Then, AutoCompare can be used with gene names, as described in § 4, taking advantage of the much larger collection of gene pathways provided with AutoCompare. The disadvantage of these conversion strategies however is that the original amount of data generally increases because of redundancy in databases and gene synonyms. Moreover, since most conversion tools do not filter results by taxonomy, this increase of non relevant data also augments the P values.

Example 2: Comparative analysis of experimental proteomes. The lymphoma cell line Karpas 299 was cultured in vitro for 48 hours in complete medium with and without the bisphosphonate drug zoledronate, the cells were isolated, their protein extract were prepared and the two resulting proteomes were analysed by mass spectrometry: briefly, the proteins were digested by trypsin, the peptides were analysed by nano-electrospray mass spectrometry and identified in SwissProt database using MASCOT (http://www.matrixscience.com/) software (unpublished results). AutoCompare allowed us to compare them to each other and to the proteomes listed the PANTHER pathways. This approach identified 52 matches between lymphoma proteins and one of the reference pathway proteomes. In control lymphoma cells for instance, Autocompare identified among others, 10 proteins of "cytoskeletal_regulation_by_Rho_GTPase" (O15144, O15145, O15511, P23528, P62736, P63261, P63267, P68032, P68133, Q5NBV3), 10 proteins involved in "inflammation mediated by chemokines and cytokines" and 6 proteins from the "Integrin_signalling_pathway". Of note, this approach also indicated that the 5 proteins P62736, P68032, P68133, Q13363 and Q969G3 expressed by the lymphoma cells in control conditions are involved in the Wnt pathway. By contrast, the proteome from cells treated with zoledronate only comprised the P68133 and Q13363 proteins from this pathway, suggesting the treatment had inhibited expression of the 3 others. Hence this example shows how Autocompare can be used to pinpoint targeting of the morphogen Wnt cascade by the bisphosphonate drug.

Protein PANTHER Pathway	Counts/Total (%)
Integrin signalling pathway	132/1175 (11%)
Inflammation mediated by chemokine and cytokine	135/1417 (10%)
Angiogenesis	118/1231 (10%)
Wnt signaling pathway	142/2085 (7%)
Interleukin signaling pathway	74/596 (12%)
Blood coagulation	51/244 (21%)
Cadherin signaling pathway	78/885 (9%)
Apoptosis signaling pathway	74/839 (9%)
PDGF signaling pathway	65/938 (7%)
p53 pathway	49/548 (9%)
Toll receptor signaling pathway	37/281 (13%)
Oxidative stress response	39/349 (11%)
B-cell activation	39/361 (11%)
TGF-beta signaling pathway	56/810 (7%)
Heterotrimeric G-protein signaling pathway	59/978 (6%)
VEGF signaling pathway	39/416 (9%)
EGF receptor signaling pathway	61/1071 (6%)
Heterotrimeric G-protein signaling pathway	48/670 (7%)
Insulin IGF pathway-mitogen activated protein kinase	32/276 (12%)
Interferon-gamma signaling pathway	29/256 (11%)
Metabotropic glutamate receptor group II pathway	36/423 (9%)
GABA-B receptor II signaling	22/126 (17%)
T-cell activation	43/631 (7%)
Cytoskeletal regulation by Rho GTPase	44/674 (7%)
FGF signaling pathway	53/978 (5%)
Enkephalin release	29/280 (10%)
Endogenous cannabinoid signaling	18/83 (22%)
Endothelin signaling pathway	45/748 (6%)
Nicotinic acetylcholine receptor signaling pathway	51/1011 (5%)
p38 MAPK pathway	17/96 (18%)

Table 1. Top rated PANTHER pathways identified by AutoCompare after conversion of follicular lymphoma genes into protein accession numbers. Counts indicate the number of genes found in the corresponding reference gene sets.

6. Conclusion

In conclusion, this example study shows how the use of Autocompare and nwCompare enables users to get fast access to multidimensional comparisons and to the corresponding analysis of large datasets such as proteomes and transcriptomes. We illustrated here this use by the determination of oncogenes and functions involved in the biology of aggressive human B cell lymphomas. Proteomics data sets (protein names, protein accession numbers) can be compared directly in nwCompare since this software performs strings comparisons.

AutoCompare is provided with a starting collection of PANTHER protein pathways for a direct analysis of proteomic datasets. Proteomics users can additionally take advantage of AutoCompare large gene starting database of about 5500 pathways by converting protein names into gene names.

7. Acknowledgements

Work in JJF's lab is supported by institutional grants from INSERM, Université de Toulouse 3 and CNRS, as well as by grants from Institut National du Cancer (contracts RITUXOP and V9V2TER). We thank L. Pasqualucci (Columbia University, NY) for kindly providing us with clinical classifications of the lymphoma samples from GSE12195 dataset.

8. References

Benjamini, Y., & Yekutieli, D. (2005). Quantitative trait loci analysis using the false discovery rate. *Genetics,* 171, pp 783-790, Print ISSN: 0016-6731; Online ISSN: 1943-2631

Benjamini, Y., & Yekutieli, D. (2001). The control of the false discovery rate in multiple testing under dependency. *Ann. Stat.,* 29, 4, pp 1165-1188, ISSN 0090-5364

Caron, G., Le Gallou, S., Lamy, T., Tarte, K. & Fest, T. (2009). CXCR4 expression functionally discriminates centroblasts versus centrocytes within human germinal center B cells. *J Immunol.* 182, pp 7595-7602.

Compagno, M., Lim, W. K., Grunn, A., Nandula, S. V., Brahmachary, M., Shen, Q., Bertoni, F., Ponzoni, M., Scandurra, M., Califano, A., Bhagat, G., Chadburn, A., Dalla-Favera, R. & Pasqualucci, L. (2009). Mutations of multiple genes cause deregulation of NF-kappaB in diffuse large B-cell lymphoma. *Nature,* 459, pp 717-721.

Côté, R.G., Jones, P., Martens, L., Kerrien, S., Reisinger, F., Lin, Q., Leinonen, R., Apweiler, R. & Hermjakob, H. (2007). The Protein Identifier Cross-Referencing (PICR) service: reconciling protein identifiers across multiple source databases. *BMC Bioinformatics,* 8, pp 401-414.

Lenz, G. and Staudt, L. (2010). Aggressive lymphomas. *The New England Journal of Medicine,.* 362, pp 1419-1429.

Mi, H., Lazareva-Ulitsky, B., Loo, R., Kejariwal, A., Vandergriff, J., Rabkin, S., Guo, N., Muruganujan, A., Doremieux, O., Campbell, M. J., Kitano, H. & Thomas* P. D. (2005). The PANTHER database of protein families, subfamilies, functions and pathways. *Nucl. Acids Res,* 33, suppl 1, D284-D288.

Pont, F. & Fournié, JJ. (2010). Sorting protein lists with nwCompare: a simple and fast algorithm for n-way comparison of proteomic data files. *Proteomics,* 10, 5, March 2010, pp 1091-1094. ISSN: 1615-9861.

Pottiez, G., Deracinois, B., Duban-Deweer, S., Cecchelli, R., Fenart, L., Karamanos, Y. & Flahaut, C. (2010). A large-scale electrophoresis- and chromatography-based determination of gene expression profiles in bovine brain capillary endothelial cells after the re-induction of blood-brain barrier properties. *Proteome Sci.,* 15, 8, November 2010, pp 57. ISSN: 1477-5956.

23

Application of Bioinformatics Tools in Gel-Based Proteomics

Kah Wai Lin, Min Jia and Serhiy Souchelnytskyi
Department of Oncology-Pathology,
Karolinska Institutet, Stockholm
Sweden

1. Introduction

Personalized medicine is the most promising approaches in the treatment of various diseases, especially cancer. The use of appropriate biomarkers for personalized treatment has advantage over conventional therapeutics approach, as it confer maximum effectiveness with minimum side effect. Personalized treatment can be achieved by implementation of omic studies in clinical practices. Application of genomic, transcriptomic, proteomic and metabolomic studies deliver a vast amount of data that lead to the discovery of novel biomarkers for diagnostic, prognostic and therapeutic purposes. Therefore, further exploration in omic study could lead to the implementation of personalized medicine as a standard therapeutic scheme in the clinic.

Proteomics is a global study of entire proteins of cell, tissue and organism in a particular condition and time point (Graves & Haystead, 2002). Proteomics is a very comprehensive discipline that includes the study of expression, function, localization, structure, modification, and protein-protein interaction (Graves & Haystead, 2002; Lim & Elenitoba-Johnson, 2004). A proteomics experiment generates vast amount of data that require further analysis, and systems biology is the main approach. Systems biology is an integrative science that studies the complex behavior of biological entities at the systems level (Kitano, 2002a, 2002b). Integrating the proteomics data into systems biology language is an important approach in understanding the behavior of the complex organisms at various levels (Souchelnytskyi, 2005). In recent years, our knowledge of proteomics and system biology is growing rapidly and create an excitement in scientific community because of its potential in novel biomarker and drug discovery (Duncan & Hunsucker, 2005).

Proteomics studies are highly dependent on the technology for protein separation and identification, and bioinformatics for data analysis. By protein separation techniques, gel-based and liquid chromatography (LC)-based approaches represent the primary stream in proteomics. In gel-based approach, that is, conventional 2D gel electrophoresis (2D-GE) and 2D differential gel electrophoresis (2D-DIGE), the proteins are separated by their molecular weight and isoelectric point. In LC-based approach, the proteins or peptides are separated by using high performance liquid chromatography (Aebersold & Mann, 2003; Cravatt et al., 2007). The identification and characterization of proteins or peptides by mass spectrometry are followed after separation (Kolker et al., 2006). In more recent years, antibody-based methods emerging as important approaches in proteomics. These approaches included the use of immunohistochemistry (IHC) on tissue microarrays (TMAs), reverse phase protein arrays (RPPAs) and serum-based diagnostic assays using antibody arrays (Borrebaeck & Wingren, 2007; Brennan, O'Connor et al., 2010; Wingren & Borrebaeck, 2004).

In the present article, we focus our discussion on the various ways of translating gel-based proteomics data into systems biology using different bioinformatics approaches. Firstly, we will discuss the dataset from gel-based proteomics, including the acquisition of primary data and type of data for bioinformatics analysis. In the subsequent section, we will discuss the several way of analyzing the data acquired from gel-based proteomics, which included the ontological-based classification, hierarchical clustering, systems and network analysis (Table 1). We will focus our discussion on the general concepts of the analysis, type of

Ontological Classification	Hierarchical Clustering	Systems and Network Analysis
Query Tools	Cluster+TreeView (Eisen et al., 1998)	Osprey (Breitkreutz et al., 2002)
GO-TermFinder (Boyle et al., 2004)	PermutMatrix (Caraux & Pinloche, 2005)	BioLayout (Enright & Ouzounis, 2001)
AmiGO (Carbon et al., 2009)	POMELO II (Morrissey & Diaz-Uriarte, 2009)	CellDesigner (Funahashi et al., 2007)
MatchMiner (Bussey et al., 2003)	Genesis (Sturn et al., 2002)	Cytoscape (Kohl et al. 2011)
Visualization Tools		
GoMiner (Bussey et al., 2003)		
FatiGO (Al-Shahrour et al., 2004, 2007)		
Onto-Express (Draghici et al., 2003; Khatri et al., 2002)		
GOSurfer (Zhong et al., 2004)		
GOTM (Zhang et al., 2004)		

Table 1. List of bioinformatics tools that are commonly used for gel-based proteomics.

datasets used and bioinformatics software. Some examples of studies and future directions are presented for each approach.

2. Dataset in gel-based proteomics

The general workflow of bioinformatics analysis of gel-based proteomics is shown in figure 1. In gel-based proteomics, various types of datasets can be generated. There can be an annotated 2D gel, mass spectra, and list of identified proteins (Taylor et al., 2003). These dataset can be qualitative or quantitative. In this review, we focus on the analysis of 2 type of datasets generated from annotated 2D gel, i.e. global expression profile and differential expression profile.

By identifying the protein spots on a 2D gel, a comprehensive, global protein expression profile can be generated. This approach can deliver a list of proteins expressed in a cell or tissue in a particular condition, which is exceptionally useful in understanding their biological characteristic. An example is a recent study on proteome profiling of breast epithelial cells with various proliferation potential. This study generate the most comprehensive 2D protein expression map with 183 proteins identified in 184A1 cells and 318 proteins identified in MCF10A cells, which lead to the understanding of their biological properties and delivered a list of potential biomarkers of early event of tumorigenesis (Bhaskaran et al., 2009).

By identifying the protein spots in 2D gels that are different in their staining intensity in different conditions, a differential expression profile can be generated. Various biological questions can be addressed by differential expression analysis. The proteome changes upon drugs treatment can be studied by comparing the 2D gel of a particular cell treated with and without drugs. For example, cellular response to histone deacetylase inhibitor in colon cancer cells was evaluated by such approach (Milli et al., 2008). Besides, various disease stages can also be compared, for example, a list of proteins were identified to be differentially regulated between normal liver tissue and hepatocellular carcinoma (Corona et al. 2010). Furthermore, the dynamic changes of proteome can also be studied. By comparing the differential expressed proteins in the neuroblastoma grown in mice in different time interval reveal the proteome changes of the disease progression and effect of host-tumor interaction (Turner et al., 2009). Therefore, differential expression analysis of 2D gels often called comparative proteomics.

By applying various systems biology analysis tools, these proteomics dataset can further improve our insight into particular biological questions. The first objective of gel-based proteomics data mining is to search for protein of biological importance, such as diagnostic biomarker and potential drug target. By comparing two or more predefined biological conditions, we can precisely define the proteins of interest among thousands of spots in the 2D gel (Meunier et al., 2007). This can be achieved by using differential expression proteome profile, or by comparative analysis of two or more global protein expression profiles. The second objective of gel-based proteomics data mining is to use clustering approach to group or classify the proteins. This is important for understanding the complex biological systems, such as classification of tumor according to the expression of proteins, for the diagnostics and therapeutics purposes (Meunier et al., 2007). This approach can be achieved by applying the bioinformatics tools on both differential expression and global expression profile. In the subsequent section, we will discuss the analysis of gel-based proteomics dataset by using various approaches, and their biological significance.

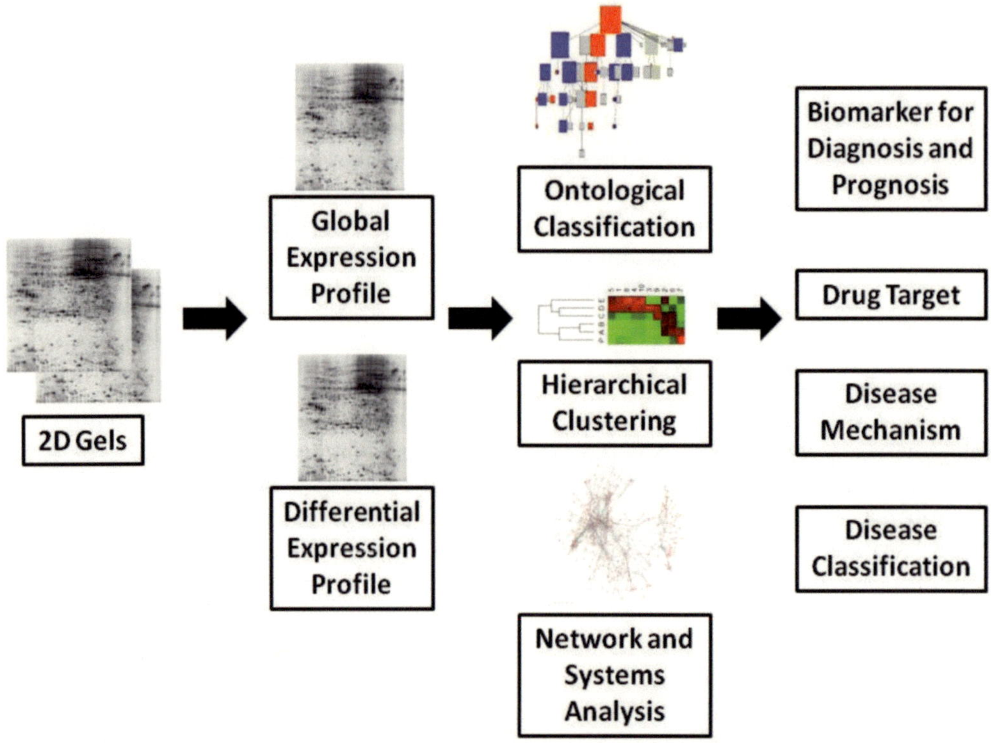

Fig. 1. General workflow of bioinformatics analysis of gel-based proteomics. Once the 2D gels are generated, 2 type of dataset can be acquired from annotated gel, i.e. global expression and differential expression profiles. These datasets can be used for further analysis by various approaches, such as ontological classification, hierarchical clustering and systems/network analysis. These analysis approaches can improve our insight into particular biological questions, such as discovery of novel disease biomarkers for diagnosis and prognosis, drug target, study of disease mechanism and disease classification.

3. Ontological classification

The postgenomic era has brought an exponential growth of biological databases. In recent years, researchers have begun to use unique identifiers to describe components of a database, and the relationship between them. The concept of unique identifiers forms the basis of ontology. Ontology can be described by a set of representative, unambiguous and non-redundant vocabulary or identifier, which define classes, relations, functions, objects and theories (Gruber, 1993). It is not only represents an individual component but also its related components. For instance, in anatomy ontology, stomach is define as an organ with cavity which continuous proximally with oesophagus and distally with small intestine; it is member of viscera of abdomen; it is part of gastrointestinal tract; it is supplied by left and right gastric artery; etc (Detwiler et al., 2003).

The Open Biomedical Ontology (OBO) consortium (http://www.obofoundry.org/) provides a resource where biomedical ontologies are presented in a standard format. Ontology-based approaches for data integration provide a platform of communication between researchers. It also allowed the retrieval/query of information across multiple resources and more efficient data mining and exploration. To gain the functional insight in a large-scale proteomics study, the traditional "literature mining" method is laborious and inefficient. Therefore, ontology-based approach is an effective solution.

In gel-based proteomics, the large dataset can be annotated and explored by application of Gene Ontology (GO) (http://www.geneontology.org/). Gene Ontology is a part of the Open Biomedical Ontologies (OBO), which is the most widely used ontology in biomedical research community (Smith et al., 2007). The main objective of GO is to produce a controlled and unified vocabulary for genes and gene products, such as proteins, that can be applied to all organisms. Furthermore, classification of these components in defined groups or classes allowed us to gain the functional insight in the large-scale proteomics data.

GO annotation organizes genes or gene products into hierarchical order based on 3 categories: cellular component, biological process and molecular function (The Gene Ontology Consortium, 2000). Cellular component describe the localization of particular active gene products in the cells or its extracellular environment. It may be particular cellular structure, e.g. mitochondrion, Golgi apparatus; or gene products groups, e.g. proteosome, ribosome. Biological process describes the biochemical reaction of gene products in the cells. Examples of higher order categories are cell death, signal transduction. Examples of lower order categories are lipid metabolism, purine metabolism. Molecular function describes the elemental activities of gene products at molecular levels. Examples of higher order categories are enzyme, cytoskeletal regulator. Examples of lower order categories are glycine dehydrogenase, apoptosis activator. Since March 2007, 25,000 unique GO identifiers have been created, these provide researchers a broad set of descriptors for cellular component, biological process and molecular function for genes and their products (Dimmer et al., 2008).

There are various GO tools available (table 1). The complete list of tools can be found in http://www.geneontology.org/. These tools belong to either query tools or visualization tools. Prior to analysis, the genes or proteins have to be converted from generic or common name into the unique identifier, i.e. GO term, by using query tools. The most commonly used query tools are GO-TermFinder (Boyle et al., 2004), AmiGO (Carbon et al., 2009), and MatchMiner (Bussey et al., 2003). For example, the GO identifier for cyclin D3 is CCND3.

Once the list of GO identifiers are generated, visualization the data are carried out, using the tools such as GoMiner (Bussey et al., 2003), FatiGO (Al-Shahrour et al., 2004, 2007), Onto-Express (Draghici et al., 2003; Khatri et al., 2002), GOSurfer (Zhong et al., 2004), and GOTM (Zhang et al., 2004). These tools provide visualization of data in the form of either AmiGo view or Direct Acyclic Graph (DAG) view (Figure 2). AmiGO view is in the form of expandable tree structures, and it is linked to external databases, such as NCBI and CGAP. DAG is similar to hierarchies but differ in that a more specialized and narrower term or "child" can be related to more than one less specialized and broader term or "parent". Each term are represented by a node and they connected by path in hierarchical order. Each node can often be reached from multiple paths, which allow the comparison of genes/gene products involved in more than one molecular function or biological processes.

In gel-based proteomics, data generated from global expression and differential expression profiles can be used for ontological-based classification. Many studies suggested that ontological classification is a powerful tool in functional characterization of the cells in gel-based proteomics studies. For instance, a study from Alfonso et al. showed the use of

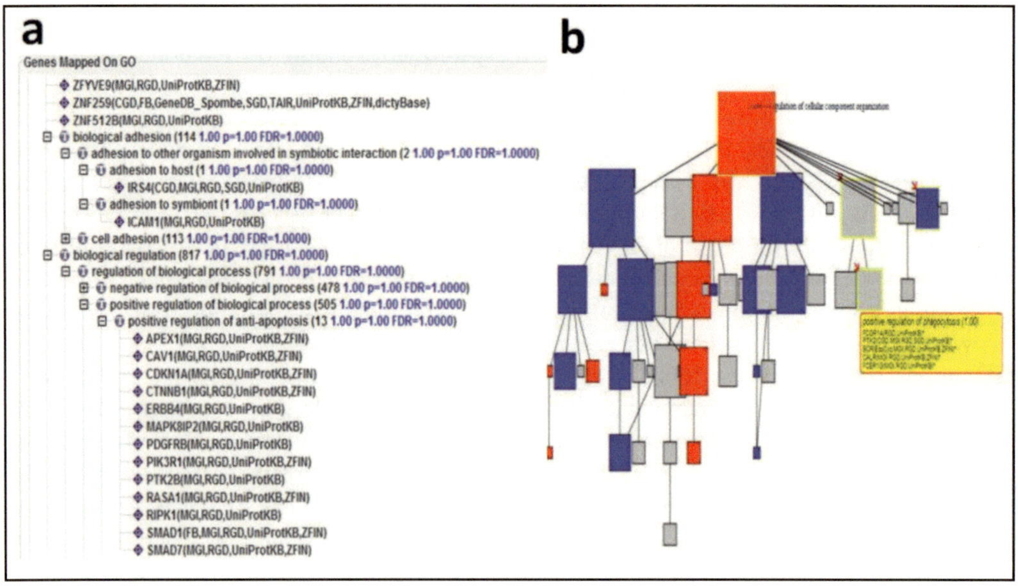

Fig. 2. Data visualization of ontological-based classification. Gene Ontology tools, such as GoMiner (Bussey et al., 2003) showed in this figure, provide visualization of data in the form of either AmiGo view or Direct Acyclic Graph (DAG) view. (a) AmiGO view is in the form of expandable tree structures, and it is linked to external databases. (b) In DAG view, each GO term are represented by a node and they connected by path in hierarchical order. Each node can often be reached from multiple paths, which allow the comparison of genes/gene products involved in more than one category.

ontological classification in a gel-based proteomics study to provide a functional insight of the colorectal cancer. In this study, 41 out of 52 analyzed proteins were unambiguously identified as being differentially expressed in colorectal cancer (Alfonso et al., 2005). An ontology analysis of these proteins revealed that they were mainly involved in regulation of transcription, cellular reorganization and cytoskeleton, cell communication and signal transduction, and protein synthesis and folding (Alfonso et al., 2005). Another example is the study of proteome changes in human T cells during peak HIV infection using 2D differential gel electrophoresis. In this study, ontological classification showed that very high proportion of differentially expressed mitochondrial and metabolic pathway proteins were identified, suggesting that metabolic reprogramming occurs upon HIV infection of T cells (Ringrose et al., 2008).

Although current proteomics study benefit from using Gene Ontology, the major drawback is that Gene Ontology does not describe and annotate the multiple forms of a gene, such as alternative slicing, proteolytic cleavage and post-translational modification. Therefore, Gene Ontology cannot describe the functional stage of the gene products. In recent year, Protein Ontology (PRO) database has been created, which provide a formal classification of proteins (Natale et al., 2007, 2011; Reeves et al., 2008). The PRO included the classification of proteins based on the basis of evolutionary relationships and the structured representation of multiple protein forms of a gene. An initial attempt in applying PRO for the annotation of TGF-beta signalling proteins showed that PRO provide a more accurate annotation and also facilitate various analysis, such as cross-species analysis, pathway analysis and disease modelling (Arighi et al., 2009). Despite of that, implementation of PRO in proteomics study is still in the infancy stage and there is no tools developed for the analysis of large-scale proteomics data. This implicates that further refinement and development of tools for PRO is needed in order to fill the gap.

4. Hierarchical clustering

Hierarchical clustering is a powerful approach for analyzing and visualizing the large proteomics dataset. Cluster analysis was initially designed for transcriptomics studies, such as analysis of microarray data, to explore the similarity between samples based on the pattern of gene expression (Eisen et al., 1998). In recent years, the hierarchical clustering has been adapted to the proteomics study. It enables the proteins to be grouped or classified blindly according to their expression profiles. It is a useful approach in understanding the interdependencies of protein in expression profile, molecular classification and protein signature discovery of diseases, and the dynamic changes of protein expression.

The major principle of hierarchical clustering is based on the dissimilarity or distance between the samples. In proteomics data analysis, this can be calculated by using Pearson correlation coefficient or Euclidean distance. Once the distant matrix is calculated, agglomerative clustering algorithm is performed. In proteomics, unweighted paired group average linkage (UPGMA), complete linkage, and Ward's methods are the most commonly used algorithms. The final results are presented as dendrogram or heat map (Meunier et al., 2007).

In dendrogram, proteins which are closely related will appear on the same branches. The length of branch represents the strength of relationship, where shorter the branch, closer the relationship. In a heat map, group of similar expression will appear as a pattern of cluster with same color. In either presentation method, the ultimate aim is to find the cluster which indicates a similar biological function related to disease mechanism for diagnosis and prognosis purpose.

There are several tools available for hierarchical clustering, for example, Cluster+TreeView (Eisen et al., 1998), PermutMatrix (Caraux & Pinloche, 2005), POMELO II (Morrissey & Diaz-Uriarte, 2009), and Genesis (Sturn et al., 2002). However, most currently available tools are mainly developed for transcriptomics study, i.e. analysis of cDNA microarray data. They are based on different algorithms, and only some of them can be well adapted to the proteomics data analysis, such as Cluster+TreeView and PermutMatrix (Eisen et al., 1998).

The general workflow of hierarchical clustering analysis using PermutMatrix, is discussed here. The proteomics data is presented in the form of standard text file that contains the data matrix: columns represent the sample, i.e. gels with various biological classes or groups, and row represent proteins of interests. Thereafter, the selection of clustering parameters for both distance and aggregation procedures, followed by the application of hierarchical clustering analysis. The result of clustering can be visualized in the form of dendrogram of gel samples and proteins, and heatmap of the clustered data matrix (Meunier et al., 2007) (Figure 3).

Many studies proven that hierarchical clustering is a powerful tool in analysis of large proteomics dataset. Hierarchical clustering can be use for analysis of differential expression protein or global protein expression profiles from the 2D gel. Studies suggested that hierarchical clustering is a powerful tool for discovery of protein signatures or cluster of proteins for molecular classification of diseases, especially cancer. These was shown in

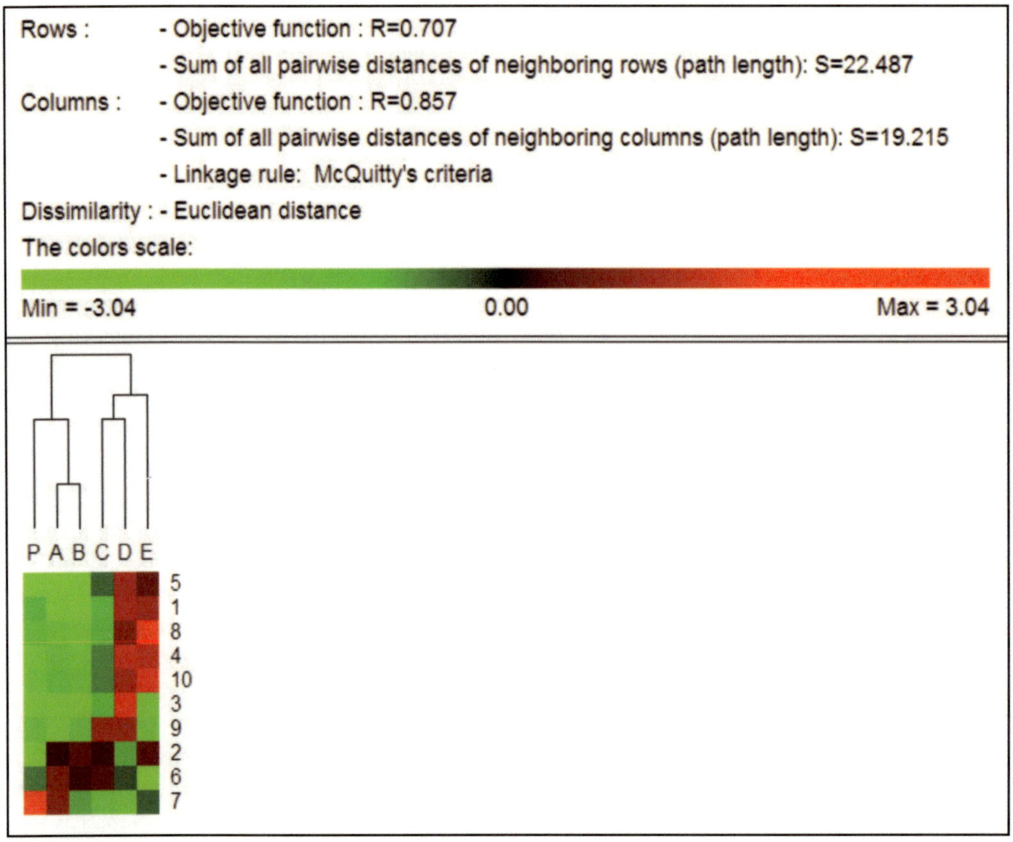

Fig. 3. Data visualization of hierarchical clustering. Using PermutMatrix (Caraux & Pinloche, 2005), hierarchical clustering are presented as dendrogram or heat map. In dendrogram, proteins which are closely related will appear on the same branches. The length of branch represents the strength of relationship, where shorter the branch, closer the relationship. In heat map, group of similar expression will appear as a pattern of cluster with same color.

several recent studies that hierarchical clustering facilitates accurate molecular classification of vaginal and cervical cancer (Hellman et al., 2004), ovarian cancer (Alaiya et al., 2002), lung cancer (Wingren & Borrebaeck, 2004), soft-tissue sarcoma (Suehara et al., 2006), based on their protein expression profile from 2D gel. These studies might lead to the discovery of tumour-specific markers among the differentially expressed proteins. Besides, hierarchical clustering facilitates the discovery of protein signature for prediction of disease progression. This was shown by the study of a set of 20 protein spots that could predict the survival of patients with lung adenocarcinoma (Chen et al., 2003).

Many studies showed the similarity in methodology between transcriptomics and proteomics data analysis using hierarchical clustering approach. As such, many bioinformatics tools that are developed for microarray study can be adapted to gel-based proteomics studies. However, special attentions are needed, as not all the algorithms used for transcriptomics study can be used in proteomics study (Meunier et al., 2007). Without strong knowledge of these algorithms, hierarchical clustering analysis of proteomics data could lead to false result and ambiguity. This implicate that the development of new tools of hierarchical clustering analysis for proteomics study is needed to fulfil the demand of ever-growing proteomics society.

5. Systems and network analysis

The behaviour of a biological system, such as cells, is the consequence of complex interaction between their individual components, such as DNAs, proteins, metabolites, and other biological active molecules. In the past decades, signalling pathway has been the only approach to understand the interaction between these components. However, it is impossible to predict the behaviour of biological systems solely from understanding of their individual component or single signalling pathway. Integration of signalling pathways into a higher order biological network is a very crucial approach for studying the complex behaviour of a biological system. These can be achieved by implementation of systems and network analysis tools. In addition, the recent success of genomics and proteomics technologies generates a vast amount of data that has increased the quest for the systems and network analysis tools.

Over the past few years, application of system and network analysis in genomics and proteomics study had showed a great promise in understanding of complex behaviours of biological systems. Global mapping of the cells or organelles using these tools enable us to discover, visualize and explore the behaviour of the biological systems relevant to our experimental design. In addition, by studying the topological, functional, and dynamic properties of biological networks, the regulatory and control mechanism of the cells underlying the changes of environment can be explored. An example is a study of the over-expression of certain signalling pathway of the tumor cells under the challenge with chemotherapeutics drug (Barabasi & Oltvai, 2004; Kwoh & Ng, 2007).

Networks are displayed as graphs, which represented by nodes and edges/links. These graphs differ from the ontological and hierarchical clustering in that each node is not a function, but a component, such as gene or protein; or a substrate/product of a reaction. Nodes are displayed in various shapes, which represent various types of molecules, such as genes, proteins, and metabolites. The nodes are connected with each other by the edges or links. Edges or links represent the biological relationships between the nodes, such as induction, activation, inhibition, post-translational modification, enzymatic-substrate reaction, and physical binding.

The interaction between the nodes can be directed or undirected. In directed network, the link between two nodes has a defined direction, for example, the induction of activation of a protein by an enzyme. In undirected network, the link does not have specific direction, for example, protein-protein interaction or physical binding. Network can provide a framework from which complex regulatory information can be extracted. Most of the biological networks are scale-free, in which most of the nodes have only a few links, while a few nodes with a very large number of links, which are called hubs (Barabasi & Oltvai, 2004).

The general principle of network construction is based on the known interaction pair of gene or protein. In brief, Swiss-Prot and GeneBank accession numbers from the experimental dataset are used to search against the external databases that contain information about the interaction between the genes or proteins. Subsequently, the genes or proteins from the experiment data were integrated and merged with their known interacting partners and pathways. This process is continued until all proteins of interest are included into the network.

There are a number of available tools for construction and analysis of networks (Thomas & Bonchev, 2010), such as Osprey, (Breitkreutz et al., 2002), BioLayout (Enright & Ouzounis, 2001), CellDesigner (Funahashi et al., 2007), and Cytoscape (Kohl et al. 2011; Smoot et al. 2011). Each tool has distinct functional features. Although most of these tools were initially designed for genomics data analysis, most of them are well adapted for proteomics data analysis. For gel-based proteomics, both global expression profile and differential expression profile can be used to construct the network, depending on the experimental design and question to be answered.

Here we show an example of workflow of network analysis in gel-based proteomics, by using Cytoscape. Cytoscape is open source software that provides basic functionality for integrating proteomics data on the network, editing and visualization of network, and also implementation of external plug-ins for network analysis. Data generated from gel-based proteomics, i.e. the list of the proteins, are integrated with the graph using tools for network construction, such as MiMi (Gao et al., 2009), cPath (Cerami et al., 2006) and BioNetBuilder (Avila-Campillo et al., 2007). Subsequently, using the annotation tools, the node and edge can be annotated with attribute and expression data, such as expression ratio obtained from 2D gel analysis. For visualization of network structure, Cytoscape supports a variety of network layout algorithms, such as spring-embedded layout, circular layout and hierarchical layout (Figure 4).

In order to reduce the complexity of a large network, user can selectively display the set of nodes and edges in the graph, using graph selection and filtering tools. Nodes and edges can be selected according to a wide variety of criteria, including selection by name or by the property of the attribute (Figure 5). Besides, Cytoscape are supported by filtering tools that includes a Minimum Neighbors filter, Local Distance filter, Differential Expression filter, or the combination filter. Minimum Neighbors filter selects nodes having a minimum number of neighbors within a specified distance in the network. Local Distance filter selects nodes within a specified distance of a group of nodes. Differential Expression filter selects nodes according to their expression data. Combination filter selects nodes by combinations of other filters (Shannon et al., 2003).

When the network construction is complete, user can implement various external plug-ins for analysis of the network. This is one of the most powerful functionality of Cytoscape for solving biological questions by mean of network exploration. There is a variety of plug-ins which is commonly used in network analysis. Several examples of Cytoscape plug-ins for network analysis, such as MCODE (Bader & Hogue, 2003), NetworkAnalyzer (Assenov et al., 2008) and Centiscape (Scardoni et al., 2009), are discussed here. MCODE is a plug-in that search for clusters or highly interconnected regions in the network (Bader & Hogue, 2003). In protein network, clusters are often attribute to a groups of proteins that represent a proteins family or protein-protein interaction networks, therefore, finding the cluster enable us to define the region of functional importance. NetworkAnalyzer is a Java plug-in that analyses and visualizes the molecular interaction networks (Assenov et al., 2008). NetworkAnalyzer computes different parameters that describe the network topology, such as diameter of a network, average number of neighbours, and numbers of connected pairs of nodes. NetworkAnalyzer also compute more complex parameters, for example, node degree distribution, topological coefficients, shortest path length distribution, closeness centrality and neighbourhood connectivity distribution. These topology parameters enable us to understanding the property of biological network, such as protein signalling network, protein-protein interaction network, that are of biological importance. Centiscape is another

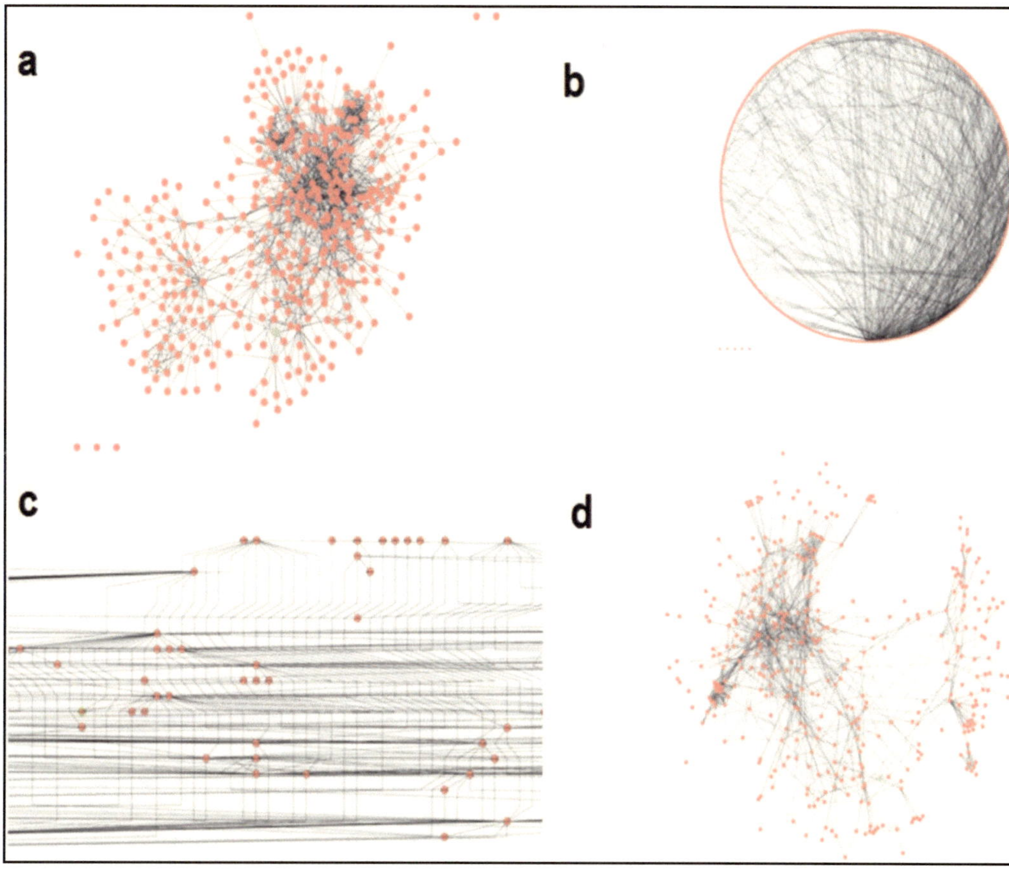

Fig. 4. Visualization of network structure using Cytoscape (Kohl et al. 2011; Smoot et al. 2011). Networks are displayed as graphs, which represented by nodes and edges. For visualization of network structure, Cytoscape supports a variety of network layout algorithms, such as (a) force-directed layout, (b) circular layout, (c) hierarchical layout, and (d) spring-embedded layout.

plug-in for analysis of complex topology of biological network (Scardoni et al., 2009). Centiscape computes centrality indexes of each node in the network, and relationship between the nodes. Thus, Centiscape provides classification of nodes according to their capability to influence the function of other nodes within the network. This may enable us to identify the critical nodes and regulatory circuits in the protein network.

In gel-based proteomics, network construction and pathway analysis are very useful in identifying novel regulatory mechanism of diseases and drug target discovery (Dudley & Butte, 2009). This was showed by a recent study that network analysis of proteomics data from clear cell renal cell carcinoma patient revealed the role of TNFα in clear cell renal cell carcinoma pathogenesis. In addition, it was suggested that clinically available TNFα

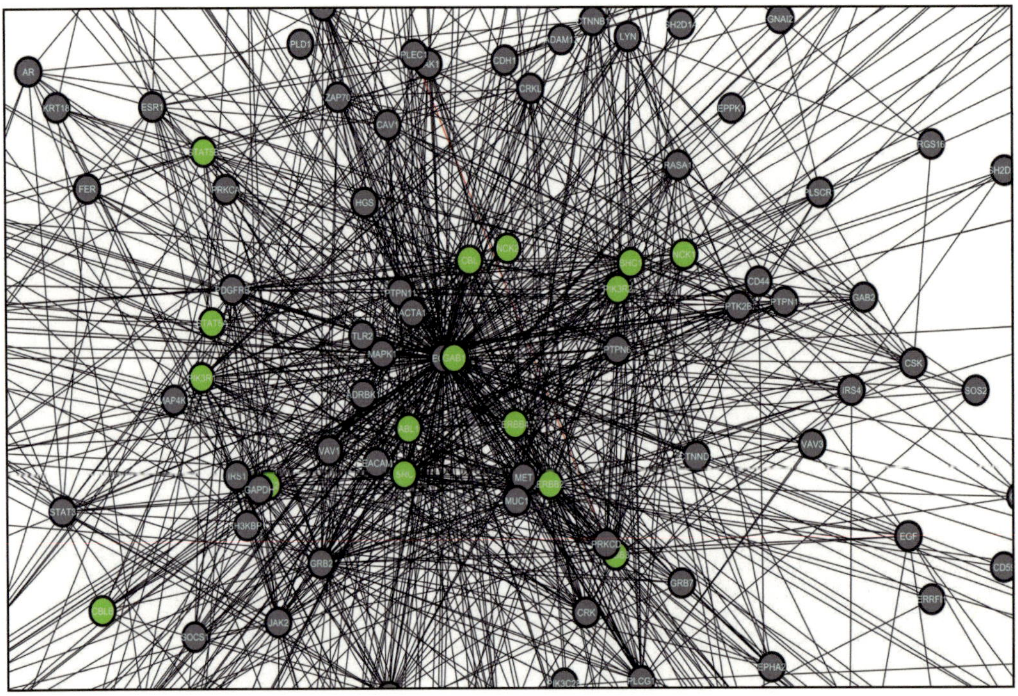

Fig. 5. Graph selection tool in Cytoscape (Kohl et al. 2011; Smoot et al. 2011). User can use graph selection tool to reduce the complexity of the graph. In this example, the components of ERBB pathway were selected and coloured (green) using selection tools.

inhibitors, such as thalidomide and etanercept can be used for the treatment of renal cell carcinoma (Perroud et al., 2006). Besides, network analysis is an indispensable tool in understanding the complex biological behaviour of the cells. A recent study showed that network analysis of gel-based proteome reveal the similarities in regulatory mechanism by MCF10A and 184A1 cells. Network analysis showed the involvement of TNF, AKT, F2 and IGF hubs in both cell types, but cell cycle regulation and mitogenic signaling networks are more representative in MCF10A cells, as compared to 184A1 cells. Study of the network also showed that enhanced expression of cell cycle and proliferation-related proteins, such as CDK4 and cyclin D3 may have an important contribution to increased proliferation rate of breast epithelial cells at the early event of tumorigenesis (Bhaskaran et al., 2009).

Network and pathway analysis is a robust approach in analyzing large proteomics dataset. However, there are several major limitations. Network analysis is unbiased and hypothesis-free because the built of network are based on known interaction sets that recruited from published data. As a consequent, network analysis is not able to uncover the new or unknown pathway and interaction. On the other hand, the qualities of network are dependent on the limitation of high-throughput experiments where the data were recruited from. For instance, protein-protein interaction studies that generate a high proportion of false-positive result will affect the quality of network based on this data (Arrell & Terzic, 2010). Nevertheless, network analysis remains a powerful tool in understanding the gel-based proteomics data, and it can serve as a good starting point for a further exploration of the dataset.

6. Concluding remarks

Tremendous effort have been made during past decade in understanding the biology of normal and diseased cells at systemic level. Proteomics is one of the most promising approaches in generating functional insight of biological systems. Recent advancement in protein separation and identification technology leads to the generation of enormous amount of data which implicate that importance of bioinformatics analysis. However, this renders a great challenge for biomedical researchers in selecting the suitable strategies in bioinformatics analysis of proteomics data.

This article gives an overview of various analysing strategies in gel-based proteomics; we hope that this will help biomedical researchers to derive more biologically meaningful information from their data. These effort will render a direct impact in the in-depth understanding of biological behaviour of cells, ultimately implemented in clinical applications.

7. References

Aebersold, R. & Mann, M. (2003). Mass spectrometry-based proteomics. *Nature,* Vol.422, No.6928, pp. 198-207, ISSN 0028-0836

Al-Shahrour, F.; Diaz-Uriarte, R. & Dopazo, J. (2004). FatiGO: a web tool for finding significant associations of Gene Ontology terms with groups of genes. *Bioinformatics,* Vol.20, No.4, pp. 578-580, ISSN 1367-4803

Al-Shahrour, F.; Minguez, P.; Tarraga, J.; Medina, I.; Alloza, E.; Montaner, D. & Dopazo, J. (2007). FatiGO +: a functional profiling tool for genomic data. Integration of functional annotation, regulatory motifs and interaction data with microarray experiments. *Nucleic Acids Research,* Vol.35, pp. W91-96, ISSN 1362-4962

Alaiya, A. A.; Franzen, B.; Hagman, A.; Dysvik, B.; Roblick, U. J.; Becker, S.; Moberger, B.; Auer, G. & Linder, S. (2002). Molecular classification of borderline ovarian tumors using hierarchical cluster analysis of protein expression profiles. *International Journal of Cancer,* Vol.98, No.6, pp. 895-899, ISSN 0020-7136

Alfonso, P.; Nunez, A.; Madoz-Gurpide, J.; Lombardia, L.; Sanchez, L. & Casal, J. I. (2005). Proteomic expression analysis of colorectal cancer by two-dimensional differential gel electrophoresis. *Proteomics,* Vol.5, No.10, pp. 2602-2611, ISSN 1615-9853

Arighi, C. N.; Liu, H.; Natale, D. A.; Barker, W. C.; Drabkin, H.; Blake, J. A.; Smith, B. & Wu, C. H. (2009). TGF-beta signaling proteins and the Protein Ontology. *BMC Bioinformatics,* Vol.10, No. S5, pp. S3, ISSN 1471-2105

Arrell, D. K. & Terzic, A. Network systems biology for drug discovery. *Clinical Pharmacology and Therapeutics,* Vol.88, No.1, pp. 120-125, ISSN 1532-6535

Assenov, Y.; Ramirez, F.; Schelhorn, S. E.; Lengauer, T. & Albrecht, M. (2008). Computing topological parameters of biological networks. *Bioinformatics,* Vol.24, No.2, pp. 282-284, ISSN 1367-4811

Avila-Campillo, I.; Drew, K.; Lin, J.; Reiss, D. J. & Bonneau, R. (2007). BioNetBuilder: automatic integration of biological networks. *Bioinformatics,* Vol.23, No.3, pp. 392-393, ISSN 1367-4811

Bader, G. D. & Hogue, C. W. (2003). An automated method for finding molecular complexes in large protein interaction networks. *BMC Bioinformatics,* Vol.4, No., pp. 2, ISSN 1471-2105

Barabasi, A. L. & Oltvai, Z. N. (2004). Network biology: understanding the cell's functional organization. *Nat Rev Genet,* Vol.5, No.2, pp. 101-113, ISSN 1471-0056

Bhaskaran, N.; Lin, K. W.; Gautier, A.; Woksepp, H.; Hellman, U. & Souchelnytskyi, S. (2009). Comparative proteome profiling of MCF10A and 184A1 human breast epithelial cells emphasized involvement of CDK4 and cyclin D3 in cell proliferation. *Proteomics - Clinical Applications,* Vol.3, No.1, pp. 68-77, ISSN 1862-8354

Borrebaeck, C. A. & Wingren, C. (2007). High-throughput proteomics using antibody microarrays: an update. *Expert Review of Molecular Diagnostics,* Vol.7, No.5, pp. 673-686, ISSN 1744-8352

Boyle, E. I.; Weng, S.; Gollub, J.; Jin, H.; Botstein, D.; Cherry, J. M. & Sherlock, G. (2004). GO::TermFinder--open source software for accessing Gene Ontology information and finding significantly enriched Gene Ontology terms associated with a list of genes. *Bioinformatics,* Vol.20, No.18, pp. 3710-3715, ISSN 1367-4803

Breitkreutz, B. J.; Stark, C. & Tyers, M. (2002). Osprey: a network visualization system. *Genome Biology,* Vol.3, No.12, pp. PREPRINT0012, ISSN 1465-6914

Brennan, D. J.; O'Connor, D. P.; Rexhepaj, E.; Ponten, F. & Gallagher, W. M (2010). Antibody-based proteomics: fast-tracking molecular diagnostics in oncology. *Nature Review Cancer,* Vol.10, No.9, pp. 605-617, ISSN 1474-1768

Bussey, K. J.; Kane, D.; Sunshine, M.; Narasimhan, S.; Nishizuka, S.; Reinhold, W. C.; Zeeberg, B.; Ajay, W. & Weinstein, J. N. (2003). MatchMiner: a tool for batch navigation among gene and gene product identifiers. *Genome Biology,* Vol.4, No.4, pp. R27, ISSN 1465-6914

Caraux, G. & Pinloche, S. (2005). PermutMatrix: a graphical environment to arrange gene expression profiles in optimal linear order. *Bioinformatics,* Vol.21, No.7, pp. 1280-1281, ISSN 1367-4803

Carbon, S.; Ireland, A.; Mungall, C. J.; Shu, S.; Marshall, B. & Lewis, S. (2009). AmiGO: online access to ontology and annotation data. *Bioinformatics,* Vol.25, No.2, pp. 288-289, ISSN 1367-4811

Cerami, E. G.; Bader, G. D.; Gross, B. E. & Sander, C. (2006). cPath: open source software for collecting, storing, and querying biological pathways. *BMC Bioinformatics,* Vol.7, pp. 497, 1471-2105

Chen, G.; Gharib, T. G.; Wang, H.; Huang, C. C.; Kuick, R.; Thomas, D. G.; Shedden, K. A.; Misek, D. E.; Taylor, J. M.; Giordano, T. J.; Kardia, S. L.; Iannettoni, M. D.; Yee, J.; Hogg, P. J.; Orringer, M. B.; Hanash, S. M. & Beer, D. G. (2003). Protein profiles associated with survival in lung adenocarcinoma. *The Proceedings of National Academy of Sciences USA,* Vol.100, No.23, pp. 13537-13542, ISSN 0027-8424

Corona, G.; De Lorenzo, E.; Elia, C.; Simula, M. P.; Avellini, C.; Baccarani, U.; Lupo, F.; Tiribelli, C.; Colombatti, A. & Toffoli, G (2010). Differential proteomic analysis of hepatocellular carcinoma. *International Journal of Oncology,* Vol.36, No.1, pp. 93-99, ISSN 1791-2423

Cravatt, B. F.; Simon, G. M. & Yates, J. R., 3rd. (2007). The biological impact of mass-spectrometry-based proteomics. *Nature,* Vol.450, No.7172, pp. 991-1000, ISSN 1476-4687

Detwiler, L. T.; Mejino Jr, J. V.; Rosse, C. & Brinkley, J. F. (2003). Efficient web-based navigation of the Foundational Model of Anatomy. *AMIA Annual Symposium Proceedings,* pp. 829, ISSN 1942-597X

Dimmer, E. C.; Huntley, R. P.; Barrell, D. G.; Binns, D.; Draghici, S.; Camon, E. B.; Hubank, M.; Talmud, P. J.; Apweiler, R. & Lovering, R. C. (2008). The Gene Ontology - Providing a Functional Role in Proteomic Studies. *Proteomics,* Vol.8 No.23-24, ISSN 1615-9861

Draghici, S.; Khatri, P.; Bhavsar, P.; Shah, A.; Krawetz, S. A. & Tainsky, M. A. (2003). Onto-Tools, the toolkit of the modern biologist: Onto-Express, Onto-Compare, Onto-Design and Onto-Translate. *Nucleic Acids Research,* Vol.31, No.13, pp. 3775-3781, ISSN 1362-4962

Dudley, J. T. & Butte, A. J. (2009). Identification of discriminating biomarkers for human disease using integrative network biology. *Pacific Symposium of Biocomputing,* pp. 27-38, ISSN 1793-5091

Duncan, M. W. & Hunsucker, S. W. (2005). Proteomics as a tool for clinically relevant biomarker discovery and validation. *Experimental Biology and Medicine (Maywood),* Vol.230, No.11, pp. 808-817, ISSN 1535-3702

Eisen, M. B.; Spellman, P. T.; Brown, P. O. & Botstein, D. (1998). Cluster analysis and display of genome-wide expression patterns. *The Proceedings of National Academy of Sciences USA,* Vol.95, No.25, pp. 14863-14868, ISSN 0027-8424

Enright, A. J. & Ouzounis, C. A. (2001). BioLayout--an automatic graph layout algorithm for similarity visualization. *Bioinformatics,* Vol.17, No.9, pp. 853-854, ISSN 1367-4803

Funahashi, A.; Jouraku, A.; Matsuoka, Y. & Kitano, H. (2007). Integration of CellDesigner and SABIO-RK. *In Silico Biology,* Vol.7, No.S2, pp. S81-90, 1386-6338

Gao, J.; Ade, A. S.; Tarcea, V. G.; Weymouth, T. E.; Mirel, B. R.; Jagadish, H. V. & States, D. J. (2009). Integrating and annotating the interactome using the MiMI plugin for cytoscape. *Bioinformatics,* Vol.25, No.1, pp. 137-138, ISSN 1367-4811

Graves, P. R. & Haystead, T. A. (2002). Molecular biologist's guide to proteomics. *Microbiology and Molecular Biology Review,* Vol.66, No.1, pp. 39-63; ISSN 1092-2172

Gruber, T. R. (1993). A translation approach to portable ontologies. *Knowledge Acquisition,* Vol.5, pp. 199-220,

Hellman, K.; Alaiya, A. A.; Schedvins, K.; Steinberg, W.; Hellstrom, A. C. & Auer, G. (2004). Protein expression patterns in primary carcinoma of the vagina. *British Journal of Cancer,* Vol.91, No.2, pp. 319-326, ISSN 0007-0920

Khatri, P.; Draghici, S.; Ostermeier, G. C. & Krawetz, S. A. (2002). Profiling gene expression using onto-express. *Genomics,* Vol.79, No.2, pp. 266-270, ISSN 0888-7543

Kitano, H. (2002). Computational systems biology. *Nature,* Vol.420, No.6912, pp. 206-210, ISSN 0028-0836

Kitano, H. (2002). Systems biology: a brief overview. *Science,* Vol.295, No.5560, pp. 1662-1664, ISSN 1095-9203

Kohl, M.; Wiese, S. & Warscheid, B. (2011). Cytoscape: software for visualization and analysis of biological networks. *Methods in Molecular Biology,* Vol.696, pp. 291-303, ISSN 1940-6029

Kolker, E.; Higdon, R. & Hogan, J. M. (2006). Protein identification and expression analysis using mass spectrometry. *Trends in Microbiology,* Vol.14, No.5, pp. 229-235, 0966-842X

Kwoh, C. K. & Ng, P. Y. (2007). Network analysis approach for biology. *Cellular and Molecular Life Sciences,* Vol.64, No.14, pp. 1739-1751, ISSN 1420-682X

Li, L. S.; Kim, H.; Rhee, H.; Kim, S. H.; Shin, D. H.; Chung, K. Y.; Park, K. S.; Paik, Y. K. & Chang, J. (2004). Proteomic analysis distinguishes basaloid carcinoma as a distinct subtype of nonsmall cell lung carcinoma. *Proteomics,* Vol.4, No.11, pp. 3394-3400, ISSN 1615-9853

Lim, M. S. & Elenitoba-Johnson, K. S. (2004). Proteomics in pathology research. *Laboratory Investigation,* Vol.84, No.10, pp. 1227-1244, ISSN 0023-6837

Meunier, B.; Dumas, E.; Piec, I.; Bechet, D.; Hebraud, M. & Hocquette, J. F. (2007). Assessment of hierarchical clustering methodologies for proteomic data mining. *Journal of Proteome Research,* Vol.6, No.1, pp. 358-366, 1535-3893

Milli, A.; Cecconi, D.; Campostrini, N.; Timperio, A. M.; Zolla, L.; Righetti, S. C.; Zunino, F.; Perego, P.; Benedetti, V.; Gatti, L.; Odreman, F.; Vindigni, A. & Righetti, P. G. (2008). A proteomic approach for evaluating the cell response to a novel histone deacetylase inhibitor in colon cancer cells. *Biochimica et Biophysica Acta,* Vol.1784, No.11, pp. 1702-1710, ISSN 0006-3002

Morrissey, E. R. & Diaz-Uriarte, R. (2009). Pomelo II: finding differentially expressed genes. *Nucleic Acids Research,* Vol.37, No.Web Server issue, pp. W581-586, ISSN 1362-4962

Natale, D. A.; Arighi, C. N.; Barker, W. C.; Blake, J.; Chang, T. C.; Hu, Z.; Liu, H.; Smith, B. & Wu, C. H. (2007). Framework for a protein ontology. *BMC Bioinformatics,* Vol.8 Suppl 9, No., pp. S1, ISSN 1471-2105

Natale, D. A.; Arighi, C. N.; Barker, W. C.; Blake, J. A.; Bult, C. J.; Caudy, M.; Drabkin, H. J.; D'Eustachio, P.; Evsikov, A. V.; Huang, H.; Nchoutmboube, J.; Roberts, N. V.; Smith, B.; Zhang, J. & Wu, C. H. (2011). The Protein Ontology: a structured representation of protein forms and complexes. *Nucleic Acids Research,* Vol.39, pp. D539-545, ISSN 1362-4962

Perroud, B.; Lee, J.; Valkova, N.; Dhirapong, A.; Lin, P. Y.; Fiehn, O.; Kultz, D. & Weiss, R. H. (2006). Pathway analysis of kidney cancer using proteomics and metabolic profiling. *Molecular Cancer,* Vol.5, pp. 64, ISSN 1476-4598

Reeves, G. A.; Eilbeck, K.; Magrane, M.; O'Donovan, C.; Montecchi-Palazzi, L.; Harris, M. A.; Orchard, S.; Jimenez, R. C.; Prlic, A.; Hubbard, T. J.; Hermjakob, H. & Thornton, J. M. (2008). The Protein Feature Ontology: a tool for the unification of protein feature annotations. *Bioinformatics,* Vol.24, No.23, pp. 2767-2772, ISSN 1367-4811

Ringrose, J. H.; Jeeninga, R. E.; Berkhout, B. & Speijer, D. (2008). Proteomic studies reveal coordinated changes in T-cell expression patterns upon infection with human immunodeficiency virus type 1. *Journal of Virology,* Vol.82, No.9, pp. 4320-4330, ISSN 1098-5514

Scardoni, G.; Petterlini, M. & Laudanna, C. (2009). Analyzing biological network parameters with CentiScaPe. *Bioinformatics,* Vol.25, No.21, pp. 2857-2859, ISSN 1367-4811

Shannon, P.; Markiel, A.; Ozier, O.; Baliga, N. S.; Wang, J. T.; Ramage, D.; Amin, N.; Schwikowski, B. & Ideker, T. (2003). Cytoscape: a software environment for integrated models of biomolecular interaction networks. *Genome Research,* Vol.13, No.11, pp. 2498-2504, ISSN 1088-9051

Smith, B.; Ashburner, M.; Rosse, C.; Bard, J.; Bug, W.; Ceusters, W.; Goldberg, L. J.; Eilbeck, K.; Ireland, A.; Mungall, C. J.; Leontis, N.; Rocca-Serra, P.; Ruttenberg, A.; Sansone, S. A.; Scheuermann, R. H.; Shah, N.; Whetzel, P. L. & Lewis, S. (2007). The OBO Foundry: coordinated evolution of ontologies to support biomedical data integration. *Nature Biotechnology,* Vol.25, No.11, pp. 1251-1255, ISSN 1087-0156

Smoot, M. E.; Ono, K.; Ruscheinski, J.; Wang, P. L. & Ideker, T. (2011). Cytoscape 2.8: new features for data integration and network visualization. *Bioinformatics,* Vol.27, No.3, pp. 431-432, ISSN 1367-4811

Souchelnytskyi, S. (2005). Bridging proteomics and systems biology: what are the roads to be traveled? *Proteomics,* Vol.5, No.16, pp. 4123-4137, ISSN 1615-9853

Sturn, A.; Quackenbush, J. & Trajanoski, Z. (2002). Genesis: cluster analysis of microarray data. *Bioinformatics,* Vol.18, No.1, pp. 207-208, ISSN 1367-4803

Suehara, Y.; Kondo, T.; Fujii, K.; Hasegawa, T.; Kawai, A.; Seki, K.; Beppu, Y.; Nishimura, T.; Kurosawa, H. & Hirohashi, S. (2006). Proteomic signatures corresponding to histological classification and grading of soft-tissue sarcomas. *Proteomics,* Vol.6, No.15, pp. 4402-4409, ISSN 1615-9853

Taylor, C. F.; Paton, N. W.; Garwood, K. L.; Kirby, P. D.; Stead, D. A.; Yin, Z.; Deutsch, E. W.; Selway, L.; Walker, J.; Riba-Garcia, I.; Mohammed, S.; Deery, M. J.; Howard, J. A.; Dunkley, T.; Aebersold, R.; Kell, D. B.; Lilley, K. S.; Roepstorff, P.; Yates, J. R., 3rd; Brass, A.; Brown, A. J.; Cash, P.; Gaskell, S. J.; Hubbard, S. J. & Oliver, S. G. (2003). A systematic approach to modeling, capturing, and disseminating proteomics experimental data. *Nature Biotechnology,* Vol.21, No.3, pp. 247-254, ISSN 1087-0156

Thomas, S. & Bonchev, D. (2010). A survey of current software for network analysis in molecular biology. *Hum Genomics,* Vol.4, No.5, pp. 353-360, ISSN 1479-7364

Turner, K. E.; Kumar, H. R.; Hoelz, D. J.; Zhong, X.; Rescorla, F. J.; Hickey, R. J.; Malkas, L. H. & Sandoval, J. A. (2009). Proteomic analysis of neuroblastoma microenvironment: effect of the host-tumor interaction on disease progression. *Journal of Surgical Research,* Vol.156, No.1, pp. 116-122, ISSN 1095-8673

Wingren, C. & Borrebaeck, C. A. (2004). High-throughput proteomics using antibody microarrays. *Expert Rev Proteomics,* Vol.1, No.3, pp. 355-364, ISSN 1744-8387

Zhang, B.; Schmoyer, D.; Kirov, S. & Snoddy, J. (2004). GOTree Machine (GOTM): a web-based platform for interpreting sets of interesting genes using Gene Ontology hierarchies. *BMC Bioinformatics,* Vol.5, No., pp. 16, ISSN 1471-2105

Zhong, S.; Storch, K. F.; Lipan, O.; Kao, M. C.; Weitz, C. J. & Wong, W. H. (2004). GoSurfer: a graphical interactive tool for comparative analysis of large gene sets in Gene Ontology space. *Applied Bioinformatics,* Vol.3, No.4, pp. 261-264, ISSN 1175-5636